T0156083

Springer Texts in Business and Economics

More information about this series at
http://www.springer.com/series/10099

Hans Peters

Game Theory

A Multi-Leveled Approach

Second Edition

 Springer

Hans Peters
Department of Quantitative Economics
Maastricht University
Maastricht
The Netherlands

ISSN 2192-4333   ISSN 2192-4341   (electronic)
Springer Texts in Business and Economics
ISBN 978-3-662-51877-9   ISBN 978-3-662-46950-7   (eBook)
DOI 10.1007/978-3-662-46950-7

Printed on acid-free paper

Springer-Verlag GmbH Berlin Heidelberg is part of Springer Science+Business Media
(www.springer.com)

Voor Lenie, Nina en Remco

Preface

This book is a compilation of much of the material I used for various game theory courses over the past decades. The first part, *Thinking Strategically*, is intended for undergraduate students in economics or business, but can also serve as an introduction for the subsequent parts of the book. The second and third parts go deeper into the various topics treated in the first part. These parts are intended for more mathematically oriented undergraduate students, or for graduate students in (for instance) economics. Part II is on noncooperative games and Part III on cooperative games. Part IV consists of a mathematical tools chapter, a chapter with review problems for Part I, and a chapter with hints and solutions to the problems of all chapters. Every chapter has a section with problems.

The material draws heavily on game theory texts developed by many others, often in collaboration. I mention in particular Jean Derks, Thijs Jansen, Andrés Perea, Ton Storcken, Frank Thuijsman, Stef Tijs, Dries Vermeulen, and Koos Vrieze. I am also seriously indebted to a large number of introductory, intermediate, and advanced texts and textbooks on game theory, and hope I have succeeded in giving sufficient credits to the authors of these works in all due places.

About the Second Edition

In this second edition, I have corrected mistakes, omissions, and typos from the first edition, and tried to improve the exposition throughout the book. I have added extra problems to some chapters, and also a chapter with review problems for Part I. In Chap. 6, I have added a few sections on auctions with incomplete information. With only few exceptions, the references to the literature are now collected in Notes sections, which conclude every chapter in the book.

This second edition has benefitted tremendously from extensive comments of Piotr Frackiewicz and Peter Wakker. The list of people from who I received comments also includes Krzysztof Apt, Maikel Bosschaert, Yukihiko Funaki, Ali Ihsan Ozkes, Mahmut Parlar, Thijs Ruijgrok, Steffen Sagave, Judith Timmer, Mark Voorneveld, and others.

How to Use This Book

Part I of the book is intended, firstly, for undergraduate students in economics and business and, secondly, as preparation and background for Parts II–IV. Part I is preceded by Chap. 1, which is a general introduction to game theory by means of examples. The first chapter of Part I, Chap. 2 of the book, is on zero-sum games. This chapter is included, not only for historical reasons—the minimax theorem of von Neumann (1928) was one of the first formal results in game theory—but also since zero-sum games (all parlor games) require basic, strictly competitive, game-theoretic thinking. The heart of Part I consists of Chaps. 3–6 on noncooperative games and applications, and Chap. 9 as a basic introduction to cooperative games. These chapters can serve as a basics course in game theory. Chapters 7 and 8 on repeated games and evolutionary games can serve as extra material, as well as Chap. 10 on cooperative game models and Chap. 11, which is an introduction to the related area of social choice theory.

Although this book can be used for self-study, it is not intended to replace the teacher. Part I is meant for students who are knowledgeable in basic calculus, and does not try to avoid the use of mathematics on that basic level. Moreover, (almost) all basic game theory models are described in a formally precise manner, although I am aware that some students may have a blind spot for mathematical notation that goes beyond simple formulas for functions and equations. This formal presentation is included especially since many students have always been asking questions about it: leaving it out may lead to confusion and ambiguities. On the other hand, a teacher may decide to drop these more formal parts and go directly to the examples of concretely specified games. For example, in Chap. 5, the game theory teacher may decide to skip the formal Sect. 5.1 and go directly to the worked out examples of games with incomplete information—and perhaps later return to Sect. 5.1.

Parts II–IV require more mathematical sophistication and are intended for graduate students in economics, or for an elective game theory course for students in (applied) mathematics. In my experience, it works well to couple the material in these parts to related chapters in Part I. In particular, one can combine Chaps. 2 and 12 on zero-sum games, Chaps. 3 and 13 on finite games, Chaps. 4, 5, and 14 on games with incomplete information and games in extensive form, and Chaps. 8 and 15 on evolutionary games. For cooperative game theory, one can combine Chap. 9 with Part III.

Each chapter contains a problems section. Moreover, Chap. 23 contains review problems for Part I. Hints, answers and solutions are provided at the end of the book in Chap. 24. For a complete set of solutions for teachers, as well as any comments, please contact me by email.[1]

Maastricht, The Netherlands   Hans Peters
January 2015

Reference

von Neumann, J. (1928). Zur Theorie der Gesellschaftsspiele. *Mathematische Annalen, 100,* 295–320.

---

[1] H.Peters@maastrichtuniversity.nl.

Contents

Introduction

1

The best introduction to game theory is by way of examples. This chapter starts with a global definition in Sect. 1.1, collects some historical facts in Sect. 1.2, and presents examples in Sect. 1.3. Section 1.4 briefly comments on the distinction between cooperative and noncooperative game theory.

1.1 A Definition

Game theory studies situations of competition and cooperation between several involved parties by using mathematical methods. This is a broad definition but it is consistent with the large number of applications. These applications range from strategic questions in warfare to understanding economic competition, from economic or social problems of fair distribution to behavior of animals in competitive situations, from parlor games to political voting systems—and this list is certainly not exhaustive.

Game theory is an official mathematical discipline (American Mathematical Society Classification code 91A) but it is developed and applied mostly by economists. In economics, articles and books on game theory and applications are found in particular under the Journal of Economic Literature codes C7x. The list of references at the end of this book contains many textbooks and other books on game theory.

1.2 Some History

In terms of applications, game theory is a broad discipline, and it is therefore not surprising that game-theoretic situations can be recognized in the Bible (Brams, 1980) or the Talmud (Aumann and Maschler, 1985). Also the literature on strategic

© Springer-Verlag Berlin Heidelberg 2015
H. Peters, *Game Theory*, Springer Texts in Business and Economics,
DOI 10.1007/978-3-662-46950-7_1

warfare contains many situations that could have been modelled using game theory: a very early reference, over 2,000 years old, is the work of the Chinese warrior-philosopher Sun Tzu (1988). Early works dealing with economic problems are Cournot (1838) on quantity competition and Bertrand (1883) on price competition. Some of the work of Dodgson—better known as Lewis Carroll, the writer of *Alice's Adventures in Wonderland*—is an early application of zero-sum games to the political problem of parliamentary representation, see Dodgson (1884) and Black (1969).

One of the first formal works on game theory is Zermelo (1913). The logician Zermelo proved that in the game of chess either White has a winning strategy (i.e., can always win), or Black has a winning strategy, or each player can always enforce a draw—see Sect. 13.2.5. Up to the present, however, it is still not known which of these three cases is the true one. A milestone in the history of game theory is Von Neumann (1928). In this article von Neumann proved the famous minimax theorem for zero-sum games. This work was the basis for the book *Theory of Games and Economic Behavior* by von Neumann and Morgenstern (1944/1947), by many regarded as the starting point of game theory. In this book the authors extended von Neumann's work on zero-sum games and laid the groundwork for the study of cooperative (coalitional) games. See Dimand and Dimand (1996) for a comprehensive study of the history of game theory up to 1945.

The title of the book of von Neumann and Morgenstern reveals the intention of the authors that game theory was to be applied to economics. Nevertheless, in the 1950s and 1960s the further development of game theory was mainly the domain of mathematicians. Seminal articles in this period were the papers by Nash[1] on Nash equilibrium (Nash, 1951) and on bargaining (Nash, 1950), and Shapley on the Shapley value and the core for games with transferable utility (Shapley, 1953, 1967). See also Bondareva (1962) on the core. Apart from these articles, the foundations of much that was to follow later were laid in the contributed volumes edited by Kuhn and Tucker (1950, 1953), Dresher et al. (1957), Luce and Tucker (1958), and Dresher et al. (1964). A classical work in game theory is Luce and Raiffa (1957): many examples still used in game theory can be traced back to this source, like the Prisoners' Dilemma and the Battle of the Sexes.

In the late 1960s and 1970s of the previous century game theory became accepted as a new formal language for economics in particular. This development was stimulated by the work of Harsanyi (1967/1968) on modelling games with incomplete information and Selten (1965, 1975) on (sub)game perfect Nash equilibrium.

In 1994, Nash, Harsanyi and Selten jointly received the Nobel prize in economics for their work in game theory. Since then, many Nobel prizes in economics have been awarded for achievements in game theory or closely related to game theory: Mirrlees and Vickrey (in 1996), Sen (in 1998), Akerlof, Spence and Stiglitz (in 2001), Aumann and Schelling (in 2005), Hurwicz, Maskin and Myerson (in 2007), and Roth and Shapley (in 2012).

---

[1] See Nasar (1998) for a biography, and the later movie with the same title *A Beautiful Mind*.

From the 1980s on, large parts of economics have been rewritten and further developed using the ideas, concepts and formal language of game theory. Articles on game theory and applications can be found in many economic journals. Journals explicitly focusing on game theory include the *International Journal of Game Theory*, *Games and Economic Behavior*, and *International Game Theory Review*. Game theorists are organized within the *Game Theory Society*, see http://www.gametheorysociety.org/.

1.3 Examples

Every example in this section is based on a *story*. Each time this story is presented first and, next, it is translated into a formal mathematical *model*. Such a mathematical model is an alternative description, capturing the essential ingredients of the story with the omission of details that are considered unimportant: the mathematical model is an abstraction of the story. After having established the model, we spend some lines on how to *solve* it: we say something about how the players should or would act. In more philosophical terms, these solutions can be normative or positive in nature, or somewhere in between, but often such considerations are left as food for thought for the reader. As a general remark, a basic distinction between optimization theory and game theory is that in optimization it is usually clear when some action or choice is optimal, whereas in game theory we deal with human (or, more generally, animal) behavior and then it may be less clear when an action is optimal or even what optimality means.[2]

Each example is concluded by further *comments*, possibly including a short preview on the treatment of the exemplified game in the book. The examples are grouped in subsections on zero-sum games, nonzero-sum games, extensive form games, cooperative games, and bargaining games.

1.3.1 Zero-Sum Games

The first example is based on a military situation staged in World War II.

1.3.1.1 The Battle of the Bismarck Sea

Story The game is set in the South-Pacific in 1943. The Japanese admiral Imamura has to transport troops across the Bismarck Sea to New Guinea, and the American admiral Kenney wants to bomb the transport. Imamura has two possible choices: a short Northern route (2 days) or a long Southern route (3 days), and Kenney must choose one of these routes to send his planes to. If he chooses the wrong route he

---

[2]Feyerabend's (1974) 'anything goes' adage reflects a workable attitude in a young science like game theory.

can call back the planes and send them to the other route, but the number of bombing days is reduced by 1. We assume that the number of bombing days represents the payoff to Kenney in a positive sense and to Imamura in a negative sense.

Model The Battle of the Bismarck Sea problem can be modelled using the following table:

$$
\begin{array}{c}
 & \begin{array}{cc} \text{North} & \text{South} \end{array} \\
\begin{array}{c} \text{North} \\ \text{South} \end{array}
\left(\begin{array}{cc} 2 & 2 \\ 1 & 3 \end{array} \right).
\end{array}
$$

This table represents a game with two players, namely Kenney and Imamura. Each player has two possible choices; Kenney (player 1) chooses a row, Imamura (player 2) chooses a column, and these choices are to be made independently and simultaneously. The numbers represent the payoffs to Kenney. For instance, the number 2 up left means that if Kenney and Imamura both choose North, the payoff to Kenney is 2 and the payoff to Imamura is -2. Thus, the convention is to let the numbers denote the payments *from* player 2 (the column player) *to* player 1 (the row player). This game is an example of a *zero-sum game* because the sum of the payoffs is always equal to zero.

Solution In this particular example, it does not seem difficult to predict what will happen. By choosing North, Imamura is always at least as well off as by choosing South, as is easily inferred from the above table of payoffs. So it is safe to assume that Imamura chooses North, and Kenney, being able to perform this same kind of reasoning, will then also choose North, since that is the best reply to the choice of North by Imamura. Observe that this game is easy to analyze because one of the players (Imamura) has a dominated choice, namely South: no matter what the opponent (Kenney) decides to do, North is at least as good as South, and sometimes better.

Another way to look at this game is to observe that the payoff 2 resulting from the combination (North, North) is maximal in its column ($2 \geq 1$) and minimal in its row ($2 \leq 2$). Such a position in the matrix is called a *saddlepoint*. In such a saddlepoint, neither player has an incentive to deviate unilaterally. (As will become clear later, this implies that the combination (North, North) is a *Nash equilibrium*.) Also observe that, in such a saddlepoint, the row player maximizes his minimal payoff (because $2 = \min\{2, 2\} \geq 1 = \min\{1, 3\}$), and the column player (who has to pay according to our convention) minimizes the maximal amount that he has to pay (because $2 = \max\{2, 1\} \leq 3 = \max\{2, 3\}$). The resulting payoff of 2 from player 2 to player 1 is called the *value* of the game.

Comments Two-person zero-sum games with finitely many choices, like the one above, are also called *matrix games* since they can be represented by a single matrix.

Matrix games are studied in Chaps. 2 and 12. The combination (North, North) in the example above corresponds to what happened in reality back in 1943.

1.3.1.2 Matching Pennies

Story In the two-player game of *matching pennies*, both players have a coin and simultaneously show heads or tails. If the coins match, player 2 gives his coin to player 1; otherwise, player 1 gives his coin to player 2.

Model This is a zero-sum game with payoff matrix

$$
\begin{array}{c}
\quad\quad\quad\text{Heads}\quad\text{Tails} \\
\begin{array}{c}
\text{Heads} \\
\text{Tails}
\end{array}
\left(
\begin{array}{cc}
1 & -1 \\
-1 & 1
\end{array}
\right).
\end{array}
$$

Solution Observe that in this game no player has a dominated choice: Heads can be better or worse than Tails, depending on the choice of the opponent. Also, there is no *saddlepoint*: there is no position in the matrix at which there is simultaneously a minimum in the row and a maximum in the column. Thus, there does not seem to be a natural way to solve the game. One way to overcome this difficulty is by allowing players to randomize between their choices: player 1 chooses Heads with a certain probability p and Tails with probability $1 - p$, and player 2 chooses Heads and Tails with probabilities q and $1 - q$ respectively. From considerations of symmetry, a good guess would be to suppose that player 1 chooses Heads or Tails both with probability $\frac{1}{2}$. As above, suppose that player 2 plays Heads with probability q and Tails with probability $1 - q$. In that case the expected payoff for player 1 is equal to

$$
\frac{1}{2}[q \cdot 1 + (1 - q) \cdot -1] + \frac{1}{2}[q \cdot -1 + (1 - q) \cdot 1]
$$

which is independent of q, namely, equal to 0. So by randomizing in this way between his two choices, player 1 can guarantee to obtain 0 in expectation (of course, the actually realized outcome is always $+1$ or -1). Analogously, player 2, by playing Heads or Tails each with probability $\frac{1}{2}$, can guarantee to pay 0 in expectation. Thus, the amount of 0 plays a role similar to that of a saddlepoint. Again, we will say that 0 is the *value* of this game.

Comments The randomized choices of the players are usually called *mixed strategies*. Randomized choices are often interpreted as *beliefs* of the other player(s) about the choice of the player under consideration. See also Sect. 3.1.

1.3.2 Nonzero-Sum Games

1.3.2.1 Prisoners' Dilemma

Story Two prisoners (players 1 and 2) have committed a crime together and are interrogated separately. Each prisoner has two possible choices: he may 'cooperate' (*C*) which means 'not betray his partner' or he may 'defect' (*D*), which means 'betray his partner'. The punishment for the crime is 10 years of prison. Betrayal yields a reduction of 1 year for the defector (traitor). If a prisoner is not betrayed, he is convicted to 1 year for a minor offense.

Model This situation can be summarized as follows:

$$\begin{array}{c} \\ C \\ D \end{array} \begin{array}{c} C \qquad\quad D \\ \begin{pmatrix} -1,-1 & -10,0 \\ 0,-10 & -9,-9 \end{pmatrix}. \end{array}$$

This table is read in the same way as before, but now there are two payoffs at each position: by convention the first number is the payoff for player 1 (the row player) and the second number is the payoff for player 2 (the column player). Observe that the game is no longer zero-sum, and we have to write down both numbers at each matrix position.

Solution For both players *C* is a strictly dominated choice: *D* is better than *C*, whatever the other player does. So it is natural to argue that the outcome of this game will be the pair of choices (D, D), leading to the payoffs $-9, -9$. Thus, due to the existence of strictly dominated choices, the Prisoners' Dilemma game is easy to analyze.

Comments The payoffs $(-9, -9)$ are inferior: they are not *Pareto optimal*, the players could obtain the higher payoff of -1 for each by cooperating, i.e., both playing *C*. There is a large literature on how to establish cooperation, e.g. by reputation effects in a repeated play of the game. If the game is played repeatedly, other (higher) payoffs are possible, see Chap. 7.

The Prisoners' Dilemma is a metaphor for many economic situations. An outstanding example is the so-called 'tragedy of the commons', see Problem 6.27 in this book.

1.3.2.2 Battle of the Sexes

Story A man and a woman want to go out together, either to a football match or to a ballet performance. They forgot to agree where they would go to that night, are in different places and have to decide on their own where to go; they have no means to communicate. Their main concern is to be together. The man has a preference for football and the woman for ballet.

Model A table reflecting the situation is as follows.

$$\begin{array}{c c} & \begin{array}{cc} \text{Football} & \text{Ballet} \end{array} \\ \begin{array}{c} \text{Football} \\ \text{Ballet} \end{array} & \left(\begin{array}{cc} 2,1 & 0,0 \\ 0,0 & 1,2 \end{array} \right). \end{array}$$

Here, the man chooses a row and the woman a column.

Solution Observe that no player has a dominated choice. The players have to coordinate without being able to communicate. Now it may be possible that the night before they discussed football at length; each player remembers this, may think that the other remembers this, and so this may serve as a focal point for both. In the absence of such considerations it is hard to give a unique prediction for this game. We can, however, say that the combinations (Football, Football) and (Ballet, Ballet) are special in the sense that the players' choices are *best replies* to each other; if the man chooses Football (Ballet), then it is optimal for the woman to choose Football (Ballet) as well, and vice versa. In the literature, such choice combinations are called *Nash equilibria*. The concept of Nash equilibrium is without doubt the main solution concept developed in game theory.

Comments The Battle of the Sexes game is metaphoric for problems of coordination.

1.3.2.3 Matching Pennies
Every zero-sum game is, trivially, a special case of a nonzero-sum game. For instance, the Matching Pennies game discussed in Sect. 1.3.1 can be represented as a nonzero-sum game as follows:

$$\begin{array}{c c} & \begin{array}{cc} \text{Heads} & \text{Tails} \end{array} \\ \begin{array}{c} \text{Heads} \\ \text{Tails} \end{array} & \left(\begin{array}{cc} 1,-1 & -1,1 \\ -1,1 & 1,-1 \end{array} \right). \end{array}$$

Clearly, no player has a dominated choice and there is no combination of a row and a column such that each player's choice is optimal given the choice of the other player—there is no Nash equilibrium. If mixed strategies are allowed, then it can be checked that if player 2 plays Heads and Tails each with probability $\frac{1}{2}$, then for player 1 it is optimal to do so too, and vice versa. Such a combination of mixed strategies is again called a Nash equilibrium. See Chaps. 3 and 13.

1.3.2.4 A Cournot Game

Story Two firms produce a similar ('homogenous') product. The market price of this product is equal to $p = 1 - Q$ or zero (whichever is larger), where Q is the total quantity produced. There are no production costs.

Model The two firms are the players, 1 and 2. Each player $i = 1, 2$ chooses a quantity $q_i \geq 0$, and makes a profit of $K_i(q_1, q_2) = q_i(1 - q_1 - q_2)$ (or zero if $q_1 + q_2 \geq 1$).

Solution Suppose player 2 produces $q_2 = \frac{1}{3}$. Then player 1 maximizes his own profit $q_1(1 - q_1 - \frac{1}{3})$ by choosing $q_1 = \frac{1}{3}$. Also the converse holds: if player 1 chooses $q_1 = \frac{1}{3}$ then $q_2 = \frac{1}{3}$ maximizes profit for player 2. This combination of strategies consists of mutual best replies and is therefore again called a Nash equilibrium.

Comments This particular Nash equilibrium is often called Cournot equilibrium. It is easy to check that the Cournot equilibrium in this example is again not Pareto optimal: if the firms each would produce $\frac{1}{4}$, then they would both be better off. The main difference between this example and the preceding ones is, that each player here has infinitely many choices, also if no mixed strategies are included. See further Chap. 6.

1.3.3 Extensive Form Games

All examples in Sects. 1.3.1 and 1.3.2 are examples of one-shot games: the players choose only once, independently and simultaneously. In parlor games as well as in games derived from real-life economic or political situations, this is often not what happens. Players may move sequentially, and observe or partially observe each others' moves. Such situations are better modelled by so-called extensive form games.

1.3.3.1 Sequential Battle of the Sexes

Story The story is similar to the one in Sect. 1.3.2, but we now assume that the man chooses first and the woman can observe the choice of the man.

Model This situation can be represented by the decision tree in Fig. 1.1.

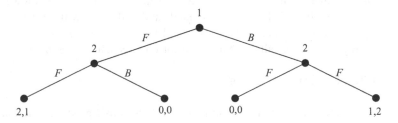

Fig. 1.1 The decision tree of sequential battle of the sexes

Player 1 (the man) chooses first, player 2 (the woman) observes player 1's choice and then makes her own choice. The first number in each pair of numbers is the payoff to player 1, and the second number is the payoff to player 2. Filled circles denote decision nodes (of a player) or end nodes (followed by payoffs).

Solution An obvious way to analyze this game is by working backwards. If player 1 chooses F, then it is optimal for player 2 to choose F as well, and if player 1 chooses B, then it is optimal for player 2 to choose B as well. Given this choice behavior of player 2 and assuming that player 1 performs this line of reasoning about the choices of player 2, player 1 should choose F.

Comments What this simple example shows is that in such a so-called extensive form game, there is a distinction between a play plan of a player and an actual move or choice of that player. Player 2 has the plan to choose F (B) if player 1 has chosen F (B). Player 2's actual choice is F—assuming as above that player 1 has chosen F. We use the word *strategy* to denote a play plan, and the word *action* to denote a particular move. In a one-shot game there is no difference between the two, and then the word 'strategy' is used.

Games in extensive form are studied in Chaps. 4, 5, and 14. The solution described above is an example of a so-called backward induction (or subgame perfect) (Nash) equilibrium. There are other Nash equilibria as well. Suppose player 1 chooses B and player 2's plan (strategy) is to choose B always, independent of player 1's choice. Observe that, given the strategy of the opponent, no player can do better, and so this combination is a Nash equilibrium, although player 2's plan is only partly 'credible': if player 1 would choose F instead of B, then player 2 would be better off by changing her choice to F.

1.3.3.2 Sequential Cournot

Story The story is similar to the one in Sect. 1.3.2, but we now assume that firm 1 chooses first and firm 2 can observe the choice of firm 1.

Model Since each player $i = 1, 2$ has infinitely many actions $q_i \geq 0$, we cannot draw a picture like Fig. 1.1 for the sequential Battle of the Sexes. Instead of straight lines we use zigzag lines to denote a continuum of possible actions. For this example we obtain Fig. 1.2.

Player 1 moves first and chooses $q_1 \geq 0$. Player 2 observes player 1's choice of q_1 and then chooses $q_2 \geq 0$.

$$1 \qquad q_1 \geq 0 \qquad 2 \qquad q_2 \geq 0$$
$$q_1(1 - q_1 - q_2), q_2(1 - q_1 - q_2)$$

Fig. 1.2 The extensive form of sequential Cournot

Solution Like in the sequential Battle of the Sexes game, an obvious way to solve this game is by working backwards. Given the observed choice q_1, player 2's optimal (profit maximizing) choice is $q_2 = \frac{1}{2}(1 - q_1)$ or $q_2 = 0$, whichever is larger. Given this *reaction function* of player 2, the optimal choice of player 1 is obtained by maximizing the profit function $q_1 \mapsto q_1 \left(1 - q_1 - \frac{1}{2}(1 - q_1)\right)$. The maximum is obtained for $q_1 = \frac{1}{2}$. Consequently, player 2 chooses $q_2 = \frac{1}{4}$.

Comments The solution described here is another example of a backward induction or subgame perfect equilibrium. It is also called *Stackelberg equilibrium*. See Chap. 6.

1.3.3.3 Entry Deterrence

Story An old question in industrial organization is whether an incumbent monopolist can maintain his position by threatening to start a price war against any new firm that enters the market. In order to analyze this question, consider the following situation. There are two players, the entrant and the incumbent. The entrant decides whether to Enter (E) or to Stay Out (O). If the entrant enters, the incumbent can Collude (C) with him, or Fight (F) by cutting the price drastically. The payoffs are as follows. Market profits are 100 at the monopoly price and 0 at the fighting price. Entry costs 10. Collusion shares the monopoly profits evenly.

Model This situation can be represented by the decision tree in Fig. 1.3.

Solution By working backward, we find that the entrant enters and the incumbent colludes.

Comments As in the sequential battle of the sexes there exists another Nash equilibrium. If the entrant stays out and the incumbent's plan is to fight if the entrant would enter, then also this is a combination where no player can do better given the strategy of the other player. Again, one might argue that the 'threat' of the incumbent firm to start a price war in case the potential entrant would enter, is not credible since the incumbent hurts himself by carrying out the threat.

Fig. 1.3 The game of entry deterrence. Payoffs: entrant, incumbent

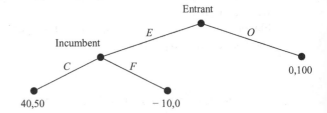

1.3.3.4 Entry Deterrence with Incomplete Information

Story Consider the following variation on the foregoing entry deterrence model. Suppose that with 50 % probability the incumbent's payoff from Fight (*F*) is equal to some amount *x* rather than the 0 above. (Here, *x* is a given number. It is not a variable but a parameter.) The entrant still moves first and knows whether this payoff is *x* or 0. The incumbent moves last and does not know whether this payoff is *x* or 0 when he moves. Both firms know the probabilities of the payoff being *x* or 0. This situation might arise, for instance, if the capacity of the entrant firm is private information. A positive value of *x* might be associated with the entrant having a capacity constraint which leaves a larger share of the market to the incumbent if he fights. The incumbent estimates the probability of the entrant having this capacity constraint at 50 %, and the entrant knows this.

Model This situation can be modelled by including a chance move in the game tree. Moreover, the tree should express the asymmetric information between the players. Consider the game tree in Fig. 1.4. First there is a chance move. The entrant learns the outcome of the chance move and decides to enter or not. If he enters, then the incumbent decides to collude or fight, without however knowing the outcome of the chance move: this is indicated by the dashed line. Put otherwise, the incumbent has two decision nodes where he should choose, but he does not know at which node he actually is. Thus, he can only choose between 'collude' and 'fight', without making this choice contingent on the outcome of the chance move.

Solution If *x* ≤ 50 then an obvious solution is that the incumbent colludes and the entrant enters. Also the combination of strategies where the entrant stays out no matter what the outcome of the chance move is, and the incumbent fights, is a Nash equilibrium. A complete analysis is more subtle and may include a consideration of the probabilistic information that the incumbent might derive from the action of the entrant in a so-called *perfect Bayesian equilibrium*, see Chaps. 5 and 14.

Fig. 1.4 Entry deterrence with incomplete information

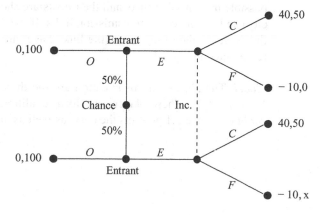

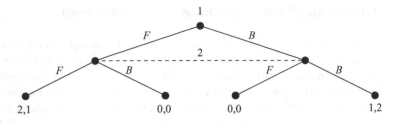

Fig. 1.5 Simultaneous battle of the sexes in extensive form

Comments The collection of the two nodes of the incumbent, connected by the dashed line, is called an *information set*. In general, information sets are used to model imperfect information. In the present example imperfect information arises since the incumbent does not know the outcome of the chance move. Imperfect information can also arise if some player does not observe some move of some other player. As a simple example, consider again the simultaneous move Battle of the Sexes game of Sect. 1.3.2. This can be modelled as a game in extensive form as in Fig. 1.5. Hence, player 2, when he moves, does not know what player 1 has chosen. This is equivalent to players 1 and 2 moving independently and simultaneously.

1.3.4 Cooperative Games

In a cooperative game the focus is on payoffs and coalitions, rather than on strategies. The prevailing analyses have an axiomatic flavor, in contrast to the equilibrium analysis of noncooperative theory. The implicit assumption is that players can make binding agreements.

1.3.4.1 Three Cooperating Cities

Story Cities 1, 2 and 3 want to be connected with a nearby power source. The possible transmission links and their costs are shown in the following figure. Each city can hire any of the transmission links. If the cities cooperate in hiring the links they save on the hiring costs (the links have unlimited capacity). The situation is represented in Fig. 1.6.

Model The players in this situation are the three cities. Denote the player set by $N = \{1, 2, 3\}$. These players can form coalitions: any subset S of N is called a *coalition*. Table 1.1 presents the costs as well as the savings of each coalition.

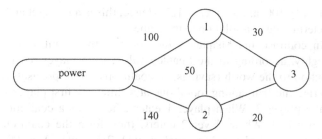

Fig. 1.6 Situation leading to the three cities game

Table 1.1 The three cities game

S	{1}	{2}	{3}	{1, 2}	{1, 3}	{2, 3}	{1, 2, 3}
$c(S)$	100	140	130	150	130	150	150
$v(S)$	0	0	0	90	100	120	220

The costs $c(S)$ are obtained by calculating the cheapest routes connecting the cities in the coalition S with the power source. The cost savings $v(S)$ are determined by

$$v(S) := \sum_{i \in S} c(\{i\}) - c(S) \text{ for each nonempty } S \subseteq N.$$

The cost savings $v(S)$ for coalition S are equal to the difference in costs corresponding to the situation where all members of S work alone and the situation where all members of S work together. The pair (N, v) is called a *cooperative game*.

Solution Basic questions in a cooperative game (N, v) are: which coalitions will actually be formed, and how should the worth (savings) of such a coalition be distributed among its members? To form a coalition the consent of every member is needed, but it is likely that the willingness of a player to participate in a coalition depends on what that player obtains in that coalition. Therefore, the second question seems to be the more basic one, and in this book attention is focussed on that question. Specifically, it is usually assumed that the *grand coalition N* of all players is formed, and the question is then reduced to the problem of distributing the amount $v(N)$ among the players. In the present example, how should the amount 220 $(= v(N))$ be distributed among the three cities? In other words, we look for vectors $\mathbf{x} = (x_1, x_2, x_3) \in \mathbb{R}^3$ such that $x_1 + x_2 + x_3 = 220$, where player $i \in \{1, 2, 3\}$ obtains x_i. One obvious candidate is to choose $x_1 = x_2 = x_3 = 220/3$, but this does not really reflect the asymmetry of the situation: some coalitions save more than others. The literature offers many quite different solutions to this distribution problem, among which are the *core*, the *Shapley value*, and the *nucleolus*.

The core consists of those payoff distributions that cannot be improved upon by any smaller coalition. For the three cities example, this means that the core consists of those vectors (x_1, x_2, x_3) such that $x_1 + x_2 + x_3 = 220, x_1, x_2, x_3 \geq 0, x_1 + x_2 \geq 90,$

$x_1 + x_3 \geq 100$, and $x_2 + x_3 \geq 120$. Hence, this is a large set and therefore it is rather indeterminate as a solution to the game.

In contrast, the Shapley value consists by definition of one point (vector). Roughly, according to the Shapley value, each player receives his average contribution to the worth (savings) of coalitions. More precisely, imagine the players entering the 'bargaining room' one at a time, say first player 1, then player 2, and finally player 3. When player 1 enters, he forms a coalition on his own, which has worth 0. When player 2 enters, they form the coalition $\{1, 2\}$, so that the contribution of player 2 is equal to $v(\{1, 2\}) - v(\{1\}) = 90 - 0 = 90$. Finally, player 3 enters and they form the grand coalition. Player 3's contribution is equal to $v(N) - v(\{1, 2\}) = 220 - 90 = 130$. Hence, this results in the payoff vector $(0, 90, 130)$. Now the Shapley value is obtained by repeating this argument for the five other possible orderings in which the players can enter the bargaining room, and then taking the average of the six payoff vectors. In this example this results in the distribution $(65, 75, 80)$.

Also the nucleolus consists of one point, in this case the vector $(56\frac{2}{3}, 76\frac{2}{3}, 86\frac{2}{3})$. The nucleolus is more complicated to define and harder to compute, and at this stage the reader should take these numbers for granted.

Formal definitions of all these concepts are provided in Chap. 9. See also Chaps. 16–20.

Comments The implicit assumptions for a game like this are, first, that a coalition which is actually formed, can make binding agreements on the distribution of its payoff and, second, that any payoff distribution which distributes (or, at least, does not exceed) the savings (or, more generally, *worth*) of the coalition is possible. For these reasons, such games are called *cooperative games with transferable utility*.

1.3.4.2 The Glove Game

Story Assume that there are three players, 1, 2, and 3. Players 1 and 2 each possess a right-hand glove, while player 3 has a left-hand glove. A pair of gloves has worth 1. The players cooperate in order to generate a profit.

Model The associated cooperative game is described by Table 1.2.

Solution The core of this game consists of exactly one vector. The Shapley value assigns 2/3 to player 3 and 1/6 to both player 1 and player 2 (see Problem 1.6). The nucleolus is the unique element of the core.

Table 1.2 The glove game

S	$\{1\}$	$\{2\}$	$\{3\}$	$\{1, 2\}$	$\{1, 3\}$	$\{2, 3\}$	$\{1, 2, 3\}$
$v(S)$	0	0	0	0	1	1	1

Table 1.3 Preferences for
dentist appointments

	Mon	Tue	Wed
Adams	2	4	8
Benson	10	5	2
Cooper	10	6	4

Table 1.4 The dentist game:
a permutation game

S	$\{1\}$	$\{2\}$	$\{3\}$	$\{1,2\}$	$\{1,3\}$	$\{2,3\}$	$\{1,2,3\}$
$v(S)$	2	5	4	14	18	9	24

1.3.4.3 A Permutation Game

Story Mr. Adams, Mrs. Benson, and Mr. Cooper have appointments with the dentist on Monday, Tuesday, and Wednesday, respectively. This schedule not necessarily matches their preferences, due to different urgencies and other factors. These preferences (expressed in numbers) are given in Table 1.3.

Model This situation gives rise to a game in which the coalitions can gain by reshuffling their appointments. For instance, Adams (player 1) and Benson (player 2) can change their appointments and obtain a total of 14 instead of 7. A complete description of the resulting game is given in Table 1.4.[3]

Solution The core of this game is the convex hull of the vectors $(15, 5, 4)$, $(14, 6, 4)$, $(8, 6, 10)$, and $(9, 5, 10)$, i.e., it is the quadrangle with these points as vertices, together with its inside. The Shapley value is the vector $(9\frac{1}{2}, 6\frac{1}{2}, 8)$, and the nucleolus is the vector $(11\frac{1}{2}, 5\frac{1}{2}, 7)$.

Comments See Chap. 20 for an analysis of permutation games.

1.3.4.4 A Voting Game

The United nations Security Council consists of five permanent members (United States, Russia, Britain, France, and China) and ten other members. Motions must be approved by at least nine members, including all the permanent members. This situation gives rise to a 15-player so-called voting game (N, v) with $v(S) = 1$ if the coalition S contains the five permanent members and at least four nonpermanent members, and $v(S) = 0$ otherwise. Games with worths 0 or 1 are also called *simple games*. Coalitions with worth equal to 1 are called winning, the other coalitions are called losing. Simple games are studied in Chap. 16.

A solution to such a voting game is interpreted as representing the power of a player, rather than payoff (money) or utility.

---

[3]The numbers in this table are the total payoffs to coalitions and not the net payoffs compared to the coalition members staying alone instead of cooperating. These would be, respectively, 0, 0, 0, 7, 12, 0, and 13.

1.3.5   Bargaining Games

Bargaining theory focusses on agreements between individual players.

1.3.5.1 A Division Problem

Story Consider the following situation. Two players have to agree on the division of
one unit of a perfectly divisible good, say a liter of wine. If they reach an agreement,
say (α, β) where $\alpha, \beta \geq 0$ and $\alpha + \beta \leq 1$, then they split up the one unit according to
this agreement; otherwise, they both receive nothing. The players have preferences
for the good, described by utility functions.

Model To fix ideas, assume that player 1 has a utility function $u_1(\alpha) = \alpha$ and player
2 has a utility function $u_2(\alpha) = \sqrt{\alpha}$. Thus, a distribution $(\alpha, 1-\alpha)$ of the good leads
to a corresponding pair of utilities $(u_1(\alpha), u_2(1 - \alpha)) = (\alpha, \sqrt{1-\alpha})$. By letting α
range from 0 to 1 we obtain all utility pairs corresponding to all distributions of
the whole unit of the good: this is the bold curve in Fig. 1.7. It is assumed that also
distributions summing to less than the whole unit are theoretically possible. This
yields the whole shaded region.

Solution According to the *Nash bargaining solution* this bargaining problem
should be solved as follows: maximize the product of the players' utilities on the
shaded area. Since this maximum will be reached on the boundary, the problem is
equivalent to

$$\max_{0 \leq \alpha \leq 1} \alpha\sqrt{1 - \alpha}.$$

The maximum is obtained for $\alpha = \frac{2}{3}$. So the solution of the bargaining problem in
utilities equals $(\frac{2}{3}, \frac{1}{3}\sqrt{3})$, which is the point z in Fig. 1.7. This implies that player
1 obtains $\frac{2}{3}$ of the 1 unit of the good, whereas player 2 obtains $\frac{1}{3}$. As described

Fig. 1.7 A bargaining game

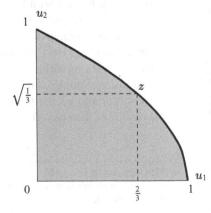

here, this solution may seem to come out of the blue, but it can be backed up axiomatically, see below.

Comments For an axiomatic characterization of the Nash bargaining solution, see Chaps. 10 and 21. The bargaining literature also includes many noncooperative, strategic approaches to the bargaining problem, notably the Rubinstein alternating offers model: see Chaps. 6 and 21. The bargaining game can be seen as a special case of a cooperative game without transferable utility. Also games with transferable utility form a subset of the more general class of games without transferable utility. See Chap. 21.

1.4   Cooperative Versus Noncooperative Game Theory

The usual distinction between cooperative and noncooperative game theory is that in a cooperative game binding agreements between players are possible, whereas this is not the case in noncooperative games. This distinction is informal and also not very clear-cut: for instance, the core of a cooperative game has a clear noncooperative flavor; a concept such as correlated equilibrium for noncooperative games (see Sect. 13.7) has a clear cooperative flavor. Moreover, quite some game-theoretic literature is concerned with viewing problems both from a cooperative and a noncooperative perspective. The latter approach is sometimes called the *Nash program*; the bargaining problem discussed above is a typical example. In a much more precise sense, the theory of *implementation* is concerned with representing outcomes from cooperative solutions as equilibrium outcomes of specific noncooperative solutions.

A workable distinction between cooperative and noncooperative games can be based on the 'modelling technique' that is used: in a noncooperative game players have explicit strategies, whereas in a cooperative game players and coalitions are characterized, more abstractly, by the outcomes and payoffs that they can reach. The examples in Sects. 1.3.1–1.3.3 are examples of noncooperative games, whereas those in Sects. 1.3.4 and 1.3.5 are examples of cooperative games.

1.5   Problems

1.1. *Battle of the Bismarck Sea*

(a) Represent the 'Battle of the Bismarck Sea' as a game in extensive form.
(b) Now assume that Imamura moves first, and Kenney observes Imamura's move and moves next. Represent this situation in extensive form and solve by working backwards.
(c) Answer the same questions as under (b) with now Kenney moving first.

1.2. *Variant of Matching Pennies*

Consider the following variant of the 'Matching Pennies' game:

$$
\begin{array}{cc}
 & \text{Heads} \quad \text{Tails} \\
\begin{array}{c} \text{Heads} \\ \text{Tails} \end{array} &
\begin{pmatrix} x & -1 \\ -1 & 1 \end{pmatrix}
\end{array}
$$

where x is a real number. For each value of x, determine all saddlepoints of the game, if any.

1.3. *Mixed Strategies*

Consider the following zero-sum game:

$$
\begin{array}{cc}
 & L \quad R \\
\begin{array}{c} T \\ B \end{array} &
\begin{pmatrix} 3 & 2 \\ 1 & 4 \end{pmatrix}
\end{array}.
$$

(a) Show that this game has no saddlepoint.
(b) Find a mixed strategy (randomized choice) of (the row) player 1 that makes his expected payoff independent of player 2's strategy.
(c) Find a mixed strategy of player 2 that makes his expected payoff independent of player 1's strategy.
(d) Consider the expected payoffs found under (b) and (c). What do you conclude about how the game could be played if randomized choices are allowed?

1.4. *Sequential Cournot*

Consider the sequential Cournot model in Sect. 1.3.3, but now based on the market price $p = 2 - 3Q$ (or zero, whichever is larger), where Q is the total quantity produced.

(a) Represent this game in extensive form, similar as in Fig. 1.2.
(b) Solve this game by working backwards.

1.5. *Three Cooperating Cities*

(a) Complete the computation of the Shapley value of the 'Three Cooperating Cities Game'. Is it an element of the core? Why or why not?
(b) Show that the nucleolus of this game is an element of the core.

1.6. *Glove Game*

(a) Compute the core of the glove game.
(b) Compute the Shapley value of this game. Is it an element of the core?

1.7. *Dentist Appointments*

(a) For the permutation (dentist appointments) game, compute the Shapley value
and check if it is an element of the core.
(b) Show that the nucleolus of this game is in the core.

1.8. *Nash Bargaining*

(a) Verify the computation of the Nash bargaining solution for the division problem
in Sect. 1.3.5.
(b) Compute the Nash bargaining outcome, both in utilities and in division of the
good, when the utility functions are $u_1(\alpha) = 2\alpha - \alpha^2$ and $u_2(\alpha) = \alpha$.

1.9. *Variant of Glove Game*
Suppose there are $n = \ell + r$ players, where ℓ players own a left-hand glove and r
players own a right-hand glove. Let N be the set of all players, let L be the subset of
N consisting of the players who own a left-hand glove, and let R be the subset of N
consisting of the players who own a right-hand glove. Let $S \subseteq N$ denote an arbitrary
coalition and let $|S|$ denote the number of players in S. For instance, $|L| = \ell$, and
$|R \cap S|$ is the number of players in S who own a right-hand glove. As before, each
pair of gloves has worth 1. Find an expression for $v(S)$, i.e., the maximal profit that
S can generate by cooperation of its members.

1.6   Notes

The 'Battle of the Bismarck Sea' example is taken from Rasmusen (1989). Also
see the memoirs of Churchill (1983): in 1953, Churchill received the Nobel prize in
literature for this work.

Von Neumann (1928) proved that every two-person finite zero-sum game in
which the players can use mixed strategies, has a value. This result is known as
the *minimax theorem*.

The Prisoners' Dilemma game has been widely studied in the literature. Axelrod
(1984) described the results of a tournament for which players could submit
strategies for repeated play of the Prisoners' Dilemma: the so-called tit-for-tat
strategy emerged as a winning strategy. For the 'tragedy of the commons', see
Hardin (1968), or Gibbons (1992, p. 27).

The concept of focal points was developed by Schelling (1960). Nash equilibria
were first explicitly proposed by Nash (1951), who proved that every game in which
each player has finitely many choices—zero-sum or nonzero-sum—has a Nash
equilibrium in mixed strategies. The basic idea of a Nash equilibrium is much older.
For instance, the Cournot equilibrium was developed in Cournot (1838).

Subgame perfect equilibria in extensive form games were first explicitly studied
by Selten (1965). Again, the basic idea occurs earlier, for instance in von Stackelberg
(1934).

The Three Cooperating Cities is an example of a minimum cost spanning tree game, see Bird (1976). The core was first introduced in Gillies (1953), the Shapley value in Shapley (1953), and the nucleolus in Schmeidler (1969). The dentist example is taken from Curiel (1997, p. 54). The United Nations Security Council game is taken from Owen (1995).

The Nash bargaining solution was proposed and axiomatically characterized in Nash (1950). Nash (1953) proposed a noncooperative game to back up the Nash bargaining solution. Rubinstein (1982) modelled the bargaining problem as an alternating offers extensive form game. Binmore et al. (1986) observed the close relationship between the Nash bargaining solution and the strategic approach of Rubinstein. See Chap. 10.

The reference list contains a number of textbooks on game theory, notably: Fudenberg and Tirole (1991a), Gardner (1995), Gibbons (1992), Maschler et al. (2013), Morris (1994), Moulin (1988), Moulin (1995), Myerson (1991), Osborne (2004), Owen (1995), Peleg and Sudhölter (2003), Perea (2012), Rasmusen (1989), Thomas (1986), and Watson (2002).

References

Aumann, R. J., & Maschler, M. (1985). Game theoretic analysis of a bankruptcy problem from the Talmud. *Journal of Economic Theory, 36*, 195–213.

Axelrod, R. (1984). *The evolution of cooperation.* New York: Basic Books.

Bertrand, J. (1883). Review of Walras's 'Théorie mathématique de la richesse sociale' and Cournot's 'Recherches sur les principes mathématiques de la théorie des richesses'. *Journal des Savants*, 499–508 [translated by M. Chevaillier and reprinted in Magnan de Bornier (1992), 646–653].

Binmore, K., Rubinstein, A., & Wolinsky, A. (1986). The Nash bargaining solution in economic modelling. *Rand Journal of Economics, 17*, 176–188.

Bird, C. G. (1976). On cost allocation for a spanning tree: a game theoretic approach. *Networks, 6*, 335–350.

Black, D. (1969). Lewis Carroll and the theory of games. In *American Economic Review, Proceedings*.

Bondareva, O. N. (1962). Theory of the core in the *n*-person game. *Vestnik Leningradskii Universitet, 13*, 141–142 (in Russian).

Brams, S. J. (1980). *Biblical games: A strategic analysis of stories in the Old Testament.* Cambridge: MIT Press.

Churchill, W. (1983). *Second world war.* Boston: Houghton Mifflin.

Cournot, A. (1838). *Recherches sur les principes mathématiques de la théorie des richesses* [English translation (1897) *Researches into the mathematical principles of the theory of wealth*]. New York: Macmillan.

Curiel, I. (1997). *Cooperative game theory and applications: Cooperative games arising from combinatorial optimization problems.* Boston: Kluwer Academic.

Dimand, M. A., & Dimand, R. W. (1996). *The history of game theory, volume 1: from the beginnings to 1945.* London: Routledge.

Dodgson, C. L. (1884). *The principles of parliamentary representation.* London: Harrison & Sons.

Dresher, M., Tucker. A. W., & Wolfe, P. (Eds.), (1957). *Contributions to the theory of games III. Annals of mathematics studies* (Vol. 39). Princeton: Princeton University Press.

Dresher, M., Shapley, L. S., & Tucker, A. W., (Eds.), (1964). *Advances in game theory. Annals of mathematics studies* (Vol. 52). Princeton: Princeton University Press.

Feyerabend, P. K. (1974). *Against method*. London: New Left Books.

Fudenberg, D., & Tirole, J. (1991a). *Game theory*. Cambridge: MIT Press.

Gardner, R. (1995). *Games for business and economics*. New York: Wiley.

Gibbons, R. (1992). *A primer in game theory*. Hertfordshire: Harvester Wheatsheaf.

Gillies, D. B. (1953). *Some theorems on n-person games*, Ph.D. Thesis. Princeton: Princeton University Press.

Hardin, G. (1968). The tragedy of the commons. *Science, 162*, 1243–1248.

Harsanyi, J. C. (1967/1968) Games with incomplete information played by "Bayesian" players I, II, and III. *Management Science, 14*, 159–182, 320–334, 486–502.

Kuhn, H. W., & Tucker, A. W. (Eds.), (1950). *Contributions to the theory of games I. Annals of mathematics studies* (Vol. 24). Princeton: Princeton University Press.

Kuhn, H. W., & Tucker, A. W. (Eds.), (1953). *Contributions to the theory of games II. Annals of mathematics studies* (Vol. 28). Princeton: Princeton University Press.

Luce, R. D., & Raiffa, H. (1957). *Games and decisions: Introduction and critical survey*. New York: Wiley.

Luce, R. D., & Tucker, A. W. (Eds.), (1958). *Contributions to the theory of games IV. Annals of mathematics studies* (Vol. 40). Princeton: Princeton University Press.

Maschler, M., Solan, E., & Zamir, S. (2013). *Game theory*. Cambridge: Cambridge University Press.

Morris, P. (1994). *Introduction to game theory*. New York: Springer.

Moulin, H. (1988). *Axioms of cooperative decision making*. Cambridge: Cambridge University Press.

Moulin, H. (1995). *Cooperative microeconomics; a game-theoretic introduction*. Hemel Hempstead: Prentice Hall/Harvester Wheatsheaf.

Myerson, R. B. (1991). *Game theory, analysis of conflict*. Cambridge: Harvard University Press.

Nasar, S. (1998). *A beautiful mind*. London: Faber and Faber Ltd.

Nash, J. F. (1950). The bargaining problem. *Econometrica, 18*, 155–162.

Nash, J. F. (1951). Non-cooperative games. *Annals of Mathematics, 54*, 286–295.

Nash, J. F. (1953). Two-person cooperative games. *Econometrica, 21*, 128–140.

Osborne, M. J. (2004). *An introduction to game theory*. New York: Oxford University Press.

Owen, G. (1995). *Game theory* (3rd ed.). San Diego: Academic.

Peleg, B., & Sudhölter, P. (2003). *Introduction to the theory of cooperative games*. Boston: Kluwer Academic.

Perea, A. (2012). *Epistemic game theory: Reasoning and choice*. Cambridge: Cambridge University Press.

Rasmusen, E. (1989). *Games and information: An introduction to game theory* (2nd ed., 1994/1995). Oxford: Basil Blackwell.

Rubinstein, A. (1982). Perfect equilibrium in a bargaining model. *Econometrica, 50*, 97–109.

Schelling, T. C. (1960). *The strategy of conflict*. Cambridge: Harvard University Press.

Schmeidler, D. (1969). The nucleolus of a characteristic function game. *SIAM Journal on Applied Mathematics, 17*, 1163–1170.

Selten, R. (1965). Spieltheoretische Behandlung eines Oligopolmodels mit Nachfragezeit. *Zeitschrift für Gesammte Staatswissenschaft, 121*, 301–324.

Selten, R. (1975). Reexamination of the perfectness concept for equilibrium points in extensive games. *International Journal of Game Theory, 4*, 25–55.

Shapley, L. S. (1953). A value for *n*-person games. In: A. W. Tucker & H. W. Kuhn (Eds.), *Contributions to the theory of games II* (pp. 307–317). Princeton: Princeton University Press.

Shapley, L. S. (1967). On balanced sets and cores. *Naval Research Logistics Quarterly, 14*, 453–460.

Sun Tzu (1988) *The art of war* [Translated by Thomas Cleary]. Boston: Shambala.

Thomas, L. C. (1986). *Games, theory and applications*. Chichester: Ellis Horwood Limited.

von Neumann, J. (1928). Zur Theorie der Gesellschaftsspiele. *Mathematische Annalen, 100*, 295–320.

von Neumann, J., & Morgenstern, O. (1944/1947) *Theory of games and economic behavior*. Princeton: Princeton University Press.

von Stackelberg, H. F. (1934). *Marktform und Gleichgewicht*. Wien: Julius Springer.

Watson, J. (2002). *Strategy, an introduction to game theory*. New York: Norton.

Zermelo, E. (1913). Über eine Anwendung der Mengenlehre auf die Theorie des Schachspiels. In *Proceedings Fifth International Congress of Mathematicians* (Vol. 2, pp. 501–504).

Part I

Thinking Strategically

Finite Two-Person Zero-Sum Games

2

This chapter deals with two-player games in which each player chooses from finitely many pure strategies or randomizes among these strategies, and the sum of the players' payoffs or expected payoffs is always equal to zero. Games like the Battle of the Bismarck Sea and Matching Pennies, discussed in Sect. 1.3.1 belong to this class.

In Sect. 2.1 the basic definitions and theory are discussed. Section 2.2 shows how to solve $2 \times n$ and $m \times 2$ games, and larger games by elimination of strictly dominated strategies.

2.1 Basic Definitions and Theory

Since all data of a finite two-person zero-sum game can be summarized in one matrix, such a game is usually called a 'matrix game'.

Definition 2.1 (Matrix Game) A *matrix game* is an $m \times n$ matrix A of real numbers, where m is the number of rows and n is the number of columns. A *(mixed) strategy* of player 1 is a probability distribution $\mathbf{p}$ over the rows of A, i.e., an element of the set

$$\Delta^m := \{\mathbf{p} = (p_1, \ldots, p_m) \in \mathbb{R}^m \mid \sum_{i=1}^{m} p_i = 1, \ p_i \geq 0 \text{ for all } i = 1, \ldots, m\} \ .$$

Similarly, a *(mixed) strategy* of player 2 is a probability distribution $\mathbf{q}$ over the columns of A, i.e., an element of the set

$$\Delta^n := \{\mathbf{q} = (q_1, \ldots, q_n) \in \mathbb{R}^n \mid \sum_{j=1}^{n} q_j = 1, \ q_j \geq 0 \text{ for all } j = 1, \ldots, n\} \ .$$

© Springer-Verlag Berlin Heidelberg 2015

H. Peters, *Game Theory*, Springer Texts in Business and Economics,

DOI 10.1007/978-3-662-46950-7_2

A strategy $\mathbf{p}$ of player 1 is called *pure* if there is a row i with $p_i = 1$. This strategy is also denoted by $\mathbf{e}^i$. Similarly, a strategy $\mathbf{q}$ of player 2 is called *pure* if there is a column j with $q_j = 1$. This strategy is also denoted by $\mathbf{e}^j$.   □

The interpretation of a matrix game A is as follows. If player 1 plays row i (i.e., pure strategy $\mathbf{e}^i$) and player 2 plays column j (i.e., pure strategy $\mathbf{e}^j$), then player 1 receives payoff a_{ij} and player 2 pays a_{ij} (and, thus, receives $-a_{ij}$), where a_{ij} is the number in row i and column j of matrix A. If player 1 plays strategy $\mathbf{p}$ and player 2 plays strategy $\mathbf{q}$, then player 1 receives the expected payoff

$$\mathbf{p}A\mathbf{q} = \sum_{i=1}^{m}\sum_{j=1}^{n} p_i q_j a_{ij} \, ,$$

and player 2 receives $-\mathbf{p}A\mathbf{q}$.

Remark 2.2 (1) Note that, according to Definition 2.1, a strategy means a mixed strategy. A pure strategy is a special case of a mixed strategy. We add the adjective 'pure' if we wish to refer to a pure strategy. (2) Since no confusion is likely to arise, we do not use transpose notations like $\mathbf{p}^T A\mathbf{q}$ or $\mathbf{p}A\mathbf{q}^T$.   □

For 'solving' matrix games, i.e., establishing what clever players would or should do, the concepts of maximin and minimax strategies are important, as will be explained below. First we give the definitions.

Definition 2.3 (Maximin and Minimax Strategies) A strategy $\mathbf{p}$ is a *maximin strategy* of player 1 in matrix game A if

$$\min\{\mathbf{p}A\mathbf{q} \mid \mathbf{q} \in \Delta^n\} \geq \min\{\mathbf{p}'A\mathbf{q} \mid \mathbf{q} \in \Delta^n\} \text{ for all } \mathbf{p}' \in \Delta^m \, .$$

A strategy $\mathbf{q}$ is a *minimax strategy* of player 2 in matrix game A if

$$\max\{\mathbf{p}A\mathbf{q} \mid \mathbf{p} \in \Delta^m\} \leq \max\{\mathbf{p}A\mathbf{q}' \mid \mathbf{p} \in \Delta^m\} \text{ for all } \mathbf{q}' \in \Delta^n \, .$$

                                                                                      □

In words: a maximin strategy of player 1 maximizes the minimal (with respect to player 2's strategies) payoff of player 1, and a minimax strategy of player 2 minimizes the maximum (with respect to player 1's strategies) that player 2 has to pay to player 1. (It can be proved by basic mathematical analysis that maximin and minimax strategies always exist.) Of course, the asymmetry in these definitions is caused by the fact that, by convention, a matrix game represents the amounts that player 2 has to pay to player 1.

In order to check if a strategy $\mathbf{p}$ of player 1 is a maximin strategy it is sufficient to check that the first inequality in Definition 2.3 holds with $\mathbf{e}^j$ for every $j = 1, \ldots, n$

instead of every $\mathbf{q} \in \Delta^n$. This is not difficult to see but the reader is referred to Chap. 12 for a formal treatment. A similar observation holds for minimax strategies. In other words, to check if a strategy is maximin (minimax) it is sufficient to consider its performance against every pure strategy, i.e., column (row).

Why would we be interested in such strategies? At first glance, these strategies seem to express a very conservative or pessimistic, worst-case scenario attitude. The reason for nevertheless considering maximin/minimax strategies is provided by the so-called *minimax theorem*, which states that for every matrix game A there is a real number $v = v(A)$ with the following properties:

(a) A strategy $\mathbf{p}$ of player 1 guarantees a payoff of at least v to player 1 (i.e., $\mathbf{p}A\mathbf{q} \geq v$ for all strategies $\mathbf{q}$ of player 2) if and only if $\mathbf{p}$ is a maximin strategy.
(b) A strategy $\mathbf{q}$ of player 2 guarantees a payment of at most v by player 2 to player 1 (i.e., $\mathbf{p}A\mathbf{q} \leq v$ for all strategies $\mathbf{p}$ of player 1) if and only if $\mathbf{q}$ is a minimax strategy.

Hence, player 1 can obtain a payoff of at least v by playing a maximin strategy, and player 2 can guarantee to pay not more than v—hence secure a payoff of at least $-v$—by playing a minimax strategy. For these reasons, the number $v = v(A)$ is also called the *value* of the game A—it represents the worth to player 1 of playing the game A—and maximin and minimax strategies are called *optimal strategies* for players 1 and 2, respectively.

Therefore, 'solving' the game A means, naturally, determining the optimal strategies and the value of the game. In the Battle of the Bismarck Sea in Sect. 1.3.1, the pure strategies N of both players guarantee the same amount 2. Therefore, this is the value of the game and N is optimal for both players. The analysis of that game is easy since it has a 'saddlepoint', namely position $(1, 1)$ with $a_{11} = 2$. The definition of a saddlepoint is as follows.

Definition 2.4 (Saddlepoint) A position (i, j) in a matrix game A is a *saddlepoint* if

$$a_{ij} \geq a_{kj} \text{ for all } k = 1, \ldots, m \text{ and } a_{ij} \leq a_{ik} \text{ for all } k = 1, \ldots, n \,,$$

i.e., if a_{ij} is maximal in its column j and minimal in its row i.   □

Clearly, if (i, j) is a saddlepoint, then player 1 can guarantee a payoff of at least a_{ij} by playing the pure strategy row i, since a_{ij} is minimal in row i. Similarly, player 2 can guarantee a payoff of at least $-a_{ij}$ by playing the pure strategy column j, since a_{ij} is maximal in column j. Hence, a_{ij} must be the value of the game A: $v(A) = a_{ij}$, $\mathbf{e}^i$ is an optimal (maximin) strategy of player 1, and $\mathbf{e}^j$ is an optimal (minimax) strategy of player 2.

2.2   Solving 2 × n Games and m × 2 Games

In this section we show how to solve matrix games where at least one of the players
has only two pure strategies. We also show how the idea of strict domination can be
of help in solving matrix games.

2.2.1   2 × n Games

We demonstrate how to solve a matrix game with 2 rows and n columns graphically,
by considering the following 2×4 example:

$$
\begin{array}{cccc}
\mathbf{e}^1 & \mathbf{e}^2 & \mathbf{e}^3 & \mathbf{e}^4
\end{array}
$$
$$
A = \begin{pmatrix} 10 & 2 & 4 & 1 \\ 2 & 10 & 8 & 12 \end{pmatrix}.
$$

We have labelled the columns of A, i.e., the pure strategies of player 2 for reference
below. Let $\mathbf{p} = (p, 1 - p)$ be an arbitrary strategy of player 1. The expected payoffs
to player 1 if player 2 plays a pure strategy are equal to:

$$
\begin{aligned}
\mathbf{p}A\mathbf{e}^1 &= 10p + 2(1 - p) = 8p + 2 \\
\mathbf{p}A\mathbf{e}^2 &= 2p + 10(1 - p) = 10 - 8p \\
\mathbf{p}A\mathbf{e}^3 &= 4p + 8(1 - p) \ \ = 8 - 4p \\
\mathbf{p}A\mathbf{e}^4 &= p + 12(1 - p) \ \ = 12 - 11p.
\end{aligned}
$$

We plot these four linear functions of p in one diagram:

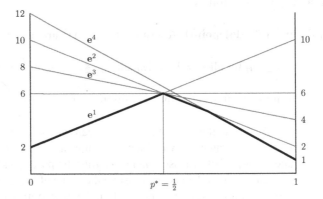

In this diagram the values of p are plotted on the horizontal axis, and the four straight
gray lines plot the payoffs to player 1 for each of the four pure strategies of player 2.
Observe that for every $0 \le p \le 1$ the minimum payoff that player 1 may obtain is
given by the lower envelope of these curves, the thick black curve in the diagram:

for any p, any combination (q_1, q_2, q_3, q_4) of the points on $\mathbf{e}^1$, $\mathbf{e}^2$, $\mathbf{e}^3$, and $\mathbf{e}^4$ with first coordinate p would lie on or above this lower envelope. Clearly, the lower envelope is maximal for $p = p^* = \frac{1}{2}$, and the maximal value is 6. Hence, we have established that player 1 has a unique optimal (maximin) strategy, namely $\mathbf{p}^* = (\frac{1}{2}, \frac{1}{2})$, and that the value of the game, $v(A)$, is equal to 6.

What are the optimal or minimax strategies of player 2? From the theory of the previous section we know that a minimax strategy $\mathbf{q} = (q_1, q_2, q_3, q_4)$ of player 2 should guarantee to player 2 to have to pay at most the value of the game. From the diagram it is clear that q_4 should be equal to zero since otherwise the payoff to player 1 would be larger than 6 if player 1 plays $(\frac{1}{2}, \frac{1}{2})$, and thus $\mathbf{q}$ would not be a minimax strategy. So a minimax strategy has the form $\mathbf{q} = (q_1, q_2, q_3, 0)$. Any such strategy, plotted in the diagram, gives a straight line that is a combination of the lines associated with $\mathbf{e}^1$, $\mathbf{e}^2$, and $\mathbf{e}^3$ and which passes through the point $(\frac{1}{2}, 6)$ since all three lines pas through this point. Moreover, for no value of p should this straight line exceed the value 6, otherwise $\mathbf{q}$ would not guarantee a payment of at most 6 by player 2. Consequently, this straight line has to be horizontal. Summarizing this argument, we look for numbers $q_1, q_2, q_3 \geq 0$ such that

$$2q_1 + 10q_2 + 8q_3 = 6 \quad \text{(left endpoint should be } (0, 6))$$
$$10q_1 + 2q_2 + 4q_3 = 6 \quad \text{(right endpoint should be } (1, 6))$$
$$q_1 + q_2 + q_3 = 1 \quad \text{(}\mathbf{q}\text{ is a probability vector)} .$$

By substitution, it is easy to reduce this system of equations to the two equations

$$3q_1 - q_2 = 1$$
$$q_1 + q_2 + q_3 = 1 .$$

In fact, one of the two first equations could have been omitted from the beginning, since we already know that any combination of the three lines passes through $(\frac{1}{2}, 6)$, and two points are sufficient to determine a straight line.

From the remaining two equations, we obtain that the set of optimal strategies of player 2 is

$$\{\mathbf{q} = (q_1, q_2, q_3, q_4) \in \Delta^4 \mid q_2 = 3q_1 - 1, \ q_4 = 0\} .$$

Note that, if $q_1 = \frac{1}{3}$, then $q_2 = 0$, and if $q_1 = \frac{1}{2}$, then $q_2 = \frac{1}{2}$. Clearly, q_1 and q_2 cannot be smaller, since then their sum would be negative, and they cannot be larger since then their sum would exceed 1. Hence, the set of optimal strategies of player 2 can alternatively be described as

$$\{\mathbf{q} = (q_1, q_2, q_3, q_4) \in \mathbb{R}^4 \mid \frac{1}{3} \leq q_1 \leq \frac{1}{2}, \ q_2 = 3q_1 - 1, \ q_3 = 1 - q_1 - q_2, \ q_4 = 0\} .$$

This means that the set of optimal strategies of player 2 in this game is one-dimensional, i.e., a line segment.

2.2.2 $m \times 2$ Games

The solution method to solve $m \times 2$ games is analogous. Consider the following example:

$$A = \begin{array}{c} \mathbf{e}^1 \\ \mathbf{e}^2 \\ \mathbf{e}^3 \\ \mathbf{e}^4 \end{array} \left(\begin{array}{cc} 10 & 2 \\ 2 & 10 \\ 4 & 8 \\ 1 & 12 \end{array} \right).$$

Let $\mathbf{q} = (q, 1 - q)$ be an arbitrary strategy of player 2. Again, we make a diagram in which now the values of q are put on the horizontal axis, and the straight lines indicated by $\mathbf{e}^i$ for $i = 1, 2, 3, 4$ are the payoffs to player 1 associated with his four pure strategies (rows) as functions of q. The equations of these lines are given by:

$$\begin{aligned} \mathbf{e}^1 A \mathbf{q} &= 10q + 2(1 - q) = 8q + 2 \\ \mathbf{e}^2 A \mathbf{q} &= 2q + 10(1 - q) = 10 - 8q \\ \mathbf{e}^3 A \mathbf{q} &= 4q + 8(1 - q) \ = 8 - 4q \\ \mathbf{e}^4 A \mathbf{q} &= q + 12(1 - q) \ = 12 - 11q \, . \end{aligned}$$

The resulting diagram is as follows.

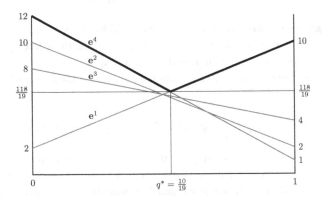

Observe that the maximum payments that player 2 has to make are now located on the upper envelope, represented by the thick black curve. The minimum is reached at the point of intersection of $\mathbf{e}^1$ and $\mathbf{e}^4$ in the diagram, which has coordinates $(\frac{10}{19}, \frac{118}{19})$. Hence, the value of the game is $\frac{118}{19}$, and the unique optimal (minimax) strategy of player 2 is $\mathbf{q}^* = (\frac{10}{19}, \frac{9}{19})$.

To find the optimal strategy or strategies $\mathbf{p} = (p_1, p_2, p_3, p_4)$ of player 1, it follows from the diagram that $p_2 = p_3 = 0$, otherwise for $q = \frac{10}{19}$ the value $\frac{118}{19}$ of the game is not reached, so that $\mathbf{p}$ is not a maximin strategy. So we look for a combination of $\mathbf{e}^1$ and $\mathbf{e}^4$ that gives at least $\frac{118}{19}$ for every q, hence it has to be equal to $\frac{118}{19}$ for every q. This gives rise to the equations $2p_1 + 12p_4 = 10p_1 + p_4 = \frac{118}{19}$ and $p_1 + p_4 = 1$, with unique solution $p_1 = \frac{11}{19}$ and $p_4 = \frac{8}{19}$. So the unique optimal strategy of player 1 is $(\frac{11}{19}, 0, 0, \frac{8}{19})$.

2.2.3 Strict Domination

The idea of strict domination can be used to first eliminate pure strategies before the graphical analysis of a matrix game. Consider the game

$$
A = \begin{pmatrix} 10 & 2 & 5 & 1 & 6 \\ 2 & 10 & 8 & 12 & 9 \end{pmatrix}.
\begin{matrix} \mathbf{e}^1 & \mathbf{e}^2 & \mathbf{e}^3 & \mathbf{e}^4 & \mathbf{e}^5 \end{matrix}
$$

This game is obtained from the game in Sect. 2.2.1 by adding a fifth column and changing a_{13} from 4 to 5.

In this game it cannot be optimal for player 2 to put positive probability on the fifth column, since the payoffs in the third column are always—that is, no matter what player 1's strategy is—better for player 2. So we can assume that the fifth column is played with zero probability: it is strictly dominated by the third column. This is a case where a pure strategy is strictly dominated by another pure strategy.

It can also happen that a pure strategy is strictly dominated by a mixed strategy. For instance, consider the third column. The payoff 5 in the first row is in between the payoffs 10 and 2 in the first row and first and second columns, respectively; and also the payoff 8 in the second row is in between the payoffs 2 and 10 in the second row and first and second columns. So it may be possible to find a combination of the first two columns that results in smaller payoffs than those in the third column. In order to see if such a combination is possible, suppose that probability α is put on the first column and $1 - \alpha$ is put on the second column. The resulting payoffs are

$$
\alpha \begin{pmatrix} 10 \\ 2 \end{pmatrix} + (1 - \alpha) \begin{pmatrix} 2 \\ 10 \end{pmatrix} = \begin{pmatrix} 8\alpha + 2 \\ 10 - 8\alpha \end{pmatrix}.
$$

We wish to have $8\alpha + 2 < 5$ and $10 - 8\alpha < 8$, and both inequalities hold as long as $\frac{1}{4} < \alpha < \frac{3}{8}$. This means that the pure strategy $\mathbf{e}^3$ is strictly dominated by any (mixed) strategy $(\alpha, 1 - \alpha, 0, 0, 0)$, as long as α is in the computed range. This implies that, in an optimal strategy, the probability q_3 put by player 2 on the third column must be zero, otherwise player 2 could guarantee to pay less by adding αq_3 to the first column and $(1 - \alpha)q_3$ to the second column, for any $\frac{1}{4} < \alpha < \frac{3}{8}$, and playing the third column with zero probability.

The preceding analysis implies that, in order to solve the above game, we can start by eliminating the third and fifth columns of the matrix. Thus, in the diagram in Sect. 2.2.1, we do not have to draw the line corresponding to $\mathbf{e}^3$. The value of the game is still 6, player 1 still has a unique optimal strategy $\mathbf{p}^* = (\frac{1}{2}, \frac{1}{2})$, and player 2 now also has a unique optimal strategy, namely the one where $q_3 = 0$, which is the strategy $(\frac{1}{2}, \frac{1}{2}, 0, 0, 0)$.

In general, strictly dominated pure strategies in a matrix game are not played with positive probability in any optimal strategy and can therefore be eliminated before solving the game. Sometimes, this idea can also be used to solve matrix games in which each player has more than two pure strategies ($m, n > 2$). Moreover, the idea can be applied iteratively, that is, after elimination of a strictly dominated pure strategy, in the smaller game perhaps another strictly dominated pure strategy can be eliminated, etc., until no more pure strategies are strictly dominated. See Example 2.7 for an illustration, and see Chap. 13 for a rigorous treatment.

We first give the formal definition of strict domination, and then discuss the announced example.

Definition 2.5 (Strict Domination) Let A be an $m \times n$ matrix game and i a row. The pure strategy $\mathbf{e}^i$ is *strictly dominated* if there is a strategy $\mathbf{p} = (p_1, \ldots, p_m) \in \Delta^m$ such that $\mathbf{p}A\mathbf{e}^j > \mathbf{e}^iA\mathbf{e}^j$ for every $j = 1, \ldots, n$. Similarly, let j be a column. The pure strategy $\mathbf{e}^j$ is *strictly dominated* if there is a strategy $\mathbf{q} = (q_1, \ldots, q_n) \in \Delta^n$ such that $\mathbf{e}^iA\mathbf{q} < \mathbf{e}^iA\mathbf{e}^j$ for every $i = 1, \ldots, m$. $\qquad\square$

Remark 2.6 It is not difficult to see that if $\mathbf{e}^i$ is strictly dominated, then it is strictly dominated by some $\mathbf{p} \in \Delta^m$ with $p_i = 0$. Similarly, if $\mathbf{e}^j$ is strictly dominated, then it is strictly dominated by some $\mathbf{q} \in \Delta^n$ with $q_j = 0$. $\qquad\square$

Example 2.7 Consider the following 3×3 matrix game:

$$A = \begin{pmatrix} 6 & 0 & 2 \\ 0 & 5 & 4 \\ 3 & 2 & 1 \end{pmatrix}.$$

For player 1, the third strategy $\mathbf{e}^3$ is strictly dominated by the strategy $\mathbf{p} = (\frac{7}{12}, \frac{5}{12}, 0)$, since

$$\mathbf{p}A = \begin{pmatrix} \frac{7}{2} & \frac{25}{12} & \frac{17}{6} \end{pmatrix} \text{ and } \mathbf{e}^3A = \begin{pmatrix} 3 & 2 & 1 \end{pmatrix}.$$

Hence, in any optimal strategy player 1 puts zero probability on the third row. Elimination of this row results in the matrix

$$B = \begin{pmatrix} 6 & 0 & 2 \\ 0 & 5 & 4 \end{pmatrix}.$$

Now, player 2's third strategy $\mathbf{e}^3$ is strictly dominated by the strategy $\mathbf{q} = (\frac{1}{4}, \frac{3}{4}, 0)$, since

$$B\mathbf{q} = \begin{pmatrix} \frac{3}{2} \\ \frac{15}{4} \end{pmatrix} \text{ and } B\mathbf{e}^3 = \begin{pmatrix} 2 \\ 4 \end{pmatrix}.$$

Hence, in any optimal strategy player 2 puts zero probability on the third column. Elimination of this column results in the matrix

$$C = \begin{pmatrix} 6 & 0 \\ 0 & 5 \end{pmatrix}.$$

This is a 2×2 matrix game, which can be solved by the method in Sect. 2.2.1 or Sect. 2.2.2. See Problem 2.1(a).   □

2.3 Problems

2.1. *Solving Matrix Games*
Solve the following matrix games, i.e., determine the optimal strategies and the value of the game. Each time, start by checking if the game has a saddlepoint.

(a)

$$\begin{pmatrix} 6 & 0 \\ 0 & 5 \end{pmatrix}$$

What are the optimal strategies in the original matrix game A in Example 2.7?

(b)

$$\begin{pmatrix} 2 & -1 & 0 & 2 \\ 2 & 0 & 0 & 3 \\ 0 & 0 & -1 & 2 \end{pmatrix}$$

(c)

$$\begin{pmatrix} 1 & 3 & 1 \\ 2 & 2 & 0 \\ 0 & 3 & 2 \end{pmatrix}$$

(d)

$$\begin{pmatrix} 16 & 12 & 2 \\ 2 & 6 & 16 \\ 8 & 8 & 6 \\ 0 & 7 & 8 \end{pmatrix}$$

(e)

$$\begin{pmatrix} 3 & 1 & 4 & 0 \\ 1 & 2 & 0 & 5 \end{pmatrix}$$

(f)

$$\begin{pmatrix} 1 & 0 & 2 \\ 4 & 1 & 1 \\ 3 & 1 & 3 \end{pmatrix}.$$

2.2. *Saddlepoints*

(a) Let A be an arbitrary $m \times n$ matrix game. Show that any two saddlepoints must have the same value. In other words, if (i,j) and (k,l) are two saddlepoints, show that $a_{ij} = a_{kl}$.
(b) Let A be a 4×4 matrix game in which $(1,1)$ and $(4,4)$ are saddlepoints. Show that A has at least two other saddlepoints.
(c) Give an example of a 4×4 matrix game with exactly three saddlepoints.

2.3. *Maximin Rows and Minimax Columns*
Row i is a *maximin row* in an $m \times n$ matrix game A if $\min_{j \in \{1,\dots,n\}} a_{ij} \geq \min_{j \in \{1,\dots,n\}} a_{kj}$ for all $k \in \{1,\dots,m\}$. Column j is a *minimax column* if $\max_{i \in \{1,\dots,m\}} a_{ij} \leq \max_{i \in \{1,\dots,m\}} a_{i\ell}$ for all $\ell \in \{1,\dots,n\}$.
Consider the following matrix game:

$$A = \begin{pmatrix} 3 & 2 & 0 \\ 1 & 2 & 2 \\ 0 & 2 & 4 \\ 0 & 3 & 1 \end{pmatrix}.$$

(a) Determine all maximin rows and minimax columns. What can you conclude from this about the value of this game?
(b) The value of this game is $\frac{12}{7}$. Use this to give an argument why player 2 will put zero probability on column 2 in any minimax strategy.
(c) Determine all minimax strategies of player 2 and all maximin strategies of player 1.

2.4. *Subgames of Matrix Games*
Consider the following matrix game:

$$A = \begin{pmatrix} 3 & 1 & 4 & 0 \\ 1 & 2 & 0 & 5 \end{pmatrix}.$$

(a) Determine all maximin rows and minimax columns. (See Problem 2.3.) What can you conclude from this about the value of this game?

(b) Consider the six different 2×2-matrix games that can be obtained by choosing two columns from A, as follows:

$$A_1 = \begin{pmatrix} 3 & 1 \\ 1 & 2 \end{pmatrix} \quad A_2 = \begin{pmatrix} 3 & 4 \\ 1 & 0 \end{pmatrix} \quad A_3 = \begin{pmatrix} 3 & 0 \\ 1 & 5 \end{pmatrix}$$

$$A_4 = \begin{pmatrix} 1 & 4 \\ 2 & 0 \end{pmatrix} \quad A_5 = \begin{pmatrix} 1 & 0 \\ 2 & 5 \end{pmatrix} \quad A_6 = \begin{pmatrix} 4 & 0 \\ 0 & 5 \end{pmatrix}.$$

Determine the values of all these games. Which one must be equal to the value of A?

(c) Determine all maximin and minimax strategies of A. [Hint: Use your answer to (b).]

2.5. *Rock-Paper-Scissors*

In the famous Rock-Paper-Scissors two-player game each player has three pure strategies: Rock, Paper, and Scissors. Here, Scissors beats Paper, Paper beats Rock, Rock beats Scissors. Assign a 1 to winning, 0 to a draw, and -1 to losing. Model this game as a matrix game, try to guess its optimal strategies, and then show that these are the unique optimal strategies. What is the value of this game?

2.4 Notes

The theory of zero-sum games was developed by von Neumann (1928), who proved the minimax theorem. In general, matrix games can be solved by Linear Programming. See Chap. 12 for details.

Reference

von Neumann, J. (1928). Zur Theorie der Gesellschaftsspiele. *Mathematische Annalen, 100,* 295–320.

Finite Two-Person Games

3

In this chapter we consider two-player games where each player chooses from finitely many pure strategies or randomizes among these strategies. In contrast to Chap. 2 it is no longer required that the sum of the players' payoffs is zero (or, equivalently, constant). This allows for a much larger class of games, including many games relevant for economic or other applications. Famous examples are the Prisoners' Dilemma and the Battle of the Sexes discussed in Sect. 1.3.2.

In Sect. 3.1 we introduce the model and the concept of Nash equilibrium. Section 3.2 shows how to compute Nash equilibria in pure strategies for arbitrary games, all Nash equilibria in games where both players have exactly two pure strategies, and how to use the concept of strict domination to facilitate computation of Nash equilibria and to compute equilibria also of larger games. The structure of this chapter thus parallels the structure of Chap. 2. For a deeper and more comprehensive analysis of finite two-person games see Chap. 13.

3.1 Basic Definitions and Theory

The data of a finite two-person game can be summarized by two matrices. Usually, these matrices are written as one matrix with two numbers at each position. Therefore, such games are often called 'bimatrix games'. The definition is as follows.

Definition 3.1 (Bimatrix Game) A *bimatrix game* is a pair of $m \times n$ matrices (A, B), where m is the number of rows and n the number of columns. □

The interpretation of a bimatrix game (A, B) is that, if player 1 (the row player) plays row i and player 2 (the column player) plays column j, then player 1 receives payoff a_{ij} and player 2 receives b_{ij}, where these numbers are the corresponding entries of A and B, respectively. Definitions and notations for pure and mixed strategies, strategy

© Springer-Verlag Berlin Heidelberg 2015
H. Peters, *Game Theory*, Springer Texts in Business and Economics,
DOI 10.1007/978-3-662-46950-7_3

sets and expected payoffs are similar to those for matrix games, see Sect. 2.1, but for easy reference we repeat them here. A *(mixed) strategy* of player 1 is a probability distribution $\mathbf{p}$ over the rows of A and B, i.e., an element of the set

$$\Delta^m := \{\mathbf{p} = (p_1, \ldots, p_m) \in \mathbb{R}^m \mid \sum_{i=1}^{m} p_i = 1, \; p_i \geq 0 \text{ for all } i = 1, \ldots, m\} .$$

Similarly, a *(mixed) strategy* of player 2 is a probability distribution $\mathbf{q}$ over the columns of A and B, i.e., an element of the set

$$\Delta^n := \{\mathbf{q} = (q_1, \ldots, q_n) \in \mathbb{R}^n \mid \sum_{j=1}^{n} q_j = 1, \; q_j \geq 0 \text{ for all } j = 1, \ldots, n\} .$$

A strategy $\mathbf{p}$ of player 1 is called *pure* if there is a row i with $p_i = 1$. This strategy is also denoted by $\mathbf{e}^i$. Similarly, a strategy $\mathbf{q}$ of player 2 is called *pure* if there is a column j with $q_j = 1$. This strategy is also denoted by $\mathbf{e}^j$. If player 1 plays $\mathbf{p}$ and player 2 plays $\mathbf{q}$ then the payoff to player 1 is the expected payoff

$$\mathbf{p}A\mathbf{q} = \sum_{i=1}^{m} \sum_{j=1}^{n} p_i q_j a_{ij} ,$$

and the payoff to player 2 is the expected payoff

$$\mathbf{p}B\mathbf{q} = \sum_{i=1}^{m} \sum_{j=1}^{n} p_i q_j b_{ij} .$$

As mentioned, the entries of A and B are usually grouped together in one (bi)matrix, by putting the pair a_{ij}, b_{ij} at position (i, j) of the matrix. Cf. the examples in Sect. 1.3.2.

Central to noncooperative game theory is the idea of best reply. It says that a rational selfish player should always maximize his (expected) payoff, given his knowledge of or conjecture about the strategies chosen by the other players.

Definition 3.2 (Best Reply) A strategy $\mathbf{p}$ of player 1 is a *best reply* to a strategy $\mathbf{q}$ of player 2 in an $m \times n$ bimatrix game (A, B) if

$$\mathbf{p}A\mathbf{q} \geq \mathbf{p}'A\mathbf{q} \text{ for all } \mathbf{p}' \in \Delta^m .$$

Similarly, $\mathbf{q}$ is a *best reply* of player 2 to $\mathbf{p}$ if

$$\mathbf{p}B\mathbf{q} \geq \mathbf{p}B\mathbf{q}' \text{ for all } \mathbf{q}' \in \Delta^n .$$

$\square$

In a Nash equilibrium, each player's strategy is a best reply to the other player's strategy.

Definition 3.3 (Nash Equilibrium)   A pair of strategies $(\mathbf{p}^*, \mathbf{q}^*)$ in a bimatrix game (A, B) is a *Nash equilibrium* if $\mathbf{p}^*$ is a best reply of player 1 to $\mathbf{q}^*$ and $\mathbf{q}^*$ is a best reply of player 2 to $\mathbf{p}^*$. A Nash equilibrium $(\mathbf{p}^*, \mathbf{q}^*)$ is called *pure* if both $\mathbf{p}^*$ and $\mathbf{q}^*$ are pure strategies. $\square$

The concept of a Nash equilibrium can be extended to arbitrary games, including games with arbitrary numbers of players, strategy sets, and payoff functions. We will see many examples in later chapters.

Every bimatrix game has a Nash equilibrium: for a proof see Sect. 13.1. Generally speaking, the main concern with Nash equilibrium is not its existence but rather the opposite, namely its abundance, as well as its interpretation. In many games, there are many Nash equilibria, and then the questions of equilibrium selection and equilibrium refinement are relevant (cf. Chap. 13). With respect to interpretation, an old question is how the players would come to play a Nash equilibrium in reality. The definition of Nash equilibrium does not say anything about this.

For a Nash equilibrium in mixed strategies as in Definition 3.3, an additional question is what the meaning of such a mixed strategy is. Does it mean that the players actually randomize when playing the game? A different and common interpretation is that a mixed strategy of a player, say player 1, represents the belief, or conjecture, of the other player, player 2, about what player 1 will do. Thus, it embodies the 'strategic uncertainty' of the players in a game.

For now, we just leave these questions aside and take the definition of Nash equilibrium at face value. We show how to compute pure Nash equilibria in general, and all Nash equilibria in games where both players have two pure strategies. Just as in Chap. 2, we also consider the role of strict domination.

3.2   Finding Nash Equilibria

To find all Nash equilibria of an arbitrary bimatrix game is a difficult task. We refer to Sect. 13.2.3 for more discussion on this problem. Here we restrict ourselves to, first, the much easier problem of finding all Nash equilibria in pure strategies of an arbitrary bimatrix game and, second, to showing how to find all Nash equilibria in 2×2 games graphically. It is also possible to solve 2×3 and 3×2 games graphically, see Sect. 13.2.2. For larger games, graphical solutions are impractical or, indeed, impossible.

3.2.1   Pure Nash Equilibria

To find the pure Nash equilibria in a bimatrix game, one can first determine the pure best replies of player 2 to every pure strategy of player 1, and next determine the pure best replies of player 1 to every pure strategy of player 2. Those pairs of pure strategies that are mutual best replies are the pure Nash equilibria of the game. To illustrate this method, consider the bimatrix game

$$
\begin{array}{c}
\,\begin{array}{cccc} W & X & Y & Z \end{array} \\
\begin{array}{c} T \\ M \\ B \end{array}
\left(\begin{array}{cccc}
2,2 & 4,0 & 1,1 & 3,2 \\
0,3 & 1,5 & 4,4 & 3,4 \\
2,0 & 2,1 & 5,1 & 1,0
\end{array}\right).
\end{array}
$$

First we determine the pure best replies of player 2 to every pure strategy of player 1, indicated by underlining the corresponding entries. This yields:

$$
\begin{array}{c}
\,\begin{array}{cccc} W & X & Y & Z \end{array} \\
\begin{array}{c} T \\ M \\ B \end{array}
\left(\begin{array}{cccc}
2,\underline{2} & 4,0 & 1,1 & 3,\underline{2} \\
0,3 & 1,\underline{5} & 4,4 & 3,4 \\
2,0 & 2,\underline{1} & 5,\underline{1} & 1,0
\end{array}\right).
\end{array}
$$

Next, we determine the pure best replies of player 1 to every pure strategy of player 2, again indicated by underlining the corresponding entries. This yields:

$$
\begin{array}{c}
\,\begin{array}{cccc} W & X & Y & Z \end{array} \\
\begin{array}{c} T \\ M \\ B \end{array}
\left(\begin{array}{cccc}
\underline{2},2 & \underline{4},0 & 1,1 & \underline{3},2 \\
0,3 & 1,5 & 4,4 & \underline{3},4 \\
\underline{2},0 & 2,1 & \underline{5},1 & 1,0
\end{array}\right).
\end{array}
$$

Putting the two results together yields:

$$
\begin{array}{c}
\,\begin{array}{cccc} W & X & Y & Z \end{array} \\
\begin{array}{c} T \\ M \\ B \end{array}
\left(\begin{array}{cccc}
\underline{2},\underline{2} & \underline{4},0 & 1,1 & \underline{3},\underline{2} \\
0,3 & 1,\underline{5} & 4,4 & \underline{3},4 \\
\underline{2},0 & 2,\underline{1} & \underline{5},\underline{1} & 1,0
\end{array}\right).
\end{array}
$$

We conclude that the game has three Nash equilibria in pure strategies, namely (T, W), (T, Z), and (B, Y). In mixed strategy notation, these are the pairs $(\mathbf{e}^1, \mathbf{e}^1)$, $(\mathbf{e}^1, \mathbf{e}^4)$, and $(\mathbf{e}^3, \mathbf{e}^3)$. In more extensive notation: $((1, 0, 0), (1, 0, 0, 0))$, $((1, 0, 0), (0, 0, 0, 1))$, and $((0, 0, 1), (0, 0, 1, 0))$.

Strictly speaking, one should also consider mixed best replies to a pure strategy in order to establish whether this pure strategy can occur in a Nash equilibrium, but it is not difficult to see that any mixed best reply is a combination of pure best replies and, thus, can never lead to a higher payoff. For instance, in the example above, any strategy of the form $(q, 0, 0, 1 - q)$ played against T yields to player 2 a payoff of $2 (= 2q + 2(1 - q))$ and is therefore a best reply, but does not yield a payoff higher than W or Z. However, the reader can check that all strategy pairs of the form $(T, (q, 0, 0, 1 - q)) \, (0 < q < 1)$ are also Nash equilibria of this game.

It is also clear from this example that a Nash equilibrium does not have to result in Pareto optimal payoffs: a pair of payoffs is *Pareto optimal* if there is no other pair of payoffs which are at least as high for both players and strictly higher for at least one player. The payoff pair $(4, 4)$, resulting from (M, Y), is better for both players than the equilibrium payoffs $(2, 2)$, resulting from (T, W). We know this phenomenon already from the Prisoners' Dilemma game in Sect. 1.3.2.

3.2.2 2 × 2 Games

We demonstrate the graphical solution method for 2×2 games by means of an example. Consider the bimatrix game

$$
\begin{array}{cc}
 & \begin{array}{cc} L & \quad R \end{array} \\
\begin{array}{c} T \\ B \end{array} & \begin{pmatrix} 2,2 & 0,1 \\ 1,1 & 3,3 \end{pmatrix}
\end{array} .
$$

Observe that this game has two Nash equilibria in pure strategies, namely (T, L) and (B, R). To find all Nash equilibria we determine the best replies of both players.

First consider the strategy $(q, 1 - q)$ of player 2. The unique best reply of player 1 to this strategy is T or, equivalently, $(1, 0)$, if the expected payoff from playing T is higher than the expected payoff from playing B, since then it is also higher than the expected payoff from playing any combination $(p, 1 - p)$ of T and B. Hence, the best reply is T if

$$
2q + 0(1 - q) > 1q + 3(1 - q) ,
$$

so if $q > \frac{3}{4}$. Similarly, we find that B is the unique best reply if $q < \frac{3}{4}$, and that T and B are both best replies if $q = \frac{3}{4}$. In the last case, since T and B yield the same payoff to player 1 against $(q, 1 - q)$, it follows that any $(p, 1 - p)$ is a best reply. Summarizing, if we denote the set of best replies of player 1 against $(q, 1 - q)$ by $\beta_1(q, 1 - q)$, we have

$$
\beta_1(q, 1 - q) = \begin{cases} \{(1, 0)\} & \text{if } \frac{3}{4} < q \leq 1 \\ \{(p, 1 - p) \mid 0 \leq p \leq 1\} & \text{if } q = \frac{3}{4} \\ \{(0, 1)\} & \text{if } 0 \leq q < \frac{3}{4} . \end{cases} \tag{3.1}
$$

By analogous arguments, we find that for a strategy $(p, 1 - p)$ the best replies $\beta_2(p, 1 - p)$ of player 2 are given by

$$\beta_2(p, 1 - p) = \begin{cases} \{(1, 0)\} & \text{if } \frac{2}{3} < p \leq 1 \\ \{(q, 1 - q) \mid 0 \leq q \leq 1\} & \text{if } p = \frac{2}{3} \\ \{(0, 1)\} & \text{if } 0 \leq p < \frac{2}{3}. \end{cases} \tag{3.2}$$

By definition, the Nash equilibria of the game are the strategy combinations $(\mathbf{p}^*, \mathbf{q}^*)$ such that $\mathbf{p}^* \in \beta_1(\mathbf{q}^*)$ and $\mathbf{q}^* \in \beta_2(\mathbf{p}^*)$, i.e., the points of intersection of the best reply functions in (3.1) and (3.2). A convenient way to find these points is by drawing the graphs of $\beta_1(q, 1 - q)$ and $\beta_2(p, 1 - p)$. We put p on the horizontal axis and q on the vertical axis and obtain the following diagram.

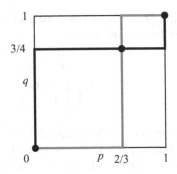

The solid black curve depicts the best replies of player 1 and the solid grey curve depicts the best replies of player 2. The solid circles indicate the three Nash equilibria of the game: $((1, 0), (1, 0))$, $((2/3, 1/3), (3/4, 1/4))$, and $((0, 1), (0, 1))$.

3.2.3 Strict Domination

The graphical method discussed in Sect. 3.2.2 is suited for 2×2 games. It can be extended to 2×3 and 3×2 games as well, see Sect. 13.2.2.

In general, for the purpose of finding Nash equilibria the size of a game can sometimes be reduced by iteratively eliminating strictly dominated strategies. We look for a strictly dominated (pure) strategy of a player, eliminate the associated row or column, and continue this procedure for the smaller game until there is no more strictly dominated strategy. It can be shown (see Sect. 13.3) that no pure strategy that is eliminated by this procedure is ever played with positive probability in a Nash equilibrium of the original game. Thus, no Nash equilibrium of the original game is eliminated. Also, no Nash equilibrium is added. It follows, in particular, that the order in which strictly dominated strategies are eliminated does not matter.

For completeness we first repeat the definition of strict domination, formulated for a bimatrix game, and then present an example.

Definition 3.4 (Strict Domination) Let (A, B) be an $m \times n$ bimatrix game and i a row. The pure strategy $\mathbf{e}^i$ is *strictly dominated* if there is a strategy $\mathbf{p} = (p_1, \ldots, p_m) \in \Delta^m$ such that $\mathbf{p}A\mathbf{e}^j > \mathbf{e}^iA\mathbf{e}^j$ for every $j = 1, \ldots, n$. Similarly, let j be a column. The pure strategy $\mathbf{e}^j$ is *strictly dominated* if there is a strategy $\mathbf{q} = (q_1, \ldots, q_n) \in \Delta^n$ such that $\mathbf{e}^i B\mathbf{q} > \mathbf{e}^i B\mathbf{e}^j$ for every $i = 1, \ldots, m$. $\square$

We observe that Remark 2.6 is still valid: if $\mathbf{e}^i$ for player 1 is strictly dominated, then it is strictly dominated by some $\mathbf{p}$ with $p_i = 0$; and similar for player 2.

Consider the following bimatrix game

$$
\begin{array}{c c}
 & \begin{array}{cccc} W & X & Y & Z \end{array} \\
\begin{array}{c} T \\ M \\ B \end{array} &
\left(\begin{array}{cccc}
2,2 & 2,1 & 2,2 & 0,0 \\
1,0 & 4,1 & 2,4 & 1,5 \\
0,4 & 3,1 & 3,0 & 3,3
\end{array} \right).
\end{array}
$$

Observe, first, that no pure strategy (row) of player 1 is strictly dominated by another pure strategy of player 1, and that no pure strategy (column) of player 2 is strictly dominated by another pure strategy of player 2. Consider the pure strategy X of player 2. Note that the payoffs in column X for player 2 are below the maximum of the payoffs in columns W and Y: $1 < \max\{2, 2\}$, $1 < \max\{0, 4\}$, and $1 < \max\{4, 0\}$. Therefore, we may try and see if X can be strictly dominated by a combination of W and Y, i.e., by a strategy of the form $(q, 0, 1 - q, 0)$. For this we need: $1 < 2q + 2(1 - q)$, $1 < 0q + 4(1 - q)$, and $1 < 4q + 0(1 - q)$. These three inequalities hold for all q with $\frac{1}{4} < q < \frac{3}{4}$. For instance, X is strictly dominated by $(\frac{1}{2}, 0, \frac{1}{2}, 0)$. So X can be eliminated, to obtain

$$
\begin{array}{c c}
 & \begin{array}{ccc} W & Y & Z \end{array} \\
\begin{array}{c} T \\ M \\ B \end{array} &
\left(\begin{array}{ccc}
2,2 & 2,2 & 0,0 \\
1,0 & 2,4 & 1,5 \\
0,4 & 3,0 & 3,3
\end{array} \right).
\end{array}
$$

Next, in this reduced game, for player 1 pure strategy M is strictly dominated by any strategy of the form $(p, 0, 1 - p)$ with $\frac{1}{2} < p < \frac{2}{3}$. So M can be eliminated to obtain

$$
\begin{array}{c c}
 & \begin{array}{ccc} W & Y & Z \end{array} \\
\begin{array}{c} T \\ B \end{array} &
\left(\begin{array}{ccc}
2,2 & 2,2 & 0,0 \\
0,4 & 3,0 & 3,3
\end{array} \right).
\end{array}
$$

Here, finally, Z can be eliminated since it is strictly dominated by W, and we are left with the 2×2 game

$$
\begin{array}{c c}
 & \begin{array}{cc} W & Y \end{array} \\
\begin{array}{c} T \\ B \end{array} &
\left(\begin{array}{cc}
2,2 & 2,2 \\
0,4 & 3,0
\end{array} \right).
\end{array}
$$

This game can be solved by the graphical method of Sect. 3.2.2. Doing so results in the diagram

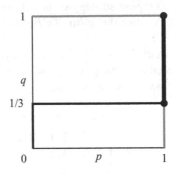

The solid black (grey) curve depicts player 1's (2's) best replies. In this case, the curves overlap in infinitely many points, resulting in the set of Nash equilibria $\{((1,0),(q,1-q)) \mid 1/3 \leq q \leq 1\}$. In the original 3×4 game, the set of all Nash equilibria is therefore equal to

$$\{((1,0,0),(q,0,1-q,0)) \mid 1/3 \leq q \leq 1\} .$$

3.3   Problems

3.1. *Some Applications*
In each of the following situations, set up the corresponding bimatrix game and solve for all Nash equilibria.

(a) *Pure coordination.* Two firms (Smith and Brown) decide whether to design the computers they sell to use large or small floppy disks. Both players will sell more computers if their disk drives are compatible. If they both choose for large disks the payoffs will be 2 for each. If they both choose for small disks the payoffs will be 1 for each. If they choose different sizes the payoffs will be -1 for each.

(b) *The welfare game.* This game models a government that wishes to aid a pauper if he searches for work but not otherwise, and a pauper who searches for work only if he cannot depend on government aid, and who may not succeed in finding a job even if the tries. The payoffs are $3,2$ (for government, pauper) if the government aids and the pauper tries to work; $-1,1$ if the government does not aid and the pauper tries to work; $-1,3$ if the government aids and the pauper does not try to work; and $0,0$ in the remaining case. [These payoffs represent the preferences of the players, rather than monetary values. For instance, the government ranks the combination Aid/Search above the combination No

Aid/No Search, which in turn is ranked above the combinations Aid/No Search and No Aid/Search, between which the government is indifferent.]

(c) *Wage game.* Each of two firms has one job opening. Suppose that firm i ($i = 1, 2$) offers wage w_i, where $0 < \frac{1}{2}w_1 < w_2 < 2w_1$ and $w_1 \neq w_2$. Imagine that there are two workers, each of whom can apply to only one firm. The workers simultaneously decide whether to apply to firm 1 or firm 2. If only one worker applies to a given firm, that worker gets the job; if both workers apply to the same firm, each worker has probability $1/2$ of getting the job while the other worker remains unemployed and has a payoff of zero.

(d) *Marketing game.* Two firms sell a similar product. Each percent of market share yields a net payoff of 1. Without advertising both firms have 50 % of the market. The cost of advertising is equal to 10 but leads to an increase in market share of 20 % at the expense of the other firm. The firms make their advertising decisions simultaneously and independently. The total market for the product is of fixed size.

(e) *Voting game.* Two political parties, I and II, each have three votes that they can distribute over three party-candidates each. A committee is to be elected, consisting of three members. Each political party would like to see as many as possible of their own candidates elected in the committee. Of the total of six candidates, those three who have most of the votes will be elected; in case of ties, tied candidates are drawn with equal probabilities.

(f) *Voting game, revisited.* Consider the situation in (e) but now assume that each party is risk averse. For instance, each party strictly prefers to have one candidate for sure in the committee over a lottery in which it has zero or two candidates in the committee each with probability 50 %. Model this by each party having a payoff of $\sqrt{c}$ for a sure number $c \in \{0, 1, 2, 3\}$ of its candidates in the committee.

3.2. *Matrix Games*

(a) Since a matrix game is a special case of a bimatrix game, it may be analyzed by the graphical method of Sect. 3.2.2. Do this for the game in Problem 2.1(a). Compare your answer with what you found previously.

(b) Argue that a pair consisting of a maximin and a minimax strategy in a matrix game is a Nash equilibrium; and that any Nash equilibrium in a matrix game must be a pair consisting of a maximin and a minimax strategy. (You may give all your arguments in words.)

(c) Define a maximin strategy for player 1 in the bimatrix game (A, B) to be a maximin strategy in the matrix game A. Which definition is appropriate for player 2 in this respect? With these definitions, find examples showing that a Nash equilibrium in a bimatrix game does not have to consist of maximin strategies, and that a maximin strategy does not have to be part of a Nash equilibrium.

3.3. *Strict Domination*

Consider the bimatrix game

$$
\begin{array}{c c c c}
 & W & X & Y & Z \\
T & \begin{pmatrix} 6,6 & 4,4 & 1,2 & 8,5 \\ B & 4,5 & 6,6 & 2,8 & 4,4 \end{pmatrix}
\end{array}.
$$

(a) Which pure strategy of player 1 or player 2 is strictly dominated by a pure strategy?

(b) Describe all combinations of strategies W and Y of player 2 that strictly dominate X.

(c) Find all Nash equilibria of this game.

3.4. *Iterated Elimination (1)*

Consider the bimatrix game

$$
\begin{array}{c c c c c}
 & W & X & Y & Z \\
A & \begin{pmatrix} 5,4 & 4,4 & 4,5 & 12,2 \\ B & 3,7 & 8,7 & 5,8 & 10,6 \\ C & 2,10 & 7,6 & 4,6 & 9,5 \\ D & 4,4 & 5,9 & 4,10 & 10,9 \end{pmatrix}
\end{array}.
$$

(a) Find a few different ways in which strictly dominated strategies can be iteratedly eliminated in this game.

(b) Find the Nash equilibria of this game.

3.5. *Iterated Elimination (2)*

Consider the bimatrix game

$$
\begin{pmatrix} 2,0 & 1,1 & 4,2 \\ 3,4 & 1,2 & 2,3 \\ 1,3 & 0,2 & 3,0 \end{pmatrix}.
$$

Find the Nash equilibria of this game.

3.6. *Weakly Dominated Strategies*

A pure strategy i of player 1 in an $m \times n$ bimatrix game (A, B) is *weakly dominated* if there a strategy $\mathbf{p} = (p_1, \ldots, p_m) \in \Delta^m$ such that $\mathbf{p}Ae^j \geq e^iAe^j$ for every $j = 1, \ldots, n$, and $\mathbf{p}Ae^j > e^iAe^j$ for at least one j. The definition of a weakly dominated strategy of player 2 is similar. In words, a pure strategy is weakly dominated if there is some pure or mixed strategy that is always at least as good, and that is better against at least one pure strategy of the opponent. Instead of iterated elimination of strictly dominated strategies one might also consider iterated elimination of weakly dominated strategies. The advantage is that in games where no strategy is strictly dominated it might still be possible to eliminate strategies that are weakly dominated. However, some Nash equilibria of the original game may be eliminated as well, and also the order of elimination may matter. These issues are illustrated by the following examples.

(a) Consider the bimatrix game

$$
\begin{array}{c c c c}
 & X & Y & Z \\
A & \begin{pmatrix} 11,10 & 6,9 & 10,9 \\ 11,6 & 6,6 & 9,6 \\ 12,10 & 6,9 & 9,11 \end{pmatrix} \\
B & & & \\
C & & &
\end{array}.
$$

First, determine the pure Nash equilibria of this game. Next, apply iterated elimination of weakly dominated strategies to reduce the game to a 2×2 game and determine the unique Nash equilibrium of this smaller game.

(b) Consider the bimatrix game

$$
\begin{array}{c c c c}
 & X & Y & Z \\
A & \begin{pmatrix} 1,1 & 0,0 & 2,0 \\ 1,2 & 1,2 & 1,1 \\ 0,0 & 1,1 & 1,1 \end{pmatrix} \\
B & & & \\
C & & &
\end{array}.
$$

Show that different orders of eliminating weakly dominated strategies may result in different Nash equilibria.

3.7. *A Parameter Game*
Consider the bimatrix game

$$
\begin{array}{c c c}
 & L & R \\
T & \begin{pmatrix} 1,1 & a,0 \\ 0,0 & 2,1 \end{pmatrix} \\
B & &
\end{array}
$$

where $a \in \mathbb{R}$. Determine the Nash equilibria of this game for every possible value of a.

3.8. *Equalizing Property of Mixed Equilibrium Strategies*

(a) Consider again the game of Problem 3.3, which has a unique Nash equilibrium in mixed strategies. In this equilibrium, player 1 puts positive probability p^* on T and $1 - p^*$ on B, and player 2 puts positive probability q^* on W and $1 - q^*$ on Y. Show that, if player 2 plays this strategy, then both T and B give player 1 the same expected payoff, equal to the equilibrium payoff. Also show that, if player 1 plays his equilibrium strategy, then both W and Y give player 2 the same expected payoff, equal to the equilibrium payoff, and higher than the expected payoff from X or from Z.

(b) Generalize the observations made in (a), more precisely, give an argument for the following statement:

Let (A, B) be an $m \times n$ bimatrix game and let $(\mathbf{p}^*, \mathbf{q}^*)$ be a Nash equilibrium. Then each row played with positive probability in this Nash equilibrium has the same expected payoff for player 1 against $\mathbf{q}^*$ and this payoff is at least as high as the payoff from any other row. Each column played with positive probability in this Nash equilibrium has the same expected payoff for player 2 against $\mathbf{p}^*$ and this payoff is at least as high as the payoff from any other column.

You may state your argument in words, without using formulas.

3.9. *Voting*

Suppose the spectrum of political positions is described by the closed interval (line segment) $[0, 5]$. Voters are uniformly distributed over $[0, 5]$. There are two candidates, who may occupy any of the positions in the set $\{0, 1, 2, 3, 4, 5\}$. Voters will always vote for the nearest candidate. If the candidates occupy the same position they each get half of the votes. The candidates simultaneously and independently choose positions. Each candidate wants to maximize the number of votes for him/herself. Only pure strategies are considered.

(a) Model this situation as a bimatrix game between the two candidates.
(b) Determine the best replies of both players.
(c) Determine all Nash equilibria (in pure strategies), if any.

We now change the situation as follows. Candidate 1 can only occupy the positions $1, 3, 5$, and candidate 2 can only occupy the positions $0, 2, 4$.

(d) Answer questions (a), (b), and (c) for this new situation.
(e) The two games above are constant-sum games. How would you turn them into zero-sum games without changing the strategic possibilities (best replies, equilibria)? What would be the value of these games and the (pure) optimal strategies?

3.10. *Guessing Numbers*

Players 1 and 2 each choose a number from the set $\{1, \ldots, K\}$. If the players choose the same number, then player 2 pays 1 Euro to player 1; otherwise no payment is made. The players can use mixed strategies and each player's preferences are determined by his or her expected monetary payoff.

(a) Show that each player playing each pure strategy with probability $\frac{1}{K}$ is a Nash equilibrium.
(b) Show that in every Nash equilibrium, player 1 must choose every number with positive probability.
(c) Show that in every Nash equilibrium, player 2 must choose every number with positive probability.

(d) Determine all (mixed strategy) Nash equilibria of the game.

(e) This is a zero-sum game. What are the optimal strategies and the value of the game?

3.11. *Bimatrix Games*

(a) Give an example of a 2 × 2-bimatrix game with exactly two Nash equilibria in pure strategies and no other Nash equilibrium. For your example, determine the players' reaction functions and make a picture, showing that your example is as desired.

(b) Consider the following bimatrix game:

$$(A, B) = \begin{pmatrix} a,b & c,d \\ e,f & g,h \end{pmatrix}.$$

Assume that $a > e$. Give necessary and sufficient further conditions on the payoffs such that the game has no pure strategy Nash equilibria. Under these conditions, determine all mixed Nash equilibria of the game.

3.4   Notes

For finite two-person games, Nash (1951) proved that every game has a Nash equilibrium in mixed strategies. The term 'strategic uncertainty' can already be found in von Neumann and Morgenstern (1944/1947). For general bimatrix games, Nash equilibria can be found by Nonlinear Programming methods, see Chap. 13.

Problems 3.1(a, b) are taken from Rasmusen (1989), and Problem 3.1(c) from Gibbons (1992). Problem 3.4 is taken from Watson (2002).

References

Gibbons, R. (1992). *A primer in game theory*. Hertfordshire: Harvester Wheatsheaf.

Nash, J. F. (1951). Non-cooperative games. *Annals of Mathematics, 54*, 286–295.

Rasmusen, E. (1989). *Games and information: An introduction to game theory* (2nd ed., 1994/1995). Oxford: Basil Blackwell.

von Neumann, J., & Morgenstern, O. (1944/1947). *Theory of games and economic behavior*. Princeton: Princeton University Press.

Watson, J. (2002). *Strategy, an introduction to game theory*. New York: Norton.

Finite Extensive Form Games

4

Most games derived from economic or political situations have in common with most parlor games (like card games and board games) that they are not 'one-shot': players move sequentially, and one and the same player may move more often than once. Such games are best described by drawing a decision tree which tells us whose move it is and what a player's information is when that player has to make a move.

In this chapter these 'games in extensive form' are studied. Attention is restricted to games with finitely many players (usually two), finitely many decision moments and finitely many moves. See Sect. 1.3.3 for a few examples. We also assume that each player has 'complete information': this means that either there is no chance move in the game or, if there is one, there is no player who still has to move after the chance move and is not informed about the outcome of the chance move. This excludes, for instance, the game of entry deterrence with incomplete information in Sect. 1.3.3, but it also excludes most card games—most card games start with cards being dealt, which is a chance move. For the analysis of games with incomplete information see Chap. 5. Chapter 14 extends the analysis of the present and the next chapter.

The first section of this chapter introduces games in extensive form. In order to avoid a load of cumbersome notation the treatment will be somewhat informal but—hopefully—not imprecise. In Sect. 4.2 we define strategies and the 'strategic form' of a game: the definition of Nash equilibrium for extensive form games is then practically implied. The focus in this chapter is on pure Nash equilibrium.

In the third section the concept of Nash equilibrium is refined by considering backward induction and subgame perfection. A further important refinement, called 'perfect Bayesian equilibrium', is treated in the fourth section.

© Springer-Verlag Berlin Heidelberg 2015
H. Peters, *Game Theory*, Springer Texts in Business and Economics,
DOI 10.1007/978-3-662-46950-7_4

4.1   The Extensive Form

A *game in extensive form* is described by a *game tree*. Such a game tree is characterized by *nodes* and *edges*. Each node is either a *decision node* of a player, or a *chance node*, or an *end node*. Each edge corresponds to either an *action* of a player or a choice made by chance, sometimes called a 'move of Nature'.

Figure 4.1 illustrates these and other concepts.

The upper node in the tree, the *root* of the tree, is a decision node of player 1 and the starting point of the game. Player 1 chooses between three actions, namely *A*, *B*, and *C*. Player 2 learns that player 1 has chosen either one of the actions *A* and *B*, or action *C*. The first event is indicated by the dashed line connecting the two left decision nodes of player 2. In that case, player 2 has two actions, namely *l* and *r*. We call the two connected nodes an *information set* of player 2: player 2 knows that the play of the game has arrived at one of these nodes but he does not know at which one. The fact that player 2 has the same set of actions at each of the two nodes in this information set is a necessary consequence: if this were not the case, player 2 would know at which node he was (i.e., would know whether player 1 would have played *A* or *B*) by simply examining the set of available actions, which would go against the interpretation of an information set. This last argument is one consequence of the more general assumption that the whole game tree is common knowledge between the players: each player knows it, knows that the other player(s) know(s) it, knows that the other player(s) know(s) that he knows it, and so on.

If player 1 plays *C*, then there is a chance move, resulting with probability 1/4 in a decision node of player 2 (following *U*) and with probability 3/4 in a decision node of player 1 (following *D*). At player 2's decision node this player has two actions, namely *L* and *R*. At player 1's decision node this player also has two actions, namely *a* and *b*. All the remaining nodes are end nodes, indicated by payoff pairs, where the upper number is the payoff to player 1 and the lower number the payoff to player 2. In this diagram, the payoffs are written as column vectors, but we also write them as row vectors, whatever is convenient in a given situation.

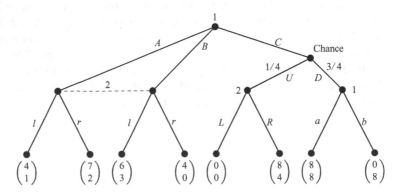

Fig. 4.1 A game in extensive form

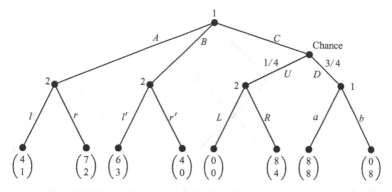

Fig. 4.2 The game of Fig. 4.1, now with perfect information

Also the singleton decision nodes are called information sets. Thus, in this game, each player has two different information sets. An information set is *nontrivial* if it consists of at least two nodes. Games with nontrivial information sets are called games with *imperfect information*. If a game has only trivial information sets, then we say that it has *perfect information*. If we change the present example by assuming that player 2 observes whether player 1 chooses A or B, then the game has perfect information. See Fig. 4.2. We have renamed the actions of player 2 after action B of player 1 for later convenience: these actions can now be regarded as different from player 2's actions following A, since player 2 knows that player 1 has chosen B.

The chance move in our example is not a particularly interesting one, since the players learn what the outcome of the chance move is. (The situation is different if at least one player is not completely informed about the outcome of a chance move and if this lack of information has strategic consequences. In that case, we talk about games with 'incomplete' information, see Chap. 5.)

As mentioned before, we do not give a formal definition of a *game in extensive form*: the examples in Figs. 4.1 and 4.2 illustrate the main ingredients of such a game.[1] An important condition is that the game tree should be a *tree* indeed: it should have a single root and no 'cycles'. This means that a situation as for instance in Fig. 4.3a is not allowed.

We also restrict attention to games in extensive form that have *perfect recall*: each player remembers what he did in the past. For instance, the situation in Fig. 4.3b, where player 1 at his lower information set does not recall which action he took earlier, is not considered.[2]

---

[1] See Chap. 14 for a formal definition.

[2] The assumption of perfect recall plays a particular role for the relation between mixed and behavioral strategies, see Chap. 14.

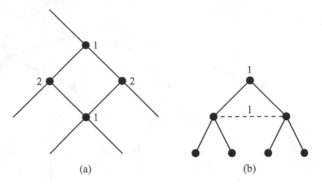

Fig. 4.3 An example of a cycle (**a**), and of a game without perfect recall (**b**)

4.2   The Strategic Form

In a game in extensive form, it is extremely important to distinguish between actions and strategies. An *action* is a possible move of a player at an information set. In the games in Figs. 4.1 and 4.2 player 1 has the actions A, B, and C, and a and b; and player 2 has the actions l and r, and L and R. In contrast,

   a *strategy* is a *complete plan to play the game*.

This is one of the most important concepts in game theory. In the games in Figs. 4.1 and 4.2, a possible strategy for player 1 is:

   Start by playing C; if the chance move of the game results in D, then play b.

Another strategy of player 1 is:

   Start by playing A; if the chance move of the game results in D, then play b.

The last strategy might look strange since player 1's first action A precludes him having to take any further action. Nevertheless, also this plan is regarded as a possible strategy.[3]
   A possible strategy for player 2 in the game of Fig. 4.1 is:

   Play l if player 1 plays A or B, and play L if player 1 plays C and the chance move results in U.

Note that player 2 cannot make his action contingent on whether player 1 plays A or B, since player 2 does not have that information. In the perfect information game

---

[3]Although there is not much lost if we would exclude such strategies—as some authors do.

of Fig. 4.2, however, player 2's strategy should tell what player 2 plays after A and what he plays after B. A possible strategy would then be:

Play l if player 1 plays A, play r' if player 1 plays B, and play L if player 1 plays C and the chance move results in U.

A formal definition of a strategy of a player is:

a strategy is *a list of actions, exactly one at each information set of that player.*

In both our examples, a strategy of player 1 is therefore a list of two actions since player 1 has two information sets. The number of possible strategies of player 1 is the number of different lists of actions. Since player 1 has three possible actions at his first information set and two possible actions at his second information set, this number is equal to $3 \times 2 = 6$. The strategy set of player 1 can be denoted as

$$\{Aa, Ab, Ba, Bb, Ca, Cb\} \, .$$

Similarly, in the imperfect information game in Fig. 4.1 player 2 has $2 \times 2 = 4$ different strategies, and his strategy set can be denoted as

$$\{lL, lR, rL, rR\} \, .$$

In the perfect information game in Fig. 4.2 player 2 has three information sets and two actions at each information set, so $2 \times 2 \times 2 = 8$ different strategies, and his strategy set can be denoted as

$$\{ll'L, ll'R, lr'L, lr'R, rl'L, rl'R, rr'L, rr'R\} \, .$$

In general, it is important to distinguish between a strategy combination and the associated *outcome*. The outcome is the induced play of the game, that is, the path followed in the game tree. For instance, the strategy combination $(Aa, ll'L)$ in the game in Fig. 4.2 induces the outcome (A, l) and the payoffs $(4, 1)$.

There are several reasons why we are interested in strategies. The main reason is that by considering strategies the extensive form game is effectively reduced to a one-shot game. Once we fix a profile (in the present example, pair) of strategies we can compute the payoffs by following the path followed in the game tree. Consider for instance the strategy pair (Cb, rL) in the game in Fig. 4.1. Then player 1 starts by playing C, and this is followed by a chance move; if the result of this move is U, then player 2 plays L; if the result is D, then player 1 plays b. Hence, with probability $1/4$ the resulting payoff pair is $(0, 0)$ and with probability $3/4$ the resulting payoff pair is $(0, 8)$. So the expected payoffs are 0 for player 1 and 6 for player 2. In this way, we can compute the payoffs in the game of Fig. 4.1 resulting from each of the 6×4 possible strategy combinations. Similarly, for the game in Fig. 4.2 we compute

$$
\begin{array}{c c c c c}
 & \textit{lL} & \textit{lR} & \textit{rL} & \textit{rR} \\
Aa & \begin{pmatrix} 4,1 & 4,1 & \underline{7,2} & 7,\underline{2} \\ 4,1 & 4,1 & \underline{7,2} & 7,\underline{2} \\ \underline{6,3} & 6,\underline{3} & 4,0 & 4,0 \\ \underline{6,3} & 6,\underline{3} & 4,0 & 4,0 \\ \underline{6,6} & \underline{8,7} & 6,6 & \underline{8,7} \\ 0,6 & 2,\underline{7} & 0,6 & 2,\underline{7} \end{pmatrix} \\
Ab \\
Ba \\
Bb \\
Ca \\
Cb
\end{array}
$$

	$ll'L$	$ll'R$	$lr'L$	$lr'R$	$rl'L$	$rl'R$	$rr'L$	$rr'R$
Aa	4,1	4,1	4,1	4,1	$\underline{7,2}$	7,$\underline{2}$	$\underline{7,2}$	7,$\underline{2}$
Ab	4,1	4,1	4,1	4,1	$\underline{7,2}$	7,$\underline{2}$	$\underline{7,2}$	7,$\underline{2}$
Ba	$\underline{6,3}$	6,$\underline{3}$	4,0	4,0	$\underline{6,3}$	6,$\underline{3}$	4,0	4,0
Bb	$\underline{6,3}$	6,$\underline{3}$	4,0	4,0	$\underline{6,3}$	6,$\underline{3}$	4,0	4,0
Ca	$\underline{6,6}$	$\underline{8,7}$	$\underline{6,6}$	$\underline{8,7}$	6,6	$\underline{8,7}$	6,6	$\underline{8,7}$
Cb	0,6	2,$\underline{7}$	0,6	2,$\underline{7}$	0,6	2,$\underline{7}$	0,6	2,$\underline{7}$

Fig. 4.4 The 6 × 4 strategic form of the game in Fig. 4.1 and the 6 × 8 strategic form of the game in Fig. 4.2

6 × 8 payoff pairs. We next write these payoff pairs in a bimatrix, as in Chap. 3. The resulting bimatrix games are presented in Fig. 4.4.

Such a bimatrix game is called the *strategic form* of the extensive form game. The definition of Nash equilibrium of an extensive form game is then almost implied:

A *Nash equilibrium* of a game in extensive form is a Nash equilibrium of the strategic form.

This definition holds for pure Nash equilibria and, more generally, Nash equilibria in mixed strategies, but in this chapter we restrict attention to pure strategies and pure strategy Nash equilibria.

The pure strategy Nash equilibria of the bimatrix games in Fig. 4.4 can be found by using the method of Sect. 3.2.1. The equilibria correspond to the double underlined entries. Thus, the imperfect information game has six different Nash equilibria in pure strategies, and the perfect information game has ten different Nash equilibria in pure strategies.

In the next two sections we examine these Nash equilibria more closely and discuss ways to distinguish between them.

4.3   Backward Induction and Subgame Perfection

We first consider the perfect information game of Fig. 4.2. This game can be analyzed using the principle of *backward induction*. This means that we start with the nodes preceding the end nodes, and turn them into end nodes with payoffs resulting from choosing the optimal action(s). Specifically, player 2 chooses *r* after

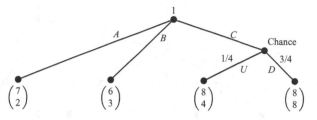

Fig. 4.5 The reduced game of Fig. 4.2

A of player 1 and l' after B of player 1; and R after C and U. Player 1 chooses a after D. Thus, we obtain the reduced game of Fig. 4.5. Note that player 2's strategy has already been completely determined by this first step: it is the strategy $rl'R$. Player 1 has chosen a at his lower information set. Next, in this reduced game, player 1 chooses the action(s) that yield(s) the highest payoff. Since A yields a payoff of 7, B a payoff of 6, and C a(n expected) payoff of $\frac{1}{4} \cdot 8 + \frac{3}{4} \cdot 8 = 8$, it is optimal for player 1 to choose C. Hence, we obtain the strategy combination $(Ca, rl'R)$ with payoffs $\frac{1}{4}(8, 4) + \frac{3}{4}(8, 8) = (8, 7)$. This is one of the ten Nash equilibria of the game (see Fig. 4.4). It is called *backward induction equilibrium*.

It can be shown that, in a game of perfect information, *applying the backward induction principle always results in a (pure) Nash equilibrium.* As a by-product, we obtain that *a game of perfect information has at least one Nash equilibrium in pure strategies, which can be obtained by backward induction.*

It is illustrative to consider backward induction equilibrium—in this game: $(Ca, rl'R)$—as opposed to the induced backward induction outcome. The latter refers to the actual play of the game or, equivalently, the equilibrium path, in this case (Ca, R). Observe that there are other Nash equilibria in this game that generate the same outcome or path, namely $(Ca, ll'R)$, $(Ca, lr'R)$, and $(Ca, rr'R)$: they all generate the path (Ca, R), but differ in the left part of the tree, where player 2 makes at least one decision that is not optimal. Hence, the principle of backward induction ensures that every player always takes an optimal action, even in parts of the game tree that are not actually reached when the game is played.

A generalization of backward induction is *subgame perfection*. The definition of a subgame is as follows:

A *subgame* is any part of the game tree, starting at an information set consisting of a single decision node (trivial information set) of a player or at a chance node, and which is not connected to the rest of the tree by any later information set.

The last part of this definition is illustrated in Fig. 4.6. Although player 2's information set consists of a single decision node, no subgame starts there: this would not make sense since player 3 does not observe any earlier move and in particular does not know at which of the three nodes in his information set he is. Formally, the subtree starting at player 2's decision node is still connected to the whole tree by player 3's information set.

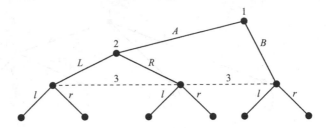

Fig. 4.6 No subgame starts at player 2's decision node

The game in Fig. 4.2 has six different subgames, namely: the entire game; the game starting from the chance move; and the four games starting from the four nodes preceding the end nodes. The definition of a subgame perfect equilibrium is as follows:

A *subgame perfect equilibrium* is a strategy combination that induces a Nash equilibrium in every subgame.

To see what this means, consider again the game in Fig. 4.2. In order for a strategy combination to be a subgame perfect equilibrium, it has to induce a Nash equilibrium in every subgame. Since the entire game is a subgame, a subgame perfect equilibrium has to be a Nash equilibrium in the entire game, and, thus, the ten Nash equilibria in this game are the candidates for a subgame perfect equilibrium. This is the case for any arbitrary game, and therefore *a subgame perfect equilibrium is always a Nash equilibrium.*

Returning to the game in Fig. 4.2, a subgame perfect equilibrium also has to induce an equilibrium in each of the four one-player subgames preceding the end nodes: although we have not defined Nash equilibria for one-person games, the only reasonable definition is that a player should choose the action that is optimal. In the example, this means that (from left to right) the actions r, l', R, and a, should be chosen. This implies that the players choose optimally also in the subgame starting from the chance node. Summarizing, we look for the Nash equilibrium or equilibria that generate the mentioned actions, and the only Nash equilibrium that does this is again $(Ca, rl'R)$. Hence, the unique subgame perfect equilibrium in this game is $(Ca, rl'R)$. It is not surprising that this is also the backward induction equilibrium: *in games of perfect information, backward induction equilibria and subgame perfect equilibria coincide.*

Let us now consider the imperfect information version of the game in Fig. 4.1. In this case, backward induction cannot be applied to the left part of the game tree: since player 2 does not know whether player 1 has played A or B when he has to choose an action in his left information set, he cannot optimally choose between l and r: l is better if player 1 has played B, but r is better if player 1 has played A. Considering subgame perfection, the only subgames are now: the entire game; the two subgames following U and D; and the subgame starting from the chance move.

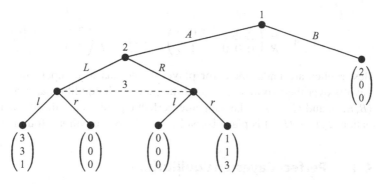

Fig. 4.7 A three-player game

Hence, the restrictions imposed by subgame perfection are that player 1 should play a, player 2 should play R, and the strategy combination should be a Nash equilibrium in the entire game. Of the six Nash equilibria of the game (see Fig. 4.4), this leaves the two equilibria (Ca, lR) and (Ca, rR). So these are the subgame perfect equilibria of the game in Fig. 4.1.

We conclude this section with an example which shows more explicitly than the preceding example that subgame perfection can be more generally applied than the backward induction principle. Consider the game in Fig. 4.7, which is a three player game (for a change). Clearly, backward induction cannot be applied here: player 3 cannot unambiguously determine his optimal action since he does not know whether player 2 has played L or has played R. For subgame perfection, notice that this game has two subgames: the entire game; and the game starting with player 2's decision node. The latter game is a game between players 2 and 3 with strategic form

$$
\begin{array}{c}
\quad\quad l \quad\quad r \\
\begin{array}{c} L \\ R \end{array}
\begin{pmatrix} 3,1 & 0,0 \\ 0,0 & 1,3 \end{pmatrix}
\end{array}
$$

where player 2 is the row player and player 3 the column player. This game has two pure Nash equilibria, namely (L, l) and (R, r). Hence, a subgame perfect equilibrium has to induce one of these two equilibria in this subgame. Note that if the first equilibrium is played, then player 1 should play A, yielding him a payoff of 3 rather than the payoff of 2 obtained by playing B. If the other equilibrium is played in the subgame, then player 1 should obviously play B since A now yields only 1. So the two subgame perfect equilibria are (A, L, l) and (B, R, r).

Alternatively, one can first compute the (pure) Nash equilibria of the entire game. The strategic form of the game can be represented as follows, where the left matrix results from player 1 playing A and the right matrix from player 1 playing B.

$$1:A \quad \begin{array}{c} L \\ R \end{array} \begin{pmatrix} \overset{l}{\underline{3,3},1} & \overset{r}{0,0,0} \\ 0,0,0 & 1,\underline{1},\underline{3} \end{pmatrix} \qquad 1:B \quad \begin{array}{c} L \\ R \end{array} \begin{pmatrix} \overset{l}{\underline{2},\underline{0},\underline{0}} & \overset{r}{\underline{2},\underline{0},\underline{0}} \\ \underline{2},\underline{0},\underline{0} & \underline{2},\underline{0},\underline{0} \end{pmatrix}.$$

Best replies are underlined (for player 1 one has to compare the corresponding payoffs over the two matrices), and the pure Nash equilibria are (A, L, l), (B, L, r), (B, R, l), and (B, R, r). The subgame perfect equilibria are those where the combination (L, l) or (R, r) is played, resulting in the two equilibria found above.

4.4   Perfect Bayesian Equilibrium

A further refinement of Nash equilibrium and of subgame perfect equilibrium is provided by the concept of 'perfect Bayesian equilibrium'. Consider an information set of a player in an extensive form game. A *belief* of that player on that information set is a probability distribution over the nodes of that information set or, equivalently, over the actions leading to that information set. Of course, if the information set is trivial (consists of a single node) then also the belief is trivial, namely attaching probability 1 to the unique node. Our somewhat informal definition of a perfect Bayesian equilibrium is as follows.

A *perfect Bayesian equilibrium* in an extensive form game is a combination of strategies and a specification of beliefs such that the following two conditions are satisfied:

(i) *Bayesian consistency*: the beliefs are consistent with the strategies under consideration;

(ii) *sequential rationality*: the players choose optimally given their beliefs.

The first condition says that the beliefs should satisfy Bayesian updating with respect to the strategies whenever possible. The second condition says that a player should maximize his expected payoff given his beliefs. In order to see what these conditions mean exactly, we consider some examples. (Formal definitions are provided in Chap. 14.)

Consider the game in Fig. 4.8, which is identical to the game in Fig. 4.1. This game has one nontrivial information set. Suppose player 2's belief at this information set is given by the probabilities α at the left node and $1 - \alpha$ at the right node, where $0 \le \alpha \le 1$. That is, if this information set is reached then player 2 attaches probability α to player 1 having played A and probability $1 - \alpha$ to player 1 having played B. All the other information sets are trivial and therefore the beliefs attach probability 1 to each of the corresponding nodes. Condition (ii), sequential rationality, means that player 2 should choose R and player 1 should choose a at the corresponding (trivial) information sets. At the nontrivial information set, player 2 should choose the action that maximizes his expected payoff, *given his belief at this information set*. The expected payoff from l is equal to $\alpha \cdot 1 + (1 - \alpha) \cdot 3 = 3 - 2\alpha$ and the expected payoff from r is $\alpha \cdot 2 + (1 - \alpha) \cdot 0 = 2\alpha$. Hence, l is optimal if $3 - 2\alpha \ge 2\alpha$, i.e., if $\alpha \le 3/4$ and r is optimal if $\alpha \ge 3/4$.

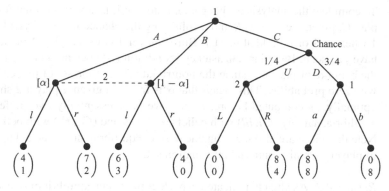

Fig. 4.8 The extensive form game of Fig. 4.1

What does condition (i), Bayesian consistency, imply for this game? It says that, whenever possible, the belief of player 2 at the left, nontrivial information set should be computed by Bayesian updating using the strategy of player 1, specifically, using the initial action of player 1. Thus, if the strategy of player 1 prescribes action A, then player 2 should indeed believe this, so $\alpha = 1$. Similarly, if the strategy of player 1 prescribes action B, then $1 - \alpha = 1$ so $\alpha = 0$. The formula behind this is the formula for conditional probability:

$$\alpha = \text{Prob}[A \text{ is played} \mid A \text{ or } B \text{ is played}]$$
$$= \frac{\text{Prob}[A \text{ and } (A \text{ or } B) \text{ is played}]}{\text{Prob}[A \text{ or } B \text{ is played}]}$$
$$= \frac{\text{Prob}[A \text{ is played}]}{\text{Prob}[A \text{ or } B \text{ is played}]}$$

where these probabilities should be computed given the strategy of player 1. Hence, indeed, if the strategy of player 1 prescribes A then $\alpha = 1/1 = 1$ and if the strategy of player 1 prescribes B then $\alpha = 0/1 = 0$. If, however, player 1's strategy prescribes C, then $\text{Prob}[A \text{ or } B \text{ is played}] = 0$ and we cannot use the formula for computing α: the belief of player 2 on his nontrivial information set is undetermined or *free*.

Remark 4.1 We do not consider mixed strategies in this chapter, but suppose we did and consider, for instance, the mixed strategy where player 1 plays Aa with probability $\frac{1}{2}$, Ab with probability $\frac{1}{6}$, and Ba, Bb, Ca and Cb each with probability $\frac{1}{12}$. Then we would have $\alpha = (\frac{1}{2} + \frac{1}{6})/(\frac{1}{2} + \frac{1}{6} + 2 \cdot \frac{1}{12}) = \frac{4}{5}$. In words, given player 1's strategy, if it turns out that player 1 has played A or B, then player 2 should believe that player 1 has played A with probability $\frac{4}{5}$ and B with probability $\frac{1}{5}$. □

To complete the analysis of this game, note that it is always optimal for player 1 to play C, given the actions R and a following the chance move: C yields 8 to player 1 whereas A or B yield at most 7. But if player 1 does not play A or B, then, as we have just seen, Bayesian consistency [condition (i)] does not put any restriction on the belief α of player 2, since the nontrivial information set of player 2 is reached with zero probability. This means that α is free, but given α player 2 should choose optimally, as computed before. Hence, we have essentially two perfect Bayesian equilibria, namely (Ca, lR) with beliefs $\alpha \le 3/4$ and (Ca, rR) with beliefs $\alpha \ge 3/4$. Note that these are also the subgame perfect equilibria, now 'backed up' by a belief of player 2 on his nontrivial information set.

Remark 4.2 As already indicated, a perfect Bayesian equilibrium is also a subgame perfect equilibrium, but in order to show this a more detailed formal definition is required. See Chap. 14.   ☐

In order to show that the perfect Bayesian equilibrium requirement can have an additional impact compared to subgame perfection, consider the variation on the game of Fig. 4.1, obtained by replacing the payoffs $(4, 1)$ after A and l by the payoffs $(4, 3)$, resulting in the game in Fig. 4.9.

One can check that the subgame perfect equilibria are still (Ca, lR) and (Ca, rR). Obviously, a rational player 2 would never play r at his nontrivial information set since l is always better, but subgame perfection does not rule this out. But clearly, there is no belief that player 2 could have at this information set that would make r optimal: if we denote player 2's belief by $(\alpha, 1 - \alpha)$ as before, then r yields 2α whereas l yields $3\alpha + 3(1 - \alpha) = 3$, which is always larger than 2α. Hence, the only perfect Bayesian equilibrium is (Ca, lR), with arbitrary free belief of player 2 at his nontrivial information set.

We conclude with another example. Consider again the game of Fig. 4.7, reproduced in Fig. 4.10 with belief $(\alpha, 1-\alpha)$ attached to the nodes in the information set of player 3. There are two ways to find the perfect Bayesian equilibria of this

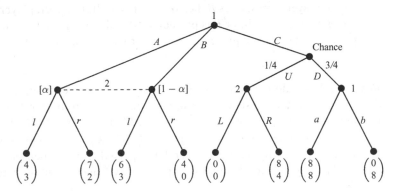

Fig. 4.9 The extensive form game of Fig. 4.1 with payoffs (4,3) after A and l

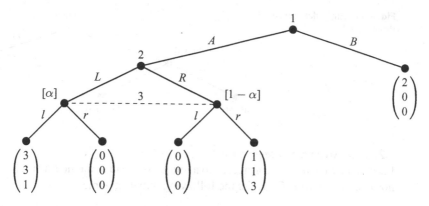

Fig. 4.10 The three-player game of Fig. 4.7 with belief of player 3

game. One can consider the subgame perfect equilibria and find appropriate beliefs. Alternatively, one can start from scratch and apply a form of backward induction. To illustrate the last method, start with player 3. If player 3 plays l then his (expected) payoff is $1\alpha + 0(1 - \alpha) = \alpha$. If player 3 plays r then his (expected) payoff is $0\alpha + 3(1 - \alpha) = 3 - 3\alpha$. Therefore, l is optimal if $\alpha \geq 3/4$ and r is optimal if $\alpha \leq 3/4$. In fact, we have just applied condition (ii), sequential rationality, in the definition of a perfect Bayesian equilibrium.

Now suppose player 3 plays l. Then it is optimal for player 2 to play L. If player 2 plays L, then condition (i) in the definition of a perfect Bayesian equilibrium, Bayesian consistency, implies $\alpha = 1$: that is, player 3 should indeed believe that player 2 has played L. Since $1 \geq 3/4$, l is the optimal action for player 3. Player 1, finally, should play A, yielding payoff 3 instead of the payoff 2 resulting from B. So we have a perfect Bayesian equilibrium (A, L, l) with belief $\alpha = 1$.

If player 3 plays r, then it is optimal for player 2 to play R, resulting in $\alpha = 0$ by Bayesian consistency; since $0 \leq 3/4$, r is the optimal action for player 3. In this case, player 1 should play B: this yields payoff 2, whereas A yields only 1. Hence, we have a perfect Bayesian equilibrium (B, R, r) with belief $\alpha = 0$.

4.5 Problems

4.1. *Counting Strategies*
Consider the following simplified chess game. White moves first (in accordance with the usual rules). Black observes White's move and then makes its move. Then the game is over and ends in a draw. Determine the strategy sets of White and Black. How many strategies does Black have? [In case you do not know the rules of chess: the only information you need is that both players have 20 possible actions.]

Fig. 4.11 Entry deterrence, Problem 4.3

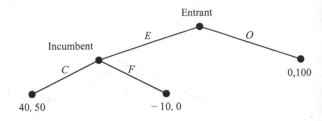

4.2. *Extensive vs. Strategic Form*

Each game in extensive form leads to a unique game in strategic form. The converse, however, is not true. Consider the following bimatrix game:

$$\begin{pmatrix} 1,0 & 1,0 & 0,1 & 0,1 \\ 2,0 & 0,2 & 2,0 & 0,2 \end{pmatrix}.$$

Find an extensive form game with perfect information and an extensive form game with imperfect information, both having this bimatrix game as their strategic form game.

4.3. *Entry Deterrence*

Consider the entry deterrence game of Chap. 1, of which the extensive form is reproduced in Fig. 4.11.

(a) Write down the strategic form of this game.
(b) Determine the Nash equilibria (in pure strategies). Which one is the backward induction equilibrium? Which one is subgame perfect? In which sense is the other equilibrium based on an 'incredible threat'?

4.4. *Choosing Objects*

Four objects O_1, O_2, O_3, and O_4 have different worths for two players 1 and 2, given by the following table:

	O_1	O_2	O_3	O_4
Worth for player 1	1	2	3	4
Worth for player 2	2	3	4	1

Player 1 starts with choosing an object. After him player 2 chooses an object, then player 1 takes his second object, and finally player 2 gets the object that is left. The payoff for a player is the sum of the worths of the objects he obtains.

(a) Draw the decision tree for this extensive form game.
(b) How many strategies does each player have?

(c) Determine the backward induction or subgame perfect equilibria (in pure strategies). How many different subgame perfect equilibria are there? What are the associated outcomes, and the resulting divisions of the objects?

(d) Is there a Nash equilibrium in this game resulting in a division of the objects different from the division in a subgame perfect equilibrium?

4.5. *A Bidding Game*

Players 1 and 2 bid for an object that has value 2 for each of them. They both have wealth 3 and are not allowed to bid higher than this amount. Each bid must be a nonnegative integer amount. Besides bidding, each player, when it is his turn, has the options to pass (P) or to match (M) the last bid, where the last bid is set at zero at the beginning of the game. If a player passes (P), then the game is over and the other player gets the object and pays the last bid. If a player matches (M), then the game is over and each player gets the object and pays the last bid with probability $\frac{1}{2}$. Player 1 starts, and the players alternate until the game is over. Each new bid must be higher than the last bid.

(a) Draw the game tree of this extensive form game.
(b) How many strategies does player 1 have? Player 2?
(c) How many subgame perfect equilibria does this game have? What is (are) the possible subgame perfect equilibrium outcome(s)?
(d) Describe all (pure strategy) Nash equilibria of the game (do not make the strategic form). Is there any Nash equilibrium that does not result in a subgame perfect equilibrium outcome?

4.6. *An Extensive Form Game*

For the game in Fig. 4.12, write down the strategic form and compute all Nash equilibria, subgame perfect equilibria, and perfect Bayesian equilibria in pure strategies.

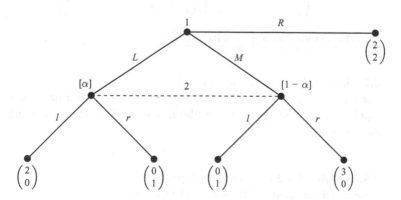

Fig. 4.12 Extensive form game of Problem 4.6

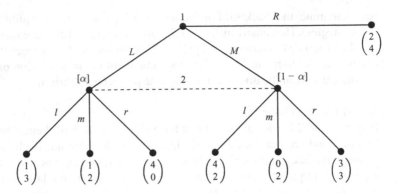

Fig. 4.13 Extensive form game of Problem 4.7

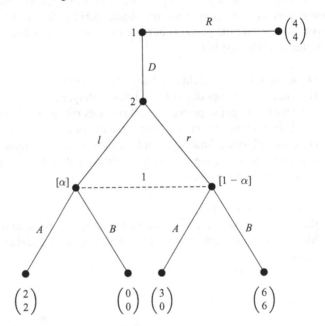

Fig. 4.14 Extensive form game of Problem 4.8

4.7. *Another Extensive Form Game*

For the game in Fig. 4.13, write down the strategic form and compute all Nash equilibria, subgame perfect equilibria, and perfect Bayesian equilibria in pure strategies.

4.8. *Still Another Extensive Form Game*

Consider the extensive form game in Fig. 4.14.

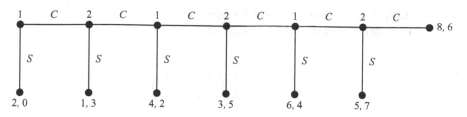

Fig. 4.15 The centipede game of Problem 4.9

(a) Determine the strategic form of this game.
(b) Determine all Nash equilibria in pure strategies.
(c) Determine all subgame perfect equilibria in pure strategies.
(d) Determine all perfect Bayesian equilibria in pure strategies.

4.9. *A Centipede Game*
In the centipede game, the two players move alternatingly. On each move, a player can stop (S) or continue (C). On any move, a player is better off stopping the game than continuing if the other player stops immediately afterward, but is worse off stopping than continuing if the other player continues, regardless of the subsequent actions. The game ends after a finite number of periods. Consider an example of this game in Fig. 4.15.

(a) Determine the backward induction or subgame perfect equilibrium of this game. What is the associated outcome?
(b) Show that there are other Nash equilibria, but that these always result in the same outcome as the subgame perfect equilibrium.

4.10. *Finitely Repeated Prisoners' Dilemma*
Consider the prisoners' dilemma game of Chap. 1:

$$
\begin{array}{cc}
 & \begin{array}{cc} C & \qquad D \end{array} \\
\begin{array}{c} C \\ D \end{array} &
\begin{pmatrix} -1,-1 & -10,0 \\ 0,-10 & -9,-9 \end{pmatrix}.
\end{array}
$$

Suppose that this game is played twice. After the first play of the game the players learn the outcome of that play. The final payoff for each player is the sum of the payoffs of the two stages.

(a) Write down the extensive form of this game. How many strategies does each player have? How many subgames does this game have?
(b) Determine the subgame perfect equilibrium or equilibria of this game. What if the game is repeated more than twice but still finitely many times?

4.11. *A Twice Repeated* 2×2 *Bimatrix Game*
Consider the following bimatrix game:

$$
\begin{array}{cc}
 & \begin{array}{cc} L & R \end{array} \\
\begin{array}{c} T \\ B \end{array} &
\left(\begin{array}{cc} 2,1 & 1,0 \\ 5,2 & 4,4 \end{array} \right)
\end{array} .
$$

Suppose that the game is played twice, and that after the first play of the game the players learn the outcome of that play. The final payoff for each player is the sum of the payoffs of the two stages.

(a) Determine the subgame perfect equilibrium or equilibria of this game. What if the game is repeated more than twice but still finitely many times?
(b) Exhibit a Nash equilibrium (of the twice repeated game) where (B, L) is played in the first round.

4.12. *Twice Repeated* 3×3 *Bimatrix Games*
Consider the following bimatrix game:

$$
\begin{array}{cc}
 & \begin{array}{ccc} L & M & R \end{array} \\
\begin{array}{c} T \\ C \\ B \end{array} &
\left(\begin{array}{ccc} 8,8 & 0,9 & 0,0 \\ 9,0 & 0,0 & 3,1 \\ 0,0 & 1,3 & 3,3 \end{array} \right)
\end{array} .
$$

Suppose that the game is played twice, and that after the first play of the game the players learn the outcome of that play. The final payoff for each player is the sum of the payoffs of the two stages.

(a) How many strategies does each player have in this game? How many subgames does the game have?
(b) Describe a subgame perfect equilibrium in which (T, L) is played in the first round.

For question (c), consider the bimatrix game

$$
\begin{array}{cc}
 & \begin{array}{ccc} L & M & R \end{array} \\
\begin{array}{c} T \\ C \\ B \end{array} &
\left(\begin{array}{ccc} 5,3 & 0,0 & 2,0 \\ 0,0 & 2,2 & 0,0 \\ 0,0 & 0,0 & 0,0 \end{array} \right)
\end{array} .
$$

(c) For the twice repeated version of this game, describe a subgame perfect equilibrium in which (B, R) is played in the first round.

4.6   Notes

The concept of subgame perfection was first formally introduced by Selten (1965, 1975). The result that each finite extensive form game of perfect information has a backward induction equilibrium in pure strategies is intuitive but nevertheless somewhat cumbersome to prove, see for instance Perea (2001, Chap. 3).

There is also a stronger version of Bayesian consistency, resulting in the concept of 'sequential equilibrium', see Kreps and Wilson (1982) and Chap. 14.

The game in Problem 4.12(c) is taken from Benoit and Krishna (1985).

References

Benoit, J.-P., & Krishna, V. (1985). Finitely repeatd games. *Econometrica, 53*, 905–922.

Kreps, D. M., & Wilson, R. B. (1982). Sequential equilibria. *Econometrica, 50*, 863–894.

Perea, A. (2001). *Rationality in extensive form games*. Boston: Kluwer Academic.

Selten, R. (1965). Spieltheoretische Behandlung eines Oligopolmodels mit Nachfragezeit. *Zeitschrift für Gesammte Staatswissenschaft, 121*, 301–324.

Selten, R. (1975). Reexamination of the perfectness concept for equilibrium points in extensive games. *International Journal of Game Theory, 4*, 25–55.

Finite Games with Incomplete Information

5

In a game of *imperfect* information players may be uninformed about the moves made by other players. Every one-shot, simultaneous move game is a game of imperfect information. In a game of *incomplete* information players may be uninformed about certain characteristics of the game or of the players. For instance, a player may have incomplete information about actions available to some other player, or about payoffs of other players. Incomplete information is modelled by assuming that every player can be of a number of different types. A type of a player summarizes all relevant information (in particular, actions and payoffs) about that player. Furthermore, it is assumed that each player knows his own type and, given his own type, has a probability distribution over the types of the other players. Often, these probability distributions are assumed to be consistent in the sense that they are the marginal probability distributions derived from a basic commonly known distribution over all combinations of player types.

In this chapter we consider games with finitely many players, finitely many types, and finitely many strategies. These games can be either static (simultaneous, one-shot) or dynamic (extensive form games). A Nash equilibrium in this context is also called 'Bayesian equilibrium', and in games in extensive form an appropriate refinement is perfect Bayesian equilibrium. As will become clear, in essence the concepts studied in Chaps. 3 and 4 are applied again. Throughout this chapter we restrict attention to pure strategies and pure strategy Nash equilibria.

In Sect. 5.1 we present a brief introduction to the concept of player types in a game, but the remainder of the chapter can also be understood without this general introduction. Section 5.2 considers static games of incomplete information, and Sect. 5.3 discusses so-called signaling games, which is the most widely applied class of extensive form games with incomplete information. Both Sects. 5.2 and 5.3 are based on examples, rather than general definitions. For the latter, see Chap. 14.

© Springer-Verlag Berlin Heidelberg 2015
H. Peters, *Game Theory*, Springer Texts in Business and Economics,
DOI 10.1007/978-3-662-46950-7_5

5.1   Player Types

Consider the set of players $N = \{1, \ldots, n\}$. For each player $i \in N$, there is a finite set T_i of *types* which that player can have. If we denote by $T = T_1 \times \ldots \times T_n$ the set

$$T = \{(t_1, \ldots, t_n) \mid t_1 \in T_1, t_2 \in T_2, \ldots, t_n \in T_n\},$$

i.e., the set of all possible combinations of types, then a *game with incomplete information* specifies a separate game for every possible combination $t = (t_1, \ldots, t_n) \in T$, in a way to be explained in the next sections. We assume that each player i knows his own type t_i and, given t_i, attaches probabilities

$$p(t_1, \ldots, t_{i-1}, t_{i+1}, \ldots, t_n \mid t_i)$$

to all type combinations $t_1 \in T_1, \ldots, t_{i-1} \in T_{i-1}, t_{i+1} \in T_{i+1}, \ldots, t_n \in T_n$ of the other players.

Often, these probabilities are derived from a common probability distribution p over T, where $p(t)$ is the probability that the type combination is t. This is also what we assume in this chapter. Moreover, we assume that every player i, apart from his own type t_i, also knows the probability distribution p. Hence, if player i has type t_i, then he can compute the probability that the type combination of the other players is the vector $(t_1, \ldots, t_{i-1}, t_{i+1}, \ldots, t_n)$. Formally, this probability is equal to the conditional probability

$$p(t_1, \ldots, t_{i-1}, t_{i+1}, \ldots, t_n \mid t_i) = \frac{p(t_1, \ldots, t_{i-1}, t_i, t_{i+1}, \ldots, t_n)}{\sum p(t'_1, \ldots, t'_{i-1}, t_i, t'_{i+1}, \ldots, t'_n)}$$

where the sum in the denominator is taken over all possible types of the other players, i.e., over all possible $t'_1 \in T_1, \ldots, t'_{i-1} \in T_{i-1}, t'_{i+1} \in T_{i+1}, \ldots, t'_n \in T_n$. Hence, the sum in the denominator is the probability that player i has type t_i.

Thus, a player in a game of incomplete information can make his actions dependent on his own type but not on the types of the other players. However, since he knows the probabilities of the other players' types, he can compute the expected payoffs from taking specific actions. In the next two sections we will see how this works in static and in extensive form games.

5.2   Static Games of Incomplete Information

We discuss a few examples.

5.2.1   Battle-of-the-Sexes with One-Sided Incomplete Information

The first example is a variant of the Battle-of-the-Sexes (see Sect. 1.3.2) in which player 1 (the man) does not know whether player 2 (the woman) wants to go out with him or avoid him. More precisely, player 1 does not know whether he plays the game y (from 'yes') or the game n (from 'no'), where these games are as follows:

$$y: \begin{array}{c} \\ F \\ B \end{array} \begin{array}{cc} F & B \\ \begin{pmatrix} 2,1 & 0,0 \\ 0,0 & 1,2 \end{pmatrix} \end{array} \qquad n: \begin{array}{c} \\ F \\ B \end{array} \begin{array}{cc} F & B \\ \begin{pmatrix} 2,0 & 0,2 \\ 0,1 & 1,0 \end{pmatrix} \end{array}.$$

Player 1 attaches probability $1/2$ to each of these games, and player 2 knows this. In the terminology of types, this means that player 1 has only one type, simply indicated by '1', and that player 2 has two types, namely y and n. So there are two type combinations, namely $(1, y)$ and $(1, n)$, each occurring with probability $1/2$. Player 2 knows player 1's type with certainty, and also knows her own type, that is, she knows which game is actually being played. Player 1 attaches probability $1/2$ to each type of player 2.

What would be a Nash equilibrium in this game? To see this, it is helpful to model the game as a game in extensive form, using the tree representation of Chap. 4. Such a tree is drawn in Fig. 5.1.

The game starts with a chance move which selects which of the two bimatrix games is going to be played. In the terminology of types, it selects the type of player 2. Player 2 is informed but player 1 is not. Player 2 has four different strategies but player 1 only two.

From this extensive form it is apparent that every Nash equilibrium is subgame perfect, since there are no nontrivial subgames.

Also, every Nash equilibrium is perfect Bayesian, since the only nontrivial information set (namely, that of player 1) is reached with positive probability (namely, equal to 1) for any strategy of player 2, and thus the beliefs are completely

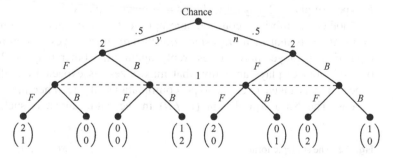

Fig. 5.1 An extensive form representation of the Battle-of-the-Sexes game with incomplete information. The upper numbers are the payoffs for player 1

determined by player 2's strategy through Bayesian updating (see Sect. 4.4). More precisely, suppose the belief of player 1 is denoted by the nonnegative vector $(\alpha_1, \alpha_2, \alpha_3, \alpha_4)$, where $\alpha_1 + \ldots + \alpha_4 = 1$, from left to right in Fig. 5.1. For instance, α_3 is the probability attached by player 1 to player 2 having type n and playing strategy F. Suppose, for instance, that player 2 plays F if she has type y and B if she has type n, and let E denote the event that player 1's information set is reached. Then

$$\alpha_1 = \text{Prob}\,[2 \text{ plays } F \text{ and has type } y \mid E]$$

$$= \frac{\text{Prob}\,[2 \text{ plays } F, \text{ has type } y, \text{ and } E]}{\text{Prob}\,[E]}$$

$$= \frac{\text{Prob}\,[2 \text{ plays } F \mid 2 \text{ has type } y]\,\text{Prob}\,[2 \text{ has type } y]\,\text{Prob}\,[E]}{\text{Prob}\,[E]}$$

$$= \frac{1 \cdot 0.5 \cdot 1}{1} = 0.5\,.$$

By a similar computation we find $\alpha_2 = 0$, $\alpha_3 = 0$, and $\alpha_4 = 0.5$. Such computations can also be made if player 2 would play mixed but we restrict attention here to pure strategies. Important is that the beliefs of player 1 are always determined by computing the conditional probabilities since his (only) information set is always reached with positive probability, namely 1. Hence, there are no *free* beliefs, in the terminology of Sect. 4.4.

The strategic form of the game is given in Fig. 5.2. There, the first letter in a strategy of player 2 says what player 2 plays if y is chosen by the Chance move, and the second letter says what player 2 plays if n is chosen. For instance, if player 1 plays F and player 2 plays FF, then the expected payoffs are equal to $0.5 \cdot (2, 1) + 0.5 \cdot (2, 0) = (2, 0.5)$. Or if player 1 plays B and player 2 plays FB, the expected payoffs are equal to $0.5 \cdot (0, 0) + 0.5 \cdot (1, 0) = (0.5, 0)$, etc. Also the best replies are indicated.

From the strategic form it is apparent that the game has a unique Nash equilibrium in pure strategies, namely (F, FB). In this equilibrium player 1 plays F, type y of player 2 plays F and type n of player 2 plays B.

Another equilibrium concept, appropriate for static games with incomplete information, is that of a *Bayesian equilibrium*. In a Bayesian equilibrium, each type of each player plays a best reply against the other players. That is, each type of a player plays an action that maximizes his expected payoff, where the expectation is taken over the type combinations of the other players and their actions. The Nash equilibrium (F, FB) in the game above is such a Bayesian

Fig. 5.2 The strategic form of the game in Fig. 5.1. Player 1 is the row player

$$\begin{array}{c@{\quad}cccc} & FF & FB & BF & BB \\ F & \begin{pmatrix} \underline{2}, 0.5 \\ B & 0, 0.5 \end{pmatrix} & \underline{1}, \underline{1.5} & \underline{1}, 0 & 0, 1 \\ & 0.5, 0 & 0.5, \underline{1.5} & \underline{1}, 1 \end{array}$$

equilibrium: the action F of type y of player 2 is a best reply against F of player 1 (player 1 has only one type), the action B of type n of player 2 is a best reply against F of player 1, and action F of player 1 maximizes player 1's expected payoff against the strategy FB of player 2. In fact, if every type of every player has positive probability (as will be the case throughout this chapter), then the Nash equilibria of the strategic form coincide with the Bayesian equilibria.

The (pure) Nash equilibrium or equilibria of a game like this can also be found without drawing the extensive form and computing the strategic form, as follows. Suppose first that player 1 plays F in an equilibrium. Then the best reply of player 2 is to play F if her type is y and B if here type is n. The expected payoff to player 1 is then 1; playing B against this strategy FB of player 2 yields only 0.5. So (F, FB) is a Nash equilibrium. If, on the other hand, player 1 plays B, then the best reply of player 2 if her type is y, is B and if her type is n it is F. This yields a payoff of 0.5 to player 1, whereas playing F against this strategy BF of player 2 yields payoff 1. Hence, there is no equilibrium where player 1 plays B. Of course, this is also apparent from the strategic form, but the argument can be made without complete computation of the strategic form.

5.2.2 Battle-of-the-Sexes with Two-Sided Incomplete Information

The next example is a further variation on the Battle-of-the-Sexes game in which neither player knows whether the other player wants to be together with him/her or not. It is based on the four bimatrix games in Fig. 5.3. These four bimatrix games correspond to the four possible type combinations of players 1 and 2. The probabilities of these four different combinations are given in Table 5.1. One way to find the Nash equilibria of this game is to draw the extensive form and compute the associated strategic form: see Problem 5.1. Alternatively, we can systematically examine the possible strategy pairs, as follows.

First observe that each player now has four strategies, namely FF, FB, BF, and BB, where the first letter is the action taken by the yes-type (y_1 or y_2), and the second letter is the action taken by the no-type (n_1 or n_2).

Fig. 5.3 Payoffs for Battle-of-the-Sexes with two types per player

$$y_1y_2 : \begin{array}{c} \\ F \\ B \end{array} \begin{array}{cc} F & B \\ \left(\begin{array}{cc} 2,1 & 0,0 \\ 0,0 & 1,2 \end{array} \right. & \left. \right) \end{array} \qquad y_1n_2 : \begin{array}{c} \\ F \\ B \end{array} \begin{array}{cc} F & B \\ \left(\begin{array}{cc} 2,0 & 0,2 \\ 0,1 & 1,0 \end{array} \right. & \left. \right) \end{array}$$

$$n_1y_2 : \begin{array}{c} \\ F \\ B \end{array} \begin{array}{cc} F & B \\ \left(\begin{array}{cc} 0,1 & 2,0 \\ 1,0 & 0,2 \end{array} \right. & \left. \right) \end{array} \qquad n_1n_2 : \begin{array}{c} \\ F \\ B \end{array} \begin{array}{cc} F & B \\ \left(\begin{array}{cc} 0,0 & 2,2 \\ 1,1 & 0,0 \end{array} \right. & \left. \right) \end{array}$$

Table 5.1 Type probabilities for Battle-of-the-Sexes with two types per player

t	y_1y_2	y_1n_2	n_1y_2	n_1n_2
$p(t)$	2/6	2/6	1/6	1/6

Next, the conditional type probabilities can easily be computed from Table 5.1. For instance,

$$p(y_2|y_1) = \frac{p(y_1 y_2)}{p(y_1)} = \frac{p(y_1 y_2)}{p(y_1 y_2) + p(y_1 n_2)} = \frac{2/6}{(2/6) + (2/6)} = 1/2 \, .$$

The other conditional probabilities are computed in the same way, yielding:

$$p(n_2|y_1) = 1/2, \; p(y_2|n_1) = 1/2, \; p(n_2|n_1) = 1/2 \, ,$$

$$p(y_1|y_2) = 2/3, \; p(n_1|y_2) = 1/3, \; p(y_1|n_2) = 2/3, \; p(n_1|n_2) = 1/3 \, .$$

We now consider the four pure strategies of player 1 one by one.

(i) Suppose that player 1 plays the strategy FF, meaning that he plays F (the first letter) if his type is y_1 and also F (the second letter) if his type is n_1. Then the expected payoff for type y_2 of player 2 if she plays F is

$$p(y_1|y_2) \cdot 1 + p(n_1|y_2) \cdot 1 = (2/3) \cdot 1 + (1/3) \cdot 1 = 1 \, .$$

If type y_2 of player 2 plays B her expected payoff is

$$p(y_1|y_2) \cdot 0 + p(n_1|y_2) \cdot 0 = (2/3) \cdot 0 + (1/3) \cdot 0 = 0 \, .$$

Hence the best reply of type y_2 of player 2 is F. Similarly, for type n_2 of player 2, playing F yields

$$p(y_1|n_2) \cdot 0 + p(n_1|n_2) \cdot 0 = (2/3) \cdot 0 + (1/3) \cdot 0 = 0$$

and playing B yields

$$p(y_1|n_2) \cdot 2 + p(n_1|n_2) \cdot 2 = (2/3) \cdot 2 + (1/3) \cdot 2 = 2 \, ,$$

so that B is the best reply. Hence, player 2's best reply against FF is FB. Suppose, now, that player 2 plays FB, so type y_2 plays F and type n_2 plays B. Then playing F yields type y_1 of player 1 an expected payoff of

$$p(y_2|y_1) \cdot 2 + p(n_2|y_1) \cdot 0 = (1/2) \cdot 2 + (1/2) \cdot 0 = 1$$

and playing B yields

$$p(y_2|y_1) \cdot 0 + p(n_2|y_1) \cdot 1 = (1/2) \cdot 0 + (1/2) \cdot 1 = 1/2 \, ,$$

so that F is the best reply for type y_1 of player 1. Similarly, for type n_1 playing F yields

$$p(y_1|n_1) \cdot 0 + p(n_2|n_1) \cdot 2 = (1/2) \cdot 0 + (1/2) \cdot 2 = 1$$

whereas playing B yields

$$p(y_1|n_1) \cdot 1 + p(n_2|n_1) \cdot 0 = (1/2) \cdot 1 + (1/2) \cdot 0 = 1/2 \,.$$

Hence, F is the best reply for type n_1. Hence, player 1's best reply against FB is FF. We conclude that (FF, FB) is a Nash equilibrium.

(ii) Suppose player 1 plays FB. Playing F yields type y_2 of player 2 a payoff of

$$p(y_1|y_2) \cdot 1 + p(n_1|y_2) \cdot 0 = 2/3 \cdot 1 + 1/3 \cdot 0 = 2/3$$

and playing B yields

$$p(y_1|y_2) \cdot 0 + p(n_1|y_2) \cdot 2 = 2/3 \cdot 0 + 1/3 \cdot 2 = 2/3$$

so that both F and B are best replies. Playing F yields type n_2 a payoff of

$$p(y_1|n_2) \cdot 0 + p(n_1|n_2) \cdot 1 = 2/3 \cdot 0 + 1/3 \cdot 1 = 1/3$$

and playing B yields

$$p(y_1|n_2) \cdot 2 + p(n_1|n_2) \cdot 0 = 2/3 \cdot 2 + 1/3 \cdot 0 = 4/3$$

so that B is the best reply. Hence, player 2 has two best replies, namely FB and BB. Against FB player 1's best reply is FF [as established in case (i)] and not FB, so this does not result in a Nash equilibrium. Against BB one can compute in the same way as hitherto that player 1's best reply is BF and not FB, so also this combination is not a Nash equilibrium.

(iii) Suppose that player 1 plays BF. Then player 2 has two best replies, namely BF and BB. Against BF the best reply of player 1 is FF and not BF, so this combination is not a Nash equilibrium. Against BB, player 1's best reply is BF, so the combination (BF, BB) is a Nash equilibrium.

(iv) Finally, suppose player 1 plays BB. Then player 2's best reply is BF. Against this, player 1's best reply is FF and not BB. So BB of player 1 is not part of a Nash equilibrium.

We conclude that the game has two Nash equilibria in pure strategies, namely: (i) both types of player 1 play F, type y_2 of player 2 also plays F but type n_2 of player 2 plays B; (ii) type y_1 of player 1 plays B, type n_1 plays F, and both types of player 2 play B. Again, these equilibria are also called Bayesian Nash equilibria.

5.3   Signaling Games

The extensive form can be used to examine a static game of incomplete information, usually by letting the game start with a chance move that picks the types of the players (see Sect. 5.2). More generally, the extensive form can be used to describe incomplete information games where players move sequentially. An important class of such games is the class of signaling games.

A (finite) signaling game starts with a chance move that picks the type of player 1. Player 1 is informed about his type but player 2 is not. Player 1 moves first, player 2 observes player 1's action and moves next, and then the game ends. Such a game is called a *signaling game* because the action of player 1 may be a signal about his type: that is, from the action of player 1 player 2 may be able to infer something about the type of player 1.

5.3.1   An Example

Consider the example in Fig. 5.4. (The numbers between square brackets at player 2's decision nodes are the beliefs of player 2, which are used in a perfect Bayesian equilibrium below.) In this game, player 1 learns the result of the chance move but player 2 does not. In the terminology of Sect. 5.1, there are two type combinations, namely $(t, 2)$ and $(\tilde{t}, 2)$, each one occurring with probability $1/2$: these notations express the fact that player 2 has only one type (called '2'). Both types of player 1 can choose between L and R. Player 2 only observes the action, L or R, and not the type of player 1. For this reason we use the same letter L for the 'left' action of each type: player 2 cannot distinguish between them. Similar for the 'right' action.

In order to analyze this game and find the (pure strategy) Nash equilibria, one possibility is to first compute the strategic form. Both players have four strategies. Player 1 has strategy set

$$\{LL, LR, RL, RR\} \, ,$$

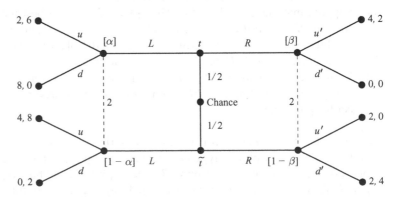

Fig. 5.4 A signaling game

Fig. 5.5 The strategic form
of the game in Fig. 5.4

$$
\begin{array}{c@{\quad}c@{\quad}c@{\quad}c}
 & uu' & ud' & du' & dd' \\
\begin{array}{c} LL \\ LR \\ RL \\ RR \end{array} &
\left(\begin{array}{cccc}
3,\underline{7} & \underline{3},\underline{7} & 4,1 & 4,1 \\
2,3 & 2,\underline{5} & \underline{5},0 & \underline{5},2 \\
\underline{4},\underline{5} & 2,4 & 2,2 & 0,1 \\
3,1 & 1,\underline{2} & 3,1 & 1,\underline{2}
\end{array}\right)
\end{array}
$$

where the first letter refers to the action of type t and the second letter to the action of type $\bar{t}$. Player 2 has strategy set

$$\{uu', ud', du', dd'\}\,.$$

The (expected) strategic form of the game can be computed in the usual way and is presented in Fig. 5.5. For instance, consider the strategy combination (LR, ud'). Then the expected payoffs are $1/2 \cdot (2, 6) + 1/2 \cdot (2, 4) = (2, 5)$, etc. The (pure) best replies are underlined. This shows that the game has two Nash equilibria, namely (RL, uu') and (LL, ud'). What else can be said about these equilibria? Observe that the only subgame of the game is the entire game, so that both equilibria are trivially subgame perfect. Are they also perfect Bayesian? That is, do they satisfy the conditions of Bayesian consistency and sequential rationality (Sect. 4.4)?

First consider the equilibrium (RL, uu'). Bayesian consistency requires

$$
\begin{aligned}
\alpha &= \text{Prob [player 1 has type } t \mid \text{player 1 plays } L] \\
&= \frac{\text{Prob [player 1 has type } t \text{ and plays } L]}{\text{Prob [player 1 plays } L]} \\
&= \frac{1/2 \cdot 0}{1/2} \\
&= 0
\end{aligned}
$$

and, similarly,

$$
\begin{aligned}
\beta &= \text{Prob [player 1 has type } t \mid \text{player 1 plays } R] \\
&= \frac{\text{Prob [player 1 has type } t \text{ and plays } R]}{\text{Prob [player 1 plays } R]} \\
&= \frac{1/2 \cdot 1}{1/2} \\
&= 1\,.
\end{aligned}
$$

Given these beliefs, playing u yields a payoff of 8 to player 2 (at the left information set), whereas playing d yields only 2, so u is optimal. Playing u' yields a payoff of 2 to player 2 (at the right information set), whereas playing d' yields 0, so u' is optimal. Hence, uu' is indeed the best reply of player 2—as we already knew from the strategic form. Thus, the pair (RL, uu') is a perfect Bayesian equilibrium with beliefs $\alpha = 0$ and $\beta = 1$.

Note that, in this case and, more generally, in case of a Nash equilibrium where each information set of player 2 is reached with positive probability (i.e., every possible action of player 1 is played by some type of player 1), the perfect Bayesian equilibrium requirement does not add anything that is essential: the beliefs of player 2 are completely determined by Bayesian consistency, and sequential rationality is automatically satisfied for these beliefs, given that we already have a Nash equilibrium.

The perfect Bayesian equilibrium (RL, uu') is called *separating*: it separates the two types of player 1, since these types play different actions. In this equilibrium, the action of player 1 is a signal for his type, and the equilibrium is 'information revealing'.

Next, consider the Nash equilibrium (LL, ud'). In this case, by Bayesian consistency we obtain

$$
\begin{aligned}
\alpha &= \text{Prob}\left[\text{player 1 has type } t \mid \text{player 1 plays } L\right] \\
&= \frac{\text{Prob}\left[\text{player 1 has type } t \text{ and plays } L\right]}{\text{Prob}\left[\text{player 1 plays } L\right]} \\
&= \frac{1/2 \cdot 1}{1} \\
&= 1/2 \,.
\end{aligned}
$$

In words: since each type of player 1 plays L, the conditional probabilities of the two decision nodes in the left information set of player 2 are both equal to $1/2$. Given $\alpha = 1/2$ it follows that u is optimal at player 2's left information set (in fact, in this particular game u is optimal for any α): we know this already from the strategic form, where both uu' and ud' are best replies of player 2 against LL.

How about the belief $(\beta, 1 - \beta)$? Since player 1 always plays L, the right information set of player 2 is reached with zero probability, and therefore we cannot compute β by the formula for conditional probability: Bayesian consistency has no bite, the belief $(\beta, 1 - \beta)$ is *free* (in the terminology of Sect. 4.4). Formally,

$$
\begin{aligned}
\beta &= \text{Prob}\left[\text{player 1 has type } t \mid \text{player 1 plays } R\right] \\
&= \frac{\text{Prob}\left[\text{player 1 has type } t \text{ and plays } R\right]}{\text{Prob}\left[\text{player 1 plays } R\right]} \,,
\end{aligned}
$$

but the probability in the denominator of the last expression is zero if player 1 plays LL. However, we still have the sequential rationality requirement: in order for (LL, ud') to be a perfect Bayesian equilibrium the belief $(\beta, 1 - \beta)$ should be such that player 2's action d' is optimal. Hence, the expected payoff to player 2 from playing d' should be at least as large as the expected payoff from playing u', so $4(1 - \beta) \geq 2\beta$, which is equivalent to $\beta \leq 2/3$. Thus, (LL, ud') is a perfect Bayesian equilibrium with beliefs $\alpha = 1/2$ and $\beta \leq 2/3$.

In this case, and more generally, in cases where in a Nash equilibrium *not* every information set of player 2 is reached with positive probability, i.e., there is some action of player 1 which is played by *no* type of player 1, the perfect Bayesian equilibrium requirement does have an impact.

The equilibrium (LL, ud') is called *pooling*, since it 'pools' the two types of player 1: both types play the same action, L in this case. In this equilibrium, the action of player 1 does not reveal any information about his type.

5.3.2 Computing Perfect Bayesian Equilibria in the Extensive Form

Perfect Bayesian equilibria can also be found without first computing the strategic form. We consider again the signaling game in Fig. 5.4.

First, assume that there is an equilibrium where player 1 plays LL. Then $\alpha = 1/2$ by Bayesian consistency, and player 2's optimal action at the left information set (following L) is u. At the right information set, player 2's optimal action is u' if $\beta \geq 2/3$ and d' if $\beta \leq 2/3$. If player 2 would play u' after R, then type t of player 1 would improve by playing R instead of L, so this cannot be an equilibrium. If player 2 plays d' after R, then no type of player 1 would want to play R instead of L. We have established that (LL, ud') with beliefs $\alpha = 1/2$ and $\beta \leq 2/3$ is a (pooling) perfect Bayesian equilibrium.

Second, assume player 1 plays LR in equilibrium. Then player 2's beliefs are given by $\alpha = 1$ and $\beta = 0$, and player 2's best reply is ud'. But then type $\bar{t}$ of player 1 would gain by playing L instead of R, so this cannot be an equilibrium.

Third, assume player 1 plays RL in equilibrium. Then $\alpha = 0$, $\beta = 1$, and player 2's best reply is uu'. Against uu', RL is player 1's best reply, so that (RL, uu) is a (separating) perfect Bayesian equilibrium with beliefs $\alpha = 0$ and $\beta = 1$.

Fourth, suppose player 1 plays RR in equilibrium. Then $\beta = 1/2$ and player 2's best reply after R is d'. After L, player 2's best reply is u for any value of α. Against ud', however, type t of payer 1 would gain by playing L instead of R. So RR is not part of an equilibrium.

Of course, these considerations can also be based on the strategic form, but we do not need the entire strategic form to find the perfect Bayesian equilibria.

5.3.3 The Intuitive Criterion

In a perfect Bayesian equilibrium, if an information set of player 2 is reached with zero probability, then the belief of player 2 on that information set is free—the only requirement is the sequential rationality requirement demanding that, given this belief, player 2 should choose the optimal action. The question is whether such a free belief is always plausible or reasonable. The so-called *intuitive criterion* (IC) puts a restriction on the plausibility of free beliefs.

It works as follows. Consider a perfect Bayesian equilibrium in a signaling game and suppose that there is an information set which is reached in the equilibrium with zero probability. In other words, there is an action, say A, of player 1 which is played by no type of player 1. Consider a type t of player 1. Suppose this type t obtains payoff x in the equilibrium under consideration. Then consider the maximal possible payoff for this type t attainable by playing A, say m. If $m < x$, then the belief of player 2 on the information set following A should assign zero probability to type t of player 1. The reason is indeed intuitive: type t could never possibly gain by playing A instead of his equilibrium action, since *at best* he would obtain m from doing so, but m is less than what t can get in equilibrium, namely x. Therefore, player 2 should not believe that type t would ever deviate to A. This comparison should be made for every *type* of player 1. This way, we may obtain some restrictions on the belief of player 2 at the information set following action A. If the original belief of player 2 on this information set, corresponding to the perfect Bayesian equilibrium under consideration, satisfies these restrictions, then we say that this perfect Bayesian equilibrium *survives* the IC. However, it could happen that this way all types of player 1 get assigned zero probability: in that case, the IC simply does *not* apply, since the probabilities in a belief have to sum up to 1 [cf. Problem 5.9(a)].

Let us apply the IC to the perfect Bayesian equilibrium (LL, ud') with $\beta \leq 2/3$ in the game in Fig. 5.4. The equilibrium payoff to type t of player 1 is equal to 2. If type t of player 1 deviates to R, then he could get maximally 4, namely if player 2 would play u' following R. It is important to notice that we consider the *maximal possible* payoff after such a deviation—of course, the payoff in the equilibrium after a deviation (in this case 0) can never be higher by the mere definition of Nash equilibrium. Since $4 \not< 2$, type t could have a reason to deviate, and so the IC puts no restriction on the belief of player 2 that player 1, if he would deviate to R, is of type t. The equilibrium payoff to type $\tilde{t}$ of player 1 is 4. Type $\tilde{t}$, however, could get at most 2 (in fact, would always get 2 in this game) by deviating to R. Since $2 < 4$, the IC now says that it is not reasonable for player 2 to assume that type $\tilde{t}$ would ever deviate to R. Thus, $1 - \beta = 0$, so that $\beta = 1$. With this belief, however, (LL, ud') can no longer be sustained as a perfect Bayesian equilibrium, since for this we need $\beta \leq 2/3$. Hence, the perfect Bayesian equilibrium with $\beta \leq 2/3$ does not survive the IC.

5.3.4 Another Example

Consider the signaling game in Fig. 5.6. We compute the pure strategy perfect Bayesian equilibria of this game by considering the strategies of player 1 one by one.

Suppose player 1 plays LL (the first letter refers to type t and the second to type $\tilde{t}$). Then $\alpha = 1/2$ by Bayesian consistency, so player 2 is indifferent between u and d. If player 2 plays u, however, then type $\tilde{t}$ obtains 0 and will therefore deviate to R, so that he obtains at least 1. Hence, player 2 should play d following L in order to get an equilibrium. Then, again to keep type $\tilde{t}$ from deviating, player 2 should

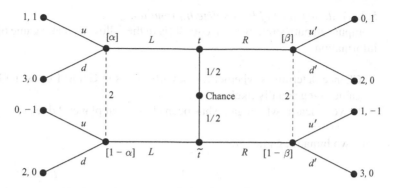

Fig. 5.6 Another signaling game

play u' following R. This is indeed optimal for player 2 for $\beta \geq 1/2$. Obviously, type t will not deviate to R, so (LL, du') with $\alpha = 1/2$ and $\beta \geq 1/2$ is a (pooling) perfect Bayesian equilibrium. In this equilibrium, type t obtains 3 and by deviating to R obtains maximally 2. Type $\tilde{t}$ obtains 2 in equilibrium and by deviating to R maximally 3. Hence, the IC prescribes $\beta = 0$, but for this belief (LL, du') does not result in a perfect Bayesian equilibrium. Hence, (LL, du') with $\alpha = 1/2$ and $\beta \geq 1/2$ does not survive the IC.

Suppose player 1 plays LR. Then the best reply of player 2 is ud', but then type t will deviate to R. So there is no perfect Bayesian equilibrium (or even Nash equilibrium) in which player 1 plays LR.

Suppose player 1 plays RL. Then the best reply of player 2 is du', but then type t will deviate to L. So there is no perfect Bayesian equilibrium (or even Nash equilibrium) in which player 1 plays RL.

Finally, suppose player 1 plays RR. Then $\beta = 1/2$ and player 2 is indifferent between u' and d' following R. If player 2 plays u' then type t of player 1 will deviate to R, where he obtains always more than 0. Hence in an equilibrium player 2 should play d'. Again to keep type t from deviating, player 2 should play u following L. This is indeed optimal for player 2 if $\alpha \geq 1/2$. Hence, (RR, ud') with beliefs $\beta = 1/2$ and $\alpha \geq 1/2$ is a (pooling) perfect Bayesian equilibrium. Type t obtains 2 in equilibrium and maximally 3 by deviating to L. Type $\tilde{t}$ obtains 3 in equilibrium and maximally 2 by deviating to L. Thus, the IC prescribes $1 - \alpha = 0$ or $\alpha = 1$, and therefore this perfect Bayesian equilibrium survives the IC for $\alpha = 1$.

5.4 Problems

5.1. *Battle-of-the-Sexes*
Draw the extensive form of the Battle-of-the-Sexes game in Sect. 5.2 with payoffs in Fig. 5.3 and type probabilities in Table 5.1. Compute the strategic form and find the pure strategy Nash equilibria of the game.

5.2. *A Static Game of Incomplete Information*
Compute all pure strategy Nash equilibria in the following static game of incomplete information:

1. Chance determines whether the payoffs are as in Game 1 or as in Game 2, each game being equally likely.
2. Player 1 learns which game has been chosen but player 2 does not.

The two bimatrix games are:

$$
\begin{array}{cc}
 & \begin{array}{cc} L & R \end{array} \\
\text{Game 1:} \quad \begin{array}{c} T \\ B \end{array} & \begin{pmatrix} 1,1 & 0,0 \\ 0,0 & 0,0 \end{pmatrix}
\end{array}
\qquad
\begin{array}{cc}
 & \begin{array}{cc} L & R \end{array} \\
\text{Game 2:} \quad \begin{array}{c} T \\ B \end{array} & \begin{pmatrix} 0,0 & 0,0 \\ 0,0 & 2,2 \end{pmatrix}
\end{array}
$$

5.3. *Another Static Game of Incomplete Information*
Player 1 has two types, t_1 and t_1', and player 2 has two types, t_2 and t_2'. The conditional probabilities of these types are:

$$p(t_2|t_1) = 1, \; p(t_2|t_1') = 3/4, \; p(t_1|t_2) = 3/4, \; p(t_1|t_2') = 0 .$$

(a) Show that these conditional probabilities can be derived from a common distribution p over the four type combinations, and determine p.

As usual suppose that each player learns his own type and knows the conditional probabilities above. Then player 1 chooses between T and B and player 2 between L and R, where these actions may be contingent on the information a player has. The payoffs for the different type combinations are given by the bimatrix games

$$
t_1 t_2 : \begin{array}{c} T \\ B \end{array} \begin{pmatrix} 2,2 & 0,0 \\ 3,0 & 1,1 \end{pmatrix}
\quad
t_1' t_2 : \begin{array}{c} T \\ B \end{array} \begin{pmatrix} 2,2 & 0,0 \\ 0,0 & 1,1 \end{pmatrix}
\quad
t_1' t_2' : \begin{array}{c} T \\ B \end{array} \begin{pmatrix} 2,2 & 0,0 \\ 0,0 & 1,1 \end{pmatrix} ,
$$

where the type combination (t_1, t_2') is left out since it has zero probability.

(b) Compute all pure strategy Nash equilibria for this game.

5.4. *Job-Market Signaling*
A worker can have either high or low ability, where the probability of high ability is equal to $2/5$. A worker knows his ability, but a firm which wants to hire the worker does not. The worker, whether a high or a low ability type, can choose between additional education or not. Choosing additional education does not enlarge the worker's productivity but may serve as a signal to the firm: a high ability worker can choose education without additional costs, whereas for a low ability worker the cost of education equals $e > 0$. The firm chooses either a high or a low wage, having observed whether the worker took additional education

or not. The payoff to the firm equals the productivity of the worker minus the wage. The payoff to the worker equals the wage minus the cost of education; if, however, this payoff is lower than the worker's reservation utility, he chooses not to work at all and to receive his reservation utility, leaving the firm with 0 payoff. Denote the productivities of the high and low ability worker by p^H and p^L, respectively, and denote the high and low wages by w^h and w^l. Finally, let r^H and r^L denote the reservation utilities of both worker types. (All these numbers are fixed.)

(a) Determine the extensive form of this game.
(b) Choose $p^H = 10$, $p^L = 8$, $w^h = 6$, $w^l = 3$, $r^H = 4$, $r^L = 0$, $e = 4$. Compute the strategic form of this game, and determine the pure strategy Nash equilibria. Also compute the perfect Bayesian equilibrium or equilibria in pure strategies, determine whether they are separating or pooling and whether they survive the IC.

5.5. *A Joint Venture*

Software Inc. and Hardware Inc. are in a joint venture together. The parts used in the joint product can be defective or not; the probability of defective parts is 0.7, and this is commonly known before the start of the game. Each can exert either high or low effort, which is equivalent to costs of 20 and 0. Hardware moves first, but software cannot observe his effort. Revenues are split equally at the end. If both firms exert low effort, total profits are 100. If the parts are defective, the total profit is 100; otherwise (i.e., if the parts are not defective), if both exert high effort, profit is 200, but if only one player does, profit is 100 with probability 0.9 and 200 with probability 0.1. Hardware discovers the truth about the parts by observation before he chooses effort, but software does not.

(a) Determine the extensive form of this game. Is this a signaling game?
(b) Determine the strategic form of this game.
(c) Compute the (pure) Nash equilibria? Which one(s) is (are) subgame perfect? Perfect Bayesian?

5.6. *Entry Deterrence*

The entry deterrence game of Chap. 1 is reproduced in Fig. 5.7. For this game, compute the pure strategy perfect Bayesian equilibria for every value of $x \in \mathbb{R}$. Which one(s) is (are) pooling or separating? Satisfy the intuitive criterion?

5.7. *The Beer-Quiche Game*

Consider the following two-player signaling game. Player 1 is either 'weak' or 'strong'. This is determined by a chance move, resulting in player 1 being 'weak' with probability 1/10. Player 1 is informed about the outcome of this chance move but player 2 is not; but the probabilities of either type of player 1 are

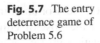

Fig. 5.7 The entry deterrence game of Problem 5.6

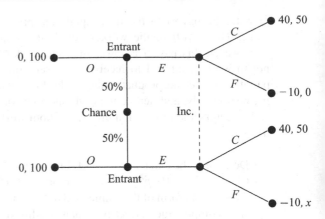

common knowledge among the two players. Player 1 has two actions: either have quiche (Q) or have beer (B) for breakfast. Player 2 observes the breakfast of player 1 and then decides to duel (D) or not to duel (N) with player 1. The payoffs are as follows. If player 1 is weak and eats quiche then D and N give him payoffs of 1 and 3, respectively; if he is weak and drinks beer, then these payoffs are 0 and 2, respectively. If player 1 is strong, then the payoffs are 0 and 2 from D and N, respectively, if he eats quiche; and 1 and 3 from D and N, respectively, if he drinks beer. Player 2 has payoff 0 from not duelling, payoff 1 from duelling with the weak player 1, and payoff −1 from duelling with the strong player 1.

(a) Draw a diagram modelling this situation.
(b) Compute all the pure strategy Nash equilibria of the game. Find out which of these Nash equilibria are perfect Bayesian equilibria. Give the corresponding beliefs and determine whether these equilibria are pooling or separating, and which ones satisfy the intuitive criterion.

5.8. *Issuing Stock*
In this story the players are a manager (M) and an existing shareholder (O). The manager is informed about the current value of the firm, a, and the NPV (net present value) of a potential investment opportunity, b, but the shareholder only knows that high values and low values each have probability $1/2$. More precisely, either $(a, b) = (\bar{a}, \bar{b})$ or $(a, b) = (\underline{a}, \underline{b})$, each with probability $1/2$, where $\underline{a} < \bar{a}$ and $\underline{b} < \bar{b}$. The manager moves first and either proposes to issue new stock E (where E is fixed) to undertake the investment opportunity, or decides not to issue new stock. The existing shareholder decides whether to approve of the new stock issue or not. The manager always acts in the interest of the existing shareholder: their payoffs in the game are always equal.

If the manager decides not to issue new stock, then the investment opportunity is foregone, and the payoff is either $\bar{a}$ or $\underline{a}$. If the manager proposes to issue new stock but this is not approved by the existing shareholder, then again the investment opportunity is foregone and the payoff is either $\bar{a}$ or $\underline{a}$. If the manager proposes to issue new stock E and the existing shareholder approves of this, then the payoff to the existing shareholder is equal to $[M/(M+E)](\bar{a}+\bar{b}+E)$ in the good state $(\bar{a},\bar{b})$ and $[M/(M+E)](\underline{a}+\underline{b}+E)$ in the bad state $(\underline{a},\underline{b})$; here, $M = (1/2)[\bar{a}+\bar{b}] + (1/2)[\underline{a}+\underline{b}]$ is the price of the existing shares if the investment is undertaken.

(a) Set up the extensive form of this signaling game.
(b) Take $\bar{a} = 150$, $\underline{a} = 50$, $\bar{b} = 20$, $\underline{b} = 10$, and $E = 100$. Compute the pure strategy perfect Bayesian equilibria of this game. Are they pooling, separating? How about the intuitive criterion? Try to interpret the results from an economic point of view.
(c) Repeat the analysis of (b) for $\bar{b} = 100$.

5.9. *More Signaling Games*

(a) Consider the signaling game in Fig. 5.4, but with payoffs $(1,2)$ instead of $(4,2)$ if type t of player 1 plays R and player 2 plays u'. Show that this game has only pooling equilibria. Which ones survive the IC?
(b) Compute the pure strategy perfect Bayesian equilibria and test for the intuitive criterion in the signaling game in Fig. 5.8.
(c) Consider the signaling game in Fig. 5.9, where the chance move is not explicitly drawn in order to keep the diagram simple. Compute the pure strategy perfect Bayesian equilibria and test for the intuitive criterion.

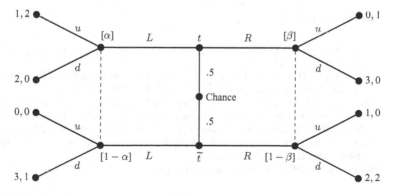

Fig. 5.8 The signaling game of Problem 5.9(b)

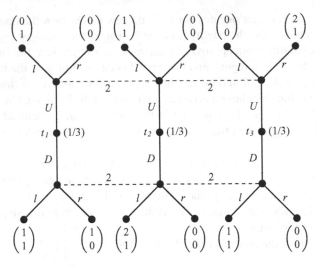

Fig. 5.9 The signaling game of Problem 5.9(c). Each type of player 1 has probability $1/3$

5.5   Notes

The type terminology and corresponding theory is due to Harsanyi (1967/1968).
The Battle of the Sexes examples in Sect. 5.2 are taken from Osborne (2004). One
of the first examples of a signaling game is the Spence (1973) job market signaling
model (see Problem 5.4). The intuitive criterion is due to Cho and Kreps (1987).

Problem 5.5 is taken from Rasmusen (1989). The beer-quiche game of Problem 5.7 is from Cho and Kreps (1987). Problem 5.8 is based on Myers and Majluf (1984).

References

Cho, I. K., & Kreps, D. M. (1987). Signalling games and stable equilibria. *Quarterly Journal of Economics, 102*, 179–221.

Harsanyi, J. C. (1967/1968). Games with incomplete information played by "Bayesian" players I, II, and III. *Management Science, 14*, 159–182, 320–334, 486–502.

Myers, S. C., & Majluf, N. S. (1984). Corporate financing and investment decisions when firms have information that investors do not have. *Journal of Financial Economics, 13*, 187–221.

Osborne, M. J. (2004). *An introduction to game theory*. New York: Oxford University Press.

Rasmusen, E. (1989). *Games and information: An introduction to game theory* (2nd ed., 1994/1995). Oxford: Basil Blackwell.

Spence, A. M. (1973). Job market signalling. *Quarterly Journal of Economics, 87*, 355–374.

Noncooperative Games: Extensions

6

In Chaps. 2–5 we have studied noncooperative games in which the players have finitely many (pure) strategies. The reason for the finiteness restriction is that in such games special results hold, such as the existence of a value and optimal strategies for two-person zero-sum games, and the existence of a Nash equilibrium in mixed strategies for finite nonzero-sum games.

The basic game-theoretical concepts discussed in these chapters can be applied to more general games. Once, in a game-theoretic situation, the players, their possible strategies, and the associated payoffs are identified, the concepts of best reply and of Nash equilibrium can be applied. Also the concepts of backward induction, subgame perfection, and perfect Bayesian equilibrium carry over to quite general extensive form games. In games of incomplete information, the concept of player types and the associated Nash equilibrium (Bayesian Nash equilibrium) can be applied also if the game has infinitely many strategies.

The bulk of this chapter consists of diverse examples verifying these claims. The main objective of the chapter is, indeed, to show how the basic game-theoretic apparatus can be applied to various different conflict situations; and, of course, to show these applications themselves.

In Sect. 6.1 we generalize some of the concepts of Chaps. 2 and 3. This section serves only as background and general framework for the examples in the following sections—most of the remainder of this chapter can also be understood without this general framework and the reader may choose to postpone reading it. Concepts specific to extensive form games and to incomplete information games are adapted later, when they are applied. In Sects. 6.2–6.7 we discuss Cournot competition with complete and incomplete information, Bertrand competition, Stackelberg equilibrium, auctions with complete and incomplete information, mixed strategies with objective probabilities, and sequential bargaining. Variations on these topics and various other topics are treated in the problem section.

© Springer-Verlag Berlin Heidelberg 2015
H. Peters, *Game Theory*, Springer Texts in Business and Economics,
DOI 10.1007/978-3-662-46950-7_6

6.1   General Framework: Strategic Games

An *n-person strategic game* is a $2n + 1$-tuple

$$G = (N, S_1, \ldots, S_n, u_1, \ldots, u_n) ,$$

where

- $N = \{1, \ldots, n\}$, with $n \geq 1$, is the set of *players*;
- for every $i \in N$, S_i is the *strategy set* of player i;
- for every $i \in N$, $u_i : S = S_1 \times \ldots \times S_n \to \mathbb{R}$ is the *payoff function* of player i; i.e., for every strategy combination $(s_1, \ldots, s_n) \in S$ where $s_1 \in S_1$, ..., $s_n \in S_n$, $u_i(s_1, \ldots, s_n) \in \mathbb{R}$ is player i's payoff.

A *best reply* of player i to the strategy combination $(s_1, \ldots, s_{i-1}, s_{i+1}, \ldots, s_n)$ of the other players is a strategy $s_i \in S_i$ such that

$$u_i(s_1, \ldots, s_{i-1}, s_i, s_{i+1}, \ldots, s_n) \geq u_i(s_1, \ldots, s_{i-1}, s'_i, s_{i+1}, \ldots, s_n)$$

for all $s'_i \in S_i$.

A *Nash equilibrium* of G is a strategy combination $(s_1^*, \ldots, s_n^*) \in S$ such that for each player i, s_i^* is a best reply to $(s_1^*, \ldots, s_{i-1}^*, s_{i+1}^*, \ldots, s_n^*)$.

A strategy $s'_i \in S_i$ of player i is *strictly dominated* by $s_i \in S_i$ if

$$u_i(s_1, \ldots, s_{i-1}, s_i, s_{i+1}, \ldots, s_n) > u_i(s_1, \ldots, s_{i-1}, s'_i, s_{i+1}, \ldots, s_n)$$

for all $(s_1, \ldots, s_{i-1}, s_{i+1}, \ldots, s_n) \in S_1 \times \ldots \times S_{i-1} \times S_{i+1} \times \ldots \times S_n$, i.e., for all strategy combinations of players other than i. Clearly, a strictly dominated strategy is never used in a Nash equilibrium.

Finally we define weak domination. A strategy $s'_i \in S_i$ of player i is *weakly dominated* by $s_i \in S_i$ if

$$u_i(s_1, \ldots, s_{i-1}, s_i, s_{i+1}, \ldots, s_n) \geq u_i(s_1, \ldots, s_{i-1}, s'_i, s_{i+1}, \ldots, s_n)$$

for all $(s_1, \ldots, s_{i-1}, s_{i+1}, \ldots, s_n) \in S_1 \times \ldots \times S_{i-1} \times S_{i+1} \times \ldots \times S_n$, such that at least once this inequality is strict.

The reader may verify that matrix games (Chap. 2) and bimatrix games (Chap. 3) are special cases of this general framework, in which the set S_i is the set of all mixed strategies of player i. The same is true for the concepts of Nash equilibrium and domination discussed in these chapters.

6.2 Cournot Quantity Competition

6.2.1 Simple Version with Complete Information

In the simplest version of the famous Cournot model, two firms producing a homogenous good compete in quantity. Each firm offers a quantity of this good on the market. The price of the good depends on the total quantity offered: the higher this quantity is, the lower the price of the good. The profit for each firm is equal to total revenue (price times quantity) minus total cost. This gives rise to a two-person game in which the players are the firms, the players' strategies are the quantities offered and the payoff functions are the profit functions. In a simple version, the price depends linearly on total quantity and marginal cost is constant and positive while there are no fixed costs. Specifically, we study the following game.

(a) The set of players is $N = \{1, 2\}$.
(b) Each player $i = 1, 2$ has set of strategies $S_i = [0, \infty)$, with typical element q_i.
(c) The payoff function of player i is $\Pi_i(q_1, q_2) = q_i P(q_1, q_2) - cq_i$, for all $q_1, q_2 \geq 0$. Here,

$$P(q_1, q_2) = \begin{cases} a - q_1 - q_2 & \text{if } q_1 + q_2 \leq a \\ 0 & \text{if } q_1 + q_2 > a \end{cases}$$

is the market price of the good, where a is a constant, and c is marginal cost, with $a > c > 0$.

A Nash equilibrium in this game is a pair (q_1^C, q_2^C), with $q_1^C, q_2^C \geq 0$, of mutually best replies, that is,

$$\Pi_1(q_1^C, q_2^C) \geq \Pi_1(q_1, q_2^C), \ \Pi_2(q_1^C, q_2^C) \geq \Pi_2(q_1^C, q_2) \text{ for all } q_1, q_2 \geq 0 .$$

This equilibrium is also called *Cournot equilibrium*. To find the equilibrium, we first compute the best reply functions, also called *reaction functions*. The reaction function $\beta_1(q_2)$ of player 1 is found by solving the maximization problem

$$\max_{q_1 \geq 0} \ \Pi_1(q_1, q_2)$$

for each given value of $q_2 \geq 0$. If $q_2 > a$ then $P(q_1, q_2) = 0$ for every q_1 and the profit of firm 1 is equal to $-cq_1$ so that, clearly, the maximum is attained by setting $q_1 = 0$. For $q_2 \leq a$ we have

$$\Pi_1(q_1, q_2) = \begin{cases} q_1(a - q_1 - q_2) - cq_1 & \text{if } q_1 \leq a - q_2 \\ -cq_1 & \text{if } q_1 > a - q_2 \end{cases}$$

so that the maximum is attained for some $q_1 \leq a - q_2$. In that case, function

$$q_1(a - q_1 - q_2) - cq_1 = q_1(a - c - q_1 - q_2)$$

has to be maximized with respect to $q_1 \geq 0$. If $q_2 > a - c$ then the maximum is attained for $q_1 = 0$. If $q_2 \leq a - c$ then we compute the maximum by setting the derivative with respect to q_1, namely the function $a - 2q_1 - q_2 - c$, equal to zero. This yields $q_1 = (a - c - q_2)/2$. (The second derivative is equal to -2 so that, indeed, we have a maximum.) Summarizing, we have

$$\beta_1(q_2) = \begin{cases} \{\frac{a-c-q_2}{2}\} & \text{if } q_2 \leq a - c \\ \{0\} & \text{if } q_2 > a - c. \end{cases}$$

Since this reaction function is single-valued for all q_2, we can omit the braces at the right-hand side and write

$$\beta_1(q_2) = \begin{cases} \frac{a-c-q_2}{2} & \text{if } q_2 \leq a - c \\ 0 & \text{if } q_2 > a - c. \end{cases} \tag{6.1}$$

By symmetric arguments we obtain for the reaction function of player 2:

$$\beta_2(q_1) = \begin{cases} \frac{a-c-q_1}{2} & \text{if } q_1 \leq a - c \\ 0 & \text{if } q_1 > a - c. \end{cases} \tag{6.2}$$

These reaction functions are drawn in Fig. 6.1. The Nash equilibrium is the point of intersection of the reaction functions. It is obtained by simultaneously solving the two equations $q_1 = (a - c - q_2)/2$ and $q_2 = (a - c - q_1)/2$, resulting in

$$(q_1^C, q_2^C) = \left(\frac{a-c}{3}, \frac{a-c}{3}\right).$$

6.2.1.1 Pareto Optimality

A pair (q_1, q_2) of strategies is *Pareto optimal* if there is no other pair (q_1', q_2') such that the associated payoffs are at least as good for both players and strictly better for at least one player. Not surprisingly, the equilibrium (q_1^C, q_2^C) is not Pareto optimal. For instance, both players can strictly benefit from joint profit maximization, attained by solving the problem

$$\max_{q_1, q_2 \geq 0} \Pi_1(q_1, q_2) + \Pi_2(q_1, q_2).$$

This amounts to solving the maximization problem

$$\max_{q_1, q_2 \geq 0} (q_1 + q_2)(a - c - q_1 - q_2)$$

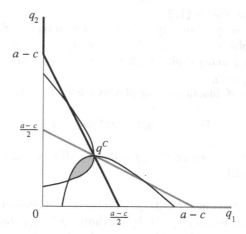

Fig. 6.1 The Cournot model: the *thick black* piecewise linear curve is the reaction function of player 1 and the *thick gray* piecewise linear curve is the reaction function of player 2. The point q^C is the Nash–Cournot equilibrium. The two isoprofit curves of the players through the Nash equilibrium are drawn. The curve intersecting the q_1-axis is the isoprofit curve of player 1: profit increases if this curve shifts downwards. The curve intersecting the q_2-axis is the isoprofit curve of player 2: profit increases if this curve shifts leftwards. The *shaded area* consists of the quantity combinations that Pareto dominate the equilibrium

or, writing $Q = q_1 + q_2$,

$$\max_{Q \geq 0} \ Q(a - c - Q)$$

which yields $Q = (a - c)/2$. Observe that this is just the monopoly quantity. Thus, joint profit maximization is attained by any pair $(q_1, q_2) \geq 0$ with $q_1 + q_2 = (a - c)/2$. Taking, in particular, $q_1 = q_2 = (a - c)/4$ yields each player a profit of $(a - c)^2/8$, whereas in the Nash equilibrium each player obtains $(a - c)^2/9$. See also Fig. 6.1, where all points in the gray-shaded area 'Pareto dominate' the Nash equilibrium: the associated payoffs are at least as good for both agents and better for at least one agent.

6.2.2 Simple Version with Incomplete Information

Consider the Cournot model of Sect. 6.2.1 but now assume that the marginal cost of firm 2 is either high, c_H, or low, c_L, where $c_H > c_L > 0$. Firm 2 knows its marginal cost but firm 1 only knows that it is c_H with probability ϑ and c_L with probability $1 - \vartheta$. The cost of firm 1 is $c > 0$ and this is commonly known. In the terminology of Sect. 5.1, player 1 has only one type but player 2 has two types, c_H and c_L. The associated game is as follows.

(a) The player set is $\{1, 2\}$.

(b) The strategy set of player 1 is $[0, \infty)$ with typical element q_1, and the strategy set of player 2 is $[0, \infty) \times [0, \infty)$ with typical element (q_H, q_L). Here, q_H is the chosen quantity if player 2 has type c_H, and q_L is the chosen quantity if player 2 has type c_L.

(c) The payoff functions of the players are the expected payoff functions. These are

$$\Pi_i(q_1, q_H, q_L) = \vartheta \Pi_i(q_1, q_H) + (1 - \vartheta)\Pi_i(q_1, q_L) \,,$$

for $i = 1, 2$, where $\Pi_i(\cdot, \cdot)$ is the payoff function from the Cournot model of Sect. 6.2.1.

To find the (Bayesian) Nash equilibrium, we first compute the best reply function or reaction function of player 1, by maximizing $\Pi_1(q_1, q_H, q_L)$ over $q_1 \geq 0$, with q_H and q_L regarded as given. Hence, we solve the problem

$$\max_{q_1 \geq 0} \ \vartheta \left[q_1(a - c - q_1 - q_H) \right] + (1 - \vartheta) \left[q_1(a - c - q_1 - q_L) \right] \,.$$

Assuming $q_H, q_L \leq a - c$ (this has to be checked later for the equilibrium), this problem is solved by setting the derivative with respect to q_1 equal to zero, which yields

$$q_1 = q_1(q_H, q_L) = \frac{a - c - \vartheta q_H - (1 - \vartheta)q_L}{2} \,. \tag{6.3}$$

Observe that, compared to (6.1), we now have the expected quantity $\vartheta q_H + (1 - \vartheta)q_L$ instead of q_2: this is due to the linearity of the model.

For player 2, we consider, for given q_1, the problem

$$\max_{q_H, q_L \geq 0} \ \vartheta \left[q_H(a - c_H - q_1 - q_H) \right] + (1 - \vartheta) \left[q_L(a - c_L - q_1 - q_L) \right] \,.$$

Since the first term in this function depends only on q_H and the second term only on q_L, solving this problem amounts to maximizing the two terms separately. In other words, we determine the best replies of types c_H and c_L separately.[1] Assuming $q_1 \leq a - c_H$ (and hence $q_1 \leq a - c_L$) this results in

$$q_H = q_H(q_1) = \frac{a - c_H - q_1}{2} \tag{6.4}$$

---

[1] This is generally so in a Bayesian, incomplete information game: maximizing the expected payoff of a player over all his types is equivalent to maximizing the payoff per type.

and

$$q_L = q_L(q_1) = \frac{a - c_L - q_1}{2} . \tag{6.5}$$

The Nash equilibrium is obtained by simultaneously solving (6.3)–(6.5) (using substitution or Gaussian elimination). The solution is the triple

$$q_1^C = \frac{a - 2c + \vartheta c_H + (1 - \vartheta)c_L}{3}$$

$$q_H^C = \frac{a - 2c_H + c}{3} + \frac{1 - \vartheta}{6}(c_H - c_L)$$

$$q_L^C = \frac{a - 2c_L + c}{3} - \frac{\vartheta}{6}(c_H - c_L) .$$

Assuming that the parameters of the game are such that these three values are nonnegative and that $q_1 \leq a - c_H$ and $q_H, q_L \leq a - c$, this is the Bayesian Nash–Cournot equilibrium of the game. This solution may be compared with the Nash equilibrium in the complete information model with asymmetric costs, see Problem 6.1. The high cost type of firm 2 produces more than it would in the complete information case: it benefits from the fact that firm 1 is unsure about the cost of firm 2 and therefore produces less than it would if it knew for sure that firm 2 had high costs. Similarly, the low cost firm 2 produces less.

6.3 Bertrand Price Competition

Consider two firms who compete in the price of a homogenous good. Specifically, assume that the demand q for the good is given by $q = q(p) = \max\{a - p, 0\}$ for every price $p \geq 0$, where a is a positive constant (the demand for the good if the price is zero). The firm with the lower price serves the whole market; if prices are equal the firms share the market equally. Each firm has the same marginal cost $0 \leq c < a$, and no fixed cost. If firm 1 sets a price p_1 and firm 2 sets a price p_2, then the profit of firm 1 is

$$\Pi_1(p_1, p_2) = \begin{cases} (p_1 - c)(a - p_1) & \text{if } p_1 < p_2 \text{ and } p_1 \leq a \\ \frac{1}{2}(p_1 - c)(a - p_1) & \text{if } p_1 = p_2 \text{ and } p_1 \leq a \\ 0 & \text{in all other cases.} \end{cases}$$

Similarly, the profit of firm 2 is

$$\Pi_2(p_1, p_2) = \begin{cases} (p_2 - c)(a - p_2) & \text{if } p_2 < p_1 \text{ and } p_2 \leq a \\ \frac{1}{2}(p_2 - c)(a - p_2) & \text{if } p_1 = p_2 \text{ and } p_2 \leq a \\ 0 & \text{in all other cases.} \end{cases}$$

Fig. 6.2 The profit function
of firm i in the monopoly
situation

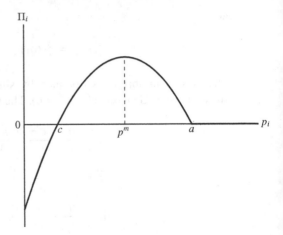

Thus, the two firms are the players in this game, and their profit functions are the
payoff functions; the strategy sets are $[0, \infty)$ for each, with typical elements p_1 and
p_2. To find a Nash equilibrium (*Bertrand equilibrium*) we first compute the best
reply functions (reaction functions). An important role is played by the price that
maximizes profit if there is only one firm in the market, i.e., the monopoly price
$p^m = (a + c)/2$, obtained by solving the problem

$$\max_{p \geq 0} \; (p - c)(a - p) \, .$$

Note that the monopoly profit function (or the profit function of each firm in the
monopoly situation) is a quadratic function, and that profit increases as the price
gets closer to the monopoly price. See Fig. 6.2.

To determine player 1's best reply function $\beta_1(p_2)$ we distinguish several cases.

If $p_2 < c$, then any $p_1 \leq p_2$ yields player 1 a negative payoff, whereas any
$p_1 > p_2$ yields a payoff of zero. Hence, the set of best replies in this case is the
interval (p_2, ∞).

If $p_2 = c$, then any $p_1 < p_2$ yields a negative payoff for player 1, and any $p_1 \geq p_2$
yields zero payoff. So the set of best replies in this case is the interval $[c, \infty)$.

If $c < p_2 \leq p^m$, then the best reply of player 1 would be a price below p_2
(to obtain the whole market) and as close to the monopoly price as possible (to
maximize payoff) but such a price does not exist: for any price $p_1 < p_2$, a price in
between p_1 and p_2 would still be better. Hence, in this case the set of best replies of
player 1 is empty.[2]

If $p_2 > p^m$ then the unique best reply of player 1 is the monopoly price p^m.

---

[2]If prices are in smallest monetary units this somewhat artificial consequence is avoided. See
Problem 6.7.

Summarizing we obtain

$$\beta_1(p_2) = \begin{cases} \{p_1 \mid p_1 > p_2\} & \text{if } p_2 < c \\ \{p_1 \mid p_1 \geq c\} & \text{if } p_2 = c \\ \emptyset & \text{if } c < p_2 \leq p^m \\ \{p^m\} & \text{if } p_2 > p^m. \end{cases}$$

For player 2, similarly,

$$\beta_2(p_1) = \begin{cases} \{p_2 \mid p_2 > p_1\} & \text{if } p_1 < c \\ \{p_2 \mid p_2 \geq c\} & \text{if } p_1 = c \\ \emptyset & \text{if } c < p_1 \leq p^m \\ \{p^m\} & \text{if } p_1 > p^m. \end{cases}$$

The point(s) of intersection of these best reply functions can be found by making a diagram or by direct inspection. We follow the latter method and leave the diagram method to the reader. If $p_2 < c$ then by $\beta_1(p_2)$ a best reply p_1 satisfies $p_1 > p_2$. But then, according to $\beta_2(p_1)$, we must have $p_2 = p^m$, a contradiction since $p^m > c$. Therefore, in equilibrium, we must have $p_2 \geq c$. If $p_2 = c$, then $p_1 \geq c$; if however, $p_1 > c$ then the only possibility is $p_2 = p^m$, a contradiction. Hence, $p_1 = c$ as well and, indeed, $p_1 = p_2 = c$ is a Nash equilibrium. If $p_2 > c$, then the only possibility is $p_1 = p^m$ but then p_2 is never a best reply. We conclude that the unique Nash equilibrium (Bertrand equilibrium) is $p_1 = p_2 = c$.

It is also possible to establish this result without completely computing the best reply functions. Suppose, in equilibrium, that $p_1 \neq p_2$, say $p_1 < p_2$. If $p_1 < p^m$ then player 1 can increase his payoff by setting a higher price still below p_2. If $p_1 \geq p^m$ then player 2 can increase his payoff by setting a price below p_1, e.g., slightly below p^m if $p_1 = p^m$ and equal to p^m if $p_1 > p^m$. Hence, we must have $p_1 = p_2$ in equilibrium. If this common price is below c then each player can improve by setting a higher price. If this common price is above c then each player can improve by setting a slightly lower price. Hence, the only possibility that remains is $p_1 = p_2 = c$, and this is indeed an equilibrium, as can be verified directly.

A few remarks on this equilibrium are in order. First, it is again Pareto inferior. For example, both firms setting the monopoly price results in higher profits. Second, each firm plays a weakly dominated strategy: any price $c < p_i < a$ weakly dominates $p_i = c$, since it always results in a positive or zero profit whereas $p_i = c$ always results in zero profit. Third, the Bertrand equilibrium is beneficial from the point of view of the consumers: it maximizes consumer surplus.

See Problem 6.3(d)–(f) for an example of price competition with heterogenous goods.

6.4   Stackelberg Equilibrium

In the Cournot model of Sect. 6.2.1, the two firms move simultaneously. Consider
now the situation where firm 1 moves first, and firm 2 observes this move and
moves next. This situation has already been discussed in Chap. 1. The corresponding
extensive form game is given in Fig. 6.3. In this game, player 1 has infinite
action/strategy set $[0, \infty)$, with typical element q_1. In the diagram, we use a zigzag
line to express the fact that the number of actions is infinite. Player 2 has the infinite
set of actions $[0, \infty)$ with typical element q_2, again represented by a zigzag line.
A strategy of player 2 assigns to each information set, hence to each decision
node—the game has perfect information—an action. Since each decision node of
player 2 follows an action q_1 of player 1, a strategy of player 2 is a function
$s_2 : [0, \infty) \rightarrow [0, \infty)$. Hence, $q_2 = s_2(q_1)$ is the quantity that firm 2 offers if
firm 1 has offered q_1. Obviously, the number of strategies of player 2 is infinite as
well.[3] The appropriate solution concept is *backward induction* or *subgame perfect*
equilibrium. The subgames of this game are the entire game and the infinite number
of one-player games starting at each decision node of player 2, i.e., following each
choice q_1 of player 1. Hence, the subgame perfect equilibrium can be found by
backward induction, as follows. In each subgame for player 2, that is, after each
choice q_1, player 2 should play optimally. This means that player 2 should play
according to the reaction function $\beta_2(q_1)$ as derived in (6.2). Then, going back
to the beginning of the game, player 1 should choose $q_1 \geq 0$ so as to maximize
$\Pi_1(q_1, \beta_2(q_1))$. In other words, player 1 takes player 2's optimal reaction into
account when choosing q_1. Assuming $q_1 \leq a - c$ (it is easy to verify that $q_1 > a - c$
is not optimal) player 1 maximizes the expression

$$q_1 \left(a - c - q_1 - \frac{a - c - q_1}{2} \right) .$$

The maximum is obtained for $q_1 = (a - c)/2$, and thus $q_2 = \beta_2((a - c)/2) =
(a - c)/4$. Hence, the subgame perfect equilibrium of the game is:

$$q_1 = (a - c)/2, \ q_2 = \beta_2(q_1) .$$

The subgame perfect equilibrium *outcome* is by definition the resulting play of the
game, that is, the actions chosen on the equilibrium path in the extensive form. In

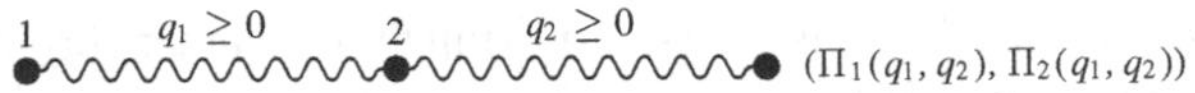

Fig. 6.3 Extensive form representation of the Stackelberg game with firm 1 as the leader

---

[3]In mathematical notation the strategy set of player 2 is the set $[0, \infty)^{[0, \infty)}$.

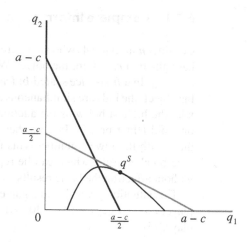

Fig. 6.4 As before, the *thick black curve* is the reaction function of player 1 and the *thick gray curve* is the reaction function of player 2. The point $q^S = (\frac{a-c}{2}, \frac{a-c}{4})$ is the Stackelberg equilibrium outcome: it is the point on the reaction curve of player 2 where player 1 maximizes profit. The associated isoprofit curve of player 1 is drawn

this case, the equilibrium outcome is:

$$q_1^S = (a-c)/2, \ q_2^S = (a-c)/4 \ .$$

The letter 'S' here is the first letter of 'Stackelberg', after whom this equilibrium is named. More precisely, this subgame perfect equilibrium (or outcome) is called the *Stackelberg equilibrium* (or *outcome*) with player 1 as the *leader* and player 2 as the *follower*. Check that player 1's profit in this equilibrium is higher and player 2's profit is lower than in the Cournot equilibrium $q_1^C = q_2^C = (a-c)/3$. See also Problem 6.9.

The Stackelberg equilibrium is depicted in Fig. 6.4. Observe that player 1, the leader, picks the point on the reaction curve of player 2 which has maximal profit for player 1. Hence, player 2 is on his reaction curve but player 1 is not.

6.5 Auctions

An auction is a procedure to sell goods among various interested parties, such that the prices are determined in the procedure. Examples range from selling a painting through an ascending bid auction (English auction) and selling flowers through a descending bid auction (Dutch auction) to tenders for public projects and selling mobile telephone frequencies.

In this section we consider a few simple, classical auction models. We start with first and second-price sealed-bid auctions under complete information, continue with a first-price sealed bid auction with incomplete information, and end with a double auction between a buyer and a seller. Some variations and extensions are discussed in Problems 6.10–6.14.

6.5.1 Complete Information

Consider n individuals who are interested in one indivisible object. Each individual i has valuation $v_i > 0$ for the object. We assume without loss of generality $v_1 \geq v_2 \geq \ldots \geq v_n$. In a *first-price sealed-bid auction* each individual submits a bid $b_i \geq 0$ for the object: the bids are simultaneous and independent ('sealed bids'). The individual with the highest bid wins the auction and obtains the object at a price equal to his own bid ('first price'). In case there are more highest bidders, the bidder among these with the lowest number wins the auction and pays his own bid—this is just a tie-breaking rule, which can be replaced by alternative tie-breaking assumptions without affecting the basic results.

This situation gives rise to a game with player set $N = \{1, 2, \ldots, n\}$, where each player i has strategy set $S_i = [0, \infty)$ with typical element b_i. The payoff function to player i is[4]

$$u_i(b_1, \ldots, b_i, \ldots, b_n) = \begin{cases} v_i - b_i & \text{if } i = \min\{k \in N \mid b_k \geq b_j \text{ for all } j \in N\} \\ 0 & \text{otherwise.} \end{cases}$$

One Nash equilibrium in this game is the strategy combination $(b_1, \ldots, b_n) = (v_2, v_2, v_3, \ldots, v_n)$. To check this one should verify that no player has a better bid, given the bids of the other players: see Problem 6.10. In this equilibrium, player 1 obtains the object and pays v_2, the second-highest valuation. Check that this is also the outcome one would approximately expect in an auction with ascending bids (English auction) or descending bids (Dutch auction).

This game has many Nash equilibria. In each of these equilibria, however, a player with a highest valuation obtains the object. Bidding one's true valuation as well as bidding higher than one's true valuation are weakly dominated strategies. Bidding lower than one's true valuation is not weakly dominated. (See Sect. 6.1 for the definition of weak domination.) Problem 6.10 is about proving all these statements.

A *second-price sealed-bid auction* differs from a first-price sealed-bid auction only in that the winner now pays the bid of the second highest bidder. In the case that two or more players have the highest bid the player with the lowest number wins and pays his own bid. The main property of this auction is that for each player i, the strategy of bidding the true valuation v_i weakly dominates all other strategies. This property and other properties are collected in Problem 6.11.

---

[4]Also here the assumption is that the players know the game. This means, in particular, that the players know each other's valuations.

6.5.2 Incomplete Information

We consider the same setting as in Sect. 6.5.1 but now assume that each bidder knows his own valuation but has only a probabilistic estimate about the valuations of the other bidders. In the terminology of types (cf. Sect. 5.1), a bidder's valuation is his true type, and each bidder holds a probability distribution over the type combinations of the other bidders. To keep things simple, we assume that every bidder's type is drawn independently from the uniform distribution over the interval [0, 1], that this is common knowledge, and that each bidder learns his true type. The auction is a first-price sealed-bid auction. Of course, we can no longer fix the ordering of the valuations, but we can still employ the same tie-breaking rule in case of more than one highest bid.

We discuss the case of two bidders and postpone the extension to $n > 2$ bidders until Problem 6.13. In the associated two-person game, a strategy of player $i \in \{1, 2\}$ should assign a bid to each of his possible types. Since the set of possible types is the interval [0, 1] and it does not make sense to ever bid more than 1, a strategy is a function $s_i : [0, 1] \rightarrow [0, 1]$. Hence, if player i's type is v_i, then $b_i = s_i(v_i)$ is his bid according to the strategy s_i. The payoff function u_i of player i assigns to each strategy pair (s_i, s_j) (where j is the other player) player i's expected payoff if these strategies are played. In a Nash equilibrium of the game, player i maximizes this payoff function given the strategy of player j, and *vice versa*. For this, it is sufficient that *each type* of player i maximizes its expected payoff given the strategy of player j, and *vice versa*; in other words (cf. Chap. 5), it is sufficient that the strategies form a Bayesian equilibrium.[5]

We claim that $s_1^*(v_1) = v_1/2$ and $s_2^*(v_2) = v_2/2$ is a Nash equilibrium of this game. To prove this, first consider type v_1 of player 1 and suppose that player 2 plays strategy s_2^*. If player 1 bids b_1, then the probability that player 1 wins the auction is equal to the probability that the bid of player 2 is smaller than or equal to b_1. This probability is equal to the probability that $v_2/2$ is smaller than or equal to b_1, i.e., to the probability that v_2 is smaller than or equal to $2b_1$. We may assume without loss of generality that $b_1 \leq 1/2$, since according to s_2^* player 2 will never bid higher than $1/2$. Since v_2 is uniformly distributed over the interval [0, 1] and $2b_1 \leq 1$, the probability that v_2 is smaller than or equal to $2b_1$ is just equal to $2b_1$. Hence, the probability that the bid b_1 of player 1 is winning is equal to $2b_1$ if player 2 plays s_2^*, and therefore the expected payoff from this bid is equal to $2b_1(v_1 - b_1)$ (if player 1 loses his payoff is zero). This is maximal for $b_1 = v_1/2$. Hence, $s_1^*(v_1) = v_1/2$ is a best reply to s_2^*. The converse is almost analogous—the only difference being that for player 2 to win player 1's bid must be strictly smaller due to the tie-breaking rule employed, but this does not change the associated probability under the uniform distribution. Hence, we have proved the claim.

---

[5]The difference is that, for instance, single types may not play a best reply in a Nash equilibrium since they have probability zero and therefore do not influence the expected payoffs.

Thus, in this equilibrium, each bidder bids half his true valuation, and a player with the highest valuation wins the auction.

How about the second-price sealed-bid auction with incomplete information? This is more straightforward since bidding one's true valuation ($s_i(v_i) = v_i$ for all $v_i \in [0, 1]$) is a strategy that weakly dominates every other strategy, for each player i. Hence, these strategies still form a (Bayesian) Nash equilibrium. See Problem 6.11.

6.5.3 Incomplete Information: A Double Auction

Assume there are two players, a buyer and a seller. The seller owns an object, for which he has a valuation v_s. The buyer has a valuation v_b for this object. These valuations are independently drawn from the uniform distribution over $[0, 1]$. The seller knows (learns) his own valuation, and also the buyer knows his own valuation, but each of them does not know the valuation of the other player, only that this is drawn from the uniform distribution over $[0, 1]$.

The auction works as follows. The buyer and the seller independently and simultaneously mention prices p_b and p_s. If $p_b \geq p_s$, then trade takes place at the average price $p = (p_b + p_s)/2$, and the payoffs are $v_b - p$ to the buyer and $p - v_s$ to the seller. If $p_b < p_s$ then no trade takes place and both have payoff 0.

This is a game of incomplete information. The buyer has infinitely many types $v_b \in [0, 1]$, and the seller has infinitely many types $v_s \in [0, 1]$. A strategy assigns a price to each type. A strategy for the buyer is therefore a function $p_b : [0, 1] \to [0, 1]$, where $p_b(v_b)$ is the price that the buyer offers if his type (valuation) is v_b. (Observe that we may indeed assume that the price is never higher than 1.) Similarly, a strategy for the seller is a function $p_s : [0, 1] \to [0, 1]$, where $p_s(v_s)$ is the price that the seller asks if his type (valuation) is v_s.

Suppose the seller plays a strategy $p_s(\cdot)$. Then the expected payoff to the buyer if his valuation is v_b and he offers price p_b is equal to

$$\left[v_b - \frac{p_b + E[p_s(v_s)|p_b \geq p_s(v_s)]}{2} \right] \text{Prob}[p_b \geq p_s(v_s)] \qquad (6.6)$$

where $E[p_s(v_s)|p_b \geq p_s(v_s)]$ denotes the expected price asked by the seller according to his strategy $p_s(\cdot)$, conditional on this price being smaller than or equal to the price p_b of the buyer.

Similarly, suppose the buyer plays a strategy $p_b(\cdot)$. Then the expected payoff to the seller if his valuation is v_s and he asks price p_s is equal to

$$\left[\frac{p_s + E[p_b(v_b)|p_s \leq p_b(v_b)]}{2} - v_s \right] \text{Prob}[p_s \leq p_b(v_b)] \qquad (6.7)$$

where $E[p_b(v_b)|p_s \leq p_b(v_b)]$ denotes the expected price offered by the buyer according to his strategy $p_b(\cdot)$, conditional on this price being larger than or equal to the price p_s of the seller.

Now the pair of strategies $(p_b(\cdot), p_s(\cdot))$ is a (Bayesian) Nash equilibrium if for each $v_b \in [0, 1]$, $p_b(v_b)$ solves

$$\max_{p_b \in [0,1]} \left[v_b - \frac{p_b + E[p_s(v_s)|p_b \geq p_s(v_s)]}{2} \right] \text{Prob}[p_b \geq p_s(v_s)] \qquad (6.8)$$

and for each $v_s \in [0, 1]$, $p_s(v_s)$ solves

$$\max_{p_s \in [0,1]} \left[\frac{p_s + E[p_b(v_b)|p_s \leq p_b(v_b)]}{2} - v_s \right] \text{Prob}[p_s \leq p_b(v_b)] . \qquad (6.9)$$

This game has many Nash equilibria. Ideally, trade should take place whenever it is efficient, i.e., whenever $v_b \geq v_s$. In Problem 6.14 we will see that not all Nash equilibria are equally efficient.

6.6   Mixed Strategies and Incomplete Information

In this section we discuss how a *mixed* strategy Nash equilibrium in a bimatrix game can be obtained as a limit of *pure* strategy Bayesian Nash equilibria in associated games of incomplete information.

Consider the bimatrix game (cf. Chap. 3)

$$G = \begin{matrix} & \begin{matrix} L & \quad R \end{matrix} \\ \begin{matrix} T \\ B \end{matrix} & \begin{pmatrix} 2,1 & 2,0 \\ 3,0 & 1,3 \end{pmatrix} \end{matrix},$$

which has a unique Nash equilibrium $((p^*, 1 - p^*), (q^*, 1 - q^*))$ with $p^* = 3/4$ and $q^* = 1/2$. The interpretation of mixed strategies and of a mixed strategy Nash equilibrium in particular is an old issue in the game-theoretic literature. One obvious interpretation is that a player actually plays according to the equilibrium probabilities. Although there is some empirical evidence that this may occur in reality,[6] this interpretation may not be entirely convincing, in particular since in a mixed strategy Nash equilibrium a player is indifferent between all pure strategies played with positive probability (cf. Problem 3.8). An alternative interpretation— also mentioned in Sect. 3.1—is that a mixed strategy of a player represents the belief(s) of the other player(s) about the strategic choice of that player. For instance, in the above equilibrium, player 2 believes that player 1 plays T with probability 3/4. The drawback of this interpretation is that these beliefs are *subjective*, and it is not explained how they are formed. In this section we discuss a way to obtain a mixed strategy Nash equilibrium as the limit of pure strategy (Bayesian) Nash equilibria in games obtained by adding some *objective* uncertainty about the payoffs. In this

---

[6]For example, Walker and Wooders (2001).

way, the strategic uncertainty of players as expressed by their beliefs is replaced by
the objective uncertainty of a chance move.

In the above example, suppose that the payoff to player 1 from (T, L) is the
uncertain amount $2 + \alpha$ and the payoff to player 2 from (B, R) is the uncertain
amount $3 + \beta$. Assume that both α and β are (independently) drawn from a uniform
distribution over the interval $[0, x]$, where $x > 0$. Moreover, player 1 learns the true
value of α and player 2 learns the true value of β, and all this is common knowledge
among the players. In terms of types, player 1 knows his type α and player 2 knows
his type β. The new payoffs are given by

$$
\begin{array}{c}
 \quad\quad\quad L \quad\quad\quad\quad R \\
\begin{array}{c} T \\ B \end{array}
\left(
\begin{array}{cc}
2+\alpha, 1 & 2, 0 \\
3, 0 & 1, 3+\beta
\end{array}
\right).
\end{array}
$$

A (pure) strategy of a player assigns an action to each of his types. Hence, for player
1 it is a map $s_1 : [0, x] \to \{T, B\}$ and for player 2 it is a map $s_2 : [0, x] \to \{L, R\}$.

To find an equilibrium of this incomplete information game, suppose that player 2
has the following rather simple strategy: play L if β is small and play R if β is large.
Specifically, let $b \in [0, x]$ such that each type $\beta \leq b$ plays L and each type $\beta > b$
plays R. Call this strategy s_2^b. What is player 1's best reply against s_2^b? Suppose the
type of player 1 is α. If player 1 plays T, then his expected payoff is equal to $2 + \alpha$
times the probability that player 2 plays L plus 2 times the probability that player 2
plays R. The probability that player 2 plays L, given the strategy s_2^b, is equal to the
probability that β is at most equal to b, and this is equal to b/x since β is uniformly
distributed over $[0, x]$. Hence, the expected payoff to player 1 from playing T is

$$
(2+\alpha) \cdot \frac{b}{x} + 2 \cdot (1 - \frac{b}{x}) = 2 + \alpha \cdot \frac{b}{x} .
$$

Similarly, the expected payoff to player 1 from playing B is

$$
3 \cdot \frac{b}{x} + 1 \cdot (1 - \frac{b}{x}) = 1 + 2 \cdot \frac{b}{x} .
$$

From this, it easily follows that T is at least as good as B if $\alpha \geq (2b - x)/b$. Hence,
the following strategy of player 1 is a best reply against strategy s_2^b of player 2: play
T if $\alpha \geq a$ and play B if $\alpha < a$, where $a = (2b - x)/b$. Call this strategy s_1^a.

For the converse, assume that player 1 plays s_1^a. To find player 2's best reply
against s_1^a we proceed similarly as above. If type β of player 2 plays L then the
expected payoff is 1 times the probability that player 1 plays T, hence 1 times $(x - a)/x$. If type β of player 2 plays R then his expected payoff is equal to $3 + \beta$ times
the probability that player 1 plays B, hence $(3 + \beta)a/x$. So L is at least as good as R
if $\beta \leq (x - 4a)/a$. Hence, a best reply of player 2 against s_1^a is the strategy s_2^b with
$b = (x - 4a)/a$.

Summarizing these arguments, we have that (s_1^a, s_2^b) is a (Bayesian) Nash equilibrium for

$$a = (2b - x)/b, \ b = (x - 4a)/a \ .$$

Solving these two equations simultaneously for solutions $a, b \in [0, x]$ yields:

$$a = (1/4)(x + 4 - \sqrt{x^2 + 16}), \ b = (1/2)(x - 4 + \sqrt{x^2 + 16}) \ .$$

In this equilibrium, the a priori probability that player 1 will play T, that is, the probability of playing T before he learns his type, is equal to $(x - a)/x$, hence to $(\sqrt{x^2 + 16} + 3x - 4)/4x$. Similarly, the a priori probability that player 2 plays L is equal to b/x, hence to $(x - 4 + \sqrt{x^2 + 16})/2x$. What happens with these probabilities as the amount of uncertainty decreases, i.e., for x approaching 0? For player 1,

$$\lim_{x \to 0} \frac{\sqrt{x^2 + 16} + 3x - 4}{4x} = \lim_{x \to 0} \frac{x/\sqrt{x^2 + 16} + 3}{4} = \frac{3}{4} \ ,$$

where the first equality follows from l'Hôpital's rule. Similarly for player 2:

$$\lim_{x \to 0} \frac{x - 4 + \sqrt{x^2 + 16}}{2x} = \lim_{x \to 0} \frac{1 + x/\sqrt{x^2 + 16}}{2} = \frac{1}{2} \ .$$

In other words, these probabilities converge to the mixed strategy Nash equilibrium of the original game.

6.7 Sequential Bargaining

In its simplest version, the bargaining problem involves two parties who have to agree on one alternative within a set of feasible alternatives. If they fail to reach an agreement, a specific 'disagreement' alternative is implemented. In the game-theoretic literature on bargaining there are two main strands, namely the cooperative, axiomatic approach, also known as the Nash bargaining problem, and the noncooperative, strategic approach, with the Rubinstein alternating offers game as main representative. In this section the focus is on the strategic approach, but in Sect. 6.7.2 we also mention the connection with the Nash bargaining solution. For an introduction to the axiomatic bargaining approach see Sect. 10.1, and see Chap. 21 for a more extensive study.

6.7.1 Finite Horizon Bargaining

Consider the example in Sect. 1.3.5, where two players bargain over the division of one unit of a perfectly divisible good, e.g., 1 liter of wine. If they do not reach an agreement, we assume that no one gets anything. To keep the problem as simple as possible, assume that the preferences of the players are represented by $u_1(\alpha) = u_2(\alpha) = \alpha$ for every $\alpha \in [0, 1]$. That is, obtaining an amount α of the good has utility α for each player. In this case the picture of the feasible set in Sect. 1.3.5 would be a triangle.

To model the bargaining process we consider the following alternating offers procedure. There are $T + 1$ rounds, where $T \in \mathbb{N}$. In round $t = 0$ player 1 makes a proposal, say $(\alpha, 1 - \alpha)$, where $\alpha \in [0, 1]$, meaning that he claims an amount α for himself, so that player 2 obtains $1 - \alpha$. Player 2 can either accept this proposal, implying that the proposal is implemented and the game is over, or reject the proposal. In the latter case the next round $t = 1$ starts, and the first round is repeated with the roles of the players interchanged: player 2 makes the proposal and player 1 accepts or rejects it. If player 1 accepts the proposal then it is implemented and the game is over; if player 1 rejects the proposal then round $t = 2$ starts, and the roles of the players are interchanged again. Thus, at even rounds, player 1 proposes; at odd rounds, player 2 proposes. The last possible round is round T: if this round is reached, then the disagreement alternative $(0, 0)$ is implemented.

We assume that utilities are discounted. Specifically, there is a discount factor $0 < \delta < 1$ such that receiving an amount α at round t has utility $\delta^t \alpha$ at round 0. Or, receiving an amount α at time t has utility $\delta\alpha$ at round $t - 1$. This reflects the fact that receiving the same amount earlier is more valuable.

In Fig. 6.5 this bargaining procedure is represented as a game in extensive form. Here, we assume that T is odd, so that the last proposal at time $t = T - 1$ is made by player 1.

We look for a subgame perfect equilibrium of this game, which can be found by backward induction. Note that subgames start at each decision node of a player, and that at each such node in Fig. 6.5 where a player has to make, accept or reject

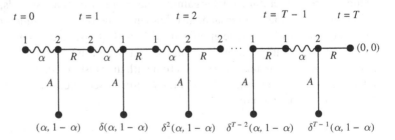

Fig. 6.5 The extensive form representation of the finite horizon bargaining procedure. The number of rounds $T + 1$ is even, α denotes the proposed amount for player 1, A is acceptance and R is rejection

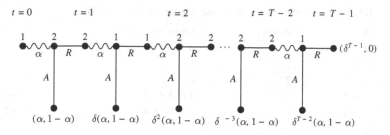

Fig. 6.6 The game of Fig. 6.5 reduced by replacing rounds $T-1$ and T by the equilibrium payoffs of the associated subgames

a proposal, actually infinitely many subgames start, since there are infinitely many possible paths leading to that node.

To start the analysis, at the final decision node, player 2 accepts if $\alpha < 1$ and is indifferent between acceptance and rejection if $\alpha = 1$: this is because rejection results in getting 0 at round T. In the subgame starting at round $T-1$ with a proposal of player 1, the only equilibrium therefore is for player 1 to propose $\alpha = 1$ and for player 2 to accept any proposal: if player 2 would reject $\alpha = 1$ then player 1 could improve by proposing $0 < \alpha < 1$; hence, we only have an equilibrium if player 2 accepts $\alpha = 1$ and, clearly, proposing $\alpha = 1$ is optimal for player 1. Hence, we can replace the part of the game from round $T-1$ on by the pair of payoffs $(\delta^{T-1}, 0)$, as in Fig. 6.6.

In this reduced game, in the subgame starting at round $T-2$, player 1 can obtain δ^{T-1} by rejecting player 2's proposal. Recall that this is the discounted utility at time $t = 0$ of receiving amount 1 at time $T-1$; the discounted utility of the amount 1 at time $T-2$ is equal to δ. Hence, in a backward induction equilibrium player 2 proposes $\alpha = \delta$ and player 1 accepts this proposal or any higher α and rejects any lower α. Hence, we can replace this whole subgame by the pair of payoffs $(\delta^{T-2}\delta, \delta^{T-2}(1-\delta)) = (\delta^{T-1}, \delta^{T-2}(1-\delta))$. Continuing this line of reasoning, in the subgame starting at round $T-3$, player 1 proposes $\alpha = 1 - \delta(1-\delta)$, which will be accepted by player 2. This results in the payoffs $(\delta^{T-3}(1-\delta(1-\delta)), \delta^{T-2}(1-\delta))$. This can be written as $(\delta^{T-3}(1-\delta+\delta^2), \delta^{T-3}(\delta-\delta^2))$. And so on and so forth. The general principle is that each player offers the other player a share equal to δ times the share the other player can expect in the next round.

By backtracking all the way to round 0 (see Table 6.1), we find that player 1 proposes $1 - \delta + \delta^2 - \ldots + \delta^{T-1}$ and player 2 accepts this proposal, resulting in the payoffs $1 - \delta + \delta^2 - \ldots + \delta^{T-1}$ for player 1 and $\delta - \delta^2 + \ldots - \delta^{T-1}$ for player 2. This is the subgame perfect equilibrium *outcome* of the game and the associated payoffs. This outcome is the path of play, induced by the following subgame perfect *equilibrium*:

- At even rounds t, player 1 proposes $\alpha = 1 - \delta + \ldots + \delta^{T-1-t}$ and player 2 accepts this proposal or any smaller α, and rejects any larger α.

Table 6.1 The proposals made in the subgame perfect equilibrium

Round	Proposer	Share for player 1	Share for player 2
T		0	0
$T-1$	1	1	0
$T-2$	2	δ	$1-\delta$
$T-3$	1	$1-\delta+\delta^2$	$\delta-\delta^2$
$T-4$	2	$\delta-\delta^2+\delta^3$	$1-\delta+\delta^2-\delta^3$
$\vdots$	$\vdots$	$\vdots$	$\vdots$
0	1	$1-\delta+\delta^2-\ldots+\delta^{T-1}$	$\delta-\delta^2+\ldots-\delta^{T-1}$

- At odd rounds t, player 2 proposes $\alpha = \delta-\delta^2+\ldots+\delta^{T-1-t}$ and player 1 accepts this proposal or any larger α, and rejects any smaller α.

In Problem 6.16 some variations on this finite horizon bargaining game are discussed.

6.7.2 Infinite Horizon Bargaining

In this subsection we consider the same bargaining problem as in the previous subsection, but now we assume $T = \infty$: the number of rounds may potentially be infinite. If no agreement is ever reached, then no player obtains anything. This game, like the finite horizon game, has many Nash equilibria: see Problem 6.16(f).

One way to analyze the game is to consider the finite horizon case and take the limit as T approaches infinity: see Problem 6.16(e). In fact, the resulting distribution is the uniquely possible outcome of a subgame perfect equilibrium, as can be seen by comparing the answer to Problem 6.16(e) with the result presented below. Of course, this claim is not proved by just taking the limit.

Note that a subgame perfect equilibrium cannot be obtained by backward induction, since the game has no final decision nodes. Here, we will just describe a pair of strategies and show that they are a subgame perfect equilibrium of the game. A proof that the associated outcome is the unique outcome resulting in any subgame perfect equilibrium can be found in the literature.

Let $\mathbf{x}^* = (x_1^*, x_2^*)$ and $\mathbf{y}^* = (y_1^*, y_2^*)$ be such that $x_1^*, x_2^*, y_1^*, y_2^* \geq 0$, $x_1^* + x_2^* = y_1^* + y_2^* = 1$, and moreover

$$x_2^* = \delta y_2^*, \quad y_1^* = \delta x_1^* . \tag{6.10}$$

It is not difficult to verify that $\mathbf{x}^* = (1/(1+\delta), \delta/(1+\delta))$ and $\mathbf{y}^* = (\delta/(1+\delta), 1/(1+\delta))$. Consider the following strategies for players 1 and 2, respectively:

(σ_1^*) At $t = 0, 2, 4, \ldots$ propose $\mathbf{x}^*$; at $t = 1, 3, 5, \ldots$ accept a proposal $\mathbf{z} = (z_1, z_2)$ of player 2 if and only if $z_1 \geq \delta x_1^*$.

(σ_2^*) At $t = 1, 3, 5, \ldots$ propose $\mathbf{y}^*$; at $t = 0, 2, 4, \ldots$ accept a proposal $\mathbf{z} = (z_1, z_2)$
of player 1 if and only if $z_2 \geq \delta y_2^*$.

These strategies are stationary: the players always make the same proposals. Moreover, a player accepts any proposal that offers him at least the discounted value of his own demand. According to (6.10), player 2 accepts the proposal $\mathbf{x}^*$ and player 1 accepts the proposal $\mathbf{y}^*$. Hence, play of the strategy pair (σ_1^*, σ_2^*) results in player 1's proposal $\mathbf{x}^* = (1/(1 + \delta), \delta/(1 + \delta))$ being accepted at round 0, so that these are also the payoffs. We will show that (σ_1^*, σ_2^*) is a subgame perfect equilibrium of the game.

To show this, note that there are two kinds of subgames: subgames where a player has to make a proposal; and subgames where a proposal is on the table and a player has to choose between accepting and rejecting the proposal.

For the first kind of subgame we may without loss of generality consider the entire game, i.e., the game starting at $t = 0$. We have to show that (σ_1^*, σ_2^*) is a Nash equilibrium in this game. First, suppose that player 1 plays σ_1^*. By accepting player 1's proposal at $t = 0$, player 2 has a payoff of $\delta/(1 + \delta)$. By rejecting this proposal, the maximum player 2 can obtain against σ_1^* is $\delta/(1 + \delta)$, by proposing $\mathbf{y}^*$ in round $t = 1$. Proposals $\mathbf{z}$ with $z_2 > y_2^*$ and thus $z_1 < y_1^*$ are rejected by player 1. Hence, σ_2^* is a best reply against σ_1^*. Similarly, if player 2 plays σ_2^*, then the best player 1 can obtain is x_1^* at $t = 0$ with payoff $1/(1 + \delta)$, since player 2 will reject any proposal that gives player 1 more than this, and also does not offer more.

For the second kind of subgame, again we may without loss of generality take $t = 0$ and assume that player 1 has made some proposal, say $\mathbf{z} = (z_1, z_2)$—the argument for t odd, when there is a proposal of player 2 on the table, is analogous. First, suppose that in this subgame player 1 plays σ_1^*. If $z_2 \geq \delta y_2^*$, then accepting this proposal yields player 2 a payoff of $z_2 \geq \delta y_2^* = \delta/(1 + \delta)$. By rejecting, the maximum player 2 can obtain against σ_1^* is $\delta/(1 + \delta)$ by proposing $\mathbf{y}^*$ at $t = 1$, which will be accepted by player 1. If, on the other hand, $z_2 < \delta y_2^*$, then player 2 can indeed better reject $\mathbf{z}$ and obtain $\delta/(1 + \delta)$ by proposing $\mathbf{y}^*$ at $t = 1$. Hence, σ_2^* is a best reply against σ_1^*. Next, suppose player 2 plays σ_2^*. Then $\mathbf{z}$ is accepted if $z_2 \geq \delta y_2^*$ and rejected otherwise. In the first case it does not matter how player 1 replies, and in the second case the game starts again with player 2 as the first proposer, and by an argument analogous to the argument in the previous paragraph, player 1's best reply is σ_1^*.

We have, thus, shown that (σ_1^*, σ_2^*) is a subgame perfect equilibrium of the game. In Problem 6.17 some variations on this game are discussed.

We conclude this section with two remarks.

Remark 6.1 Nothing in the whole analysis changes if we view the number δ not as a discount factor but as the probability that the game continues to the next round. Specifically, if a proposal is rejected, then assume that with probability $1 - \delta$ the game stops and each player receives 0, and with probability δ the game continues in the usual way. Under this alternative interpretation, the game ends with probability 1 [Problem 6.17(e)], which makes the infinite horizon assumption more acceptable.   □

Remark 6.2 In the subgame perfect equilibrium of the infinite horizon game the shares are $1/(1 + \delta)$ for player 1 and $\delta/(1 + \delta)$ for player 2. For δ converging to 1, i.e., the players becoming more patient, these shares converge to $1/2$ for each, which is the Nash bargaining outcome of the game, arising from maximizing the product $\alpha(1 - \alpha)$ for $0 \le \alpha \le 1$ (cf. Sect. 1.3.5). This is true more generally, i.e., also if the utility functions are more general. See Sects. 10.1.2 and 21.4.   □

6.8   Problems

6.1. *Cournot with Asymmetric Costs*
Consider the Cournot model of Sect. 6.2.1 but now assume that the firms have different marginal costs $c_1, c_2 \ge 0$. Assume $0 \le c_1, c_2 < a$ and $a \ge 2c_1 - c_2$, $a \ge 2c_2 - c_1$. Compute the Nash equilibrium.

6.2. *Cournot Oligopoly*
Consider the Cournot model of Sect. 6.2.1 but now assume that there are n firms instead of two. Each firm $i = 1, \ldots, n$ offers $q_i \ge 0$ and the market price is

$$P(q_1, \ldots, q_n) = \max\{a - q_1 - \ldots - q_n, 0\} \, .$$

Each firm still has marginal cost c with $a > c \ge 0$ and no fixed costs.

(a) Formulate the game corresponding to this situation. In particular, write down the payoff functions.
(b) Derive the reaction functions of the players.
(c) Derive a Nash equilibrium of the game by trying equal quantities offered. What happens if the number of firms becomes large?
(d) Show that the Nash equilibrium found in (c) is unique.

6.3. *Quantity Competition with Heterogenous Goods*
Suppose, in the Cournot model, that the firms produce heterogenous goods, which have different market prices. Specifically, suppose that these market prices are given by:

$$p_1 = \max\{5 - 3q_1 - 2q_2, 0\}, \; p_2 = \max\{4.5 - 1.5\, q_1 - 3q_2, 0\} \, .$$

The firms still compete in quantities and have equal constant marginal costs c, with $c < 4.5$.

(a) Formulate the game corresponding to this situation. In particular, write down the payoff functions.
(b) Solve for the reaction functions and the Nash equilibrium of this game. Also compute the corresponding prices and profits.

(c) Compute the quantities at which joint profit is maximized. Also compute the corresponding prices.

In (d)–(f), we assume that the firms compete in prices.

(d) Derive the demands q_1 and q_2 as a function of the prices. Set up the associated game where the prices p_1 and p_2 are now the strategic variables.

(e) Solve for the reaction functions and the Nash equilibrium of this game. Also compute the corresponding quantities and profits.

(f) Compute the prices at which joint profit is maximized. Also compute the corresponding quantities.

(g) Compare the results found under (b) and (c) with those under (e) and (f).

6.4. *A Numerical Example of Cournot Competition with Incomplete Information*
Redo the model of Sect. 6.2.2 for the following values of the parameters: $a = 1$, $c = 0$, $\vartheta = 1/2$, $c_L = 0$, $c_H = 1/4$. Compute the Nash equilibrium and compare with what was found in the text. Also compare with the complete information case by using the answer to Problem 6.1.

6.5. *Cournot Competition with Two-Sided Incomplete Information*
Consider the Cournot game of incomplete information of Sect. 6.2.2 and assume that also firm 1 can have high costs or low costs, say c_h with probability π and c_l with probability $1 - \pi$. Set up the associated game and compute the (four) reaction functions. (Assume that the parameters of the game are such that the Nash equilibrium quantities are positive and the relevant parts of the reaction functions can be found by differentiating the payoff functions (i.e., no corner solutions).) How can the Nash equilibrium be computed? (You do not actually have to compute it explicitly.)

6.6. *Incomplete Information About Demand*
Consider the Cournot game of incomplete information of Sect. 6.2.2 but now assume that the incomplete information is not about the cost of firm 2 but about market demand. Specifically, assume that the number a can be either high, a_H, with probability ϑ, or low, a_L, with probability $1 - \vartheta$. Firm 2 knows the value for sure but firm 1 only knows these probabilities. Set up the game and compute the reaction functions and the Nash equilibrium. (Make appropriate assumptions on the parameters a_H, a_L, ϑ, and c to avoid corner solutions.)

6.7. *Variations on Two-Person Bertrand*

(a) Assume that the two firms in the Bertrand model of Sect. 6.3 have different marginal costs, say $c_1 < c_2 < a$. Derive the best reply functions and find the Nash–Bertrand equilibrium or equilibria, if any.

(b) Reconsider the questions in (a) for the case where prices and costs are restricted to integer values, i.e., $p_1, p_2, c_1, c_2 \in \{0, 1, 2, \ldots\}$. (This reflects the assumption that there is a smallest monetary unit.) Specifically, consider two cases: (i) $a = 6$, $c_1 = c_2 = 2$ and (ii) $a = 6$, $c_1 = 1$, $c_2 = 2$.

6.8. *Bertrand with More Than Two Firms*
Suppose that there are $n > 2$ firms in the Bertrand model of Sect. 6.3. Assume again that all firms have equal marginal cost c, and that the firm with the lowest price gets the whole market. In case of a tie, each firm with the lowest price gets an equal share of the market. Set up the associated game and find all its Nash equilibria.

6.9. *Variations on Stackelberg*

(a) Suppose, in the model in Sect. 6.4, that the firms have different marginal costs c_1 and c_2 (cf. Problem 6.1). Compute the Stackelberg equilibrium and outcome with firm 1 as a leader and with firm 2 as a leader.
(b) Give a logical argument why the payoff of the leader in a Stackelberg equilibrium is always at least as high as his payoff in the Cournot equilibrium. Can you generalize this to arbitrary games?
(c) Consider the situation in Sect. 6.4, but now assume that there are n firms, firm 1 moves first, firm 2 second, etc. Assume again perfect information, and compute the subgame perfect equilibrium.

6.10. *First-Price Sealed-Bid Auction*
Consider the game associated with the first-price sealed-bid auction in Sect. 6.5.1, with $v_1 \geq \ldots \geq v_n > 0$ as there.

(a) Show that $(b_1, b_2, b_3, \ldots, b_n) = (v_2, v_2, v_3, \ldots, v_n)$ is a Nash equilibrium in this game.
(b) Show that, in any Nash equilibrium of the game, a player with the highest valuation obtains the object. Exhibit at least two other Nash equilibria in this game, apart from the equilibrium in (a).
(c) Show that bidding one's true valuation as well as bidding higher than one's true valuation are weakly dominated strategies. Also show that any positive bid lower than one's true valuation is not weakly dominated. (Note: to show that a strategy is weakly dominated one needs to exhibit some other strategy that is always—that is, whatever the other players do—at least as good as the strategy under consideration and at least once—that is, for at least one strategy combination of the other players—strictly better.)
(d) Show that, in any Nash equilibrium of this game, at least one player plays a weakly dominated strategy.

6.11. *Second-Price Sealed-Bid Auction*
Consider the game associated with the second price sealed bid auction in Sect. 6.5.1.

(a) Formulate the payoff functions in this game.
(b) Show that $(b_1, \ldots, b_n) = (v_1, \ldots, v_n)$ is a Nash equilibrium in this game.
(c) Show, for each player, that bidding one's true valuation weakly dominates any other action. (Show that this holds even if each player only knows his own valuation.)

(d) Show that $(b_1, \ldots, b_n) = (v_2, v_1, 0, \ldots, 0)$ is a Nash equilibrium in this game. What about $(b_1, \ldots, b_n) = (v_1, 0, 0, \ldots, 0)$?

(e) Determine *all* Nash equilibria in the game with two players ($n = 2$). (Hint: compute the best reply functions and make a diagram.)

6.12. *Third-Price Sealed-Bid Auction*
In the auction of Sect. 6.5.1, assume that there are at least three bidders and that the highest bidder wins and pays the third highest bid.

(a) Show that for any player i bidding v_i weakly dominates any lower bid but does not weakly dominate any higher bid.
(b) Show that the strategy combination in which each player i bids his true valuation v_i is in general not a Nash equilibrium.
(c) Find some Nash equilibria of this game.

6.13. *n-Player First-Price Sealed-Bid Auction with Incomplete Information*
Consider the setting of Sect. 6.5.2 but now assume that the number of bidders/players is $n \geq 2$. Show that $(s_1^*, \ldots, s_n^*)$ with $s_i^*(v_i) = (1 - 1/n)v_i$ for every player i is a Nash equilibrium of this game. (Hence, for large n, each bidder almost bids his true valuation.)

6.14. *Double Auction*
This problem is about the auction in Sect. 6.5.3.

(a) Fix a number $x \in [0, 1]$ and consider the following strategies $p_b(\cdot)$ and $p_s(\cdot)$:

$$p_b(v_b) = \begin{cases} x & \text{if } v_b \geq x \\ 0 & \text{if } v_b < x \end{cases} \quad \text{and} \quad p_s(v_s) = \begin{cases} x & \text{if } v_s \leq x \\ 1 & \text{if } v_s > x \end{cases}.$$

Show that these strategies constitute a Nash equilibrium.
(b) For the equilibrium in (a), compute the probability that trade takes place conditional on $v_b \geq v_s$. For which value of x is this probability maximal?
(c) Consider linear strategies of the form $p_b(v_b) = a_b + c_b v_b$ and $p_s(v_s) = a_s + c_s v_s$, where a_b, c_b, a_s, c_s are positive constants. Determine the values of the four constants so that these strategies constitute a Nash equilibrium.
(d) Answer the same question as in (b) for the equilibrium in (c). Which of the equilibria in (b) and (c) is the most efficient, i.e., has the highest probability of resulting in trade conditional on $v_b \geq v_s$?

6.15. *Mixed Strategies and Objective Uncertainty*
Consider the bimatrix game

$$\begin{array}{cc} & \begin{array}{cc} L & R \end{array} \\ \begin{array}{c} T \\ B \end{array} & \begin{pmatrix} 4,1 & 1,3 \\ 1,2 & 3,0 \end{pmatrix} \end{array}.$$

(a) Compute the Nash equilibrium of this game.
(b) Add some uncertainty to the payoffs of this game and find a pure (Bayesian) Nash equilibrium of the resulting game of incomplete information, such that the induced a priori mixed strategies converge to the Nash equilibrium of the original game as the amount of uncertainty shrinks to 0.

6.16. *Variations on Finite Horizon Bargaining*

(a) Adapt the arguments and the results of Sect. 6.7.1 for the case where T is even and the case where player 2 proposes at even rounds.
(b) Let $T = 3$ in Sect. 6.7.1 and suppose that the players have different discount factors δ_1 and δ_2. Compute the subgame perfect equilibrium and the subgame perfect equilibrium outcome.
(c) Consider again the model of Sect. 6.7.1, let $T = 3$, but now assume that the utility function of player 2 is $u_2(\alpha) = \sqrt{\alpha}$ for all $\alpha \in [0, 1]$. Hence, the utility of receiving α at time t for player 2 is equal to $\delta^t \sqrt{\alpha}$. Compute the subgame perfect equilibrium and the subgame perfect equilibrium outcome.
(d) Suppose, in the model of Sect. 6.7.1, that at time T the 'disagreement' distribution is $\mathbf{s} = (s_1, s_2)$ with $s_1, s_2 \geq 0$ and $s_1 + s_2 \leq 1$, rather than $(0, 0)$. Compute the subgame perfect equilibrium and the subgame perfect equilibrium outcome.
(e) In (d), compute the limits of the equilibrium shares for T going to infinity. Do these limits depend on $\mathbf{s}$?
(f) Show, in the game in Sect. 6.7.1, that subgame perfection really has a bite. Specifically, for every $\mathbf{s} = (s_1, s_2)$ with $s_1, s_2 \geq 0$ and $s_1 + s_2 = 1$, exhibit a Nash equilibrium of the game in Fig. 6.5 resulting in the distribution s.

6.17. *Variations on Infinite Horizon Bargaining*

(a) Determine the subgame perfect equilibrium outcome and subgame perfect equilibrium strategies in the game in Sect. 6.7.2 when the players have different discount factors δ_1 and δ_2.
(b) Determine the subgame perfect equilibrium outcome and subgame perfect equilibrium strategies in the game in Sect. 6.7.2 when player 2 proposes at even rounds and player 1 at odd rounds.
(c) Determine the subgame perfect equilibrium outcome and subgame perfect equilibrium strategies in the game in Sect. 6.7.2 when the 'disagreement' distribution is $\mathbf{s} = (s_1, s_2)$ with $s_1, s_2 \geq 0$ and $s_1 + s_2 \leq 1$, rather than $(0, 0)$, in case the game never stops.
(d) Consider the game in Sect. 6.7.2, but now assume that the utility function of player 2 is $u_2(\alpha) = \sqrt{\alpha}$ for all $\alpha \in [0, 1]$. Hence, the utility of receiving α at time t for player 2 is equal to $\delta^t \sqrt{\alpha}$. Compute the subgame perfect equilibrium and the subgame perfect equilibrium outcome. [Hint: first determine for this situation the values for $\mathbf{x}^*$ and $\mathbf{y}^*$ analogous to (6.10).]

(e) Interpret, as at the end of Sect. 6.7.2, the discount factor as the probability that
    the game continues to the next round. Show that the game ends with probability
    equal to 1.

6.18. *A Principal-Agent Game*

There are two players: a worker (the agent) and an employer (the principal). The
worker has three choices: either reject the contract offered to him by the employer,
or accept this contract and exert high effort, or accept the contract and exert low
effort. If the worker rejects the contract then the game ends with a payoff of zero
to the employer and a payoff of 2 to the worker (his reservation payoff). If the
worker accepts the contract he works for the employer: if he exerts high effort the
revenues for the employer will be 12 with probability 0.8 and 6 with probability
0.2; if he exerts low effort then these revenues will be 12 with probability 0.2 and
6 with probability 0.8. The employer can only observe the revenues but not the
effort exerted by the worker: in the contract he specifies a high wage w_H in case the
revenues equal 12 and a low wage w_L in case the revenues are equal to 6. These
wages are the respective choices of the employer. The final payoff to the employer
if the worker accepts the contract will be equal to revenues minus wage. The worker
will receive his wage; his payoff equals this wage minus 3 if he exerts high effort
and this wage minus 0 if he exerts low effort.

(a) Set up the extensive form of this game. Does this game have incomplete or
    imperfect information? What is the associated strategic form?
(b) Determine the subgame perfect equilibrium or equilibria of the game.

6.19. *The Market for Lemons*

A buyer wants to buy a car but does not know whether the particular car he is
interested in has good or bad quality (a lemon is a car of bad quality). About half
of the market consists of good quality cars. The buyer offers a price p to the seller,
who is informed about the quality of the car; the seller may then either accept of
reject this price. If he rejects, there is no sale and the payoff will be 0 to both. If he
accepts, the payoff to the seller will be the price minus the value of the car, and to
the buyer it will be the value of the car minus the price. A good quality car has a
value of 15,000, a lemon has a value of 5,000.

(a) Set up the extensive as well as strategic form of this game.
(b) Compute the subgame perfect equilibrium or equilibria of this game.

6.20. *Corporate Investment and Capital Structure*

Consider an entrepreneur who has started a company but needs outside financing to
undertake an attractive new project. The entrepreneur has private information about
the profitability of the existing company, but the payoff of the new project cannot be
disentangled from the payoff of the existing company—all that can be observed is
the aggregate profit of the firm. Suppose the entrepreneur offers a potential investor
an equity stake in the firm in exchange for the necessary financing. Under what

circumstances will the new project be undertaken, and what will the equity stake be? In order to model this as a game, assume that the profit of the existing company can be either high or low: $\pi = L$ or $\pi = H$, where $H > L > 0$. Suppose that the required investment for the new project is I, the payoff will be R, the potential investor's alternative rate of return is r, with $R > I(1 + r)$. The game is played as follows.

1. Nature determines the profit of the existing company. The probability that $\pi = L$ is p.
2. The entrepreneur learns π and then offers the potential investor an equity stake s, where $0 \le s \le 1$.
3. The investor observes s (but not π) and then decides either to accept or to reject the offer.
4. If the investor rejects then the investor's payoff is $I(1 + r) - I$ and the entrepreneur's payoff is π. If he accepts his payoff is $s(\pi + R) - I$ and the entrepreneur's is $(1 - s)(\pi + R)$.

(a) Set up the extensive form and the strategic form of this signaling game.
(b) Compute the perfect Bayesian Nash equilibrium or equilibria, if any.

6.21. A Poker Game
Consider the following game. There are two players, I and II. Player I deals II one of three cards—Ace, King, or Queen—at random and face down. II looks at the card. If it is an Ace, II must say "Ace", if a King he can say "King" or "Ace", and if a Queen he can say "Queen" or "Ace". If II says "Ace" player I can either believe him and give him $1 or ask him to show his card. If it is an Ace, I must pay II $2, but if it is not, II pays I $2. If II says "King" neither side looses anything, but if he says "Queen" II must pay player I $1.

(a) Set up the extensive form and the strategic form of this zerosum game.
(b) Determine its value and optimal strategies (cf. Chap. 2).

6.22. A Hotelling Location Problem
Consider n players each choosing a location in the interval $[0, 1]$. One may think of n shops choosing locations in a street, n firms choosing product characteristics on a continuous scale from 0 to 1, or n political parties choosing positions on the ideological scale. We assume that customers or voters are uniformly distributed over the interval, with a total of 1. The customers go to (voters vote for) the nearest shop (candidate). For example, if $n = 2$ and the chosen positions are $x_1 = 0.2$ and $x_2 = 0.6$, then 1 obtains 0.4 and 2 obtains 0.6 customers (votes). In case two or more players occupy the same position they share the customers or voters for that position equally.

In the first scenario, the players care only about winning or loosing in terms of the number of customers or votes. This scenario may be prominent for presidential elections, as an example. For each player the best alternative is to be the unique winner, the second best alternative is to be one of the winners, and the worst alternative is not to win. For this scenario, answer questions (a) and (b).

(a) Show that there is a unique Nash equilibrium for $n = 2$.
(b) Exhibit a Nash equilibrium for $n = 3$.

In the second scenario, the payoffs of the players are given by the total numbers of customers (or voters) they acquire. For this scenario, answer questions (c) and (d).

(c) Show that there is a unique Nash equilibrium for $n = 2$.
(d) Is there a Nash equilibrium for $n = 3$? How about $n = 4$?

6.23. *Median Voting*
Of the n persons in a room, each person i has a most favorite room temperature t_i, and the further away (lower or higher) the room temperature is from t_i, the worse it is for this person. Specifically, if the room temperature is x, then person i's utility is equal to $-|x - t_i|$. In order to find a compromise, the janitor asks each person to propose a room temperature, and based on the proposed temperatures a compromise is determined. The proposed temperatures are not necessarily equal to the favorite temperatures. Only temperatures (proposed and favorite) in the interval 0–30 °C are possible.

(a) Suppose the janitor announces that he will take the average of the proposed temperatures as the compromise temperature. Formulate this situation as an n-person game, that is, give the strategy sets of the players and the payoff functions. Does this game have a Nash equilibrium?
(b) Suppose n is odd, and suppose the janitor announces that he will take the median of the proposed temperatures as the compromise temperature. Formulate this situation as an n-person game, that is, give the strategy sets of the players and the payoff functions. Show that, for each player, proposing his ideal temperature weakly dominates any other strategy: thus, in particular, $(t_1, \ldots, t_n)$ is a Nash equilibrium of this game. Does the game have any other Nash equilibria?

6.24. *The Uniform Rule*
An amount $M \geq 0$ of a good (labor, green pea soup, ...) is to be distributed completely among n persons. Each person i considers an amount $t_i \geq 0$ as the ideal amount, and the further away the allocated amount is from this ideal, the worse it is. Specifically, if the amount allocated to person i is x_i, then person i's utility is equal to $-|x - t_i|$. In order to find a compromise, each person is asked to report an amount, and based on the reported amounts a compromise is determined. Let the ideal amounts be given by $t_1 \leq t_2 \leq \ldots \leq t_n$. The reported amounts are not necessarily equal to the ideal amounts.

(a) Suppose M is distributed proportionally to the reported amounts, that is, if the reported amounts are $(r_1, \ldots, r_n)$, then person i receives $x_i = \left(r_i / \sum_{j=1}^{n} r_j \right) M$.

(If all r_j are zero then take $x_i = M/n$.) Formulate this situation as a game. Does this game have a Nash equilibrium?

Consider the following division rule, called the *uniform rule*. Let $(r_1, \ldots, r_n)$ denote the reported amounts. If $M \leq \sum_{j=1}^{n} r_j$, then each person i receives

$$x_i = \min\{r_i, \lambda\},$$

where λ is such that $\sum_{j=1}^{n} x_j = M$. If $M \geq \sum_{j=1}^{n} r_j$, then each person i receives

$$x_i = \max\{r_i, \lambda\},$$

where, again, λ is such that $\sum_{j=1}^{n} x_j = M$.

(b) Suppose that $n = 3$ and $r_1 = 1$, $r_2 = 2$, and $r_3 = 3$. Apply the uniform rule for $M = 4, M = 5, M = 5.5, M = 6, M = 6.5, M = 7, M = 8, M = 9$.
(c) Suppose, for the general case, that the uniform rule is used to distribute the amount M. Formulate this situation as a game. Show that reporting one's ideal amount weakly dominates any other strategy: thus, in particular, $(t_1, \ldots, t_n)$ is a Nash equilibrium of this game. Does the game have any other Nash equilibria?

6.25. *Reporting a Crime*
There are n individuals who witness a crime. Everybody would like the police to be called. If this happens, each individual derives satisfaction $v > 0$ from it. Calling the police has a cost of c, where $0 < c < v$. The police will come if at least one person calls. Hence, this is an n-person game in which each player chooses from $\{C, N\}$: C means 'call the police' and N means 'do not call the police'. The payoff to person i is 0 if nobody calls the police, $v - c$ if i (and perhaps others) call the police, and v if the police is called but not by person i.

(a) What are the Nash equilibria of this game in pure strategies? In particular, show that the game does not have a symmetric Nash equilibrium in pure strategies (a Nash equilibrium is symmetric if every player plays the same strategy).
(b) Compute the symmetric Nash equilibrium or equilibria in mixed strategies. (Hint: suppose, in such an equilibrium, every person plays C with probability $0 < p < 1$. Use the fact that each player must be indifferent between C and N.)
(c) For the Nash equilibrium/equilibria in (b), compute the probability of the crime being reported. What happens to this probability if n becomes large?

6.26. *Firm Concentration*
Consider a market with ten firms. Simultaneously and independently, the firms choose between locating downtown and locating in the suburbs. The profit of each firm is influenced by the number of other firms that locate in the same area. Specifically, the profit of a firm that locates downtown is given by $5n - n^2 + 50$, where n denotes the number of firms that locate downtown. Similarly, the profit of

a firm that locates in the suburbs is given by $48 - m$, where m denotes the number of firms that locate in the suburbs. In equilibrium how many firms locate in each region and what is the profit of each?

6.27. *Tragedy of the Commons*

There are n farmers, who use a common piece of land to graze their goats. Each farmer i chooses a number of goats g_i—for simplicity we assume that goats are perfectly divisible. The value to a farmer of grazing a goat when the total number of goats is G, is equal to $v(G)$ per goat. We assume that there is a number $\bar{G}$ such that $v(G) > 0$ for $G < \bar{G}$ and $v(G) = 0$ for $G \geq \bar{G}$. Moreover, v is continuous, and twice differentiable at all $G \neq \bar{G}$, with $v'(G) < 0$ and $v''(G) < 0$ for $G < \bar{G}$. The payoff to farmer i if each farmer j chooses g_j, is equal to

$$g_i v(g_1 + \ldots + g_{i-1} + g_i + g_{i+1} + \ldots + g_n) - cg_i ,$$

where $c \geq 0$ is the cost per goat.

(a) Interpret the conditions on the function v.
(b) Show that the total number of goats in a Nash equilibrium $(g_1^*, \ldots, g_n^*)$ of this game, $G^* = g_1^* + \ldots + g_n^*$, satisfies

$$v(G^*) + (1/n)G^* v'(G^*) - c = 0 .$$

(c) The socially optimal number of goats G^{**} is obtained by maximizing $Gv(G) - cG$ over $G \geq 0$. Show that G^{**} satisfies

$$v(G^{**}) + G^{**} v'(G^{**}) - c = 0 .$$

(d) Show that $G^* > G^{**}$. (Hence, in a Nash equilibrium too many goats are grazed.)

6.9   Notes

The Cournot model in Sect. 6.2 dates back from Cournot (1838), and the Bertrand model in Sect. 6.3 from Bertrand (1883). The occurrence of the Bertrand price equilibrium is often referred to as the *Bertrand paradox*. On Cournot versus Bertrand, see Magnan de Bornier (1992). The Stackelberg equilibrium (Sect. 6.4) is named after von Stackelberg (1934).

Our coverage of auction theory is limited, and based on Osborne (2004) and Gibbons (1992). For more extensive overviews and treatments see Milgrom (2004) and Krishna (2002). The second price auction is also called *Vickrey auction* (Vickrey, 1961). The condition that bidders bid their true valuation is an example of the *incentive compatibility* requirement.

Section 6.6 is based on Harsanyi (1973), see also Gibbons (1992).

Axiomatic bargaining theory was initiated by Nash (1950). The noncooperative, strategic approach in Sect. 6.7 is based on Rubinstein (1982). See also Nash (1953) for a noncooperative approach to the Nash bargaining solution.

The goods in the model of Problem 6.3 are *strategic substitutes*. In duopoly models such as this the distinction between strategic substitutes and *strategic complements* is important for the differences between quantity and price competition. See, e.g., Tirole (1988).

Problem 6.19 (market for lemons), exhibiting the *adverse selection* problem, is based on Akerlof (1970). Problem 6.20 is taken from Gibbons (1992). Problem 6.21 (poker game) is taken from Thomas (1986). For an axiomatization of the median voting method in Problem 6.23 see Moulin (1980), and for an axiomatization of the uniform rule in Problem 6.24 see Sprumont (1991).

Problem 6.26 is taken from Watson (2002). For the tragedy of the commons situation in Problem 6.27 see Hardin (1968) and Gibbons (1992).

References

Akerlof, G. (1970). The market for lemons: Quality uncertainty and the market mechanism. *Quarterly Journal of Economics, 84*, 488–500.
Bertrand, J. (1883). Review of Walras's 'Théorie mathématique de la richesse sociale' and Cournot's 'Recherches sur les principes mathématiques de la théorie des richesses'. *Journal des Savants*, 499–508 [translated by M. Chevaillier and reprinted in Magnan de Bornier (1992), 646–653].
Cournot, A. (1838). *Recherches sur les principes mathématiques de la théorie des richesses* [English translation (1897) *Researches into the mathematical principles of the theory of wealth*]. New York: Macmillan.
Gibbons, R. (1992). *A primer in game theory*. Hertfordshire: Harvester Wheatsheaf.
Hardin, G. (1968). The tragedy of the commons. *Science, 162*, 1243–1248.
Harsanyi, J. C. (1973). Games with randomly disturbed payoffs: A new rationale of mixed strategy equilibrium points. *International Journal of Game Theory, 2*, 1–23.
Krishna, V. (2002). *Auction theory*. San Diego: Academic.
Magnan de Bornier, J. (1992). The 'Cournot-Bertrand debate': A historical perspective. *History of Political Economy, 24*, 623–656.
Milgrom, P. (2004). *Putting auction theory to work*. Cambridge: Cambridge University Press.
Moulin, H. (1980). On strategy-proofness and single-peakedness. *Public Choice, 35*, 437–455.
Nash, J. F. (1950). The bargaining problem. *Econometrica, 18*, 155–162.
Nash, J. F. (1953). Two-person cooperative games. *Econometrica, 21*, 128–140.
Osborne, M. J. (2004). *An introduction to game theory*. New York: Oxford University Press.
Rubinstein, A. (1982). Perfect equilibrium in a bargaining model. *Econometrica, 50*, 97–109.
Sprumont, Y. (1991). The division problem with single-peaked preferences: A characterization of the uniform allocation rule. *Econometrica, 59*, 509–520.
Thomas, L. C. (1986). *Games, theory and applications*. Chichester: Ellis Horwood Limited.
Tirole, J. (1988). *The theory of industrial organization*. Cambridge: MIT Press.
Vickrey, W. (1961). Counterspeculation, auctions, and competitive sealed tenders. *Journal of Finance, 16*, 8–37.
von Stackelberg, H. F. (1934). *Marktform und Gleichgewicht*. Wien: Julius Springer.
Walker, M., & Wooders, J. (2001). Minimax play at Wimbledon. *American Economic Review, 91*, 1521–1538.
Watson, J. (2002). *Strategy, an introduction to game theory*. New York: Norton.

Repeated Games

7

In the famous prisoners' dilemma game the bad (Pareto inferior) outcome, resulting from each player playing his dominant action, cannot be avoided in a Nash equilibrium or subgame perfect Nash equilibrium even if the game is repeated a finite number of times, cf. Problem 4.10. As we will see in this chapter, this bad outcome can be avoided if the game is repeated an infinite number of times. This, however, is coming at a price, namely the existence of a multitude of outcomes attainable in equilibrium. Such an *embarrassment of riches* is expressed by a so-called *folk theorem*.

As was illustrated in Problem 4.11, also *finite* repetitions of a game may sometimes lead to outcomes that are better than (repeated) Nash equilibria of the original game.

In this chapter we consider two-person *infinitely* repeated games and formulate folk theorems both for subgame perfect and for Nash equilibrium. The approach is somewhat informal, and mainly based on examples. In Sect. 7.1 we consider subgame perfect equilibrium and in Sect. 7.2 we consider Nash equilibrium.

7.1 Subgame Perfect Equilibrium

7.1.1 The Prisoners' Dilemma

Consider the prisoners' dilemma game (in the form of the 'marketing game' of Problem 3.1(d))

$$G = \begin{array}{c} \\ C \\ D \end{array} \begin{array}{c} \\ \begin{array}{cc} C & D \end{array} \\ \begin{pmatrix} 50,50 & 30,60 \\ 60,30 & 40,40 \end{pmatrix} \end{array}.$$

© Springer-Verlag Berlin Heidelberg 2015
H. Peters, *Game Theory*, Springer Texts in Business and Economics,
DOI 10.1007/978-3-662-46950-7_7

In G each player has a strictly dominated action, namely C, and (D, D) is the unique Nash equilibrium of the game, also if mixed strategies are allowed.

We assume now that G is played infinitely many times, at times $t = 0, 1, 2, \ldots$, and that after each play of G the players learn what has been played, i.e., they learn which element of the set $\{(C, C), (C, D), (D, C), (D, D)\}$ has occurred. For instance, in the marketing game, one can think of the game being played once per period—a week, month—each player observing in each period whether his opponent has advertised or not. Note that a player does not learn the exact, possibly mixed, action of his opponent, but only its realization. These realizations induce an infinite stream of associated payoffs, and we assume that there is a common discount factor $0 < \delta < 1$ such that the final payoff to each player is the δ-discounted value of the infinite stream. That is, player i $(i = 1, 2)$ obtains

$$\sum_{t=0}^{\infty} (\text{payoff from play of the stage game } G \text{ at time } t) \cdot \delta^t \, .$$

Here, the expression *stage game* is used for the one-shot game G, in order to distinguish the one-shot game from the repeated game.

As always, a strategy of a player is a complete plan to play the game. This means that, at each moment t, this plan should prescribe an action of a player—a mixed or pure strategy in the stage game G—for each possible history of the game up to time t, that is, an action for each sequence of length t (namely, at $0, \ldots, t - 1$) of elements from the set $\{(C, C), (C, D), (D, C), (D, D)\}$. Clearly, such a strategy can be quite complicated and the number of possible strategies is enormous. We will be able, however, to restrict attention to relatively simple strategies.

The infinite extensive form game just defined is denoted by $G^\infty(\delta)$. A natural solution concept for this game is subgame perfect (Nash) equilibrium. Each subgame in $G^\infty(\delta)$ is, basically, equal to the game $G^\infty(\delta)$ itself: the difference between two subgames is the difference between the two histories leading to those subgames. For instance, at $t = 6$, there are 4^6 possible histories of play and therefore there are 4^6 different subgames; each of these subgames, however, looks exactly like $G^\infty(\delta)$.

We will now exhibit a few subgame perfect equilibria of $G^\infty(\delta)$. First consider the simple strategy

D^∞: play D at each moment $t = 0, 1, 2, \ldots$, independent of the history of the game, i.e., independent of what was played before t.

Then D^∞ is a well-defined strategy. If both players play D^∞ then the resulting payoff is

$$\sum_{t=0}^{\infty} 40 \, \delta^t = 40/(1 - \delta)$$

for each player. We claim that (D^∞, D^∞) is a subgame perfect equilibrium in $G^\infty(\delta)$. Consider any $t = 0, 1, \ldots$ and any subgame starting at time t. Then (D^∞, D^∞) induces a Nash equilibrium in this subgame: given that player 2 always plays D, player 1 cannot do better than always play D as well, and *vice versa*. Hence, (D^∞, D^∞) is a subgame perfect equilibrium. In this subgame perfect equilibrium, the players just play the Nash equilibrium of the stage game at every time t.

We next exhibit another subgame perfect equilibrium. Consider the following strategy:

$Tr(C)$: at $t = 0$ and at every time t such that in the past only (C, C) has occurred in the stage game: play C. Otherwise, play D.

Strategy $Tr(C)$ is an example of a so-called *trigger strategy*. In general, if the players play trigger strategies, they follow some fixed pattern of play until a deviation occurs: then a Nash equilibrium action of the stage game is played forever. In the present example, $Tr(C)$, a player starts by playing C and keeps on playing C as long as both players have only played C in the past, i.e., as long as the history of play is $(C, C), \ldots, (C, C)$; after any deviation from this, i.e., if the history of play is *not* $(C, C), \ldots, (C, C)$, the player plays D and keeps on playing D forever. Again, $Tr(C)$ is a well-defined strategy, and if both players play $Tr(C)$, then each player obtains the payoff

$$\sum_{t=0}^{\infty} 50 \, \delta^t = 50/(1 - \delta) \, .$$

Is $(Tr(C), Tr(C))$ also a subgame perfect equilibrium? The answer is a qualified yes: if δ is large enough, then it is. The crux of the argument is as follows. At each stage of the game, a player has an incentive to deviate from C and play his dominant action D, thereby obtaining a momentary gain of 10. Deviating, however, triggers eternal 'punishment' by the other player, who is going to play D forever. The best reply to this punishment is to play D as well, entailing a loss of 10 at each moment from the next moment on. The discounted value, at the moment of deviation, of this loss is equal to $10\delta/(1 - \delta)$, and to keep a player from deviating this loss should be at least as large as the momentary gain of 10. This is the case if and only if $\delta \geq 1/2$.

More formally, we can distinguish two kinds of subgames that are relevant for the strategy combination $(Tr(C), Tr(C))$. One kind are those subgames where *not* always (C, C) has been played in the past. In such a subgame, $Tr(C)$ tells a player to play D forever, and therefore the best reply of the other player is to do so as well, which means indeed to play according to $Tr(C)$. Thus, in this kind of subgame, $(Tr(C), Tr(C))$ is a Nash equilibrium.

In the other kind of subgame, no deviation has occurred so far: in the past always (C, C) has been played. Consider this subgame at some time T and suppose that player 2 plays $Tr(C)$. If player 1 plays $Tr(C)$ as well, his payoff is equal to

$$\sum_{t=0}^{T-1} 50\, \delta^t + \sum_{t=T}^{\infty} 50\, \delta^t .$$

If, instead, he deviates at time T to D, he obtains maximally

$$\sum_{t=0}^{T-1} 50\, \delta^t + 60\, \delta^T + \sum_{t=T+1}^{\infty} 40\, \delta^t .$$

The first term in this expression is the discounted payoff from (C, C) at $t = 0, \ldots, T-1$. The second term is the discounted payoff from (D, C) at time T. The third term is the discounted payoff from (D, D) from time $t = T + 1$ on. Note that from $t = T + 1$ on player 2 plays D, according to his strategy $Tr(C)$, and the best that the (deviating) player 1 can do is to play D as well.

Hence, to avoid deviation (and make $Tr(C)$ player 1's best reply in the subgame) we need that the first payoff is at least as high as the second one, resulting in the inequality

$$50\, \delta^T/(1 - \delta) \geq 60\, \delta^T + 40\, \delta^{T+1}/(1 - \delta)$$

or, equivalently, $\delta \geq 1/2$—as found before. The arguments for the roles of the players reversed are exactly equal. We conclude that for every $\delta \geq 1/2$, $(Tr(C), Tr(C))$ is a subgame perfect equilibrium of the game $G^\infty(\delta)$. The existence of this equilibrium is a major reason to study infinitely repeated games. In popular terms, it shows that cooperation is sustainable if deviations can be credibly punished, which is the case if the future is sufficiently important (i.e., δ large enough).

To exhibit yet another subgame perfect equilibrium, different from (D^∞, D^∞) and $(Tr(C), Tr(C))$, consider the following strategies for players 1 and 2, respectively.

Tr_1: As long as the sequence (C, D), (D, C), (C, D), (D, C), (C, D), (D, C), $\ldots$ has occurred in the past from time 0 on, play C at $t \in \{0, 2, 4, 6, \ldots\}$; play D at $t \in \{1, 3, 5, 7, \ldots\}$. Otherwise, play D.

Tr_2: As long as the sequence (C, D), (D, C), (C, D), (D, C), (C, D), (D, C), $\ldots$ has occurred in the past from time 0 on, play D at $t \in \{0, 2, 4, 6, \ldots\}$; play C at $t \in \{1, 3, 5, 7, \ldots\}$. Otherwise, play D.

Note that these are again 'trigger strategies': the players 'tacitly' agree on a certain sequence (pattern) of play, but revert to playing D forever after a deviation. If player

1 plays Tr_1 and player 2 plays Tr_2, then the sequence $(C, D), (D, C), (C, D), (D, C),$
..., results. To see why (Tr_1, Tr_2) might be a subgame perfect equilibrium, note that
on average a player obtains 45 per stage, which is more than the 40 that would be
obtained from deviating from this sequence and playing D forever. More precisely,
suppose player 2 plays Tr_2 and suppose player 1 considers a deviation from Tr_1. It
is optimal to deviate at an even moment, say at $t = 0$, since then player 1 receives a
payoff of 30, and can obtain 40 by deviating to D. Since, after this deviation, player
2 plays D forever, the maximal total discounted payoff to player 1 from deviating is
obtained by also playing D forever after his deviation to D at time $t = 0$, and this
payoff is equal to

$$40 + 40(\delta + \delta^2 + \ldots) = 40/(1 - \delta) .$$

If player 1 does not deviate and sticks to the strategy Tr_1 he obtains

$$30(1 + \delta^2 + \delta^4 + \ldots) + 60(\delta + \delta^3 + \delta^5 + \ldots) = (30 + 60\delta)/(1 - \delta^2) .$$

To keep player 1 from deviating we need $40/(1 - \delta) \leq (30 + 60\delta)/(1 - \delta^2)$, which
yields $\delta \geq 1/2$. The argument if the roles of the players are reversed is similar, and
we conclude that for each $\delta \geq 1/2$, (Tr_1, Tr_2) is a subgame perfect equilibrium in
$G^\infty(\delta)$.

More generally, by playing appropriate sequences of elements from the set of
possible outcomes $\{(C, C), (C, D), (D, C), (D, D)\}$ of the stage game G, the players
can on average reach any convex combination of the associated payoffs in the long
run. That is, take any such combination

$$\alpha_1(50, 50) + \alpha_2(30, 60) + \alpha_3(60, 30) + \alpha_4(40, 40) , \qquad (*)$$

where $\alpha_i \in \mathbb{R}$, $\alpha_i \geq 0$ for every $i = 1, \ldots, 4$, and $\sum_{i=1}^{4} \alpha_i = 1$. By choosing
a sequence of possible outcomes such that (C, C) occurs on average in a fraction
α_1 of the stages, (C, D) in a fraction α_2, (D, C) in a fraction α_3, and (D, D) in a
fraction α_4, then the payoffs $(*)$ are reached as averages in the limit, i.e., as $t \to \infty$.
These are indeed average payoffs, independent of the discount factor δ. If these
limit average payoffs exceed 40 (the Nash equilibrium payoff of the stage game) for
each player, associated trigger strategies can be formulated that result in these (limit
average) payoffs and that trigger eternal play of (D, D) after a deviation, similar to
the strategies $Tr(C), Tr_1$ and Tr_2 above.

Note that for $\alpha_1 = 1$ we can take the strategy pair $(Tr(C), Tr(C))$. For $\alpha_2 = \alpha_3 =
1/2$ we can take the strategy pair (Tr_1, Tr_2).

To exhibit yet another example, consider the average payoff pair $(42, 48)$, which
is equal to

$$\frac{1}{5}(50, 50) + \frac{2}{5}(30, 60) + \frac{1}{5}(60, 30) + \frac{1}{5}(40, 40) .$$

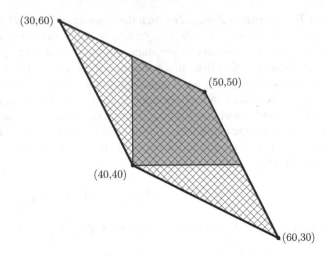

Fig. 7.1 For every payoff pair in the *shaded area* there is a δ large enough such that this payoff pair can be obtained as the limit average in a subgame perfect equilibrium of $G^\infty(\delta)$, where G is the prisoners' dilemma game of Sect. 7.1.1

This means that the payoffs $(42, 48)$ can be obtained on average by playing (C, C) at $t = 0, 5, 10, \ldots$; (C, D) at $t = 1, 2, 6, 7, 11, 12, \ldots$; (D, C) at $t = 3, 8, 13, \ldots$; and (D, D) at $t = 4, 9, 14, \ldots$ Translated to trigger strategies we have

> Tr_1^*: As long as the sequence (C, C), (C, D), (C, D), (D, C), (D, D); (C, C), (C, D), (C, D), (D, C), (D, D); $\ldots$ has occurred in the past from time 0 on, play C at $t \in \{0, 1, 2, 5, 6, 7, 10, 11, 12, \ldots\}$; play D at $t \in \{3, 4, 8, 9, 13, 14, \ldots\}$. Otherwise, play D.

> Tr_2^*: As long as the sequence (C, C), (C, D), (C, D), (D, C), (D, D); (C, C), (C, D), (C, D), (D, C), (D, D); $\ldots$ has occurred in the past from time 0 on, play C at $t \in \{0, 3, 5, 8, 10, 13, \ldots\}$; play D at $t \in \{1, 2, 4, 6, 7, 9, 11, 12, 14, \ldots\}$. Otherwise, play D.

For δ sufficiently high, these strategies form a subgame perfect equilibrium of $G^\infty(\delta)$. (See Problem 7.5.)

Figure 7.1 shows all limit average payoffs that can be reached in this way.

7.1.2 Some General Observations

For the prisoners' dilemma game we have established that each player playing always D is a subgame perfect equilibrium of $G^\infty(\delta)$ for every $0 < \delta < 1$. The logic is simple. If player 2 *always* plays the Nash equilibrium action D in the stage game, then player 1 can never do better than playing a best reply action in the stage game, i.e., playing D—'never' means: independent of the history, i.e., independent of the play in the stage game thus far, i.e., in any subgame. The same logic holds for *any* stage game, that is, with finitely or infinitely many actions, with any arbitrary number of players, and for *any* Nash equilibrium of the stage game. The following proposition merely states this more formally.[1]

Proposition 7.1 *Let G be any arbitrary (not necessarily finite) n-person game, and let the strategy combination* $\mathbf{s} = (s_1, \ldots, s_i, \ldots, s_n)$ *be a Nash equilibrium in G. Let* $0 < \delta < 1$. *Then each player i playing* s_i *at every moment t is a subgame perfect equilibrium in* $G^\infty(\delta)$.

In particular, this proposition holds for any bimatrix game (see Definition 3.1) and any (not necessarily pure) Nash equilibrium in this bimatrix game. But it also holds, for instance, for the Cournot or Bertrand games (cf. Chap. 6) with two or more players (see Problems 7.6 and 7.7).

Let $G = (A, B)$ be an $m \times n$-bimatrix game with $A = (a_{ij})$ and $B = (b_{ij})$. Let $P(G)$ be the convex hull of the set $\{(a_{ij}, b_{ij}) \in \mathbb{R}^2 \mid i = 1, \ldots, m, \ j = 1, \ldots, n\}$. That is,

$$P(G) = \left\{ \sum_{i=1}^{m} \sum_{j=1}^{n} \alpha_{ij}(a_{ij}, b_{ij}) \ \middle| \ \sum_{i=1}^{m} \sum_{j=1}^{n} \alpha_{ij} = 1, \ \alpha_{ij} \geq 0 \text{ for all } i, j \right\}.$$

Equivalently, to obtain the set $P(G)$, just plot all $m \times n$ payoff pairs in $\mathbb{R}^2$, and take the smallest convex polygon containing all these points.

For the prisoners' dilemma game G, $P(G)$ is the quadrangle with vertices $(40, 40)$, $(30, 60)$, $(60, 30)$, and $(50, 50)$, see Fig. 7.1. The elements (payoff pairs) of $P(G)$ can be obtained as limit average payoffs in the infinitely repeated game G by an appropriate sequence of play, as demonstrated before in Sect. 7.1.1. The following proposition says that every payoff pair in $P(G)$ that strictly dominates the payoffs associated with a Nash equilibrium of G can be obtained as limit average payoffs in a subgame perfect equilibrium of $G^\infty(\delta)$ for δ large enough. Such a proposition is known as a *folk theorem*. Its proof (omitted here) is somewhat technical but basically consists of formulating trigger strategies in a similar way as for the prisoners' dilemma game above. In these strategies, after a deviation from the pattern leading to the desired limit average payoffs, players revert to the Nash equilibrium under consideration of the stage game forever.

Proposition 7.2 (Folk Theorem for Subgame Perfect Equilibrium) *Let* $(\mathbf{p}^*, \mathbf{q}^*)$ *be a Nash equilibrium of* $G = (A, B)$, *and let* $\mathbf{x} = (x_1, x_2) \in P(G)$ *such that* $x_1 > \mathbf{p}^* A \mathbf{q}^*$ *and* $x_2 > \mathbf{p}^* B \mathbf{q}^*$. *Then there is a* $0 < \delta^* < 1$ *and a subgame perfect equilibrium in* $G^\infty(\delta^*)$ *with limit average payoffs* $\mathbf{x}$.

---

[1]In this proposition it is assumed that $G^\infty(\delta)$ is well-defined, in particular that the discounted payoff sums are finite.

Remark 7.3 For the purpose of this chapter, a limit average payoff pair is just a payoff pair in the set $P(G)$, i.e., in the polygon with the payoff pairs in (A, B) as vertices. More formally, if $\xi_0, \xi_1, \xi_2, \ldots$ is a sequence of real numbers, then the limit average of this sequence is the number $\lim_{T \to \infty} \frac{1}{T+1} \sum_{t=0}^{T} \xi_t$, assuming that this limit exists. In other words, we take the average of the first $T + 1$ numbers of the sequence and let T go to infinity.   □

7.1.3 Another Example

In order to illustrate Propositions 7.1 and 7.2 we consider another example. Let the bimatrix game (stage game) $G = (A, B)$ be given by

$$
G = \begin{array}{c} \\ U \\ D \end{array} \begin{array}{c} L \quad\quad R \\ \begin{pmatrix} 4, 4 & 0, 2 \\ 3, 6 & 1, 8 \end{pmatrix} \end{array}.
$$

This game has two pure strategy Nash equilibria (U, L) and (D, R) and a mixed equilibrium $((\frac{1}{2}, \frac{1}{2}), (\frac{1}{2}, \frac{1}{2}))$, with associated payoffs respectively $(4, 4)$, $(1, 8)$, and $(2, 5)$. The set of limit average payoff pairs $P(G)$ is depicted in Fig. 7.2.

Proposition 7.1 applies to the three Nash equilibria of the stage game:

- Player 1 always playing U and player 2 always L is a subgame perfect Nash equilibrium of $G^\infty(\delta)$ for any value of δ. The payoffs are $4/(1-\delta)$ for each. The limit average payoffs are 4 for each.

Fig. 7.2 The *crosshatched area* is the set $P(G)$ of the game G in Sect. 7.1.3. The *shaded area*—all points strictly above $(2, 5)$—as well as $(2, 5)$, $(4, 4)$, and $(1, 8)$, are the limit average payoffs pairs attainable in a subgame perfect equilibrium of $G^\infty(\delta)$ according to Propositions 7.1 and 7.1, for a sufficiently high value of δ

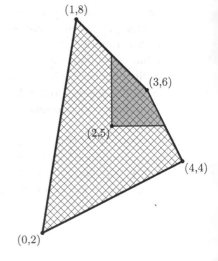

- Player 1 always playing D and player 2 always L is a subgame perfect Nash equilibrium of $G^\infty(\delta)$ for any value of δ. The payoff is $1/(1-\delta)$ for player 1 and $8/(1-\delta)$ for player 2. The limit average payoffs are 1 and 8, respectively.
- Player 1 always playing $(\frac{1}{2}, \frac{1}{2})$ and player 2 always $(\frac{1}{2}, \frac{1}{2})$ is a subgame perfect Nash equilibrium of $G^\infty(\delta)$ for any value of δ. The payoff is $2/(1-\delta)$ for player 1 and $5/(1-\delta)$ for player 2. The limit average payoffs are 2 and 5, respectively.

Proposition 7.2 says that all payoffs in the shaded area strictly above $(2, 5)$ in Fig. 7.2 can be obtained as limit average payoffs in $G^\infty(\delta)$, provided δ is sufficiently high.

As a first example we consider the payoffs $(3, 6)$. Clearly, these payoffs can be obtained as limit average payoffs if the players play (D, L) always. In order to obtain them in a subgame perfect equilibrium we can use the mixed Nash equilibrium of the stage game as a punishment after a deviation. Specifically, consider the following trigger strategy pair (S_1^*, S_2^*).

S_1^*: Start with D and keep playing D as long as (D, L) has been played so far. After any deviation, play $(\frac{1}{2}, \frac{1}{2})$ forever.

S_2^*: Start with L and keep playing L as long as (D, L) has been played so far. After any deviation, play $(\frac{1}{2}, \frac{1}{2})$ forever.

In this case, the maximal payoff to player 1 from deviating to U, given that player 2 plays S_2^*, is equal to $4 + 2 \cdot \delta/(1-\delta)$, since player 2 switches to $(\frac{1}{2}, \frac{1}{2})$ forever: to this, player 1's best reply is to also play $(\frac{1}{2}, \frac{1}{2})$ forever, resulting in the payoff 2 at each stage. This maximal payoff is smaller than or equal to $3/(1-\delta)$ if and only if $\delta \geq 1/2$. Similarly, the maximal payoff to player 2 from deviating to R, given that player 1 plays S_1^*, is equal to $8 + 5 \cdot \delta/(1-\delta)$, and this is smaller than or equal to $6/(1-\delta)$ if and only if $\delta \geq 2/3$. We conclude that (S_1^*, S_2^*) is a subgame perfect equilibrium of $G^\infty(\delta)$ for $\delta \geq 2/3$.

As a second example, consider the limit average payoff pair $(3, 5\frac{1}{3})$. From Proposition 7.2 we can conclude that these payoffs can be obtained as limit average payoffs in a subgame perfect equilibrium of $G^\infty(\delta)$. Note that we can write $(3, 5\frac{1}{3}) = \frac{2}{3}(4, 4) + \frac{1}{3}(1, 8)$, so that these limit averages can be achieved by the sequence of play

$$(U, L), (U, L), (D, R), (U, L), (U, L), (D, R), \ldots$$

Consider the following strategies.

$\bar{S}_1$: Play U at times $t = 0, 1, 3, 4, 6, 7, \ldots$ and play D at times $t = 2, 5, 8, \ldots$.

$\bar{S}_2$: Play L at times $t = 0, 1, 3, 4, 6, 7, \ldots$ and play R at times $t = 2, 5, 8, \ldots$.

Observe that, with these strategies, the players play a Nash equilibrium of the stage game at every time t, which implies that they can only loose by deviating. Thus, we do not need trigger strategies to punish deviations. The pair $(\bar{S}_1, \bar{S}_2)$ is a subgame perfect equilibrium of $G^\infty(\delta)$ for every value of $\delta \in (0, 1)$.

As a final example, consider following the strategies in $G^\infty(\delta)$ for players 1 and 2, respectively.

$\hat{S}_1$: Play D at times $t = 4, 9, 14, 19, \ldots$ and play U at all other times.

$\hat{S}_2$: Play R at times $t = 4, 9, 14, 19, \ldots$ and play L at all other times.

These strategies result in the sequence of play

$$(U,L), (U,L), (U,L), (U,L), (D,R), (U,L), (U,L), (U,L), (U,L), (D,R), \ldots$$

resulting in limit average payoffs $\frac{4}{5}(4, 4) + \frac{1}{5}(1, 8) = (3\frac{2}{5}, 4\frac{4}{5})$. This pair is outside the shaded region in Fig. 7.2, i.e., does not dominate the pair $(2, 5)$. This means that the mixed equilibrium $((\frac{1}{2}, \frac{1}{2}), (\frac{1}{2}, \frac{1}{2}))$ of the stage game cannot serve as a punishment for deviations by player 2, since the payoff 5 to player 2 from this equilibrium is larger than $4\frac{4}{5}$. Nevertheless, by the same logic as for the pair $(\bar{S}_1, \bar{S}_2)$, the strategy pair $(\hat{S}_1, \hat{S}_2)$ is a subgame perfect equilibrium of $G^\infty(\delta)$ for every value of $\delta \in (0, 1)$: at each time t the players play a Nash equilibrium of the stage game. This example shows that Proposition 7.2 is not exhaustive: it does not necessarily give *all* limit average payoff pairs attainable in a subgame perfect equilibrium.

7.2 Nash Equilibrium

In this section we consider the consequences of relaxing the subgame perfection requirement for a Nash equilibrium in an infinitely repeated game. When thinking of trigger strategies as in Sect. 7.1, this means that deviations can be punished more severely, since the equilibrium does not have to induce a Nash equilibrium in the 'punishment subgame'.

For the infinitely repeated prisoners' dilemma game of Sect. 7.1 this has no consequences. In this game each player can guarantee to obtain at least 40 at each stage, so that more severe punishments are not possible. In the following subsection we consider a different example.

7.2.1 An Example

Consider the bimatrix game

$$G = (A, B) = \begin{matrix} & L & R \\ U & \\ D & \end{matrix} \begin{pmatrix} 1,1 & 0,0 \\ 0,0 & -1,4 \end{pmatrix}.$$

The set $P(G)$ (see Sect. 7.1.2) is the triangle with vertices $(1, 1)$, $(0, 0)$, and $(-1, 4)$. In the game G the strategy D is a strictly dominated strategy for player 1. The unique Nash equilibrium of the stage game is (U, L). Player 1 always playing U and player 2 always playing L is a subgame perfect equilibrium in $G^\infty(\delta)$ for every $0 < \delta < 1$, cf. Proposition 7.1. Note that Proposition 7.2 does not add anything to this observation, since $P(G)$ does not contain any payoff pair larger than $(1, 1)$ for each player.

Now consider the following strategy pair (N_1, N_2) in the infinitely repeated game $G^\infty(\delta)$.

N_1: At $t = 0$ play D. After a history where (D, R) was played at $t = 0, 4, 8, 12, \ldots$ and (U, L) at all other times: play D at $t = 0, 4, 8, 12, \ldots$ and play U at all other times. After any other history play the mixed action $(\frac{4}{5}, \frac{1}{5})$, that is, play U with probability $\frac{4}{5}$ and D with probability $\frac{1}{5}$.

N_2: At $t = 0$ play R. After a history where (D, R) was played at $t = 0, 4, 8, 12, \ldots$ and (U, L) at all other times: play R at $t = 0, 4, 8, 12, \ldots$ and play L at all other times. After any other history play R.

These strategies are again trigger strategies. They induce a sequence of play in which within each four times, (D, R) is played once and (U, L) is played thrice. After a deviation player 1 plays the mixed action $(\frac{4}{5}, \frac{1}{5})$ and player 2 the pure action R forever. Thus, in a subgame following a deviation the players do not play a Nash equilibrium: if player 2 plays R always, then player 1's best reply is to play U always. Hence, (N_1, N_2) is not a subgame perfect equilibrium.

We claim, however, that (N_1, N_2) is a Nash equilibrium if δ is sufficiently large.

First observe that player 2 can never achieve a momentary gain from deviating since, if player 1 plays N_1, then N_2 requires player 2 to play a best reply in the stage game at every moment t. Moreover, after any deviation player 1 plays $(\frac{4}{5}, \frac{1}{5})$ at any moment t, so that both L and R have an expected payoff of $\frac{4}{5}$ for player 2, which is less than 1 and less than 4.

Suppose player 2 plays N_2. If player 1 wants to deviate from N_1, the best moment to do so is one where he is supposed to play D, so at $t = 0, 4, \ldots$. Without loss of generality suppose player 1 deviates at $t = 0$. Then (U, R) is played at $t = 0$, yielding payoff 0 to player 1. After that, player 2 plays R forever, and the best reply of player 1 to this is to play U forever, again yielding 0 each time. So his total payoff from deviating is 0. Without deviation player 1's total discounted payoff is equal to

$$-1(\delta^0 + \delta^4 + \delta^8 + \ldots) + 1(\delta^1 + \delta^2 + \delta^3 + \delta^5 + \delta^6 + \delta^7 + \ldots).$$

In order to keep player 1 from deviating this expression should be at least 0, i.e.

$$\frac{-1}{1 - \delta^4} + \left[\frac{\delta}{1 - \delta} - \frac{\delta^4}{1 - \delta^4} \right] \geq 0$$

which holds if and only if $\delta \geq \delta^*$ with $\delta^* \approx 0.54$. Hence, for these values of δ, (N_1, N_2) is a Nash equilibrium in $G^\infty(\delta)$. The limit average payoffs in this equilibrium are equal to $\frac{3}{4}(1, 1) + \frac{1}{4}(-1, 4)$, hence to $(\frac{1}{2}, \frac{7}{4})$.

The actions played in this equilibrium after a deviation are, in fact, the actions that keep the opponent to his maximin payoff. To see this, first consider the action of player 2, R. The payoff matrix of player 1 is the matrix A with

$$A = \begin{array}{c} \\ U \\ D \end{array} \overset{\begin{array}{cc} L & R \end{array}}{\begin{pmatrix} 1 & 0 \\ 0 & -1 \end{pmatrix}}.$$

The value of the *matrix game A* (cf. Chap. 2) is equal to 0—in fact, (U, R) is a saddlepoint of A—and, thus, player 1 can always obtain at least 0. By playing R, which is player 2's optimal strategy in A, player 2 can hold player 1 down to 0. Hence, this is the most severe punishment that player 2 can inflict upon player 1 after a deviation.

Similarly, if we view the payoff matrix B for player 2 as a zerosum game with payoffs to player 2 and, following convention, convert this to a matrix game giving the payoffs to player 1, we obtain

$$-B = \begin{array}{c} \\ U \\ D \end{array} \overset{\begin{array}{cc} L & R \end{array}}{\begin{pmatrix} -1 & 0 \\ 0 & -4 \end{pmatrix}}.$$

In this game, $(\frac{4}{5}, \frac{1}{5})$ is an (the) optimal strategy for player 1, yielding the value of the game, which is equal to $-\frac{4}{5}$. Hence, player 2 can guarantee to obtain a payoff of $\frac{4}{5}$, but player 1 can make sure that player 2 does not obtain more than this by playing $(\frac{4}{5}, \frac{1}{5})$. Again, this is the most severe punishment that player 1 can inflict upon player 2 after a deviation.

By using these punishments in a trigger strategy, the same logic as in Sect. 7.1 tells us that any pair of payoffs in $P(G)$ that strictly dominates the pair $\mathbf{v} = (v(A), -v(-B)) = (0, \frac{4}{5})$ can be obtained as limit average payoffs in a Nash equilibrium of the game $G^\infty(\delta)$ for δ sufficiently large. This is illustrated in Fig. 7.3.

7.2.2 A Folk Theorem for Nash Equilibrium

Let $G = (A, B)$ be an arbitrary $m \times n$ bimatrix game. Let $v(A)$ be the value of the matrix game A and let $v(-B)$ be the value of the matrix game $-B$. Let the set $P(G)$ be defined as in Sect. 7.1.2. The following proposition generalizes what we found above.

Fig. 7.3 For every payoff pair in the *shaded area* there is a δ large enough such that this payoff pair can be obtained as the pair of limit averages in a Nash equilibrium of $G^\infty(\delta)$

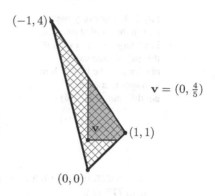

$(-1, 4)$

$\mathbf{v} = (0, \frac{4}{5})$

$(1, 1)$

$(0, 0)$

Proposition 7.4 (Folk Theorem for Nash Equilibrium) *Let* $\mathbf{x} = (x_1, x_2) \in P(G)$ *such that* $x_1 > v(A)$ *and* $x_2 > -v(-B)$. *Then there is a* $0 < \delta^* < 1$ *and a Nash equilibrium in* $G^\infty(\delta^*)$ *with limit average payoffs* $\mathbf{x}$.

The set of payoff pairs that can be reached as limit average payoff pairs in a Nash equilibrium (Proposition 7.4) contains the set of payoff pairs that can be obtained this way in a subgame perfect equilibrium (Proposition 7.2). This follows because the payoffs in a Nash equilibrium of the stage game (A, B) are at least as large as the payoffs $(v(A), -v(-B))$: if not then a player could improve by playing an optimal strategy in the associated matrix game, guaranteeing $v(A)$ (player 1) or $-v(-B)$ (player 2).

7.2.3 Another Example

Consider the bimatrix game

$$G = (A, B) = \begin{array}{c} U \\ D \end{array} \begin{pmatrix} \overset{L}{5,1} & \overset{R}{1,2} \\ 4,2 & 2,4 \end{pmatrix}.$$

The matrix game A has a saddlepoint at (D, R). Thus, $v(A) = 2$ and R is the unique optimal strategy of player 2 in A. The matrix game $-B$, given by

$$-B = \begin{array}{c} U \\ D \end{array} \begin{pmatrix} \overset{L}{-1} & \overset{R}{-2} \\ -2 & -4 \end{pmatrix},$$

has a saddlepoint at (U, R). Its value is $v(-B) = -2$, and U is the unique optimal strategy of player 1 in $-B$. The set of limit average payoff pairs and the point $\mathbf{v} = (v(A), -v(-B)) = (2, 2)$ are depicted in Fig. 7.4.

Fig. 7.4 For every payoff
pair in the *shaded area* there
is a δ large enough such that
this payoff pair can be
obtained as the pair of limit
averages in a Nash
equilibrium of $G^{\infty}(\delta)$

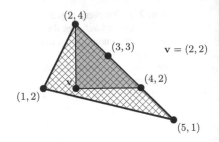

Consider, for instance, the limit average payoffs $(3, 3)$, and the following strategy
pair in $G^{\infty}(\delta)$.

N_1^*: Start by playing D and keep playing D as long as (D, L) was played at even
moments and (D, R) at odd moments. After any deviation from this, play U
forever.

N_2^*: Play L at even moments and R at odd moments as long as (D, L) was played at
even moments and (D, R) at odd moments. After any deviation from this, play
R forever.

These strategies result in the limit average payoffs $\frac{1}{2}(4, 2) + \frac{1}{2}(2, 4) = (3, 3)$.

Suppose player 2 plays N_2^*. If player 1 deviates from N_1^*, it is optimal to do so at
even moments, say at $t = 0$. His payoff from deviating is $5 + 2 \cdot \delta/(1 - \delta)$, since
after this deviation player 2 plays R forever and player 1's best reply is to play D
forever. If player 1 plays according to N_1^* he obtains $4/(1 - \delta^2) + 2 \cdot \delta/(1 - \delta^2)$. To
obtain an equilibrium, this should be at least as large as the maximal payoff from
deviating, hence

$$4/(1 - \delta^2) + 2 \cdot \delta/(1 - \delta^2) \geq 5 + 2 \cdot \delta/(1 - \delta) ,$$

which holds if and only if $\delta \geq \frac{1}{3}\sqrt{3} \approx 0.58$.

Suppose player 1 plays N_1^*. If player 2 deviates from N_2^*, it is optimal to do so at
even moments, say at $t = 0$. His payoff from deviating is $4 + 2 \cdot \delta/(1 - \delta)$, since
after this deviation player 1 plays U forever and player 2's best reply is to play R
forever. If player 2 plays according to N_2^* he obtains $2/(1 - \delta^2) + 4 \cdot \delta/(1 - \delta^2)$. To
obtain an equilibrium, this should be at least as large as the maximal payoff from
deviating, hence

$$2/(1 - \delta^2) + 4 \cdot \delta/(1 - \delta^2) \geq 4 + 2 \cdot \delta/(1 - \delta) ,$$

which holds if and only if $\delta \geq \frac{1}{2}\sqrt{5} - \frac{1}{2} \approx 0.62$.

Thus, (N_1^*, N_2^*) is a Nash equilibrium of $G^{\infty}(\delta)$ for $\delta \geq 0.62$. Note that this
equilibrium is not subgame perfect: in a subgame following a deviation the players
end up playing (U, R) forever, which is not a Nash equilibrium in such a subgame.
Also note that the only Nash equilibrium in the stage game is (D, R). Hence,

Propositions 7.1 and 7.2 only imply that the limit average payoffs $(2, 4)$ can be obtained in a subgame perfect Nash equilibrium of $G^\infty(\delta)$.

7.3 Problems

7.1. *Nash and Subgame Perfect Equilibrium in a Repeated Game (1)*
Consider the following bimatrix game:

$$G = (A, B) = \begin{array}{c} \\ U \\ D \end{array} \begin{array}{c} L \quad\quad R \\ \left(\begin{array}{cc} 2,3 & 1,5 \\ 0,1 & 0,1 \end{array} \right) \end{array}.$$

(a) Determine all Nash equilibria of this game. Also determine the value $v(A)$ of the matrix game A and the value $v(-B)$ of the matrix game $-B$. Determine the optimal strategies of player 2 in A and of player 1 in $-B$.
(b) Consider the repeated game $G^\infty(\delta)$. Which limit average payoffs can be obtained in a subgame perfect equilibrium of this repeated game according to Proposition 7.1 or Proposition 7.2? Does this depend on δ?
(c) Which limit average payoffs can be obtained in a Nash equilibrium in $G^\infty(\delta)$ according to Proposition 7.4?
(d) Describe a pair of Nash equilibrium strategies in $G^\infty(\delta)$ that result in the limit average payoffs $(2, 3)$. What is the associated minimum value of δ?

7.2. *Nash and Subgame Perfect Equilibrium in a Repeated Game (2)*
Consider the following bimatrix game:

$$G = (A, B) = \begin{array}{c} \\ U \\ D \end{array} \begin{array}{c} L \quad\quad R \\ \left(\begin{array}{cc} 2,1 & 0,0 \\ 0,0 & 1,2 \end{array} \right) \end{array}.$$

(a) Which payoffs can be reached as limit average payoffs in a subgame perfect equilibrium of the infinitely repeated game $G^\infty(\delta)$ for suitable choices of δ according to Propositions 7.1 and 7.2?
(b) Which payoffs can be reached as limit average payoffs in a Nash equilibrium of the infinitely repeated game $G^\infty(\delta)$ for suitable choices of δ according to Proposition 7.4?
(c) Describe a subgame perfect Nash equilibrium of $G^\infty(\delta)$ resulting in the limit average payoffs $(\frac{3}{2}, \frac{3}{2})$. Also give the corresponding restriction on δ.
(d) Describe a subgame perfect Nash equilibrium of $G^\infty(\delta)$ resulting in the limit average payoffs $(1, 1)$. Also give the corresponding restriction on δ.

7.3. *Nash and Subgame Perfect Equilibrium in a Repeated Game (3)*
Consider the following bimatrix game:

$$G = (A, B) = \begin{matrix} & L & R \\ U \\ D \end{matrix}\begin{pmatrix} 3,2 & 8,0 \\ 4,0 & 6,2 \end{pmatrix}$$

(a) Which payoffs can be reached as limit average payoffs in a subgame perfect Nash equilibrium of the infinitely repeated discounted game $G^\infty(\delta)$ for suitable choices of δ according to Propositions 7.1 and 7.2?
(b) Which payoffs can be reached as limit average payoffs in a Nash equilibrium of the infinitely repeated game $G^\infty(\delta)$ for suitable choices of δ according to Proposition 7.4?
(c) Describe a Nash equilibrium of $G^\infty(\delta)$ resulting in the limit average payoffs $(4\frac{1}{2}, 2)$. Is there any value of δ for which this equilibrium is subgame perfect? Why or why not?

7.4. *Subgame Perfect Equilibrium in a Repeated Game*
Consider the following bimatrix game:

$$G = \begin{matrix} & L & C & R \\ T \\ M \\ B \end{matrix}\begin{pmatrix} 6,4 & 0,7 & 0,0 \\ 8,0 & 4,6 & 0,0 \\ 0,0 & 0,0 & 1,1 \end{pmatrix}.$$

(a) What are the pure strategy Nash equilibria of this game?
(b) Which limit average payoffs can be obtained in a subgame perfect equilibrium of $G^\infty(\delta)$ according to your answer to (a) and Propositions 7.1 and 7.2?
(c) Describe a subgame perfect equilibrium in the infinitely repeated game which results in the limit average payoffs of 5 for each player. Also give the minimum value of δ for which your strategy combination is a subgame perfect equilibrium.

7.5. *The Strategies Tr_1^* and Tr_2^**
Give the inequalities that determine the lower bound on δ for the strategy combination (Tr_1^*, Tr_2^*) to be a subgame perfect equilibrium in the infinitely repeated prisoners' dilemma game in Sect. 7.1.1.

7.6. *Repeated Cournot and Bertrand*

(a) Reconsider the duopoly (Cournot) game of Sect. 6.2.1. Suppose that this game is repeated infinitely many times, and that the two firms discount the streams of payoffs by a common discount factor δ. Describe a subgame perfect Nash equilibrium of the repeated game that results in each firm receiving half of the

monopoly profits at each time. Also give the corresponding restriction on δ. What could be meant by the expression 'tacit collusion'?
(b) Answer the same questions as in (a) for the Bertrand game of Sect. 6.3.

7.7. *Repeated Duopoly*
Two firms (1 and 2) offer heterogenous goods at prices

$$p_1 = \max\{10 - 2q_1 + q_2, 0\}$$
$$p_2 = \max\{10 - 2q_2 + q_1, 0\}$$

where q_1, q_2 are the quantities offered. All costs are assumed to be zero, and the firms are engaged in price competition.

(a) Show that the market clearing quantity of firm 1 at prices p_1 and p_2 is given by

$$q_1 = \max\{10 - \frac{2}{3}p_1 - \frac{1}{3}p_2, 0\} .$$

Also derive the quantity q_2 of firm 2 as a function of the prices, and set up the profit functions of the two firms.
(b) Derive the reaction functions of the firms and use these to compute the Nash equilibrium prices. Compute the associated profits of both firms.
(c) Compute the prices at which joint profit is maximized. Compute the associated profits of both firms.
(d) Now suppose that this price competition game is played infinitely many times, and that the firms' payoffs are discounted by a common factor δ. Describe a subgame perfect equilibrium in which joint profit is maximized in each period. Also give the associated lower bound for the discount factor δ.

7.8. *On Discounting*
In a repeated game, interpret the discount factor $0 < \delta < 1$ as the probability that the game will continue, i.e., that the stage game will be played again. Show that, with this interpretation, the repeated game will end with probability 1. (Cf. Problem 6.17(e).)

7.9. *On Limit Average*
Can you give an example in which the limit that defines the long run average payoffs, does not exist? (Cf. Remark 7.3.)

7.4   Notes

For finitely repeated games as in Problems 4.10–4.12 see also Benoit and Krishna (1985) and Friedman (1985).

Fudenberg and Maskin (1986) show that Proposition 7.4—the folk theorem for Nash equilibrium in infinitely repeated games—also holds for subgame perfect Nash equilibrium if the dimension of the 'cooperative payoff space' (the set $P(G)$ in the text) is equal to the number of players. This, however, requires more sophisticated strategies. See Fudenberg and Tirole (1991a) for further references.

The expression 'folk theorem' refers to the fact that results like this had been known among game theorists even before they were formulated and written down explicitly. They belonged to the *folklore* of game theory.

For more advanced and elaborate treatments of repeated games see, e.g., Fudenberg and Tirole (1991a), Mailath and Samuelson (2006), or Myerson (1991).

References

Benoit, J.-P., & Krishna, V. (1985). Finitely repeatd games. *Econometrica, 53*, 905–922.

Friedman, J. W. (1985). Cooperative equilibria in finite horizon noncooperative supergames. *Journal of Economic Theory, 35*, 390–398.

Fudenberg, D., & Maskin, E. (1986). The folk theorem in repeated games with discounting or with incomplete information. *Econometrica, 54*, 533–556.

Fudenberg, D., & Tirole, J. (1991a). *Game theory*. Cambridge: MIT Press.

Mailath, G. J., & Samuelson, L. (2006). *Repeated games and reputations*. New York: Oxford University Press.

Myerson, R. B. (1991). *Game theory, analysis of conflict*. Cambridge: Harvard University Press.

An Introduction to Evolutionary Games

8

In an evolutionary game, players are interpreted as populations—of animals or individuals. The probabilities in a mixed strategy of a player in a bimatrix game are interpreted as shares of the population. Individuals within the same part of the population play the same pure strategy. The main 'solution' concept is the concept of an evolutionary stable strategy.

Evolutionary game theory originated in biology. The developed evolutionary biological concepts were later applied to boundedly rational human behavior, and a connection was established with dynamic systems and with game-theoretic concepts such as Nash equilibrium.

This chapter presents a short introduction to evolutionary game theory. For a more advanced continuation see Chap. 15.

In Sect. 8.1 we consider symmetric two-player games and evolutionary stable strategies. Evolutionary stability is meant to capture the idea of *mutation* from the theory of evolution. We also establish that an evolutionary stable strategy is part of a symmetric Nash equilibrium. In Sect. 8.2 the connection with the so-called replicator dynamics is studied. Replicator dynamics intends to capture the evolutionary idea of *selection* based on fitness. In Sect. 8.3 asymmetric games are considered. Specifically, a connection between replicator dynamics and strict Nash equilibrium is discussed.

8.1 Symmetric Two-Player Games and Evolutionary Stable Strategies

A famous example from evolutionary game theory is the *Hawk-Dove game*:

$$
\begin{array}{c} \\ \text{Hawk} \\ \text{Dove} \end{array}
\begin{array}{cc} \text{Hawk} & \text{Dove} \\ \left(\begin{array}{cc} 0,0 & 3,1 \\ 1,3 & 2,2 \end{array} \right) \end{array} .
$$

© Springer-Verlag Berlin Heidelberg 2015
H. Peters, *Game Theory*, Springer Texts in Business and Economics,
DOI 10.1007/978-3-662-46950-7_8

This game models the following situation. Individuals of the same large population meet randomly, in pairs, and behave either aggressively (Hawk) or passively (Dove)—the fight is about nest sites or territories, for instance. This behavior is genetically determined, so an individual does not really choose between the two modes of behavior. The payoffs reflect (Darwinian) fitness, e.g., the number of offspring. In this context, players 1 and 2 are just two different members of the same population who meet: indeed, the game is symmetric—see below for the formal definition. A mixed strategy $\mathbf{p} = (p_1, p_2)$ (of player 1 or player 2) is naturally interpreted as expressing the population shares of individuals characterized by the same type of behavior. In other words, $p_1 \times 100\%$ of the population are Hawks and $p_2 \times 100\%$ are Doves. In view of this interpretation, in what follows we are particularly interested in symmetric Nash equilibria, i.e., Nash equilibria in which the players use the same strategy. The Hawk-Dove game has three Nash equilibria, only one of which is symmetric namely $((\frac{1}{2}, \frac{1}{2}), (\frac{1}{2}, \frac{1}{2}))$.

Remark 8.1 The Hawk-Dove game can also be interpreted as a *Game of Chicken*. Two car drivers approach each other on a road, each one driving in the middle. The driver who is the first to return to his own lane (Dove) 'loses' the game, the one who stays in the middle 'wins' (Hawk). With this interpretation also the asymmetric equilibria are of interest. The asymmetric equilibria can also be of interest within the evolutionary approach: the Hawk-Dove game can be interpreted as modelling competition between two species, represented by the row and the column player. Within each species, there are again two types of behavior. See Sect. 8.3.   □

The definitions of a symmetric game and a symmetric Nash equilibrium are as follows.

Definition 8.2 Let $G = (A, B)$ be an $m \times n$ bimatrix game. Then G is *symmetric* if $m = n$ and $b_{ij} = a_{ji}$ for all $i, j = 1, \ldots, m$. A Nash equilibrium $(\mathbf{p}^*, \mathbf{q}^*)$ of G is *symmetric* if $\mathbf{p}^* = \mathbf{q}^*$.   □

In other words, a bimatrix game (A, B) is symmetric if both players have the same number of pure strategies and the payoff matrix of player 2 is the transpose of the payoff matrix of player 1, i.e., we obtain B by interchanging the rows and columns of A. This will also be denoted by $B = A^T$, where 'T' stands for 'transpose'. A Nash equilibrium is symmetric if both players play the same strategy.

We state the following fact without a proof (see Chap. 15).

Proposition 8.3 *Every symmetric bimatrix game G has a symmetric Nash equilibrium.*

With the interpretation above—different types of behavior within one and the same population—it is only meaningful to consider symmetric Nash equilibria. But in fact, we will require more.

Let $G = (A, B)$ be a symmetric game. Knowing that the game is symmetric, it is sufficient to know the payoff matrix A, since $B = A^T$. In what follows, when we consider a symmetric game A we mean the game $G = (A, A^T)$. Let A be an $m \times m$ matrix. Recall (Chaps. 2 and 3) that Δ^m denotes the set of mixed strategies (for player 1 or player 2).

The main concept in evolutionary game theory is that of an *evolutionary stable strategy*. The original concept will be formally introduced in Chap. 15. Here, we give an equivalent but easier to handle definition. With some abuse of language we give it the same name.

Definition 8.4 Let A be an $m \times m$ matrix. A strategy $\mathbf{x} \in \Delta^m$ is an *evolutionary stable strategy (ESS)* if the following two conditions hold.

(a) $(\mathbf{x}, \mathbf{x})$ is a Nash equilibrium in (A, A^T).
(b) For every $\mathbf{y} \in \Delta^m$ with $\mathbf{y} \neq \mathbf{x}$ we have:

$$\mathbf{x}A\mathbf{x} = \mathbf{y}A\mathbf{x} \Rightarrow \mathbf{x}A\mathbf{y} > \mathbf{y}A\mathbf{y} . \tag{8.1}$$

□

To interpret this definition, think again of $\mathbf{x}$ as shares of one and the same large population. The first condition says that this population is in equilibrium: $\mathbf{x}$ is one of the possible distributions of shares that maximize average fitness against $\mathbf{x}$. The second condition concerns *mutations*. Suppose there is another distribution of shares $\mathbf{y}$ (a mutation) that fares equally well against $\mathbf{x}$ as $\mathbf{x}$ itself does: $\mathbf{y}$ is an alternative 'best reply' to $\mathbf{x}$. Then (8.1) says that $\mathbf{x}$ fares better against $\mathbf{y}$ than $\mathbf{y}$ does against itself. Hence, $\mathbf{y}$ does not take over: the 'mutation' $\mathbf{y}$ is not successful. The original definition of *ESS* is phrased in terms of *small* mutations, but this turns out to be equivalent to the definition above (Chap. 15).

The evolutionary stable strategies for an $m \times m$ matrix A can be found as follows. First, compute the symmetric Nash equilibria of the game $G = (A, B)$ with $B = A^T$. This can be done using the methods developed in Chap. 3. Second, for each such equilibrium $(\mathbf{x}, \mathbf{x})$, check whether (8.1) holds. If it does, then $\mathbf{x}$ is an evolutionary stable strategy.

We apply this method to the Hawk-Dove game. For this game,

$$
A = \begin{array}{c} \\ \text{Hawk} \\ \text{Dove} \end{array}
\begin{array}{cc} \text{Hawk} & \text{Dove} \end{array}
\left(\begin{array}{cc} 0 & 3 \\ 1 & 2 \end{array} \right) .
$$

The unique symmetric equilibrium strategy was $\mathbf{x} = (\frac{1}{2}, \frac{1}{2})$. The condition $\mathbf{x}A\mathbf{x} = \mathbf{y}A\mathbf{x}$ in (8.1) is satisfied for every $\mathbf{y} = (y, 1 - y)$. This can be seen by direct computation but it also follows from the fact that $(\mathbf{x}, \mathbf{x})$ is a Nash equilibrium and $\mathbf{x}$

has all coordinates positive (how?). Hence, we have to check if

$$\mathbf{x}A\mathbf{y} > \mathbf{y}A\mathbf{y}$$

for all $\mathbf{y} = (y, 1 - y) \neq \mathbf{x}$. This inequality reduces (check!) to:

$$(2y - 1)^2 > 0 \,,$$

which is true for all $y \neq \frac{1}{2}$. Thus, $\mathbf{x} = (\frac{1}{2}, \frac{1}{2})$ is the unique ESS in A.

8.2   Replicator Dynamics and Evolutionary Stability

Central in the theory of evolution are the concepts of *mutation* and *selection*. While the idea of mutation is meant to be captured by the concept of evolutionary stability, the idea of selection is captured by the so-called replicator dynamics. We illustrate the concept of replicator dynamics by considering again the Hawk-Dove game

$$
\begin{array}{cc}
 & \begin{array}{cc} \text{Hawk} & \text{Dove} \end{array} \\
\begin{array}{c} \text{Hawk} \\ \text{Dove} \end{array} &
\left(\begin{array}{cc} 0,0 & 3,1 \\ 1,3 & 2,2 \end{array} \right)
\end{array} \,.
$$

Consider a mixed strategy or, in the present context, vector of population shares $\mathbf{x} = (x, 1 - x)$. Consider an arbitrary individual of the population. Playing 'Hawk' against the population $\mathbf{x}$ yields an expected payoff or 'fitness' of

$$0 \cdot x + 3 \cdot (1 - x) = 3(1 - x)$$

and playing 'Dove' yields

$$1 \cdot x + 2 \cdot (1 - x) = 2 - x \,.$$

Hence, the *average fitness* of the population is

$$x \cdot 3(1 - x) + (1 - x) \cdot (2 - x) = 2 - 2x^2 \,.$$

We now assume that the population shares develop over time, i.e., that x is a function of time t, and that the change in x, described by the time derivative $\dot{x} = \dot{x}(t) = dx(t)/dt$, is proportional to the difference between Hawk's fitness and average fitness. In particular, if Hawk's fitness is larger than average fitness, then the percentage of Hawks increases, and if Hawk's fitness is smaller than average fitness, then the percentage of Hawks decreases. In case of equality the population is at rest.

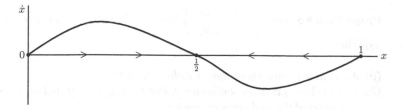

Fig. 8.1   Replicator dynamics for the Hawk-Dove game

Formally, we assume that $\dot{x}$ is given by the following equation.

$$\dot{x}(t) = dx(t)/dt = x(t)\left[3(1 - x(t)) - (2 - 2x(t)^2)\right] . \tag{8.2}$$

Equation (8.2) is the *replicator dynamics* for the Hawk-Dove game. The equation says that the population of Hawks changes continuously (described by $dx(t)/dt$), and that this change is proportional to the difference between the fitness at time t—which is equal to $3(1 - x(t))$—and the average fitness of the population—which is equal to $2 - 2x(t)^2$. Simplifying (8.2) and writing x instead of $x(t)$ yields

$$\dot{x} = dx/dt = x(x - 1)(2x - 1).$$

This makes it possible to draw a diagram of dx/dt as a function of x (a so-called *phase diagram*). See Fig. 8.1. We see that this replicator dynamics has three different roots, the so-called *rest points*[1] $x = 0$, $x = \frac{1}{2}$, and $x = 1$. For these values of x, the derivative dx/dt is equal to zero, so the population shares do not change: the system is at rest. In case $x = 0$ all members of the species are Doves, their fitness is equal to the average fitness, and so nothing changes. This rest point, however, is not stable. A slight disturbance, e.g., a genetic mutation resulting in a Hawk, makes the number of Hawks increase because dx/dt becomes positive. This increase will go on until the rest point $x = \frac{1}{2}$ is reached. A similar story holds for the rest point $x = 1$, where the population consists of only Hawks. Now suppose the system is at the rest point $x = \frac{1}{2}$. Note that, after a disturbance in either direction, the system will move back again to the state where half the population consists of Doves. Thus, of the three rest points, only $x = \frac{1}{2}$ is *stable*. (A formal definition of stability of a rest point is provided in Chap. 15.)

Recall from the previous section that $\mathbf{x} = (\frac{1}{2}, \frac{1}{2})$ is also the unique evolutionary stable strategy of the Hawk-Dove game. That this is no coincidence follows from the next proposition, which we state here without a proof (see Chap. 15 for a proof). (The definition of replicator dynamics is analogous to the one in the Hawk-Dove game.)

---

[1] In the literature also called equilibrium points, critical points, stationary points.

Proposition 8.5 *Let* $A = \begin{pmatrix} a_{11} & a_{12} \\ a_{21} & a_{22} \end{pmatrix}$ *be a* 2×2 *matrix with* $a_{11} \neq a_{21}$ *and* $a_{12} \neq a_{22}$. *Then:*

(a) *A has at least one evolutionary stable strategy.*
(b) $\mathbf{x} = (x, 1 - x)$ *is an evolutionary stable strategy of A if and only if* $\mathbf{x}$ *is a stable rest point of the replicator dynamics.*

Remark 8.6 For general $m \times m$ matrices the set of completely mixed (i.e., with all coordinates positive) rest points of the replicator dynamics coincides with the set of completely mixed strategies in symmetric Nash equilibria. There are also connections between stability of rest points and further properties of Nash equilibria. See Chap. 15 for details.   □

Example 8.7 As another example, consider the matrix

$$A = \begin{matrix} & V & W \\ V \\ W \end{matrix}\begin{pmatrix} 3 & 1 \\ 1 & 2 \end{pmatrix}.$$

The bimatrix game (A, A^T) has three Nash equilibria all of which are symmetric, namely: (V, V), (W, W), and $((1/3, 2/3), (1/3, 2/3))$. Against V the unique best reply is V, so that V is an *ESS*: (8.1) is satisfied trivially. By a similar argument, W is an *ESS*.

Against $(1/3, 2/3)$, any $\mathbf{y} = (y, 1 - y)$ is a best reply. For $(1/3, 2/3)$ to be an *ESS* we therefore need

$$\begin{pmatrix} 1/3 & 2/3 \end{pmatrix} A \begin{pmatrix} y \\ 1-y \end{pmatrix} > \begin{pmatrix} y & 1-y \end{pmatrix} A \begin{pmatrix} y \\ 1-y \end{pmatrix}$$

for all $0 \leq y \leq 1$ with $y \neq 1/3$. The inequality simplifies (check!) to the inequality $(3y - 1)^2 < 0$, which never holds. Hence, $(1/3, 2/3)$ is not an ESS.

We now investigate the replicator dynamics. The expected payoff of V against $(x, 1 - x)$ is equal to $3x + 1 \cdot (1 - x) = 2x + 1$. The expected payoff of W against $(x, 1 - x)$ is equal to $x + 2 \cdot (1 - x) = 2 - x$. The average payoff is $x(2x + 1) + (1 - x)(2 - x) = 3x^2 - 2x + 2$. Hence, the replicator dynamics is

$$dx/dt = x(2x + 1 - (3x^2 - 2x + 2)) = -x(x - 1)(3x - 1).$$

Figure 8.2 presents the phase diagram, which shows that $x = 0$ and $x = 1$ are stable rest points, and that the rest point $x = 1/3$ is not stable, in accordance with Proposition 8.5.   □

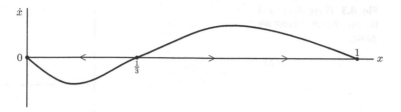

Fig. 8.2 Replicator dynamics for Example 8.7

8.3 Asymmetric Games

The evolutionary approach to game theory is not necessarily restricted to symmetric situations, i.e., bimatrix games of the form (A, A^T) in which the row and column players play identical strategies. In biology as well as economics one can find many examples of asymmetric situations. Think, for instance, of two different species competing about territory in biology; and see Problem 8.6 for an example from economics.

Consider the 2×2-bimatrix game

$$(A, B) = \begin{array}{c} \\ U \\ D \end{array} \begin{pmatrix} \overset{L}{0,0} & \overset{R}{2,2} \\ 1,5 & 1,5 \end{pmatrix} .$$

Think of two populations, the row population and the column population. In each population there are two different types: U and D in the row population and L and R in the column population. Individuals of one population are continuously and randomly matched with individuals of the other population, and we are interested again in the development of the population shares. To start with, assume the shares of U and D types in the row population are x and $1 - x$, respectively, and the shares of L and R types in the column population are y and $1 - y$. The expected payoff of a U type individual is given by:

$$0 \cdot y + 2 \cdot (1 - y) = 2 - 2y .$$

For a D type individual it is

$$1 \cdot y + 1 \cdot (1 - y) = 1 .$$

For an L type individual it is

$$0 \cdot x + 5 \cdot (1 - x) = 5 - 5x .$$

Fig. 8.3 Phase diagram of
the asymmetric evolutionary
game

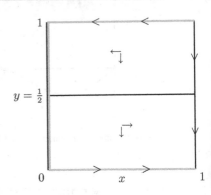

And for an R type individual:

$$2 \cdot x + 5 \cdot (1 - x) = 5 - 3x .$$

The average of the row types is therefore:

$$x[2(1 - y)] + (1 - x) \cdot 1$$

and the replicator dynamics for the population share $x(t)$ of U individuals is given by

$$dx/dt = x[2(1 - y) - x[2(1 - y)] - (1 - x)] = x(1 - x)(1 - 2y) . \qquad (8.3)$$

Here, we write x and y instead of $x(t)$ and $y(t)$. Similarly one can calculate the replicator dynamics for the column population (check this result!):

$$dy/dt = y(1 - y)(-2x) . \qquad (8.4)$$

We are interested in the rest points of the dynamic system described by Eqs. (8.3) and (8.4), and, in particular, by the stable rest points. Figure 8.3 presents a diagram of the possible values of x and y. In this diagram, the black lines are the values of x and y for which the derivative in (8.3) is equal to 0, i.e., for which the row population is at rest. The gray lines are the values of x and y for which the derivative in (8.4) is equal to 0: there, the column population is at rest. The points of intersection are the points where the whole system is at rest; this is the set

$$\{(0, y) \mid 0 \le y \le 1\} \cup \{(1, 0)\} \cup \{(1, 1)\} .$$

In Fig. 8.3, arrows indicate the direction in which x and y move. For instance, if $1 > y \ge \frac{1}{2}$ and $0 < x < 1$ we have $dx/dt < 0$ and $dy/dt < 0$, so that in that region x as well as y decrease. A *stable rest point* is a rest point such that, if the system is slightly disturbed and moves to some point close to the rest point in question, then it should move back again to this rest point. In terms of the arrows in Fig. 8.3

this means that a stable rest point is one where all arrows in the neighborhood point towards that point. It is obvious that in our example the point $(1, 0)$ is the only such point. So the situation where the row population consists only of U type individuals $(x = 1)$ and the column population consists only of R type individuals $(y = 0)$ is the only stable situation with respect to the replicator dynamics.

Is there a relation with Nash equilibrium? One can check (!) that the set of Nash equilibria in this example is the set:

$$\{(U, R), (D, L)\} \cup \{(D, (q, 1 - q)) \mid \frac{1}{2} \le q \le 1\} .$$

So the stable rest point (U, R) is a Nash equilibrium. Furthermore, it has a special characteristic, namely, it is the only strict Nash equilibrium of the game. A *strict Nash equilibrium* in a game is a Nash equilibrium where each player not only does not gain but in fact strictly looses by deviating. For instance, if the row player deviates from U in the Nash equilibrium (U, R) then he obtains strictly less than 2. All the other equilibria in this game do not have this property. For instance, if the column player deviates from L to R in the Nash equilibrium (D, L), then he still obtains 5.

The observation that the stable rest point of the replicator dynamics coincides with a strict Nash equilibrium is not a coincidence. The following proposition is stated here without a proof.

Proposition 8.8 *In a 2×2 bimatrix game a pair of strategies is a stable rest point of the replicator dynamics if and only if it is a strict Nash equilibrium. For larger games, any stable rest point of the replicator dynamics is a strict Nash equilibrium, but the converse does not necessarily hold.*

Remark 8.9 A strict Nash equilibrium in a bimatrix game must be a pure Nash equilibrium, for the following reason. If a player plays two or more pure strategies with positive probability in a Nash equilibrium, then he must be indifferent between these pure strategies and, thus, can deviate to any of them while keeping the same payoff. This holds true in any arbitrary game, not only in bimatrix games.   □

8.4   Problems

8.1. *Symmetric Games*
Compute the evolutionary stable strategies for the following payoff matrices A.

(a) $A = \begin{pmatrix} 4 & 0 \\ 5 & 3 \end{pmatrix}$ (Prisoners' Dilemma)

(b) $A = \begin{pmatrix} 2 & 0 \\ 0 & 1 \end{pmatrix}$ (Coordination game)

8.2. *More Symmetric Games*
For each of the following two matrices, determine the replicator dynamics, rest points and stable rest points, and evolutionary stable strategies. Include phase diagrams for the replicator dynamics. For the evolutionary stable strategies, provide independent arguments to show evolutionary stability by using Definition 8.4.

(a) $A = \begin{pmatrix} 0 & 1 \\ 1 & 0 \end{pmatrix}$

(b) $A = \begin{pmatrix} 2 & 0 \\ 1 & 0 \end{pmatrix}$

8.3. *Asymmetric Games*
For each of the following two asymmetric situations (i.e., row and column populations are assumed to be different and we do not only consider symmetric population shares), determine the replicator dynamics, rest points and stable rest points, including phase diagrams. Also determine all Nash and strict Nash equilibria.

(a) $(A, A^T) = \begin{pmatrix} 0,0 & 1,1 \\ 1,1 & 0,0 \end{pmatrix}$

(b) $(A, A^T) = \begin{pmatrix} 2,2 & 0,1 \\ 1,0 & 0,0 \end{pmatrix}$

8.4. *More Asymmetric Games*
For each of the following two bimatrix games, determine the replicator dynamics and all rest points and stable rest points. Also compute all Nash equilibria, and discuss the relation using Proposition 8.8.

(a) $(A, B) = \begin{pmatrix} 3,2 & 8,0 \\ 4,0 & 6,2 \end{pmatrix}$

(b) $(A, B) = \begin{pmatrix} 4,3 & 3,4 \\ 5,5 & 2,4 \end{pmatrix}$

8.5. *Frogs Call For Mates*
Consider the following game played by male frogs who Call or Don't Call their mates.

$$\begin{array}{cc} & \begin{array}{cc} \text{Call} & \text{Don't Call} \end{array} \\ \begin{array}{c} \text{Call} \\ \text{Don't Call} \end{array} & \begin{pmatrix} P-z, P-z & m-z, 1-m \\ 1-m, m-z & 0,0 \end{pmatrix} \end{array}$$

The payoffs are in units of 'fitness', measured by the frog's offspring. Here z denotes the cost of Calling (danger of becoming prey, danger of running out of energy); and m is the probability that the male who calls in a pair of males, the other of whom is not calling, gets a mate. Typically, $m \geq \frac{1}{2}$. Next, if no male calls then no female is attracted, and if both call returns diminish and they each attract P females with $0 < P < 1$.

(a) Show that there are several possible evolutionary stable strategies for this game, depending on the parameters (m, z, P).
(b) Set $m = 0.6$ and $P = 0.8$. Find values for z for each of the following situations: (i) Don't Call is an evolutionary stable strategy (ESS); (ii) Call is an ESS; (iii) A mixture of Call and Don't Call is an ESS.
(c) Suppose there are two kinds of frogs in *Frogs Call For Mates*. Large frogs have a larger cost of calling (z_1) than do small frogs (z_2). Determine the corresponding asymmetric bimatrix game. Determine the possible stable rest points of the replicator dynamics.

8.6. *Video Market Game*
Two boundedly rational video companies are playing the following asymmetric game:

	Open system	Lockout system
Open system	6, 4	5, 5
Lockout system	9, 1	10, 0

Company I (the row company) has to decide whether to have an open system or a lockout system. Company II (the column company) has to decide whether to create its own system or copy that of company I. What is a rest point of the replicator dynamics for this system?

8.5 Notes

Evolutionary game theory originated in the work of the biologists Maynard Smith and Price (1973). Taylor and Jonker (1978) and Selten (1983), among others, played an important role in applying the developed evolutionary biological concepts to boundedly rational human behavior, and in establishing the connection with dynamic systems and with game-theoretic concepts such as Nash equilibrium. A comprehensive treatment is Weibull (1995).

The original definition of evolutionary stable strategy (Chap. 15) is due to Maynard Smith and Price (1973). Taylor and Jonker (1978) introduced the replicator dynamics.

For more economic applications of asymmetric evolutionary games see for instance Gardner (1995). In the literature the concept of evolutionary stable strategy is extended to asymmetric games. See Selten (1980) or Hofbauer and Sigmund (1988) for details, also for the relation between stable rest points and strict Nash equilibrium.

Problems 8.5 and 8.6 are taken from Gardner (1995).

References

Gardner, R. (1995). *Games for business and economics*. New York: Wiley.

Hofbauer, J., & Sigmund, K. (1988). *The theory of evolution and dynamical systems*. Cambridge: Cambridge University Press.

Maynard Smith, J., & Price, G. R. (1973). The logic of animal conflict. *Nature, 246*, 15–18.

Selten, R. (1980). A note on evolutionary stable strategies in asymmetric animal conflicts. *Journal of Theoretical Biology, 84*, 93–101.

Selten, R. (1983). Evolutionary stability in extensive-form two-person games. *Mathematical Social Sciences, 5*, 269–363.

Taylor, P., & Jonker, L. (1978). Evolutionary stable strategies and game dynamics. *Mathematical Biosciences, 40*, 145–156.

Weibull, J. W. (1995). *Evolutionary game theory*. Cambridge: MIT Press.

Cooperative Games with Transferable Utility

9

The implicit assumption in a cooperative game is that players can form coalitions and make binding agreements on how to distribute the proceeds of these coalitions. A cooperative game is more abstract than a noncooperative game in the sense that strategies are not explicitly modelled: rather, the game describes what each possible coalition can earn by cooperation. In a cooperative game with *transferable utility* it is assumed that the earnings of a coalition can be expressed by one number. One may think of this number as an amount of money, which can be distributed among the players in any conceivable way—including negative payments—if the coalition is actually formed. More generally, it is an amount of *utility* and the implicit assumption is that it makes sense to transfer this utility among the players— for instance, due to the presence of a medium like money, assuming that individual utilities can be expressed in monetary terms.

This chapter presents a first acquaintance with the theory of cooperative games with transferable utility. A few important solution concepts—the core, the Shapley value, and the nucleolus—are briefly discussed in Sects. 9.2–9.4. We start with examples and preliminaries in Sect. 9.1.

9.1   Examples and Preliminaries

In Chap. 1 we have seen several examples of cooperative games with transferable utility: the three cities game, a glove game, a permutation game, and a voting game. For the stories giving rise to these games the reader is referred to Sect. 1.3.4. Here we reconsider the resulting games.

In the three cities game, cooperation between cities leads to cost savings expressed in amounts of money, as in Table 9.1. In the first line of this table all possible coalitions are listed. It important to note that the term 'coalition' is used for any subset of the set of players. So a coalition is not necessarily formed. The empty subset (empty coalition) has been added for convenience: it is assigned the number

© Springer-Verlag Berlin Heidelberg 2015
H. Peters, *Game Theory*, Springer Texts in Business and Economics,
DOI 10.1007/978-3-662-46950-7_9

Table 9.1 The three cities game

S	$\emptyset$	$\{1\}$	$\{2\}$	$\{3\}$	$\{1,2\}$	$\{1,3\}$	$\{2,3\}$	$\{1,2,3\}$
$v(S)$	0	0	0	0	90	100	120	220

Table 9.2 A glove game

S	$\emptyset$	$\{1\}$	$\{2\}$	$\{3\}$	$\{1,2\}$	$\{1,3\}$	$\{2,3\}$	$\{1,2,3\}$
$v(S)$	0	0	0	0	0	1	1	1

Table 9.3 A permutation game

S	$\emptyset$	$\{1\}$	$\{2\}$	$\{3\}$	$\{1,2\}$	$\{1,3\}$	$\{2,3\}$	$\{1,2,3\}$
$v(S)$	0	2	5	4	14	18	9	24

0 by convention. The numbers in the second line of the table are called the 'worths' of the coalitions. For instance, coalition $S = \{1,2\}$ has worth 90. In this particular example, 90 are the costs saved by cities 1 and 2 if they cooperate. It is assumed that this amount can be split between the two players (cities) if the coalition is actually formed: that is, player 1 may receive $x_1 \in \mathbb{R}$ and player 2 may receive $x_2 \in \mathbb{R}$ such that $x_1 + x_2 = 90$ or, more generally, $x_1 + x_2 \leq 90$.

In the glove game in Sect. 1.3.4 coalitions may make pairs of gloves. The game is described in Table 9.2. In this game the worth 1 of the 'grand coalition' $\{1,2,3\}$, for instance, means that this coalition can earn 1 by producing one pair of gloves. One can think of this number as expressing the monetary value of this pair of gloves. Alternatively, one can think of one pair of gloves having 'utility' equal to 1. Again, it is assumed that the players can split up this amount in any way they like. So a possible distribution of the worth of the grand coalition takes the form $(x_1, x_2, x_2) \in \mathbb{R}^3$ such that $x_1 + x_2 + x_3 \leq 1$. For $i = 1,2,3$, the number x_i may represent the money that player i receives, or (if nonnegative) the percentage of time that player i is allowed to wear the gloves.

The permutation game (dentist game) of Sect. 1.3.4 is reproduced in Table 9.3. In this game, one could think of the worth of a coalition as expressing, for instance, savings of opportunity costs by having dentist appointments on certain days. What is important is that, again, these worths can be distributed in any way among the players of the coalitions.

For the voting game related to the UN Security Council, a table could be constructed as well, but this table would be huge: there are $2^{15} = 32,768$ possible coalitions (cf. Problem 9.1). Therefore, it is convenient to describe this game as follows. Let the permanent members be the players $1, \ldots, 5$ and let the other members be the players $6, \ldots, 15$. Denote by $N = \{1, 2, \ldots, 15\}$ the grand coalition of all players and by $v(S)$ the worth of a coalition $S \subseteq N$. Then

$$v(S) := \begin{cases} 1 \text{ if } \{1, \ldots, 5\} \subseteq S \text{ and } |S| \geq 9 \\ 0 \text{ otherwise} \end{cases}$$

where $|S|$ denotes the number of players in S. In this case the number 1 indicates that the coalition is 'winning' and the number 0 that the coalition is 'losing'.

In analyzing games like this the resulting numbers—e.g., nonnegative numbers $x_1, \ldots, x_{15}$ summing to 1—are usually interpreted as power indices, expressing the power of a player in some way or another.

We summarize the concepts introduced informally in the preceding examples, formally in the following definition.

Definition 9.1 A *cooperative game with transferable utility* or *TU-game* is a pair (N, v), where $N = \{1, \ldots, n\}$ with $n \in \mathbb{N}$ is the set of *players*, and v is a function assigning to each *coalition* S, i.e., to each subset $S \subseteq N$ a real number $v(S)$, such that $v(\emptyset) = 0$. The function v is called the *characteristic function* and $v(S)$ is called the *worth* of S. The coalition N is called the *grand coalition*. A *payoff distribution* or *payoff vector* for coalition S is a vector of real numbers $(x_i)_{i \in S}$. $\qquad\square$

When analyzing a TU-game there are two important questions to answer: which coalitions are formed; and how are the worths of these coalitions distributed among their members? In this chapter we assume that the grand coalition is formed and we concentrate on the second question. This is less restrictive than it may seem at first sight, since coalition formation depends, naturally, on how the proceeds of a coalition are going to be distributed among its members. Thus, also if smaller coalitions are formed the distribution question has to be considered for these coalitions.

9.2 The Core

Consider the three cities game in Table 9.1, suppose that the grand coalition gets together, and suppose that there is a proposal $x_1 = 40$, $x_2 = 40$, and $x_3 = 140$ for distribution of the savings $v(N) = 220$ on the bargaining table. One can imagine, for instance, that player 3 made such a proposal. In that case, players 1 and 2 could protest successfully, since they can save $v(\{1, 2\}) = 90 > 80 = x_1 + x_2$ without player 3. We express this by saying that $\mathbf{x} = (x_1, x_2, x_3)$ is not in the "core" of this game. More generally, the *core* of the three cities game is the set of payoff distributions for $N = \{1, 2, 3\}$ such that the sum of the payoffs is equal to $v(N) = 220$ and each nonempty coalition S obtains at least its own worth. Thus, it is the set

$$C = \{(x_1, x_2, x_3) \in \mathbb{R}^3 \mid x_1, x_2, x_3 \geq 0,$$
$$x_1 + x_2 \geq 90, \ x_1 + x_3 \geq 100, \ x_2 + x_3 \geq 120,$$
$$x_1 + x_2 + x_3 = 220\}.$$

To obtain a better idea of what this set looks like, we can make a diagram. Although C is a subset of $\mathbb{R}^3$, the constraint $x_1 + x_2 + x_3 = 220$ makes that the set C is contained in a two-dimensional subset of $\mathbb{R}^3$, i.c., the plane through the points $(220, 0, 0)$, $(0, 220, 0)$, and $(0, 0, 220)$. The triangle formed by these three points is represented in Fig. 9.1. By the constraints $x_i \geq 0$ for every $i = 1, 2, 3$, the set C must be a

Fig. 9.1 The set C (*shaded*) is the core of the three cities game. Line segment a corresponds to the constraint $x_1 + x_2 \geq 90$; it consists of the payoff vectors in the *triangle* with $x_1 + x_2 = 90$ or, equivalently, $x_3 = 130$. Line segment b corresponds to the constraint $x_1 + x_3 \geq 100$, and line segment c corresponds to the constraint $x_2 + x_3 \geq 120$

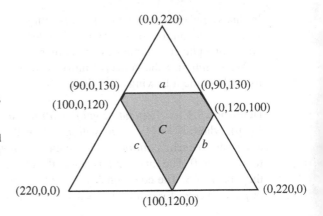

subset of this triangle. The set C is further restricted by the three constraints for the two-person coalitions: it is the shaded area in Fig. 9.1.

Hence, the core of the three cities game is the polygon and its inside with vertices $(100, 120, 0)$, $(0, 120, 100)$, $(0, 90, 130)$, $(90, 0, 130)$, and $(100, 0, 120)$.

We now give the formal definition of the core and of some other related concepts. We write $x(S) := \sum_{i \in S} x_i$ for a payoff distribution $\mathbf{x} = (x_1, \ldots, x_n) \in \mathbb{R}^n$ and a nonempty coalition $S \subseteq N = \{1, \ldots, n\}$. Hence, $x(S)$ is what the members of the coalition S obtain together if the payoff vector is $\mathbf{x}$.

Definition 9.2 For a TU-game (N, v), a payoff distribution $\mathbf{x} = (x_1, \ldots, x_n) \in \mathbb{R}^n$ is

- *efficient* if $x(N) = v(N)$,
- *individually rational* if $x_i \geq v(\{i\})$ for all $i \in N$,
- *coalitionally rational* if $x(S) \geq v(S)$ for all nonempty coalitions S.

The *core* of (N, v) is the set

$$C(N, v) = \{\mathbf{x} \in \mathbb{R}^n \mid x(N) = v(N) \text{ and } x(S) \geq v(S) \text{ for all } \emptyset \neq S \subseteq N\} \, .$$

Thus, the core of (N, v) is the set of all efficient and coalitionally rational payoff distributions.   □

The core of a game can be a large set, as in the three cities game; a small set, as in the glove game (see Problem 9.2); or it can be empty (see again Problem 9.2). In general, core elements can be computed by linear programming. For games with two or three players the core can be computed graphically, as we did for the three cities game. Sometimes, the core can be computed by using the special structure of the specific game under consideration.

We conclude this section with analyzing the core of the general glove game (cf. Problem 1.9).

Example 9.3 In the general glove game, there are $\ell > 0$ players who own a left-hand glove and $r > 0$ players who own a right-hand glove. In total there are n players, so $n = \ell + r$. The coalition N of all players is the grand coalition. The worth of a coalition S is equal to the number of pairs of gloves that the members of the coalition can make. Hence, it is equal to the *minimum* of two numbers: the number of left-hand glove owners in S and the number of right-hand glove owners in S. If we denote the set of all left-hand glove owners by L and the set of all right-hand glove owners by R, then we can also write this as

$$v(S) = \min\{|S \cap L|, |S \cap R|\} .$$

Here, $S \cap L$ is the intersection of S and L, i.e., the set of left-hand glove owners in S, and as before $|\cdot|$ denotes the number of elements in a set, in this case the number of players. What is the core of this game?

To answer this question we cannot just make a diagram as before. Even if we take specific numbers for ℓ and r, as soon as $\ell + r$ is larger than three we cannot make a picture. (This is not quite true: we could still make a picture for four players, but that would be a three-dimensional picture, which is not easy to draw.) Therefore, we have to argue in a different way.

Let us assume, first, that $\ell > r$. Then $v(N) = r$: we can use the above formula or simply observe that the grand coalition can make r pairs. Since each single player can make zero pairs, we also have $v(\{i\}) = 0$ for each player $i \in N$. Hence, for a payoff vector $\mathbf{x} = (x_1, \ldots, x_n)$ to be in the core, we already need $x_1 + \ldots + x_n = r$ and $x_i \geq 0$ for each $i \in N$. Now take a player $j \in L$, that is, j owns a left-hand glove. Consider the coalition S that consists of all right-hand glove owners and of at least r left-hand glove owners but not player j: this is possible since $\ell > r$. Then we still have $v(S) = r$, since S can still make r pairs, but this means that the members of S together should obtain at least r. In turn, this implies that they obtain exactly r, and thus player j obtains zero: $x_j = 0$. Since this argument holds for every arbitrary left-hand glove owner, every such player obtains zero. So far we have derived: all right-hand glove owners together obtain r, each one of them obtains at least zero, and every left-hand glove owner obtains zero.

Now let i be a right-hand glove owner and j a left-hand glove owner. Then $v(\{i,j\}) = 1$, hence $x_i + x_j \geq 1$. Since $x_j = 0$, as already established, we have $x_i \geq 1$. But this holds for every right-hand glove owner. Since there are r of them and together they obtain r, we must have that $x_i = 1$ for every right-hand glove owner. So we have found that $\mathbf{x}$ is unique: it assigns 1 to right-hand glove owners and 0 to left-hand glove owners. Thus, we have found that *if* $\mathbf{x}$ is in the core, *then* it can only be this specific payoff vector.

Conversely, this payoff vector is indeed in the core. All players together receive r, which is indeed equal to $v(N)$. An arbitrary coalition S receives an amount which is equal to the number of right-hand glove owners in S: this is indeed at least equal to the worth of S, i.e., the number of glove pairs that S can make.

Altogether, in one formula, we have for the core of the general glove game with $\ell > r$:

$$C(v) = \{\mathbf{x}\} \text{ with } \mathbf{x} \in \mathbb{R}^n, x_i = 1 \text{ for } i \in R \text{ and } x_i = 0 \text{ for } i \in L.$$

Similarly we find for $r > \ell$:

$$C(v) = \{\mathbf{x}\} \text{ with } \mathbf{x} \in \mathbb{R}^n, x_i = 1 \text{ for } i \in L \text{ and } x_i = 0 \text{ for } i \in R.$$

The case $\ell = r$ is left as Problem 9.4.   □

9.3   The Shapley Value

The Shapley value is a solution concept for TU-games that is quite different from the core. Whereas the core is a (possibly empty) *set*, the Shapley value assigns a *unique payoff distribution* for the grand coalition to every TU-game. The Shapley value is not so much based on strategic considerations but, rather, assigns to each player his "average marginal contribution" in the game. For three-player games we already explained the definition of the Shapley value in Chap. 1. Here, we repeat and extend this definition.

Consider again the three cities game of Table 9.1. Imagine a setting where the players enter a bargaining room one by one, and upon entering each player demands and obtains what he contributes to the worth of the coalition present in the room. Suppose that player 1 enters first, player 2 enters next, and player 3 enters last. Player 1 enters an empty room and can take his "marginal contribution" $v(\{1\}) - v(\emptyset) = 0 - 0 = 0$. When player 2 enters, player 1 is already present, and player 2 obtains his marginal contribution $v(\{1,2\}) - v(\{1\}) = 90 - 0 = 90$. When, finally, player 3 enters, then the coalition $\{1,2\}$ is already present. So player 3 obtains his marginal contribution $v(\{1,2,3\}) - v(\{1,2\}) = 220 - 90 = 130$. Hence, this procedure results in the payoff distribution $(0, 90, 130)$, which is called a *marginal vector*. Of course, this payoff distribution does not seem fair since it depends on the order in which the players enter the room, and this order is arbitrary: there are five other possible orders. The Shapley value takes the marginal vectors of all six orders into consideration, and assigns to a TU-game their average. See Table 9.4.

For an arbitrary TU-game (N, v) with player set $N = \{1, \ldots, n\}$ the Shapley value can be computed in the same way. There are $n \cdot (n - 1) \cdot \ldots \cdot 2 \cdot 1 = n!$ possible orders of the players. First compute the marginal vectors corresponding to these $n!$ different orders, and then take the average—that is, sum all marginal vectors and divide the result by $n!$. If the number of players is large, then this is a huge task. In the UN security council voting game of Sect. 9.1, for instance, this would mean

Table 9.4 Computation of
the Shapley value for the
three cities game. The
Shapley value is obtained by
dividing the totals of the
marginal contributions by 6

Order of entry	1	2	3
1, 2, 3	0	90	130
1, 3, 2	0	120	100
2, 1, 3	90	0	130
2, 3, 1	100	0	120
3, 1, 2	100	120	0
3, 2, 1	100	120	0
Total	390	450	480
Shapley value	65	75	80

computing $15! > 13 \times 10^{11}$ marginal vectors. Fortunately, there is a more clever
way to compute the total marginal contribution of a player.

For instance, let (N, v) be a TU-game with ten players. Consider player 7 and the
coalition $\{3, 5, 9\}$. The marginal contribution $v(\{3, 5, 9, 7\}) - v(\{3, 5, 9\})$ accruing
to player 7 occurs in more than one marginal vector. In how many marginal vectors
does it occur? To compute this, note that first players 3, 5, and 9 must enter, and this
can happen in 3! different orders. Then player 7 enters. Finally, the other six players
enter, and this can happen in 6! different orders. Therefore, the total number of
marginal vectors in which player 7 obtains the marginal contribution $v(\{3, 5, 9, 7\}) -$
$v(\{3, 5, 9\})$ is equal to $3! \times 6!$. By counting in this way the number of computations
is greatly reduced.

We now repeat this argument for an arbitrary TU-game (N, v), an arbitrary player
$i \in N$, and an arbitrary coalition S that does not contain player i. By the same
argument as in the preceding paragraph, the total number of marginal vectors in
which player i receives the marginal contribution $v(S \cup \{i\}) - v(S)$ is equal to the
number of different orders in which the players of S can enter first, $|S|!$, multiplied
by the number of different orders in which the players not in $S \cup \{i\}$ can enter after
player i, which is $(n - |S| - 1)!$. Hence, the total contribution obtained by player
i by entering after the coalition S is equal to $|S|!(n - |S| - 1)![v(S \cup \{i\}) - v(S)]$.
The Shapley value for player i is then obtained by summing over all coalitions S *not
containing* player i, and dividing by $n!$. In fact, we use this alternative computation
as the definition of the Shapley value.

Definition 9.4 The *Shapley value* of a TU-game (N, v) is denoted by $\Phi(N, v)$. Its
i-th coordinate, i.e., the Shapley value payoff to player $i \in N$, is given by

$$\Phi_i(N, v) = \sum_{S \subseteq N: \, i \notin S} \frac{|S|!(n - |S| - 1)!}{n!} [v(S \cup \{i\}) - v(S)] .$$

$\square$

Especially for larger TU-games it is easier to work with the formula in Definition 9.4 than to use the definition based on marginal vectors. For some purposes, however, it is easier to use the latter definition (Problem 9.7).

The Shapley value of the three cities game is an element of the core of that game (check this). In general, however, this does not have to be the case even if the core is nonempty (Problem 9.8).

Example 9.5 Consider the general glove game of Example 9.3. We assume $\ell = 4$ and $r = 2$. The worth $v(N)$ of the grand coalition is equal to 2. First note that, for reasons of symmetry, in the Shapley value all left-hand glove owners receive the same payoff and also all right-hand glove owners receive the same payoff. (This is intuitive, but can also be made more precise: see Problem 9.17 or Chap. 16.) Since the total payoff is 2, this means that it is sufficient to compute the Shapley value of either one left-hand glove owner or one right-hand glove owner: if, say, the payoff of a left-hand glove owner in the Shapley value is α, then every left-hand glove owner receives α and every right-hand glove owner receives $(2-4\alpha)/2$. Let us compute α.

Suppose i is a left-hand glove owner. Then i makes a contribution of 1 to any coalition with strictly less left-hand players than right-hand players. To all other coalitions i's contribution is zero, so we do not have to take those into consideration. Take a coalition S with k right-hand glove owners and j left-hand glove owners, such that player i is not in S and such that $j < k$, where k is equal to 1 or 2. Then indeed $v(S \cup \{i\}) - v(S) = 1$, since $v(S) = j$ and $v(S \cup \{i\}) = j + 1$. There are $\binom{2}{k} \cdot \binom{3}{j}$ coalitions with k right-hand glove owners and j left-hand glove owners.[1] Since $|S| = k + j$, the weight of S as in the formula in Definition 9.4 is $(k+j)!(5-k-j)!$. Hence, summing over all possible values of k and j we obtain

$$\Phi_i(N, v) = \alpha = \sum_{k=1}^{2} \sum_{j=0}^{k-1} \frac{(k+j)!(5-k-j)!}{6!} \binom{2}{k} \cdot \binom{3}{j} .$$

This is readily computed and yields $\alpha = 2/15$. Hence, every left-hand glove owner receives $2/15$ in the Shapley value, which implies that for the two right-hand glove owners $2 - 4 \cdot 2/15 = 22/15$ is left. This implies

$$\Phi_i(N, v) = \begin{cases} 2/15 & \text{if } i \text{ is a left-hand glove owner} \\ 11/15 & \text{if } i \text{ is a right-hand glove owner.} \end{cases}$$

Observe that the Shapley value of this game is *not* in the core (see Example 9.3).

---

[1] In general, $\binom{p}{q} = \frac{p!}{q!(p-q)!}$ is the number of ways in which we can choose a set of q elements from a set of p elements.

The same arguments can be used to compute the Shapley value for any number of left-hand and right-hand glove owners. If $\ell = r$, then for reasons of symmetry, all players receive the same payoff in the Shapley value, namely $1/2$. □

The definition of the Shapley value as assigning to each player in a game his average marginal contribution, can be regarded as a justification of this solution concept by itself. In the literature there are, moreover, a number of axiomatic characterizations of the Shapley value. In an axiomatic characterization one proceeds as follows. Consider an arbitrary map, which (like the Shapley value) assigns to each game with player set N a payoff vector. Next, define "reasonable" properties or *axioms* for this map. Such axioms limit the possible maps (i.e., solution concepts), and if the axioms are strong enough, they admit only one solution concept. This so-called *axiomatic approach* is common in cooperative game theory. Problem 9.17 preludes to this. For details, see Chap. 17.

9.4 The Nucleolus

The last concept we discuss in this introduction to TU-games is the *nucleolus*. Like the Shapley value it assigns a unique payoff distribution to a game. An advantage compared to the Shapley value is that the nucleolus assigns a payoff distribution in the core of a game, provided the core is nonempty. The nucleolus is defined for games which possess the following property. A TU-game (N, v) is *essential* if $v(N) \geq \sum_{i \in N} v(\{i\})$. For an essential game there are payoff distributions for the grand coalition that are both efficient and individually rational. Such payoff distributions are called *imputations*. The set

$$I(N, v) = \{\mathbf{x} \in \mathbb{R}^N \mid x(N) = v(N), \; x_i \geq v(\{i\}) \text{ for all } i \in N\}$$

is called the *imputation set* of the TU-game (N, v). Hence, a game (N, v) is essential if and only if $I(N, v) \neq \emptyset$ (check!).

Let (N, v) be an essential TU-game, let $\mathbf{x} \in I(N, v)$, and let S be a nonempty coalition unequal to N. The *excess of S at* $\mathbf{x}$, denoted by $e(S, \mathbf{x})$, is defined by

$$e(S, \mathbf{x}) = v(S) - x(S) .$$

The excess of a coalition S at a payoff distribution $\mathbf{x}$ is, thus, the difference between what S can acquire on its own and what it receives in total from $\mathbf{x}$. The excess can be seen as a measure of the dissatisfaction of the coalition S with the imputation $\mathbf{x}$: the smaller the total payoff of S at $\mathbf{x}$, the larger $e(S, \mathbf{x})$. In particular, if this excess is positive then S obtains less than its own worth.

In words, the *nucleolus* of an essential TU-game (N, v) is defined as follows.

1. First, find all imputations for which the maximal excess among all coalitions (not equal to N or the empty set) is as small as possible. If there is a unique such imputation, then that is the nucleolus.
2. If not, then determine those coalitions for which the maximal excess found in (1) cannot be decreased any further. Then continue with the remaining coalitions and among the imputations found in (1) find those imputations for which the maximal excess among these remaining coalitions is as small as possible. If there is a unique such imputation, then that is the nucleolus.
3. If not, then determine those coalitions for which the maximal excess found in (2) cannot be decreased any further. Then continue with the remaining coalitions and among the imputations found in (2) find those imputations for which the maximal excess among these remaining coalitions is as small as possible. If there is a unique such imputation, then that is the nucleolus.
4. Etc.

Thus, the idea behind the nucleolus is to make the largest dissatisfaction as small as possible. If there is more than one possibility to do this, then we also make the second largest dissatisfaction as small as possible, and so on, until a unique distribution is reached.

A formal definition of the nucleolus can be found in Chap. 19. Here we content ourselves with the given verbal description and some examples.

Our first illustration of this procedure is its application to the three cities game, reproduced in Table 9.5. The third line of the table gives the excesses at the imputation $(70, 70, 80)$. The choice of this particular imputation is arbitrary: we use it as a starting point to find the nucleolus. The largest excess at this imputation is -30, namely for the coalition $\{2, 3\}$. Clearly, we can decrease this excess by giving players 2 and 3 more at the expense of player 1. Doing so implies that the excesses of $\{1, 2\}$ or of $\{1, 3\}$ or of both will increase. These excesses are equal to -50. We can increase the payoffs of players 2 and 3 by $6\frac{2}{3}$ and decrease the payoff of player 1 by $2 \cdot 6\frac{2}{3} = 13\frac{1}{3}$, so that these three excesses become equal. Thus we obtain the imputation $(56\frac{2}{3}, 76\frac{2}{3}, 86\frac{2}{3})$, at which the excesses of the three two-player coalitions are all equal to $-43\frac{1}{3}$, and these are also the maximal excesses. Now first observe that at this imputation the maximal excess is as small as possible. This follows since

Table 9.5 Heuristic determination of the nucleolus of the three cities game

S	$\{1\}$	$\{2\}$	$\{3\}$	$\{1,2\}$	$\{1,3\}$	$\{2,3\}$	$\{1,2,3\}$
$v(S)$	0	0	0	90	100	120	220
$e(S, (70,70,80))$	-70	-70	-80	-50	-50	-30	
$e(S, (56\frac{2}{3},76\frac{2}{3},86\frac{2}{3}))$	$-56\frac{2}{3}$	$-76\frac{2}{3}$	$-86\frac{2}{3}$	$-43\frac{1}{3}$	$-43\frac{1}{3}$	$-43\frac{1}{3}$	

the sum of the excesses of the three two-player coalitions at *any* imputation must be the same, namely equal to -130, as follows from

$$e(\{1,2\}, \mathbf{x}) + e(\{1,3\}, \mathbf{x}) + e(\{2,3\}, \mathbf{x}) = v(\{1,2\}) + v(\{1,3\}) + v(\{2,3\})$$
$$-2(x_1 + x_2 + x_3)$$
$$= 310 - 2 \cdot 220$$
$$= -130 .$$

This implies that none of these excesses can be decreased without increasing at least one other excess. Second, the imputation at which these three excesses are equal is unique, since the system

$$90 - x_1 - x_2 = 100 - x_1 - x_3$$
$$100 - x_1 - x_3 = 120 - x_2 - x_3$$
$$x_1 + x_2 + x_3 = 220$$
$$x_1, x_2, x_3 \geq 0$$

has a unique solution—namely, indeed, $(56\frac{2}{3}, 76\frac{2}{3}, 86\frac{2}{3})$. So this imputation must be the nucleolus of the three cities game.

This example suggests that, at least for a three player TU-game, it is easy to find the nucleolus, namely simply by equating the excesses of the three two-player coalitions. Unfortunately, this is erroneous. It works if the worths of the two-player coalitions are large relative to the worths of the single player coalitions, but otherwise it may fail to result in the nucleolus. Consider the three-player TU-game in Table 9.6, which is identical to the three cities game except that now $v(\{1\}) = 20$. The third line of the table shows the excesses at $(56\frac{2}{3}, 76\frac{2}{3}, 86\frac{2}{3})$ in this TU-game. (This vector is still an imputation.) The maximal excess is now $-36\frac{2}{3}$ for the single-player coalition $\{1\}$. Clearly, $(56\frac{2}{3}, 76\frac{2}{3}, 86\frac{2}{3})$ is no longer the nucleolus: the excess of $\{1\}$ can be decreased by giving player 1 more at the expense of players 2 and/or 3. Suppose we equalize the excesses of $\{1\}$ and $\{2,3\}$ by solving the equation $20 - x_1 = 120 - x_2 - x_3$. Together with $x_1 + x_2 + x_3 = 220$ this yields $x_1 = 60$ and $x_2 + x_3 = 160$. Trying the imputation $(60, 75, 85)$, obtained by taking away the same amount $1\frac{2}{3}$ from players 2 and 3, yields the excesses in the fourth line of

Table 9.6 Heuristic determination of the nucleolus in the three cities game with the worth of coalition $\{1\}$ changed to 20

S	$\{1\}$	$\{2\}$	$\{3\}$	$\{1,2\}$	$\{1,3\}$	$\{2,3\}$	$\{1,2,3\}$
$v(S)$	20	0	0	90	100	120	220
$e(S, (56\frac{2}{3}, 76\frac{2}{3}, 86\frac{2}{3}))$	$-36\frac{2}{3}$	$-76\frac{2}{3}$	$-86\frac{2}{3}$	$-43\frac{1}{3}$	$-43\frac{1}{3}$	$-43\frac{1}{3}$	
$e(S, (60, 75, 85))$	-40	-75	-85	-45	-45	-40	

Table 9.6. We claim that $(60, 75, 85)$ is the nucleolus of this TU-game. The maximal excess is -40, reached by the coalitions $\{1\}$ and $\{2, 3\}$, and this cannot be decreased: decreasing the excess for one of those two coalitions implies increasing the excess for the other coalition. Hence, x_1 has to be equal to 60 in the nucleolus. The second maximal excess is -45, reached by the coalitions $\{1, 2\}$ and $\{1, 3\}$. Since x_1 has already been fixed at 60, a decrease in the excess for one of these two coalitions implies an increase of the excess for the other coalition. Hence, also x_2 and x_3 are fixed, at 75 and 85, respectively.

These two examples indicate that it may not be easy to compute the nucleolus. For three-player games the heuristic method above works well. In general, it can be computed by solving a series of linear programs. The following example illustrates this for the three-player games considered above.

Example 9.6 The maximal excess at the nucleolus of the three-cities game in Table 9.5 can be found by solving the following linear minimization problem.

Minimize α subject to

$$x_1 + x_2 + x_3 = 220$$

$$x_1, x_2, x_3 \geq 0$$

$$0 - x_1 \leq \alpha$$

$$0 - x_2 \leq \alpha$$

$$0 - x_3 \leq \alpha$$

$$90 - x_1 - x_2 \leq \alpha$$

$$100 - x_1 - x_3 \leq \alpha$$

$$120 - x_2 - x_3 \leq \alpha$$

Here, α is the maximal excess to be minimized. The first two constraints make sure that α is minimized over the set of imputations. The next three constraints are those for the excesses of the single-player coalitions, and the last three constraints are those for the excesses of the two-player coalitions: these six inequalities ensure that α will be the maximal excess. By using the efficiency constraint $x_1 + x_2 + x_3 = 220$ we can rewrite the problem into a minimization problem with three variables (α and for instance x_1, x_2) which could be solved graphically (by making a three-dimensional diagram); in general, however, we can use a computer program, based on for instance the simplex method, to solve the problem. In this case, from the preceding analysis we already know that the optimal solution of the problem is $\alpha = -43\frac{1}{3}$, attained at a unique point $(x_1, x_2, x_3) = (56\frac{2}{3}, 76\frac{2}{3}, 86\frac{2}{3})$.

For the modified three-cities game in Table 9.6 the linear program becomes:

Minimize α subject to

$$x_1 + x_2 + x_3 = 220$$

$$x_1, x_2, x_3 \geq 0$$

$$20 - x_1 \leq \alpha$$

$$0 - x_2 \leq \alpha$$

$$0 - x_3 \leq \alpha$$

$$90 - x_1 - x_2 \leq \alpha$$

$$100 - x_1 - x_3 \leq \alpha$$

$$120 - x_2 - x_3 \leq \alpha$$

From the preceding analysis the solution of this problem is $\alpha = -40$, and at all points at which this value is attained we have $x_1 = 60$ and, consequently, $x_2 + x_3 = 160$. To determine the second maximal excess we need to solve the following linear program:

Minimize α subject to

$$x_2 + x_3 = 160$$

$$x_2, x_3 \geq 0$$

$$-x_2 \leq \alpha$$

$$-x_3 \leq \alpha$$

$$30 - x_2 \leq \alpha$$

$$40 - x_3 \leq \alpha$$

The optimal solution to this problem is $\alpha = -45$, attained at $(x_2, x_3) = (75, 85)$— this follows from the preceding analysis, or from solving the problem graphically after reducing it to a problem of two variables using the constraint $x_2 + x_3 = 160$. Thus, the nucleolus of the modified three-cities problem is $(60, 75, 85)$, as found earlier.   □

Although the nucleolus is not easy to compute it is an attractive solution for the following reasons. It assigns a unique imputation to every essential game and if a game has a nonempty core, the nucleolus assigns a core element.

Proposition 9.7 *Let (N, v) be a game with nonempty core. Then the nucleolus of (N, v) is a payoff distribution in the core.*

Table 9.7 The game in Example 9.8

S	$\{1\}$	$\{2\}$	$\{3\}$	$\{1,2\}$	$\{1,3\}$	$\{2,3\}$	$\{1,2,3\}$
$v(S)$	0	0	0	10	0	20	20
$e(S,(0,10,10))$	0	-10	-10	0	-10	0	
$e(S,(0,15,5))$	0	-15	-5	-5	-5	0	

Proof Take any $\mathbf{x}$ in the core of (N,v). Then for every nonempty coalition $S \neq N$ we have $x(S) \geq v(S)$, hence $e(S,\mathbf{x}) = v(S) - x(S) \leq 0$. Suppose that $\mathbf{z}$ is the nucleolus of (N,v). Since the nucleolus minimizes the maximal excess over all imputations, and we already have an imputation, namely $\mathbf{x}$, at which all excesses and thus also the maximal excess are non-positive, we must have that the maximal excess at $\mathbf{z}$ is non-positive and, hence, all excesses at $\mathbf{z}$ are non-positive. This means that for every nonempty coalition $S \neq N$ we have $e(S,\mathbf{z}) = v(S) - z(S) \leq 0$, hence $z(S) \geq v(S)$. But this means that $\mathbf{z}$ is in the core of (N,v). ∎

In view of Proposition 9.7, if the core of a game is known then this may be very helpful in finding the nucleolus since we can restrict consideration to the core. We illustrate this by the next examples.

Example 9.8 Consider the three-person game in Table 9.7. This game has a nonempty core, and player 1 obtains 0 in every core distribution. Starting with the core distribution $(0,10,10)$, we find that the maximal excess is equal to 0, reached for the coalitions $\{1\}$, $\{1,2\}$, and $\{2,3\}$. Clearly, this excess cannot be decreased any further since a decrease of the excess for $\{1\}$ implies an increase for $\{2,3\}$, and conversely. Hence, if $\mathbf{z}$ is the nucleolus, then $z_1 = 0$ and $z_2 + z_3 = 20$. Note, however, that the excess for $\{1,2\}$ can be decreased by increasing the payoff for player 2 at the expense of player 3. For the payoff distribution $(0,15,5)$, we find that the second maximal excess is reached by $\{1,2\}$, $\{1,3\}$, and $\{3\}$. We cannot decrease the excess of $\{1,2\}$ without increasing it for $\{1,3\}$ and $\{3\}$, given that z_1 is already fixed at 0. Hence, we have obtained the nucleolus: $\mathbf{z} = (0,15,5)$. □

Example 9.9 Consider again the general glove game of Example 9.3. For $\ell > r$, the core of this game consists of the unique payoff distribution where the left-hand glove owners obtain 0 and the right-hand glove owners obtain 1. By Proposition 9.7, this is also the nucleolus of this game. □

We conclude with an example illustrating again the core, Shapley value and nucleolus of a game.

Example 9.10 Consider the following six-player cooperative game. The player set is $N = \{1,\ldots,6\}$. A coalition S has worth 1 in exactly two cases: either it contains player 1 and at least one other player, or it is the coalition $\{2,\ldots,6\}$. All other coalitions have worth 0. (Such a game is called an *apex game*, player 1 is the *apex player* and the other players are called *minor players*.)

First observe that the core of this game is empty. To see this, suppose $\mathbf{x} = (x_1,\ldots,x_6)$ is in the core. Then $x_1 + \ldots + x_5 \geq v(\{1,\ldots,5\}) = 1$. Since $x_6 \geq v(\{6\}) = 0$ and $x_1 + \ldots + x_6 = v(N) = 1$, we must have $x_1 + \ldots + x_5 = 1$ and $x_6 = 0$. Similarly, one derives $x_2 = \ldots = x_5 = 0$. Hence, $x_2 + \ldots + x_6 = 0$, but then $1 = v(\{2,\ldots,6\}) > x_2 + \ldots + x_6$, so that the core constraint for coalition $\{2,\ldots,6\}$ is violated. From this contradiction, we conclude that the core of this game is empty.

To find the nucleolus, for reasons of symmetry we may assume that it is of the form $(1 - 5\alpha, \alpha,\ldots,\alpha)$, where $0 \leq \alpha \leq 1/5$ in order to make it an imputation. The excess of a coalition of the form $\{1,i\}$ for every $i \in \{2,\ldots,6\}$ is equal to $1 - (1 - 5\alpha) - \alpha = 4\alpha$, and the excess of the coalition $\{2,\ldots,6\}$ is $1 - 5\alpha$. Clearly, these will be the maximal excesses. By increasing α the excesses of the $\{1,i\}$ coalitions will increase and the excess of the coalition $\{2,\ldots,6\}$ will decrease; by decreasing α the effects will be opposite. Therefore, we find the nucleolus by equating these excesses. Setting $4\alpha = 1 - 5\alpha$ yields $\alpha = 1/9$, hence the nucleolus is $(4/9, 1/9,\ldots, 1/9)$.

To find the Shapley value, we use Definition 9.4. (There are $6! = 720$ possible orderings of the six players, so listing all of them is an inefficient method to compute the Shapley value.) It is sufficient to compute the Shapley value for player 2 (for instance) since then we also know the Shapley value for the other players. Player 2 makes a nonzero contribution, equal to 1, in exactly two cases: either to $S = \{1\}$ or to $S = \{3,\ldots,6\}$. So we obtain

$$\Phi_2(N,v) = \frac{1!(6-1-1)!}{6!} \cdot 1 + \frac{4!(6-4-1)!}{6!} \cdot 1 = \frac{1}{15}.$$

Thus, the Shapley value of this game is

$$\Phi(N,v) = (1 - 5 \cdot \frac{1}{15}, \frac{1}{15},\ldots,\frac{1}{15}) = (\frac{2}{3}, \frac{1}{15},\ldots,\frac{1}{15}).$$

$\square$

9.5 Problems

9.1. *Number of Coalitions*
Show that a set of $n \in \mathbb{N}$ elements has 2^n different subsets.

9.2. *Computing the Core*

(a) Compute the core of the glove game of Table 9.2 by making a diagram.
(b) Compute the core of the dentist game of Table 9.3 by making a diagram.
(c) Compute the core of the UN security council voting game in Sect. 9.1.

9.3. *The Core of a Two-Person Game*
Consider the two-person game $(\{1,2\}), v)$ given by $v(\{1\}) = a$, $v(\{2\}) = b$, and $v(\{1,2\}) = c$, where $a, b, c \in \mathbb{R}$. Give a necessary and sufficient condition on a, b, and c for the core of $(\{1,2\}, v)$ to be nonempty. Make a diagram and compute the core.

9.4. *The Core of the General Glove Game*
Compute the core of the general glove game in Example 9.3 for the case where the numbers of left-hand glove owners and right-hand glove owners are equal. Is the Shapley value (cf. Example 9.5) in the core?

9.5. *A Condition for Nonemptiness of the Core of a Three-Person Game*
Let $(\{1,2,3\}, v)$ be a three-person game which has a nonempty core. Show that $2v(\{1,2,3\}) \geq v(\{1,2\}) + v(\{1,3\}) + v(\{2,3\})$. (Hint: Take a core element $\mathbf{x} = (x_1, x_2, x_3)$ and write down the core constraints.)

9.6. *Non-monotonicity of the Core*
Consider the following four-person game: $v(\{i\}) = 0$ for every $i = 1, \ldots, 4$, $v(\{1,2\}) = v(\{3,4\}) = 0$, $v(S) = 1$ for all other two-person coalitions and for all three-person coalitions, and $v(N) = 2$.

(a) Show that $C(N, v) = \{(\alpha, \alpha, 1 - \alpha, 1 - \alpha) \in \mathbb{R}^4 \mid 0 \leq \alpha \leq 1\}$.
(b) Consider the game (N, v') equal to (N, v) except for $v'(\{1,3,4\}) = 2$. Show that the core of (N, v') consists of a single element. What about the payoff to player 1 if core elements in (N, v) and (N, v') are compared? Conclude that the core is not "monotonic" (consider player 1).

9.7. *Efficiency of the Shapley Value*
Let (N, v) be an arbitrary TU-game. Show that the Shapley value $\Phi(N, v)$ is efficient. [Hint: take an order $i_1, i_2, \ldots, i_n$ of the players and show that the sum of the coordinates of the corresponding marginal vector is equal to $v(N)$; use this to conclude that $\Phi(N, v)$ is efficient.]

9.8. *Computing the Shapley Value*

(a) Compute the Shapley value of the glove game of Table 9.2. Is it an element of the core?
(b) Compute the Shapley value of the dentist game of Table 9.3. Is it an element of the core?
(c) Compute the Shapley value of the UN security council voting game in Sect. 9.1. (Hint: observe the—more or less—obvious fact that the Shapley value assigns the same payoff to all permanent members and also to all nonpermanent members. Use the formula in Definition 9.4.) Is it an element of the core?

9.9. *The Shapley Value and the Core*
For every real number a the three-player TU-game v_a is given by: $v_a(\{i\}) = 0$ for $i = 1, 2, 3$, $v_a(\{1,2\}) = 3$, $v_a(\{1,3\}) = 2$, $v_a(\{2,3\}) = 1$, $v_a(\{1,2,3\}) = a$.

(a) Determine the minimal value of a so that the TU-game v_a has a nonempty core.
(b) Calculate the Shapley value of v_a for $a = 6$.
(c) Determine the minimal value of a so that the Shapley value of v_a is a core distribution.

9.10. *Shapley Value in a Two-Player Game*
Let (N, v) be a two-player TU-game, i.e., $N = \{1, 2\}$. Compute the Shapley value (expressed in $v(\{1\})$, $v(\{2\})$, and $v(\{1,2\})$), and show that it is in the core of the game provided the core is nonempty. Make a diagram.

9.11. *Computing the Nucleolus*

(a) Compute the nucleolus of the glove game of Table 9.2.
(b) Compute the nucleolus of the dentist game of Table 9.3.
(c) Compute the nucleolus of the UN security council voting game in Sect. 9.1. (Hint: use Proposition 9.7.)
(d) Compute the nucleolus of the games (N, v) and (N, v') in Problem 9.6.

9.12. *Nucleolus of Two-Player Games*
Let (N, v) be an essential two-player TU-game. Compute the nucleolus.

9.13. *Computing the Core, the Shapley Value, and the Nucleolus*

(a) Compute the Shapley value and the nucleolus in the three-player TU-game given by: $v(\{i\}) = 1$ for $i = 1, 2, 3$, $v(\{1,2\}) = 2$, $v(\{1,3\}) = 3$, $v(\{2,3\}) = 4$, $v(\{1,2,3\}) = 6$. Is the Shapley value a core element in this game?
(b) Compute the core of this game. Make a picture.
(c) Suppose we increase $v(\{1\})$. What is the maximal value of $v(\{1\})$ such that the game still has a nonempty core?

9.14. *Voting (1)*
Suppose in Parliament there are four parties A, B, C, D with numbers of votes equal to $40, 30, 20, 10$, respectively. To pass any law a two-third majority is needed.

(a) Formulate this situation as a four-person cooperative game where winning coalitions have worth 1 and losing coalitions worth 0. Determine the Shapley value of this game.
(b) Determine also the core and the nucleolus of this game.

9.15. *Voting (2)*

A voting committee consists of five members: members 1 and 2 belong to party I, while members 3, 4, and 5 belong to party II. In order to pass a decision at least a weak majority of each party (that is, at least 50 % of the votes of each party) is required. A coalition that has a weak majority of both parties is called *winning*. We model this situation as a so-called *simple game*: winning coalitions obtain worth 1, all other coalitions worth 0.

(a) Show that there are six winning coalitions of minimal size: list all of them.
(b) Use your answer to (a) to give a concise description of the game (i.e., without listing all 32 coalitions).
(c) Compute the Shapley value of this game. According to the Shapley value, which players (members) are most powerful?
(d) Compute the nucleolus of this game. According to the nucleolus, which players (members) are most powerful?
(e) Compute the core of this game.

9.16. *Two Buyers and a Seller*

Players 1 and 2 are buyers, while player 3 is a seller. The seller owns an object that is worth nothing to him, but has value 1 for buyer 1 and value 2 for buyer 2. These are the prices that the buyers are willing to pay to the seller in order to get the object.

(a) Model this situation as a three-person TU-game, where the worth of each coalition is the maximal surplus it can create by a transaction between a buyer and the seller.
(b) Compute the core of this game.
(c) Compute the Shapley value of this game. Is it in the core?
(d) Compute the nucleolus of this game.

9.17. *Properties of the Shapley Value*

The properties of the Shapley value described in (a)–(c) below are called symmetry, additivity, and dummy property, respectively. It can be shown (Chap. 17) that the Shapley value is the unique solution concept that assigns exactly one payoff vector to each TU-game and has these three properties together with efficiency (cf. Problem 9.7). In other words, a solution concept has these four properties if, and only if, it is the Shapley value. In this exercise you are asked to show the "easy" part of this statement, namely the if-part. (Hint: in each case, decide which of the two formulas for the Shapley value—the one in Definition 9.4 or the formula based on marginal vectors—is most convenient to use.)

(a) Let (N, v) be a TU-game, and suppose players i and j are *symmetric* in this game, i.e., $v(S \cup \{i\}) = v(S \cup \{j\})$ for all coalitions S which do not contain i and j. Show that i and j obtain the same payoff from the Shapley value.

(b) Let (N, v) and (N, w) be two TU-games with the same player set N. Define the sum of these TU-games as the TU-game with player set N where the worth of each coalition S is given by $v(S) + w(S)$; denote this TU-game by $(N, v + w)$. Show that the Shapley value assigns to this sum TU-game the payoff vector which is the sum of the Shapley values of (N, v) and (N, w).

(c) Call player i a *dummy* in the TU-game (N, v) if $v(S \cup \{i\}) = v(S) + v(\{i\})$ for every coalition S to which player i does not belong. Show that the Shapley value assigns exactly the payoff $v(\{i\})$ to player i.

9.6   Notes

For a more advanced treatment of cooperative game theory, see Chaps. 16–20. Games with nonempty cores were characterized in Bondareva (1962) and Shapley (1967), see Chap. 16.

The Shapley value was introduced and axiomatically characterized in Shapley (1953)—see Chap. 17.

Imputations were first introduced by von Neumann and Morgenstern (1944/1947). They introduced cooperative games in order to cope with multi-person zero-sum games, which in general do not have a value.

The nucleolus was introduced in Schmeidler (1969). The nucleolus is similar in spirit to the main principle of distributive justice proposed in Rawls (1971), namely to maximize the lot of the worst off people in society.

The arguments used above to show that a particular imputation is indeed the nucleolus implicitly use a general property of the nucleolus called the Kohlberg criterion. See Chap. 19 for a detailed study of the nucleolus.

Problem 9.6 is taken from Moulin (1988).

References

Bondareva, O. N. (1962). Theory of the core in the n-person game. *Vestnik Leningradskii Universitet, 13*, 141–142 (in Russian).

Moulin, H. (1988). *Axioms of cooperative decision making*. Cambridge: Cambridge University Press.

Rawls, J. (1971). *A theory of justice*. Cambridge: Harvard University Press.

Schmeidler, D. (1969). The nucleolus of a characteristic function game. *SIAM Journal on Applied Mathematics, 17*, 1163–1170.

Shapley, L. S. (1953). A value for n-person games. In A.W. Tucker, & H. W. Kuhn (Eds.), *Contributions to the theory of games II* (pp. 307–317). Princeton: Princeton University Press.

Shapley, L. S. (1967). On balanced sets and cores. *Naval Research Logistics Quarterly, 14*, 453–460.

von Neumann, J., & Morgenstern, O. (1944/1947). *Theory of games and economic behavior*. Princeton: Princeton University Press.

Cooperative Game Models

10

The common features of a *cooperative game model*—such as the model of a game with transferable utility in Chap. 9—include: the abstraction from a detailed description of the strategic possibilities of a player; instead, a detailed description of what players and coalitions can attain in terms of outcomes or utilities; solution concepts based on strategic considerations and/or considerations of fairness, equity, efficiency, etc.; if possible, an axiomatic characterization of such solution concepts. For instance, one can argue that the core for TU-games is based on strategic considerations whereas the Shapley value is based on a combination of efficiency and symmetry or fairness with respect to contributions. The latter is made precise by an axiomatic characterization as in Problem 9.17.

In this chapter a few other cooperative game models are discussed: bargaining problems in Sect. 10.1, exchange economies in Sect. 10.2, matching problems in Sect. 10.3, and house exchange in Sect. 10.4.

10.1 Bargaining Problems

An example of a bargaining problem is the division problem in Sect. 1.3.5. A noncooperative, strategic approach to such a bargaining problem can be found in Sect. 6.7, see also Problems 6.16 and 6.17. In this section we treat the bargaining problem from a cooperative, axiomatic perspective. Surprisingly, there is a close relation between this approach and the strategic approach, as we will see. In Sect. 10.1.1 we discuss the Nash bargaining solution and in Sect. 10.1.2 its relation with the Rubinstein bargaining procedure of Sect. 6.7.

© Springer-Verlag Berlin Heidelberg 2015
H. Peters, *Game Theory*, Springer Texts in Business and Economics,
DOI 10.1007/978-3-662-46950-7_10

10.1.1 The Nash Bargaining Solution

We start with the definition of a two-person bargaining problem.[1]

Definition 10.1 A two-person *bargaining problem* is a pair (S, d), where

(i) $S \subseteq \mathbb{R}^2$ is a convex, closed and bounded set,[2]
(ii) $\mathbf{d} = (d_1, d_2) \in S$ such that there is some point $\mathbf{x} = (x_1, x_2) \in S$ with $x_1 > d_1$
    and $x_2 > d_2$.

S is the *feasible set* and $\mathbf{d}$ is the *disagreement point*.   ☐

The interpretation of a bargaining problem $(S, \mathbf{d})$ is as follows. The two players
bargain over the feasible outcomes in S. If they reach an agreement $\mathbf{x} = (x_1, x_2) \in S$,
then player 1 receives utility x_1 and player 2 receives utility x_2. If they do not reach
an agreement, then the game ends in the disagreement point $\mathbf{d}$, yielding utility d_1
to player 1 and d_2 to player 2. This is an interpretation, and the actual bargaining
procedure is not spelled out.

For the example in Sect. 1.3.5, the feasible set and the disagreement point are
given by

$$S = \{\mathbf{x} \in \mathbb{R}^2 \mid 0 \le x_1, x_2 \le 1,\ x_2 \le \sqrt{1 - x_1}\},\ d_1 = d_2 = 0 .$$

See also Fig. 1.7. In general, a bargaining problem may look as in Fig. 10.1. The set
of all such bargaining problems is denoted by $\mathcal{B}$.

Fig. 10.1 A two-person
bargaining problem

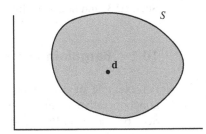

---

[1] We restrict attention here to two-person bargaining problems. For n-person bargaining problems
and, more generally, NTU-games, see the Notes section at the end of the chapter and Chap. 21.
[2] A subset of $\mathbb{R}^k$ is convex if with each pair of points in the set also the line segment connecting these
points is in the set. A set is closed if it contains its boundary or, equivalently, if for every sequence
of points in the set that converges to a point that limit point is also in the set. It is bounded if there
is a number $M > 0$ such that $|x_i| \le M$ for all points $\mathbf{x}$ in the set and all coordinates i.

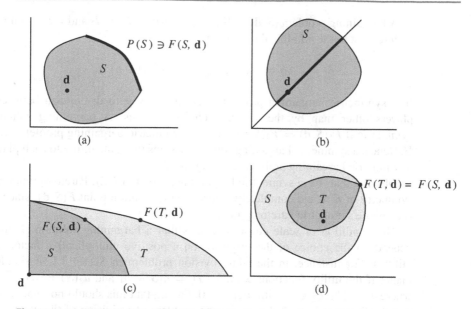

Fig. 10.2 Illustration of the four conditions ('axioms') determining the Nash bargaining solution—cf. Theorem 10.2. In (**a**) the Pareto optimal subset of S is the *thick black curve*. The bargaining problem $(S, \mathbf{d})$ in (**b**) is symmetric, and symmetry of F means that F should assign a point on the *thick black line* segment. In (**c**), which illustrates scale covariance, we took $\mathbf{d}$ to be the origin, and T results from S by multiplying all first coordinates by 2: then scale covariance implies that $F_1(T, \mathbf{d}) = 2F_1(S, \mathbf{d})$. The independence of irrelevant alternatives axiom is illustrated in (**d**)

We consider the following question: for any given bargaining problem $(S, \mathbf{d})$, what is a good compromise? We answer this question by looking for a map $F : \mathcal{B} \rightarrow \mathbb{R}^2$ which assigns a feasible point to every bargaining problem, i.e., satisfies $F(S, \mathbf{d}) \in S$ for every $(S, \mathbf{d}) \in \mathcal{B}$. Such a map is called a (two-person) *bargaining solution*. According to Nash (1950), a bargaining solution should satisfy four conditions, namely: Pareto optimality, symmetry, scale covariance, and independence of irrelevant alternatives. We discuss each of these conditions in detail. The conditions are illustrated in Fig. 10.2a–d.

For a bargaining problem $(S, \mathbf{d}) \in \mathcal{B}$, the Pareto optimal points of S are those where the utility of no player can be increased without decreasing the utility of the other player. Formally,

$$P(S) = \{\mathbf{x} \in S \mid \text{for all } \mathbf{y} \in S \text{ with } y_1 \geq x_1, y_2 \geq x_2, \text{ we have } \mathbf{y} = \mathbf{x}\}$$

is the *Pareto optimal* (sub)set of S. The bargaining solution F is *Pareto optimal* if $F(S, \mathbf{d}) \in P(S)$ for all $(S, \mathbf{d}) \in \mathcal{B}$. Hence, a Pareto optimal bargaining solution assigns a Pareto optimal point to each bargaining problem. See Fig. 10.2a for an illustration.

A bargaining problem $(S, \mathbf{d}) \in \mathcal{B}$ is *symmetric* if $d_1 = d_2$ and if S is symmetric with respect to the $45°$-line through $\mathbf{d}$, i.e., if

$$S = \{(x_2, x_1) \in \mathbb{R}^2 \mid (x_1, x_2) \in S\} .$$

In a symmetric bargaining problem there is no way to distinguish between the players other than by the arbitrary choice of axes. A bargaining solution is *symmetric* if $F_1(S, \mathbf{d}) = F_2(S, \mathbf{d})$ for each symmetric bargaining problem $(S, \mathbf{d}) \in \mathcal{B}$. Hence, a symmetric bargaining solution assigns the same utility to each player in a symmetric bargaining problem. See Fig. 10.2b.

Observe that, for a symmetric bargaining problem $(S, \mathbf{d})$, Pareto optimality and symmetry of F would completely determine the solution point $F(S, \mathbf{d})$, since there is a unique symmetric Pareto optimal point in S.

The condition of scale covariance says that a bargaining solution should not depend on the choice of the origin or on a positive multiplicative factor in the utilities. For instance, in the wine division problem in Sect. 1.3.5, it should not matter if the utility functions were $\bar{u}_1(\alpha) = a_1\alpha + b_1$ and $\bar{u}_2(\alpha) = a_2\sqrt{\alpha} + b_2$, where $a_1, a_2, b_1, b_2 \in \mathbb{R}$ with $a_1, a_2 > 0$. Saying that this should not matter means that the final outcome of the bargaining problem, the division of the wine, should not depend on this. One can think of $\bar{u}_1, \bar{u}_2$ expressing the same preferences about wine as u_1, u_2 in different units.[3] Formally, a bargaining solution F is *scale covariant* if for all $(S, \mathbf{d}) \in \mathcal{B}$ and all $a_1, a_2, b_1, b_2 \in \mathbb{R}$ with $a_1, a_2 > 0$ we have:

$$F\left(\{(a_1x_1 + b_1, a_2x_2 + b_2) \in \mathbb{R}^2 \mid (x_1, x_2) \in S\}, (a_1d_1 + b_1, a_2d_2 + b_2)\right)$$

$$= (a_1F_1(S, \mathbf{d}) + b_1, a_2F_2(S, \mathbf{d}) + b_2) .$$

For a simple case, this condition is illustrated in Fig. 10.2c.

The final condition is regarded as the most controversial one. Consider a bargaining problem $(S, \mathbf{d})$ with solution outcome $\mathbf{z} = F(S, \mathbf{d}) \in S$. In a sense, $\mathbf{z}$ can be regarded as the best compromise in S according to F. Now consider a smaller bargaining problem $(T, \mathbf{d})$ with $T \subseteq S$ and $\mathbf{z} \in T$. Since $\mathbf{z}$ was the best compromise in S, it is should certainly be regarded as the best compromise in T: $\mathbf{z}$ is available in T and every point of T is also available in S. Thus, we should conclude that $F(T, \mathbf{d}) = \mathbf{z} = F(S, \mathbf{d})$. As a less abstract example, suppose that in the wine division problem the wine is split fifty-fifty, with utilities $(1/2, \sqrt{1/2})$. Suppose now that no player wants to drink more than $3/4$ liter of wine: more wine does not increase utility. In that case, the new feasible set is

$$T = \{\mathbf{x} \in \mathbb{R}^2 \mid 0 \le x_1 \le 3/4, \ 0 \le x_2 \le \sqrt{3/4}, \ x_2 \le \sqrt{1 - x_1}\} .$$

---

[3]The usual assumption is that the utility functions are expected utility functions, which uniquely represent preferences up to choice of origin and scale.

According to the argument above, the wine should still be split fifty-fifty: $T \subseteq S$ and $(1/2, \sqrt{1/2}) \in T$. This may seem reasonable but it is not hard to change the example in such a way that the argument is, at the least, debatable. For instance, suppose that player 1 still wants to drink as much as possible but player 2 does not want to drink more than 1/2 L. In that case, the feasible set becomes

$$T' = \{\mathbf{x} \in \mathbb{R}^2 \mid 0 \le x_1 \le 1,\ 0 \le x_2 \le \sqrt{1/2},\ x_2 \le \sqrt{1 - x_1}\},$$

and we would still split the wine fifty-fifty. In this case player 2 would obtain his maximal feasible utility, and $(1/2, \sqrt{1/2})$ no longer seems a reasonable compromise since only player 1 makes a concession.

Formally, a bargaining solution F is *independent of irrelevant alternatives* if for all $(S, \mathbf{d}), (T, \mathbf{d}) \in \mathcal{B}$ with $T \subseteq S$ and $F(S, \mathbf{d}) \in T$, we have $F(T, \mathbf{d}) = F(S, \mathbf{d})$. See Fig. 10.2d for an illustration.

The theorem below says that these four conditions determine a unique bargaining solution F^{Nash}, defined as follows. For $(S, \mathbf{d}) \in \mathcal{B}$, $F^{\text{Nash}}(S, \mathbf{d})$ is equal to the unique point $\mathbf{z} \in S$ with $z_i \ge d_i$ for $i = 1, 2$ and such that

$$(z_1 - d_1)(z_2 - d_2) \ge (x_1 - d_1)(x_2 - d_2) \text{ for all } \mathbf{x} \in S \text{ with } x_i \ge d_i, i = 1, 2 .$$

The solution F^{Nash} is called the *Nash bargaining solution*. The result of Nash is as follows.

Theorem 10.2 *The Nash bargaining solution F^{Nash} is the unique bargaining solution which is Pareto optimal, symmetric, scale covariant, and independent of irrelevant alternatives.*

For a proof of this theorem and the fact that F^{Nash} is well defined—i.e., the point $\mathbf{z}$ above exists and is unique—see Chap. 21.

Example 10.3 In this example we illustrate the role of the conditions in Theorem 10.2. In fact, we show how the proof of this theorem (cf. Chap. 21) works in an example. Consider the bargaining problem $(S, \mathbf{d})$, where $S = \{(x_1, x_2) \in \mathbb{R}^2 \mid 0 \le x_1 \le 2,\ 0 \le x_2 \le 4 - x_1^2\}$ and $\mathbf{d} = (0, 1)$. The Nash bargaining solution outcome is obtained by solving the problem $\max_{0 \le x_1 \le 2} x_1(3 - x_1^2)$, which yields the point $(1, 3)$. Alternatively, consider the bargaining problem $(S', \mathbf{d}')$, obtained by subtracting 1 from the second coordinates of the points in S, including $\mathbf{d}$, yielding

$$S' = \{(x_1, x_2) \in \mathbb{R}^2 \mid 0 \le x_1 \le 2,\ -1 \le x_2 \le 3 - x_1^2\},\ \mathbf{d}' = (0, 0) .$$

Next, consider the bargaining problem $(S'', \mathbf{d}'')$, obtained from $(S', \mathbf{d}')$ by dividing all second coordinates by 2, yielding

$$S'' = \{(x_1, x_2) \in \mathbb{R}^2 \mid 0 \le x_1 \le 2,\ -\frac{1}{2} \le x_2 \le \frac{3}{2} - \frac{1}{2}x_1^2\},\ \mathbf{d}'' = (0, 0) .$$

Fig. 10.3 Illustrating
Example 10.3

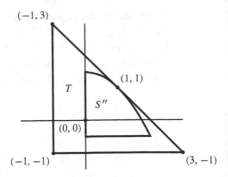

The Pareto optimal boundary of S'' is described by the function $f(x_1) = \frac{3}{2} - \frac{1}{2}x_1^2$ for $0 \le x_1 \le 2$. At $x_1 = 1$ the derivative of this function is equal to -1, so that the straight line through the point $(1, 1)$ with slope -1 is tangential to S'', i.e., the set S'' is below this line. Now consider the bargaining problem $(T, (0,0))$ with T the triangle and inside with vertices $(-1,-1)$, $(3,-1)$, and $(-1,3)$; see Fig. 10.3. The bargaining problem $(T, (0,0))$ is symmetric, so that by symmetry and Pareto optimality of the Nash bargaining solution, the outcome is the point $(1, 1)$. This point is also in S'', and moreover, S'' is a subset of T, so that by independence of irrelevant alternatives the point $(1, 1)$ is also the Nash bargaining solution outcome of $(S'', (0,0))$. By scale covariance, this implies that the Nash bargaining solution outcome of $(S', (0,0))$ is the point $(1, 2)$. Again by scale covariance, we obtain that the Nash bargaining solution outcome of $(S, (0, 1))$ is the point $(1, 3)$. We have reached this result by using only the properties of the Nash bargaining solution in Theorem 10.2, and not the formula. Observe that the result is in accordance with what we established by direct computation.   □

10.1.2 Relation with the Rubinstein Bargaining Procedure

In the Rubinstein bargaining procedure the players make alternating offers. See Sect. 6.7.2 for a detailed discussion of this noncooperative game, and Problem 6.17(d) for the application to the wine division problem of Sect. 1.3.5. Here, we use this example to illustrate the relation with the Nash bargaining solution.

The Nash bargaining solution assigns to this bargaining problem the point $\mathbf{z} = (2/3, \sqrt{1/3})$. This means that player 1 obtains 2/3 of the wine and player 2 obtains 1/3. According to the Rubinstein infinite horizon bargaining game with discount factor $0 < \delta < 1$ the players make proposals $\mathbf{x} = (x_1, x_2) \in P(S)$ and $\mathbf{y} = (y_1, y_2) \in P(S)$ such that

$$x_2 = \delta y_2, \ y_1 = \delta x_1 \, , \tag{10.1}$$

and these proposals are accepted in (subgame perfect) equilibrium. From (10.1) we derive

$$\sqrt{1-x_1} = x_2 = \delta y_2 = \delta\sqrt{1-y_1} = \delta\sqrt{1-\delta x_1}$$

hence $1 - x_1 = \delta^2(1 - \delta x_1)$, which implies

$$x_1 = \frac{1-\delta^2}{1-\delta^3} = \frac{1+\delta}{1+\delta+\delta^2}.$$

If we let δ increase to 1, we obtain

$$\lim_{\delta \to 1} \frac{1+\delta}{1+\delta+\delta^2} = \frac{2}{3},$$

which coincides with the payoff to player 1 according to the Nash bargaining solution. This is not a coincidence. It follows, in general, from (10.1) that

$$(x_1 - d_1)(x_2 - d_2) = x_1 x_2 = (y_1/\delta)(\delta y_2) = y_1 y_2 = (y_1 - d_1)(y_2 - d_2)$$

hence the Rubinstein proposals **x** and **y** have the same 'Nash product'. For our wine division example, the proposals **x** and **y** for $\delta = 0.5$ are represented in Fig. 10.4.

As δ increases to 1, this level curve shifts up until it passes through the point $\mathbf{z} = F^{\text{Nash}}(S, d)$, since this point maximizes the product $x_1 x_2$ on the set S: see again Fig. 10.4.

Fig. 10.4 The wine division problem. The disagreement point is the origin, and **z** is the Nash bargaining solution outcome. The points **x** and **y** are the proposals of players 1 and 2, respectively, in the subgame perfect equilibrium of the Rubinstein bargaining game for $\delta = 0.5$

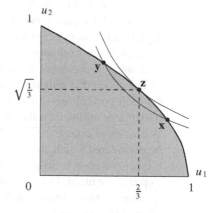

We conclude that the subgame perfect equilibrium payoffs of the infinite horizon Rubinstein bargaining game converge to the Nash bargaining solution outcome as the discount factor δ approaches 1.

Example 10.4 In the bargaining game $(S', (0,0))$ of Example 10.3 the Rubinstein proposals are determined by the equations in (10.1) together with $x_2 = 3 - x_1^2$ and $y_2 = 3 - y_1^2$, yielding

$$3 - x_1^2 = x_2 = \delta y_2 = \delta(3 - y_1^2) = \delta(3 - \delta^2 x_1^2)$$

hence

$$x_1 = \sqrt{\frac{3(1-\delta)}{1-\delta^3}} = \sqrt{\frac{3}{1+\delta+\delta^2}} \; .$$

For δ approaching 1 we obtain

$$\lim_{\delta \to 1} \sqrt{\frac{3}{1+\delta+\delta^2}} = 1$$

which is the Nash bargaining solution payoff to player 1.   □

10.2 Exchange Economies

In an exchange economy with n agents and k goods, each agent is endowed with a bundle of goods. Each agent has preferences over different bundles of goods, expressed by some utility function over these bundles. By exchanging goods among each other, it is in general possible to increase the utilities of all agents. One way to arrange this exchange is to introduce prices. For given prices the endowment of each agent represents the agent's income, which can be spent on buying a bundle of the goods that maximizes the agent's utility. If prices are such that the market for each good clears—total demand is equal to total endowment—while each agent maximizes utility, then the prices are in equilibrium: such an equilibrium is called Walrasian or competitive equilibrium. Alternatively, reallocations of the goods can be considered which are in the *core* of the exchange economy. A reallocation of the total endowment is in the core of the exchange economy if no coalition of agents can improve the utilities of its members by, instead, reallocating the total endowment of its own members among each other. It is well known that a competitive equilibrium allocation is an example of a core allocation.

This section is a first acquaintance with exchange economies. Attention is restricted to exchange economies with two agents and two goods. We work out an example of such an economy. Some variations are considered in Problem 10.4.

There are two agents, A and B, and two goods, 1 and 2. Agent A has an endowment $\mathbf{e}^A = (e_1^A, e_2^A) \in \mathbb{R}_+^2$ of the goods, and a utility function $u^A : \mathbb{R}_+^2 \to \mathbb{R}$, representing the preferences of A over bundles of goods.[4] Similarly, agent B has an endowment $\mathbf{e}^B = (e_1^B, e_2^B) \in \mathbb{R}_+^2$ of the goods, and a utility function $u^B : \mathbb{R}_+^2 \rightarrow \mathbb{R}$. (We use superscripts to denote the agents and subscripts to denote the goods.) This is a complete description of the exchange economy.

For our example we take $\mathbf{e}^A = (2,3)$, $\mathbf{e}^B = (4,1)$, $u^A(x_1, x_2) = x_1^2 x_2$ and $u^B(x_1, x_2) = x_1 x_2^2$. Hence, the total endowment in the economy is $\mathbf{e} = (6,4)$, and the purpose of the exchange is to reallocate this bundle of goods such that both agents are better off.

Let $\mathbf{p} = (p_1, p_2)$ be a vector of positive prices of the goods. Given these prices, both agents want to maximize their utilities. Agent A has an income of $p_1 e_1^A + p_2 e_2^A$, i.e., the monetary value of his endowment. Then agent A solves the maximization problem

$$\text{maximize } u^A(x_1, x_2)$$
$$\text{subject to } p_1 x_1 + p_2 x_2 = p_1 e_1^A + p_2 e_2^A, \ x_1, x_2 \geq 0 \,. \tag{10.2}$$

The income constraint is called the *budget equation*. The solution of this maximization problem is a bundle $\mathbf{x}^A(\mathbf{p}) = (x_1^A(\mathbf{p}), x_2^A(\mathbf{p}))$, called agent A's *demand function*. Maximization problem (10.2) is called the *consumer problem* (of agent A). Similarly, agent B's consumer problem is:

$$\text{maximize } u^B(x_1, x_2)$$
$$\text{subject to } p_1 x_1 + p_2 x_2 = p_1 e_1^B + p_2 e_2^B, \ x_1, x_2 \geq 0 \,. \tag{10.3}$$

For our example, (10.2) becomes

$$\text{maximize } x_1^2 x_2$$
$$\text{subject to } p_1 x_1 + p_2 x_2 = 2p_1 + 3p_2, \ x_1, x_2 \geq 0 \,,$$

which can be solved by using Lagrange's method or by substitution. By using the latter method the problem reduces to

$$\text{maximize } x_1^2 \left((2p_1 + 3p_2 - p_1 x_1)/p_2\right)$$

subject to $x_1 \geq 0$ and $2p_1 + 3p_2 - p_1 x_1 \geq 0$. Setting the derivative with respect to x_1 equal to 0 yields

$$2x_1 \left(\frac{2p_1 + 3p_2 - p_1 x_1}{p_2}\right) - x_1^2 \left(\frac{p_1}{p_2}\right) = 0 \,,$$

---

[4] $\mathbb{R}_+^2 := \{\mathbf{x} = (x_1, x_2) \in \mathbb{R}^2 \mid x_1, x_2 \geq 0\}$.

which after some simplifications yields the demand function $x_1 = x_1^A(\mathbf{p}) = (4p_1 + 6p_2)/3p_1$. By using the budget equation, $x_2^A(\mathbf{p}) = (2p_1 + 3p_2)/3p_2$. Similarly, solving (10.3) for our example yields $x_1^B(\mathbf{p}) = (4p_1 + p_2)/3p_1$ and $x_2^B(\mathbf{p}) = (8p_1 + 2p_2)/3p_2$ (check this!).

The prices $\mathbf{p}$ are *Walrasian equilibrium* prices if the markets for both goods clear. For the general model, this means that $x_1^A(\mathbf{p}) + x_1^B(\mathbf{p}) = e_1^A + e_1^B$ and $x_2^A(\mathbf{p}) + x_2^B(\mathbf{p}) = e_2^A + e_2^B$. For the example, this means

$$(4p_1 + 6p_2)/3p_1 + (4p_1 + p_2)/3p_1 = 6 \text{ and}$$

$$(2p_1 + 3p_2)/3p_2 + (8p_1 + 2p_2)/3p_2 = 4 \ .$$

Both equations result in the same condition, namely $10p_1 - 7p_2 = 0$. That there is only one equation left is no coincidence, since prices are only relative, as is easily seen from the budget equations. In fact, the prices represent the rate of exchange between the two goods, and are meaningful even if money does not exist in the economy. Thus, $\mathbf{p} = (7, 10)$ (or any positive multiple thereof) are the equilibrium prices in this exchange economy. The associated equilibrium demands are $\mathbf{x}^A(7, 10) = (88/21, 22/15)$ and $\mathbf{x}^B(7, 10) = (38/21, 38/15)$.

We now turn to the *core* of an exchange economy. A reallocation of the total endowments is in the core if no coalition can improve upon it. This definition is in the spirit as the corresponding definition for TU-games (Definition 9.2). In a two-person exchange economy, there are only three coalitions (excluding the empty coalition), namely $\{A\}$, $\{B\}$, and $\{A, B\}$. Consider an allocation $(\mathbf{x}^A, \mathbf{x}^B)$ with $x_1^A + x_1^B = e_1^A + e_1^B$ and $x_2^A + x_2^B = e_2^A + e_2^B$. To prevent that agents A or B can improve upon $(\mathbf{x}^A, \mathbf{x}^B)$ we need that

$$u^A(\mathbf{x}^A) \geq u^A(\mathbf{e}^A), \ u^B(\mathbf{x}^B) \geq u^B(\mathbf{e}^B) , \tag{10.4}$$

which are the *individual rationality* constraints. To prevent that the grand coalition $\{A, B\}$ can improve upon $(\mathbf{x}^A, \mathbf{x}^B)$ we need that

For *no* $(\mathbf{y}^A, \mathbf{y}^B)$ with $y_1^A + y_1^B = e_1^A + e_1^B$ and $y_2^A + y_2^B = e_2^A + e_2^B$ we have: $u^A(\mathbf{y}^A) \geq u^A(\mathbf{x}^A)$ and $u^B(\mathbf{y}^B) \geq u^B(\mathbf{x}^B)$ with at least one inequality strict.

$$\tag{10.5}$$

In words, (10.5) says that there should be no other reallocation of the total endowments such that no agent is worse off and at least one agent is strictly better off. This is the *efficiency* or *Pareto optimality* constraint.

We apply (10.4) and (10.5) to our example. The individual rationality constraints are

$$(x_1^A)^2 x_2^A \geq 12, \ x_1^B(x_2^B)^2 \geq 4 \ .$$

The Pareto optimal allocations, satisfying (10.5), can be computed as follows. Fix the utility level of one of the agents, say B, and maximize the utility of A subject to the utility level of B being fixed. By varying the fixed utility level of B we find all Pareto optimal allocations. In the example, we solve the following maximization problem for $c \in \mathbb{R}$:

maximize $(x_1^A)^2 x_2^A$
subject to $x_1^A + x_1^B = 6,\ x_2^A + x_2^B = 4,\ x_1^B(x_2^B)^2 = c,\ x_1^A, x_2^A, x_1^B, x_2^B \geq 0$.

By substitution this problem reduces to

maximize $(x_1^A)^2 x_2^A$
subject to $(6 - x_1^A)(4 - x_2^A)^2 = c,\ x_1^A, x_2^A \geq 0$.

The associated Lagrange function is $(x_1^A)^2 x_2^A - \lambda[(6 - x_1^A)(4 - x_2^A)^2 - c]$ and the first-order conditions are

$$2 x_1^A x_2^A + \lambda(4 - x_2^A)^2 = 0,\quad (x_1^A)^2 + 2\lambda(6 - x_1^A)(4 - x_2^A) = 0 .$$

Extracting λ from both equations and simplifying yields

$$x_2^A = \frac{4 x_1^A}{24 - 3 x_1^A} .$$

Thus, for any value of x_1^A between 0 and 6 this equation returns the corresponding value of x_2^A, resulting in a Pareto optimal allocation with $x_1^B = 6 - x_1^A$ and $x_2^B = 4 - x_2^A$.

It is straightforward to check by substitution that the Walrasian equilibrium allocation $\mathbf{x}^A(7, 10) = (88/21, 22/15)$ and $\mathbf{x}^B(7, 10) = (38/21, 38/15)$ found above, is Pareto optimal. This is no coincidence: the *First Welfare Theorem* states that in an exchange economy such as the one under consideration, a Walrasian equilibrium allocation is Pareto optimal.

Combining the individual rationality constraint for agent A with the Pareto optimality constraint yields $4(x_1^A)^3 / (24 - 3 x_1^A) \geq 12$, which holds for x_1^A larger than approximately 3.45. For agent B, similarly, the individual rationality and Pareto optimality constraints imply

$$(6 - x_1^A)\left(\frac{96 - 16 x_1^A}{24 - 3 x_1^A}\right)^2 \geq 4 ,$$

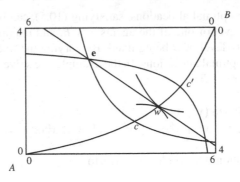

Fig. 10.5 The contract curve is the curve through c and c'. The point c is the point of intersection of the contract curve and the indifference curve of agent A through the endowment point **e**. The point c' is the point of intersection of the contract curve and the indifference curve of agent B through the endowment point **e**. The core consists of the allocations on the contract curve between c and c'. The *straight line* ('budget line') through **e** is the graph of the budget equation for A at the equilibrium prices, i.e., $7x_1 + 10x_2 = 44$, and its point of intersection with the contract curve, w, is the Walrasian equilibrium allocation. At this point the indifference curves of the two agents are both tangential to the budget line

which holds for x_1^A smaller than approximately 4.88. Hence, the core of the exchange economy in the example is, approximately, the set

$$\{(x_1^A, x_2^A, x_1^B, x_2^B) \in \mathbb{R}^4 \mid 3.45 \le x_1^A \le 4.88,$$
$$x_2^A = \frac{4x_1^A}{24 - 3x_1^A}, \ x_1^B = 6 - x_1^A, \ x_2^B = 4 - x_2^A\} \ .$$

Clearly, the Walrasian equilibrium allocation is in the core, since $3.45 \le 88/21 \le 4.88$, and also this holds more generally. Thus, decentralization of the reallocation process through prices leads to an allocation that is in the core.

For an exchange economy with two agents and two goods a very useful pictorial device is the *Edgeworth box*, see Fig. 10.5. The Edgeworth box consists of all possible reallocations of the two goods. The origin for agent A is the South West corner and the origin for agent B the North East corner. In the diagram, the indifference curves of the agents through the endowment point are plotted, as well as the *contract curve*, i.e., the set of Pareto optimal allocations. The core is the subset of the contract curve between the indifference curves of the agents through the endowment point.

10.3   Matching Problems

In a matching problem there is a group of agents that have to form couples. Examples are: students who have to be coupled with schools; hospitals that have to be coupled with doctors; workers who have to be coupled with firms; men who

Table 10.1 A matching problem

m_1	m_2	m_3	w_1	w_2	w_3
w_2	w_1	w_1	m_1	m_2	m_1
w_1	w_2	w_2	m_3	m_1	m_3
	w_3		m_2	m_3	m_2

have to be coupled with women; etc. In this section we consider so-called one-to-one matching problems.

The agents are divided in two equally large (finite and nonempty) sets, denoted M and W. Each agent in M has a strict preference over those agents in W which he prefers over staying single. Similarly, each agent in W has a strict preference over those agents in M which she prefers over staying single. In such a *matching problem*, a *matching* assigns to each agent in M at most one agent in W, and vice versa; thus, no two agents in M are assigned the same agent in W, and vice versa.

Such matching problems are also called *marriage problems*, and the agents of M and W are called *men* and *women*, respectively. While the problem may indeed refer to the 'marriage market', this terminology is of course adopted for convenience. Other examples are matching tasks and people, or matching rooms and people.

As an example, consider the matching problem in Table 10.1. The set of men is $M = \{m_1, m_2, m_3\}$ and the set of women is $W = \{w_1, w_2, w_3\}$. The columns in the table represent the preferences. For instance m_1 prefers w_2 over w_1 over staying single, but prefers staying single over w_3. An example of a matching in this particular matching problem is (m_1, w_1), (m_3, w_2), m_2, w_3, meaning that m_1 is married to w_1 and m_3 to w_2, while m_2 and w_3 stay single.[5] Observe that this matching does not seem very 'stable': m_1 and w_2 would prefer to be married to each other instead of to their partners in the given matching. Moreover, m_2 and w_3 would prefer to be married to each other instead of being single. Also, for instance, any matching in which m_1 would be married to w_3 would not be plausible, since m_1 would prefer to stay single.

The obvious way to formalize these considerations is to require that a matching should be in the core of the matching problem. A matching is in the *core* of there is no subgroup (coalition) of men and/or women who can do better by marrying (or staying single) among each other. For a matching to be in the core, the following two requirements are certainly necessary:

(c1) each person prefers his/her partner over being single;
(c2) if $m \in M$ and $w \in W$ are not matched to each other, then it is *not* the case that both m prefers w over his current partner if m is married or over being single if m is not married; and w prefers m over her current partner if w is married or over being single if w is not married.

---

[5]Check that there are 34 possible different matchings for this problem.

Obviously, if (c1) were violated then the person in question could improve by divorcing and becoming single; if (c2) were violated then m and w would both be better off by marrying each other. A matching satisfying (c1) and (c2) is called *stable*. Hence, any matching in the core must be stable. Interestingly, the converse is also true: any stable matching is in the core. To see this, suppose there is a matching outside the core and satisfying (c1) and (c2). Then there is a coalition of agents each of whom can improve by marrying or staying single within that coalition. If a member of the coalition improves by becoming single, then (c1) is violated. If two coalition members improve by marrying each other, then (c2) is violated. This contradiction establishes the claim that stable matchings must be in the core. Thus, the core of a matching problem is the set of all stable matchings.

How can stable matchings be computed? A convenient procedure is the *deferred acceptance procedure*, developed by Gale and Shapley. In this procedure, the members of one of the two parties propose and the members of the other party accept or reject proposals. Suppose men propose. In the first round, each man proposes to his favorite woman (or stays single if he prefers that) and each woman, if proposed to at least once, chooses her favorite man among those who have proposed to her (which may mean staying single). This way, a number of couples may form, and the involved men and women are called 'engaged'. In the second round, the rejected men propose to their second-best woman (or stay single); then each woman again picks here favorite among the men who proposed to her including possibly the man to whom she is currently engaged. The procedure continues until all proposals are accepted. Then all currently engaged couples marry and a matching is established.

It is not hard to verify that this matching is stable. A man who stays single was rejected by all women he preferred over staying single and therefore can find no woman who prefers him over her husband or over being single. A woman who stays single was never proposed to by any man whom she prefers over staying single. Consider, finally, an $m \in M$ and a $w \in W$ who are married but not to each other. If m prefers w over his current wife, then w must have rejected him for a better partner somewhere in the procedure. If w prefers m over her current husband, then m has never proposed to her and, thus, prefers his wife over her.

Of course, the deferred acceptance procedure can also be applied with women as proposers, resulting in a stable matching that is different in general.

Table 10.2 shows how the deferred acceptance procedure with the men proposing works, applied to the matching problem in Table 10.1.

There may be other stable matchings than those found by applying the deferred acceptance procedure with the men and with the women proposing. It can be shown

Table 10.2 The deferred acceptance procedure applied to the matching problem of Table 10.1

	Stage 1	Stage 2	Stage 3	Stage 4
m_1	$\to w_2$	rejected	$\to w_1$	
m_2	$\to w_1$ rejected	$\to w_2$		
m_3	$\to w_1$		rejected	$\to w_2$ rejected

The resulting matching is $(m_1, w_1), (m_2, w_2), m_3, w_3$

that the former procedure—with the men proposing—results in a stable matching that is optimal, among all stable matchings, from the point of view of the men, whereas the latter procedure—with the women proposing—produces the stable matching optimal from the point of view of the women. See also Problems 10.6 and 10.7.

10.4 House Exchange

In a house exchange problem each one of finitely many agents owns a house, and has a preference over all houses. The purpose of the exchange is to make the agents better off. A house exchange problem is an exchange economy with as many goods as there are agents, and where each agent is endowed with one unit of a different, indivisible good.

Formally, the set of agents is $N = \{1, \ldots, n\}$, and each agent $i \in N$ owns house h_i, and has a strict preference of the set of all (n) houses. In a *core* allocation, each agent obtains exactly one house, and there is no coalition that can make each of its members strictly better off by exchanging their *initially owned* houses among themselves.[6]

As an example, consider the house exchange problem in Table 10.3. In this problem there are six possible different allocations of the houses. Table 10.4 lists these allocations and also which coalitions could improve by exchanging their own houses.

Especially for larger problems, the 'brute force' analysis as in Table 10.4 is rather cumbersome. A different and more convenient way is to use the top trading cycle procedure. In a given house exchange problem a *top trading cycle* is a sequence $i_1, i_2, \ldots, i_k$ of agents, with $k \geq 1$, such that the favorite house of agent i_1 is house h_{i_2}, the favorite house of agent i_2 is house h_{i_3}, ..., and the favorite house of agent i_k is house h_{i_1}. If $k = 1$, then this simply means that agent i_1 already owns his favorite house. In the top trading cycle procedure, we look for a top trading cycle, assign houses within the cycle, and next the involved agents and their houses leave the scene. Then we repeat the procedure for the remaining agents, etc., until no agent is left.

Table 10.3 A house exchange problem with three agents. For instance, agent 1 prefers the house of agent 3 over the house of agent 2 over his own house

Agent 1	Agent 2	Agent 3
h_3	h_1	h_2
h_2	h_2	h_3
h_1	h_3	h_1

---

[6]Hence, by definition players in coalitions can only possibly improve by exchanging their initially owned houses, not the houses they acquired after the exchange has taken place.

Table 10.4 Analysis of the house exchange problem of Table 10.3. There are two core allocations

Agent 1	Agent 2	Agent 3	Improving coalition(s)
h_1	h_2	h_3	$\{1,2\}, \{1,2,3\}$
h_1	h_3	h_2	$\{2\}, \{1,2\}$
h_2	h_1	h_3	None: core allocation
h_2	h_3	h_1	$\{2\}, \{3\}, \{2,3\}, \{1,2,3\}$
h_3	h_1	h_2	None: core allocation
h_3	h_2	h_1	$\{3\}$

In the example in Table 10.3 there is only one top trading cycle, namely $1, 3, 2$, resulting in the allocation $1 : h_3, 3 : h_2, 2 : h_1$, a core allocation: in fact, each agent obtains his top house. In general, it is true that *for strict preferences the top trading cycle procedure results in a core allocation*. The reader should check the validity of this claim (Problem 10.8).

What about the other core allocation found in Table 10.4? In this allocation, the grand coalition could *weakly improve*: by the allocation $1 : h_3, 3 : h_2, 2 : h_1$ agents 1 and 3 would be strictly better off, while agent 2 would not be worse off. We define the *strong core* as consisting of those allocations on which no coalition could even weakly improve, that is, make all its members at least as well off and at least one member strictly better off. In the example, only the allocation $1 : h_3, 3 : h_2, 2 : h_1$ is in the strong core. In general, one can show that *the strong core of a house exchange problem with strict preferences consists of the unique allocation produced by the top trading cycle procedure*.

10.5 Problems

10.1. *A Division Problem (1)*
Suppose two players (bargainers) bargain over the division of one unit of a perfectly divisible good. Player 1 has utility function $u_1(\alpha) = \alpha$ and player 2 has utility function $u_2(\beta) = 1 - (1 - \beta)^2$ for amounts $\alpha, \beta \in [0, 1]$ of the good.

(a) Determine the set of feasible utility pairs. Make a picture.
(b) Determine the Nash bargaining solution outcome, in terms of utilities as well as of the distribution of the good.
(c) Suppose the players' utilities are discounted by a factor $\delta \in [0, 1)$. Calculate the Rubinstein bargaining outcome, i.e., the subgame perfect equilibrium outcome of the infinite horizon alternating offers bargaining game.
(d) Determine the limit of the Rubinstein bargaining outcome, for δ approaching 1, in two ways: by using the result of (b) and by using the result of (c).

10.2. *A Division Problem (2)*

Suppose that two players (bargainers) bargain over the division of one unit of a perfectly divisible good. Assume that player 1 has utility function $u(\alpha)$ $(0 \leq \alpha \leq 1)$ and player 2 has utility function $v(\alpha) = 2u(\alpha)$ $(0 \leq \alpha \leq 1)$.

Determine the distribution of the good according to the Nash bargaining solution. Can you say something about the resulting utilities? (Hint: use the relevant properties of the Nash bargaining solution.)

10.3. *A Division Problem (3)*

Suppose that two players (bargainers) bargain over the division of two units of a perfectly divisible good. Assume that player 1 has a utility function $u(\alpha) = \frac{\alpha}{2}$ $(0 \leq \alpha \leq 2)$ and player 2 has utility function $v(\alpha) = \sqrt[3]{\alpha}$ $(0 \leq \alpha \leq 2)$.

(a) Determine the distribution of the good according to the Rubinstein bargaining procedure, for any discount factor $0 < \delta < 1$.
(b) Use the result to determine the Nash bargaining solution distribution.
(c) Suppose player 1's utility function changes to $w(\alpha) = \alpha$ for $0 \leq \alpha \leq 1.6$ and $w(\alpha) = 1.6$ for $1.6 \leq \alpha \leq 2$. Determine the Nash bargaining solution outcome, both in utilities and in distribution, for this new situation.

10.4. *An Exchange Economy*

Consider an exchange economy with two agents A and B and two goods. The agents are endowed with initial bundles $e^A = (3,1)$ and $e^B = (1,3)$. Their preferences are represented by the utility functions $u^A(x_1, x_2) = \ln(x_1 + 1) + \ln(x_2 + 2)$ and $u^B(x_1, x_2) = 3\ln(x_1 + 1) + \ln(x_2 + 1)$.

(a) Compute the demand functions of the agents.
(b) Compute Walrasian equilibrium prices and the equilibrium allocation.
(c) Compute the contract curve and the core. Sketch the Edgeworth box.
(d) Show that the Walrasian equilibrium allocation is in the core.
(e) How would you set up a two-person bargaining problem associated with this economy? Would it make sense to consider the Nash bargaining solution in order to compute an allocation? Why or why not?

10.5. *The Matching Problem of Table 10.1 Continued*

(a) Apply the deferred acceptance procedure to the matching problem of Table 10.1 with the women proposing.
(b) Are there any other stable matchings in this example?

10.6. *Another Matching Problem*

Consider the matching problem with three men, three women, and preferences as in Table 10.5.

Table 10.5 The matching
problem of Problem 10.6

m_1	m_2	m_3	w_1	w_2	w_3
w_1	w_1	w_1	m_1	m_1	m_1
w_2	w_2	w_3	m_2	m_3	m_2
w_3	w_3	w_2	m_3	m_2	m_3

Table 10.6 The matching
problem of Problem 10.7

m_1	m_2	m_3	w_1	w_2	w_3
w_2	w_1	w_1	m_1	m_3	m_1
w_1	w_3	w_2	m_3	m_1	m_3
w_3	w_2	w_3	m_2	m_2	m_2

(a) Compute the two matchings produced by the deferred acceptance procedure
   with the men and with the women proposing.
(b) Are there any other stable matchings?
(c) Verify the claim made in the text about the optimality of the matchings in (a) for
   the men and the women, respectively.

10.7. *Yet Another Matching Problem: Strategic Behavior*
Consider the matching problem with three men, three women, and preferences as in
Table 10.6.

(a) Compute the two matchings produced by the deferred acceptance procedure
   with the men and with the women proposing.
(b) Are there any other stable matchings?

Now consider the following noncooperative game. The players are w_1, w_2, and w_3.
The strategy set of a player is simply the set of all possible preferences over the men.
(Thus, each player has 16 different strategies.) The outcomes of the game are the
matchings produced by the deferred acceptance procedure with the men proposing,
assuming that each man uses his true preference given in Table 10.6.

(c) Show that the following preferences form a Nash equilibrium: w_2 and w_3
   use their true preferences, as given in Table 10.6; w_1 uses the preference
   (m_1, m_2, m_3). Conclude that sometimes it may pay off to lie about one's true
   preference. (Hint: in a Nash equilibrium, no player can gain by deviating.)

10.8. *Core Property of Top Trading Cycle Procedure*
Show that for strict preferences the top trading cycle results in a core allocation.

10.9. *House Exchange with Identical Preferences*
Consider the n-agent house exchange problem where all agents have identical strict
preferences. Find the house allocation(s) in the core.

Table 10.7 The house
exchange problem of
Problem 10.10

Player 1	Player 2	Player 3	Player 4
h_3	h_4	h_1	h_3
h_2	h_1	h_4	h_2
h_4	h_2	h_3	h_1
h_1	h_3	h_2	h_4

10.10. *A House Exchange Problem*
Consider the house exchange problem with four agents in Table 10.7.
Compute all core allocations and all strong core allocations.

10.11. *Cooperative Oligopoly*
Consider the Cournot oligopoly game with n firms with different costs $c_1, c_2, \ldots, c_n$.
(This is the game of Problem 6.2 with heterogenous costs.) As before, each firm i
offers $q_i \geq 0$, and the price-demand function is $p = \max\{0, a - \sum_{i=1}^{n} q_i\}$, where
$0 < c_i < a$ for all i.

(a) Show that the reaction function of player i is

$$q_i = \max\{0, \frac{a - c_i - \sum_{j \neq i} q_i}{2}\} .$$

(b) Show that the unique Nash equilibrium of the game is $\mathbf{q}^* = (q_1^*, \ldots, q_n^*)$ with

$$q_i^* = \frac{a - nc_i + \sum_{j \neq i} c_j}{n + 1} ,$$

for each i, assuming that this quantity is positive.
(c) Derive that the corresponding profits are

$$\frac{(a - nc_i + \sum_{j \neq i} c_j)^2}{(n + 1)^2}$$

for each player i.

Let the firms now be the players in a cooperative TU-game with player set $N = \{1, 2, \ldots, n\}$, and consider a coalition $S \subseteq N$. What is the total profit that S can
make on its own? This depends on the assumptions that we make on the behavior of
the players outside S. Very pessimistically, one could solve the problem

$$\max_{q_i : i \in S} \min_{q_j : j \notin S} \sum_{i \in S} P_i(q_1, \ldots, q_n) ,$$

which is the profit that S can guarantee independent of the players outside S. This view is very pessimistic because it presumes maximal resistance of the outside players, even if this means that these outside players hurt themselves. In the present case it is not hard to see that this results in zero profit for S.

Two alternative scenarios are: S plays a Cournot-Nash equilibrium in the $(n - |S| + 1)$-player oligopoly game against the outside firms as separate firms, or S plays a Cournot-Nash equilibrium in the duopoly game against $N \setminus S$.

In the first case we in fact have an oligopoly game with costs c_j for every player $j \notin S$ and with cost $c_S := \min\{c_i : i \in S\}$ for 'player' (coalition) S.

(d) By using the results of (a)–(c) show that coalition S obtains a profit of

$$v_1(S) = \frac{[a - (n - |S| + 1)c_S + \sum_{j \notin S} c_j]^2}{(n - |S| + 2)^2}$$

in this scenario. Thus, this scenario results in a cooperative TU-game (N, v_1).

(e) Assume $n = 3$, $a = 7$, and $c_i = i$ for $i = 1, 2, 3$. Compute the core, the Shapley value, and the nucleolus for the TU-game (N, v_1).

(f) Show that in the second scenario, coalition S obtains a profit of

$$v_2(S) = \frac{(a - 2c_S + c_{N-S})^2}{9},$$

resulting in a cooperative game (N, v_2).

(g) Assume $n = 3$, $a = 7$, and $c_i = i$ for $i = 1, 2, 3$. Compute the core, the Shapley value, and the nucleolus for the TU-game (N, v_2).

10.6 Notes

The Nash bargaining model and solution were proposed and characterized in Nash (1950). Critique on the independence of irrelevant alternatives axiom was formalized in Kalai and Smorodinsky (1975), see Chap. 21.

For a proof of the First Welfare Theorem, see for instance Jehle and Reny (2001). There one can also find a proof of the fact that the Walrasian equilibrium results in a core allocation.

The deferred acceptance procedure for matching problems was first proposed in Gale and Shapley (1962). Our introduction to matching problems and house exchange is largely based on Osborne (2004). Also Problems 10.5–10.7 are from that source. Problem 10.10 is from Moulin (1995).

Two-person bargaining problems and TU-games (Chap. 9) are both special cases of the general model of *cooperative games without transferable utility*, so-called NTU-games. In an NTU-game, a *set* of feasible utility vectors $V(T)$ is assigned to each coalition T of players. For a TU-game (N, v) and a coalition T, this set takes the special form $V(T) = \{\mathbf{x} \in \mathbb{R}^n \mid \sum_{i \in T} x_i \leq v(T)\}$, i.e., a coalition T can attain any vector of utilities such that the sum of the utilities for the players in T does not exceed the worth of the coalition. In a two-player bargaining problem $(S, \mathbf{d})$, one can set $V(\{1, 2\}) = S$ and $V(\{i\}) = \{\alpha \in \mathbb{R} \mid \alpha \leq d_i\}$ for $i = 1, 2$. See also Chap. 21.

References

Gale, D., & Shapley, L. S. (1962). College admissions and the stability of marriage. *American Mathematical Monthly, 69*, 9–15.

Jehle, G. A., & Reny, P. J. (2001). *Advanced microeconomic theory*. Boston: Addison Wesley.

Kalai, E., & Smorodinsky, M. (1975). Other solutions to Nash's bargaining problem. *Econometrica, 43*, 513–518.

Moulin, H. (1995). *Cooperative microeconomics; a game-theoretic introduction*. Hemel Hempstead: Prentice Hall/Harvester Wheatsheaf.

Nash, J. F. (1950). The bargaining problem. *Econometrica, 18*, 155–162.

Osborne, M. J. (2004). *An introduction to game theory*. New York: Oxford University Press.

Social Choice

11

Social choice theory studies the aggregation of individual preferences into a common or social preference. It overlaps with several social science disciplines, such as political theory (e.g., voting for Parliament, or for a president) and game theory (e.g., voters may vote strategically, or candidates may choose positions strategically).

In the classical model of social choice, there is a finite number of agents who have preferences over a finite number of alternatives. These preferences are either aggregated into a social preference according to a so-called social welfare function, or result in a common alternative according to a so-called social choice function.

The main purpose of this chapter is to review two classical results, namely Arrow's Theorem and the Gibbard-Satterthwaite Theorem. The first theorem applies to social welfare functions and says that, if the social preference between any two alternatives should only depend on the individual preferences between these alternatives and, thus, not on individual preferences involving other alternatives, then basically the social welfare function must be dictatorial. The second theorem applies to social choice functions and says that, basically, the only social choice functions that are invulnerable to strategic manipulation are the dictatorial ones. These results are often referred to as 'impossibility theorems' since dictatorships are generally regarded undesirable.

Section 11.1 is introductory. Section 11.2 discusses Arrow's Theorem and Sect. 11.3 the Gibbard-Satterthwaite Theorem.

11.1 Introduction and Preliminaries

11.1.1 An Example

Suppose there are three agents (individuals, voters) who have strict preferences over a set of five alternatives $(a_1, \ldots, a_5)$, as given in Table 11.1.

© Springer-Verlag Berlin Heidelberg 2015
H. Peters, *Game Theory*, Springer Texts in Business and Economics,
DOI 10.1007/978-3-662-46950-7_11

Table 11.1 Borda scores

Agent	a_1	a_2	a_3	a_4	a_5
1	5	1	3	2	4
2	1	2	3	4	5
3	3	4	5	2	1

In this table the preferences of the players are represented by the *Borda scores*: the best alternative of an agent obtains 5 points, the second best 4 points, etc., until the worst alternative which obtains 1 point. For instance, agent 1 has the preference $a_1 P_1 a_5 P_1 a_3 P_1 a_4 P_1 a_2$ in the notation to be introduced below. We use the Borda scores as a convenient way to represent these preferences and, more importantly, to obtain an example of a social welfare as well as a social choice function.

First, suppose that we want to extract a common social ranking of the alternatives from the individual preferences. One way to do this is to add the Borda scores per alternative. In the example this results in $9, 7, 11, 8, 10$ for a_1, a_2, a_3, a_4, a_5, respectively, resulting in the social ranking $a_3 P a_5 P a_1 P a_4 P a_2$. If we just want to single out one alternative, then we could take the one with the maximal Borda score, in this case alternative a_3. In the terminology to be introduced formally below, Borda scores give rise to a social welfare as well as a social choice function.[1]

One potential drawback of using Borda scores to obtain a social ranking is, that the ranking between two alternatives may not just depend on the individual preferences between these two alternatives. For instance, suppose that agent 1's preference would change to $a_1 P_1 a_4 P_1 a_5 P_1 a_3 P_1 a_2$. Then the Borda scores would change to $9, 7, 10, 10, 9$ for a_1, a_2, a_3, a_4, a_5, respectively, resulting in the social ranking $a_3 I a_4 P a_1 I a_5 P a_2$ (where I denotes indifference). Observe that no agent's preference between a_1 and a_4 has changed, but that socially this preference is reversed. This is not a peculiarity of using Borda scores: Arrow's Theorem, to be discussed in Sect. 11.2, states that under some reasonable additional assumptions the only way to avoid this kind of preference reversal is to make one agent the dictator, i.e., to have the social preference coincide with the preference of one fixed agent.

Another potential drawback of using the Borda scores in order to single out a unique alternative is that this method is vulnerable to strategic manipulation. For instance, suppose that agent 1 would lie about his true preference given in Table 11.1 and claim that his preference is $a_1 P_1 a_5 P_1 a_2 P_1 a_4 P_1 a_3$ instead. Then the Borda scores would change to $9, 9, 9, 8, 10$ for a_1, a_2, a_3, a_4, a_5, respectively, resulting in the chosen alternative a_5 instead of a_3. Since agent 1 prefers a_5 over a_3 according to his *true* preference, he gains by this strategic manipulation. Again, this phenomenon

---

[1] Ties may occur, but this need not bother us here.

is not a peculiarity of the Borda method: the Gibbard-Satterthwaite Theorem in Sect. 11.3 shows that again under some reasonable additional assumptions the only way to avoid it is to make one fixed agent a dictator.

11.1.2 Preliminaries

Let $A = \{a_1, \ldots, a_m\}$ be the set of *alternatives*. To keep things interesting we assume $m \geq 3$.[2] The set of *agents* is denoted by $N = \{1, \ldots, n\}$. We assume $n \geq 2$.

A *binary relation* on A is a subset of $A \times A$. In our context, for a binary relation R on A we usually write aRb instead of $(a, b) \in R$ and interpret this as an agent or society (weakly) preferring a over b. Well-known conditions for a binary relation R on A are:

(a) *Reflexivity*: aRa for all $a \in A$.
(b) *Completeness*: aRb or bRa for all $a, b \in A$ with $a \neq b$.
(c) *Antisymmetry*: For all $a, b \in A$, if aRb and bRa, then $a = b$.
(d) *Transitivity*: For all $a, b, c \in A$, aRb and bRc imply aRc.

A *preference* on A is a reflexive, complete and transitive binary relation on A. For a preference R on A we write aPb if aRb and not bRa; and aIb if aRb and bRa. The binary relations P and I are called the *asymmetric* and *symmetric parts* of R, respectively, and are interpreted as strict preference and indifference. Check (Problem 11.1) that P is antisymmetric and transitive but not reflexive and not necessarily complete, and that I is reflexive and transitive but not necessarily antisymmetric and not necessarily complete. By $\mathcal{L}^*$ we denote the set of all preferences on A, and by $\mathcal{L} \subseteq \mathcal{L}^*$ the set of all antisymmetric (i.e., *strict*) preferences on A. In plain words, elements of $\mathcal{L}^*$ order the elements of A but allow for indifferences, while elements of $\mathcal{L}$ order the elements of A strictly.[3]

In what follows, it is assumed that agents have strict preferences while social preferences may have indifferences. A strict preference *profile* is a list $(R_1, \ldots, R_i, \ldots, R_n)$, where R_i is the strict preference of agent i. Hence, $\mathcal{L}^N$ denotes the set of all strict preference profiles. A *social choice function* is a map $f : \mathcal{L}^N \to A$, i.e., it assigns a unique alternative to every profile of strict preferences. A *social welfare function* is a map $F : \mathcal{L}^N \to \mathcal{L}^*$, i.e., it assigns a (possibly non-strict) preference to every profile of strict preferences.

---

[2]See Problem 11.7 for the case $m = 2$.
[3]Elements of $\mathcal{L}$ are usually called *linear orders* and those of $\mathcal{L}^*$ *weak orders*.

11.2   Arrow's Theorem

In this section the focus is on social welfare functions. We formulate three properties for a social welfare function $F : \mathcal{L}^N \to \mathcal{L}^*$. We call F:

(a) *Pareto Efficient* (PE) if for each profile $(R_1, \ldots, R_n) \in \mathcal{L}^N$ and all $a, b \in A$, if $a \neq b$ and aR_ib for all $i \in N$, then aPb, where P is the asymmetric part of $R = F(R_1, \ldots, R_n)$.

(b) *Independent of Irrelevant Alternatives* (IIA) if for all $(R_1, \ldots, R_n) \in \mathcal{L}^N$ and $(R'_1, \ldots, R'_n) \in \mathcal{L}^N$ and all $a, b \in A$, if $aR_ib \Leftrightarrow aR'_ib$ for all $i \in N$, then $aRb \Leftrightarrow aR'b$, where $R = F(R_1, \ldots, R_n)$ and $R' = F(R'_1, \ldots, R'_n)$.

(c) *Dictatorial* (D) if there is an $i \in N$ such that $F(R_1, \ldots, R_n) = R_i$ for all $(R_1, \ldots, R_n) \in \mathcal{L}^N$.

Pareto Efficiency requires that, if all agents prefer an alternative a over an alternative b, then the social ranking should also put a above b. Independence of Irrelevant Alternatives says that the social preference between two alternatives should only depend on the agents' preferences between these two alternatives and not on the position of any other alternative.[4] Dictatoriality says that the social ranking is always equal to the preference of a fixed agent, the *dictator*. Clearly, there are exactly n dictatorial social welfare functions.

The first two conditions are usually regarded as desirable but the third clearly not. Unfortunately, Arrow's Theorem implies that the first two conditions imply the third one.[5]

Theorem 11.1 (Arrow's Theorem) *Let F be a Pareto Efficient and IIA social welfare function. Then F is dictatorial.*

Proof

Step 1 Consider a profile in $\mathcal{L}^N$ and two distinct alternatives $a, b \in A$ such that every agent ranks a on top and b at bottom. By Pareto Efficiency, the social ranking assigned by F must also rank a on top and b at bottom.

Now change agent 1's ranking by raising b in it one position at a time. By IIA, a is ranked socially (by F) on top as long as b is still below a in the preference of agent 1. In the end, if agent 1 ranks b first and a second, we have a or b on top of the social ranking by Pareto efficiency of F. If a is still on top in the social ranking, then continue the same process with agents 2, 3, etc., until we reach some agent k such that b is on top of the social ranking after moving b above a in agent k's preference.

---

[4] Although there is some similarity in spirit, this condition is not in any formal sense related to the IIA condition in bargaining, see Sect. 10.1 or Chap. 21.

[5] For this reason the theorem is often referred to as Arrow's Impossibility Theorem.

Table 11.2 Step 1 of the proof of Theorem 11.1, agent k ranks a above b

R_1	$\cdots$	R_{k-1}	R_k	R_{k+1}	$\cdots$	R_n	F	f
b	$\cdots$	b	a	a	$\cdots$	a	a	a
a	$\cdots$	a	b	$\cdot$	$\cdots$	$\cdot$	$\cdot$	
$\cdot$	$\cdots$	$\cdot$	$\cdot$	$\cdot$	$\cdots$	$\cdot$	$\cdot$	
$\cdot$	$\cdots$	$\cdot$	$\cdot$	$\cdot$	$\cdots$	$\cdot$		b
$\cdot$	$\cdots$	$\cdot$	$\cdot$	$\cdot$	$\cdots$	$\cdot$	$\cdot$	
$\cdot$	$\cdots$	$\cdot$	$\cdot$	b	$\cdots$	b	$\cdot$	

Table 11.3 Step 1 of the proof of Theorem 11.1, agent k ranks b above a

R_1	$\cdots$	R_{k-1}	R_k	R_{k+1}	$\cdots$	R_n	F	f
b	$\cdots$	b	b	a	$\cdots$	a	b	b
a	$\cdots$	a	a	$\cdot$	$\cdots$	$\cdot$		a
$\cdot$	$\cdots$	$\cdot$	$\cdot$	$\cdot$	$\cdots$	$\cdot$	$\cdot$	
$\cdot$	$\cdots$	$\cdot$	$\cdot$	$\cdot$	$\cdots$	$\cdot$	$\cdot$	
$\cdot$	$\cdots$	$\cdot$	$\cdot$	$\cdot$	$\cdots$	$\cdot$	$\cdot$	
$\cdot$	$\cdots$	$\cdot$	$\cdot$	b	$\cdots$	b	$\cdot$	

Table 11.4 Step 2 of the proof of Theorem 11.1, arising from Table 11.2

R_1	$\cdots$	R_{k-1}	R_k	R_{k+1}	$\cdots$	R_n	F	f
b	$\cdots$	b	a	$\cdot$	$\cdots$	$\cdot$	a	a
$\cdot$	$\cdots$	$\cdot$	b	$\cdot$	$\cdots$	$\cdot$	b	
$\cdot$	$\cdots$	$\cdot$	$\cdot$	$\cdot$	$\cdots$	$\cdot$	$\cdot$	
$\cdot$	$\cdots$	$\cdot$	$\cdot$	a	$\cdots$	a	$\cdot$	
a	$\cdots$	a	$\cdot$	b	$\cdots$	b	$\cdot$	

Table 11.5 Step 2 of the proof of Theorem 11.1, arising from Table 11.3

R_1	$\cdots$	R_{k-1}	R_k	R_{k+1}	$\cdots$	R_n	F	f
b	$\cdots$	b	b	$\cdot$	$\cdots$	$\cdot$	b	b
$\cdot$	$\cdots$	$\cdot$	a	$\cdot$	$\cdots$	$\cdot$	$\cdot$	
$\cdot$	$\cdots$	$\cdot$	$\cdot$	$\cdot$	$\cdots$	$\cdot$		a
$\cdot$	$\cdots$	$\cdot$	$\cdot$	a	$\cdots$	a	$\cdot$	
a	$\cdots$	a	$\cdot$	b	$\cdots$	b	$\cdot$	

Tables 11.2 and 11.3 give the situations just before and just after b is placed above a in agent k's preference.[6]

Step 2 Now consider Tables 11.4 and 11.5.
The profile in Table 11.4 arises from the one in Table 11.2 by moving a to the last position for agents $i < k$ and to the second last position for agents $i > k$. In exactly the same way, the profile in Table 11.5 arises from the one in Table 11.3. Then IIA applied to Tables 11.3 and 11.5 implies that b is socially top-ranked in Table 11.5. Next, IIA applied to the transition from Table 11.5 to Table 11.4 implies that in

---

[6]In these tables and also the ones below, we generically denote all preferences by $R_1, \ldots, R_n$. The last column in every table will be used in Sect. 11.3.

Table 11.6 Step 3 of the proof of Theorem 11.1

R_1	$\cdots$	R_{k-1}	R_k	R_{k+1}	$\cdots$	R_n	F	f
·	$\cdots$	·	a	·	$\cdots$	·	a	a
·	$\cdots$	·	c	·	$\cdots$	·	·	
·	$\cdots$	·	b	·	$\cdots$	·	·	
c	$\cdots$	c	·	c	$\cdots$	c	·	
b	$\cdots$	b	·	a	$\cdots$	a	·	
a	$\cdots$	a	·	b	$\cdots$	b	·	

Table 11.7 Steps 4 and 5 of the proof of Theorem 11.1

R_1	$\cdots$	R_{k-1}	R_k	R_{k+1}	$\cdots$	R_n	F	f
·	$\cdots$	·	a	·	$\cdots$	·	a	a
·	$\cdots$	·	c	·	$\cdots$	·	·	
·	$\cdots$	·	b	·	$\cdots$	·	c	
c	$\cdots$	c	·	c	$\cdots$	c	·	
b	$\cdots$	b	·	b	$\cdots$	b	b	
a	$\cdots$	a	·	a	$\cdots$	a	·	

Table 11.4 b must still be socially ranked above every alternative except perhaps a. But IIA applied to the transition from Table 11.2 to Table 11.4 implies that in Table 11.4 a must still be socially ranked above every alternative. This proves that the social rankings in Tables 11.4 and 11.5 are correct.

Step 3 Consider a third alternative c distinct from a and b. The social ranking in Table 11.6 is obtained by from Table 11.4 by applying IIA.

Step 4 Consider the profile in Table 11.7, obtained from the profile in Table 11.6 by switching a and b for agents $i > k$. By IIA applied to the transition from Table 11.6 to Table 11.7, we have that a must still be socially ranked above every alternative except possibly b. However, b must be ranked below c by Pareto efficiency, which shows that the social ranking in Table 11.7 is correct.

Step 5 Consider any arbitrary profile in which agent k prefers a to b. Change the profile by moving c between a and b for agent k and to the top of every other agent's preference (if this is not already the case). By IIA this does not affect the social ranking of a vs. b. Since the preference of every agent concerning a and c is now as in Table 11.7, IIA implies that a is socially ranked above c, which itself is socially ranked above b by Pareto Efficiency. Hence, by transitivity of the social ranking we may conclude that a is socially ranked above b whenever it is preferred by agent k over b. By repeating the argument with the roles of b and c reversed, and recalling that c was an arbitrary alternative distinct from a and b, we may conclude that the social ranking of a is above some alternative whenever agent k prefers a to that alternative: k is a 'dictator' for a. Since a was arbitrary, we can repeat the whole argument to conclude that there must be a dictator for every alternative. Since there cannot be distinct dictators for distinct alternatives, there must be a single dictator for all alternatives. ■

11.3   The Gibbard-Satterthwaite Theorem

The Gibbard-Satterthwaite Theorem applies to social choice functions. We start by listing the following possible properties of a social choice function $f : \mathcal{L}^N \to A$. Call f:

(a) *Unanimous* (UN) if for each profile $(R_1, \ldots, R_n) \in \mathcal{L}^N$ and each $a \in A$, if aR_ib for all $i \in N$ and all $b \in A \setminus \{a\}$, then $f(R_1, \ldots, R_n) = a$.

(b) *Monotonic* (MON) if for all profiles $(R_1, \ldots, R_n) \in \mathcal{L}^N$ and $(R'_1, \ldots, R'_n) \in \mathcal{L}^N$ and all $a \in A$, if $f(R_1, \ldots, R_n) = a$ and $aR_ib \Rightarrow aR'_ib$ for all $b \in A \setminus \{a\}$ and $i \in N$, then $f(R'_1, \ldots, R'_n) = a$.

(c) *Dictatorial* (D) if there is an $i \in N$ such that $f(R_1, \ldots, R_n) = a$ where aR_ib for all $b \in A \setminus \{a\}$, for all $(R_1, \ldots, R_n) \in \mathcal{L}^N$.

(d) *Strategy-Proof* (SP) if for all profiles $(R_1, \ldots, R_n) \in \mathcal{L}^N$ and $(R'_1, \ldots, R'_n) \in \mathcal{L}^N$ and all $i \in N$, if $R'_j = R_j$ for all $j \in N \setminus \{i\}$, then $f(R_1, \ldots, R_n) \, R_i \, f(R'_1, \ldots, R'_n)$.

Unanimity requires that, if all agents have the same top alternative, then this alternative should be chosen. Monotonicity says that, if some alternative a is chosen and the profile changes in such a way that a is still preferred by every agent over all alternatives over which it was originally preferred, then a should remain to be chosen.[7] Dictatoriality means that there is a fixed agent whose top element is always chosen. Strategy-Proofness says that no agent can obtain a better chosen alternative by lying about his true preference.

In accordance with mathematical parlance, call a social choice function $f : \mathcal{L}^N \to A$ *surjective* if for every $a \in A$ there is some profile $(R_1, \ldots, R_n) \in \mathcal{L}^N$ such that $f(R_1, \ldots, R_n) = a$. Hence, each a is chosen at least once.[8] The Gibbard-Satterthwaite Theorem is as follows.

Theorem 11.2 (Gibbard-Satterthwaite Theorem) *Let $f : \mathcal{L}^N \to A$ be a surjective and strategy-proof social choice function. Then f is dictatorial.*

Since surjectivity is implied by unanimity, Theorem 11.2 also holds with unanimity instead of surjectivity.

We will prove the Gibbard-Satterthwaite Theorem by using the next theorem, which is a variant of the Muller-Satterthwaite Theorem.

Theorem 11.3 (Muller-Satterthwaite) *Let $f : \mathcal{L}^N \to A$ be a unanimous and monotonic social choice function. Then f is dictatorial.*

Proof of Theorem 11.2 We prove that f is unanimous and monotonic. The result then follows from Theorem 11.3.

---

[7]This property is also called *Maskin Monotonicity*, after Maskin (1999).

[8]In the social choice literature this property is sometimes called *citizen-sovereignty*.

Suppose that $f(R_1, \ldots, R_n) = a$ for some profile $(R_1, \ldots, R_n) \in \mathcal{L}^N$ and some alternative $a \in A$. Let $i \in N$ and let $(R'_1, \ldots, R'_n) \in \mathcal{L}^N$ be a profile such that for all $j \in N \setminus \{i\}$ we have $R'_j = R_j$ and for all $b \in A \setminus \{a\}$ we have aR'_ib if aR_ib. We wish to show that $f(R'_1, \ldots, R'_n) = a$. Suppose, to the contrary, that $f(R'_1, \ldots, R'_n) = b \neq a$. Then SP implies aR_ib, and hence aR'_ib. Again by SP, however, bR'_ia, hence, by antisymmetry of R'_i, $a = b$, a contradiction. This proves $f(R'_1, \ldots, R'_n) = a$.

Now suppose that $(R'_1, \ldots, R'_n) \in \mathcal{L}^N$ is a profile such that for all $i \in N$ and all $b \in A \setminus \{a\}$ we have aR'_ib if aR_ib. By applying the argument in the preceding paragraph n times, it follows that $f(R'_1, \ldots, R'_n) = a$. Hence, f is monotonic.

To prove unanimity, suppose that $(R_1, \ldots, R_n) \in \mathcal{L}^N$ and $a \in A$ such that aR_ib for all $i \in N$ and $b \in A \setminus \{a\}$. By surjectivity there is $(R'_1, \ldots, R'_n) \in \mathcal{L}^N$ with $f(R'_1, \ldots, R'_n) = a$. By monotonicity we may move a to the top of each agent's preference and still have a chosen. Next, again by monotonicity, we may change each agent i's preference to R_i without changing the chosen alternative, i.e., $f(R_1, \ldots, R_n) = a$. Hence, f is unanimous. ∎

Proof of Theorem 11.3 The proof parallels the proof of Theorem 11.1 and uses analogous steps and the same tables.

Step 1 Consider a profile in $\mathcal{L}^N$ and two distinct alternatives $a, b \in A$ such that every agent ranks a on top and b at bottom. By unanimity, f chooses a.
Now change agent 1's ranking by raising b in it one position at a time. By MON, a is chosen by f as long as b is still below a in the preference of agent 1. In the end, if agent 1 ranks b first and a second, we have a or b chosen by f, again by MON. If a is still chosen, then continue the same process with agents 2, 3, etc., until we reach some agent k such that b is chosen after moving b above a in agent k's preference. Tables 11.2 and 11.3 give the situations just before and just after b is placed above a in agent k's preference.

Step 2 Now consider Tables 11.4 and 11.5. The profile in Table 11.4 arises from the one in Table 11.2 by moving a to the last position for agents $i < k$ and to the second last position for agents $i > k$. In exactly the same way, the profile in Table 11.5 arises from the one in Table 11.3.
Then MON applied to Tables 11.3 and 11.5 implies that b is chosen in Table 11.5. Next, MON applied to the transition from Table 11.5 to Table 11.4 implies that in Table 11.4 the choice must be either b or a. Suppose b would be chosen. Then MON applied to the transition from Table 11.4 to Table 11.2 implies that in Table 11.2 b must be chosen as well, a contradiction. Hence, a is chosen in Table 11.4. This proves that the choices by f in Tables 11.4 and 11.5 are correct.

Step 3 Consider a third alternative c distinct from a and b. The choice in Table 11.6 is obtained by from Table 11.4 by applying MON.

Step 4 Consider the profile in Table 11.7, obtained from the profile in Table 11.6 by switching a and b for agents $i > k$. If the choice in Table 11.7 were some d unequal to a or b, then by MON it would also be d in Table 11.6, a contradiction. If it were b, then by MON it would remain b even if c would be moved to the top of every agent's preference, contradicting unanimity. Hence, it must be a.

Step 5 Consider any arbitrary profile with a at the top of agent k's preference. Such a profile can always be obtained from the profile in Table 11.7 without worsening the position of a with respect to any other alternative in any agent's preference. By MON therefore, a must be chosen whenever it is at the top of agent k's preference, so k is a 'dictator' for a. Since a was arbitrary, we can find a dictator for every other alternative but, clearly, these must be one and the same agent. Hence, this agent is the dictator. ∎

There is a large literature that tries to escape the negative conclusions of Theorems 11.1–11.3 by adapting the model and/or restricting the domain. Examples are provided in Problems 6.23 and 6.24.

11.4 Problems

11.1. *Preferences*
Let R be a preference on A, with symmetric part I and asymmetric part P.

(a) Prove that P is antisymmetric and transitive but not reflexive and not necessarily complete.
(b) Prove that I is reflexive and transitive but not necessarily complete and not necessarily antisymmetric.

11.2. *Pairwise Comparison*
For a profile $r = (R_1, \ldots, R_n) \in \mathcal{L}^N$ and $a, b \in A$ define

$$N(a, b, r) = \{i \in N \mid aR_ib\} \,,$$

i.e., $N(a, b, r)$ is the set of agents who (strictly) prefer a to b in the profile r. With r we can associate a binary relation $C(r)$ on A by defining $aC(r)b :\Leftrightarrow |N(a, b, r)| \geq |N(b, a, r)|$ for all $a, b \in A$. If $aC(r)b$ we say that 'a beats b by pairwise majority'.

(a) Is $C(r)$ reflexive? Complete? Antisymmetric?
(b) Show that $C(r)$ is not transitive, by considering the famous Condorcet profile for $N = \{1, 2, 3\}$ and $A = \{a, b, c\}$: aR_1bR_1c, bR_2cR_2a, cR_3aR_3b.
(c) Call a a *Condorcet winner* if $|N(a, b, r)| > |N(b, a, r)|$ for all $b \in A \setminus \{a\}$. Is there a Condorcet winner in the example in Sect. 11.1?

11.3. *Independence of the Conditions in Theorem 11.1*
Show that the conditions in Theorem 11.1 are independent. That is, exhibit a social welfare function that is Pareto efficient and does not satisfy IIA or dictatoriality, and one that satisfies IIA and is not dicatorial nor Pareto efficient.

11.4. *Independence of the Conditions in Theorem 11.2*
Show that the conditions in Theorem 11.2 are independent.

11.5. *Independence of the Conditions in Theorem 11.3*
Show that the conditions in Theorem 11.3 are independent.

11.6. *Copeland Score and Kramer Score*
The *Copeland score* of an alternative $a \in A$ at a profile $r = (R_1, \ldots, R_n) \in \mathcal{L}^N$ is defined by

$$c(a, r) = |\{b \in A \mid N(a, b, r) \geq N(b, a, r)\}| \, ,$$

i.e., the number of alternatives that a beats (cf. Problem 11.2). The *Copeland ranking* is obtained by ranking the alternatives according to their Copeland scores.

(a) Is the Copeland ranking a preference? Is it antisymmetric? Does the derived social welfare function satisfy IIA? Pareto efficiency?

The *Kramer score* of an alternative $a \in A$ at a profile $r = (R_1, \ldots, R_n) \in \mathcal{L}^N$ is defined by

$$k(a, r) = \min_{b \in A \setminus \{a\}} |N(a, b, r)| \, ,$$

i.e., the worst score among all pairwise comparisons. The *Kramer ranking* is obtained by ranking the alternatives according to their Kramer scores.

(b) Is the Kramer ranking a preference? Is it antisymmetric? Does the derived social welfare function satisfy IIA? Pareto efficiency?

11.7. *Two Alternatives*
Show that Theorems 11.1–11.3 no longer hold if there are just two alternatives, i.e., if $m = 2$.

11.5   Notes

For a general overview of social choice theory see Arrow et al. (2002, 2011). For Arrow's Theorem see Arrow (1963). For the Gibbard-Satterthwaite Theorem see Gibbard (1973) and Satterthwaite (1975). Both these theorems are closely related:

indeed, the proof of the Gibbard-Satterthwaite Theorem in Gibbard (1973) uses Arrow's Theorem. The presentation in this chapter closely follows that in Reny (2001), which is both simple and elegant, and which shows the close relation between the two results.

For the Borda scores and the so-called Borda rule see de Borda (1781). The Muller-Satterthwaite Theorem is from Muller and Satterthwaite (1977).

De Condorcet (1785) was the first to explicitly discuss the notion of a Condorcet winner; see Gehrlein (2006) for a comprehensive study of the so-called *Condorcet paradox* in Problem 11.2(b).

References

Arrow, K. J. (1963). *Social choice and individual values* (2nd ed.). New York: Wiley.

Arrow, K. J., Sen, A. K., & Suzumura, K. (Eds.), (2002). *Handbook of social choice and welfare* (Vol. 1). Amsterdam: North-Holland.

Arrow, K. J., Sen, A. K., & Suzumura, K. (Eds.), (2011). *Handbook of social choice and welfare* (Vol. 2). Amsterdam: North-Holland.

de Borda, J. C. (1781). *Mémoires sur les élections au scrutin*. Paris: Histoire de l'Académie Royale des Sciences.

de Condorcet, M. (1785). *Essai sur l'application de l'analyse à la probabilité des décisions rendues à la pluralité des voix*. Paris: Imprimerie Royale.

Gehrlein, W. V. (2006). *Condorcet's paradox*. Berlin: Springer.

Gibbard, A. (1973). Manipulation of voting schemes: a general result. *Econometrica, 41*, 587–601.

Maskin, E. (1999). Nash equilibrium and welfare optimality. *Review of Economic Studies, 66*, 23–38.

Muller, E., & Satterthwaite, M. A. (1977). The equivalence of strong positive association and strategy-proofness. *Journal of Economic Theory, 14*, 412–418.

Reny, P. J. (2001). Arrow's theorem and the Gibbard-Satterthwaite theorem: a unified approach. *Economics Letters, 70*, 99–105.

Satterthwaite, M. A. (1975). Strategy-proofness and Arrow's conditions: existence and correspondence theorems for voting procedures and social welfare functions. *Journal of Economic Theory, 10*, 187–217.

Part II

Noncooperative Games

Matrix Games

12

In this chapter we study finite two-person zero-sum games—matrix games—more rigorously. In particular, von Neumann's Minimax Theorem is proved. The chapter extends Chap. 2 in Part I. Although it is self-contained, it may be useful to (re)read Chap. 2 first.

Section 12.1 presents a proof of the Minimax Theorem, and Sect. 12.2 shows how a matrix game can be solved—i.e., optimal strategies and the value of the game can be found—by solving an associated linear programming problem.

12.1 The Minimax Theorem

A two-person zero-sum game is completely determined by a single matrix. We repeat Definition 2.1.

Definition 12.1 (Matrix Game) A *matrix game* is an $m \times n$ matrix A of real numbers, where m is the number of rows and n is the number of columns. A (*mixed*) *strategy* of player 1 is a probability distribution $\mathbf{p}$ over the rows of A, i.e., an element of the set

$$\Delta^m := \{\mathbf{p} = (p_1, \ldots, p_m) \in \mathbb{R}^m \mid \sum_{i=1}^m p_i = 1, \ p_i \geq 0 \text{ for all } i = 1, \ldots, m\} \ .$$

Similarly, a (*mixed*) *strategy* of player 2 is a probability distribution $\mathbf{q}$ over the columns of A, i.e., an element of the set

$$\Delta^n := \{\mathbf{q} = (q_1, \ldots, q_n) \in \mathbb{R}^n \mid \sum_{j=1}^n q_j = 1, \ q_j \geq 0 \text{ for all } j = 1, \ldots, n\} \ .$$

© Springer-Verlag Berlin Heidelberg 2015
H. Peters, *Game Theory*, Springer Texts in Business and Economics,
DOI 10.1007/978-3-662-46950-7_12

A strategy $\mathbf{p}$ of player 1 is called *pure* if there is a row i with $p_i = 1$. This strategy is also denoted by $\mathbf{e}^i$. Similarly, a strategy $\mathbf{q}$ of player 2 is called *pure* if there is a column j with $q_j = 1$. This strategy is also denoted by $\mathbf{e}^j$. $\qquad\square$

Let A be an $m \times n$ matrix game. For any strategy $\mathbf{p} \in \Delta^m$ of player 1, let $v_1(\mathbf{p}) = \min_{\mathbf{q}\in\Delta^n} \mathbf{p}A\mathbf{q}$. It is easy to see that $v_1(\mathbf{p}) = \min_{j\in\{1,\dots,n\}} \mathbf{p}A\mathbf{e}^j$, since $\mathbf{p}A\mathbf{q}$ is a convex combination of the numbers $\mathbf{p}A\mathbf{e}^j$. In the matrix game A player 1 can guarantee a payoff of at least

$$v_1(A) := \max_{\mathbf{p}\in\Delta^m} v_1(\mathbf{p}) .$$

Similarly, for any strategy $\mathbf{q} \in \Delta^n$ of player 2 let $v_2(\mathbf{q}) = \max_{\mathbf{p}\in\Delta^m} \mathbf{p}A\mathbf{q} = \max_{i\in\{1,\dots,m\}} \mathbf{e}^iA\mathbf{q}$, then player 2 can guarantee to have to pay at most

$$v_2(A) := \min_{\mathbf{q}\in\Delta^n} v_2(\mathbf{q}) .$$

Intuitively, player 1 should not be able to guarantee to obtain more than what player 2 can guarantee to pay maximally. Indeed, we have the following lemma.

Lemma 12.2 *For any $m \times n$ matrix game, $v_1(A) \le v_2(A)$.*

Proof Problem 12.2. $\qquad\blacksquare$

The following theorem is due to von Neumann. The proof is based on Lemma 22.3, which is equivalent to Farkas' Lemma.

Theorem 12.3 (Minimax Theorem for Matrix Games) *For any $m \times n$ matrix game A, $v_1(A) = v_2(A)$.*

Proof Let A be an $m \times n$ matrix game. In view of Lemma 12.2 it is sufficient to prove that $v_1(a) \ge v_2(A)$. Suppose, to the contrary, that $v_1(A) < v_2(A)$. We derive a contradiction, which completes the proof of the theorem.

Let B be any arbitrary $m \times n$ matrix game. Then either (i) or (ii) in Lemma 22.3 has to hold for B, i.e., exactly one of the following holds:

(i) There are $\mathbf{y} \in \mathbb{R}^n$ and $\mathbf{z} \in \mathbb{R}^m$ with $(\mathbf{y},\mathbf{z}) \ge \mathbf{0}$, $(\mathbf{y},\mathbf{z}) \ne \mathbf{0}$ and $B\mathbf{y} + \mathbf{z} = \mathbf{0}$.
(ii) There is an $\mathbf{x} \in \mathbb{R}^m$ with $\mathbf{x} > \mathbf{0}$ and $\mathbf{x}B > \mathbf{0}$.

First suppose that (i) holds and let $\mathbf{y} \in \mathbb{R}^n$ and $\mathbf{z} \in \mathbb{R}^m$ with $(\mathbf{y},\mathbf{z}) \ge \mathbf{0}$, $(\mathbf{y},\mathbf{z}) \ne \mathbf{0}$ and $B\mathbf{y} + \mathbf{z} = \mathbf{0}$. It cannot be the case that $\mathbf{y} = \mathbf{0}$, since that would imply that also $\mathbf{z} = \mathbf{0}$, a contradiction. Hence $\sum_{k=1}^n y_k > 0$. Define $\mathbf{q} \in \Delta^n$ by $q_j = y_j/\sum_{k=1}^n y_k$ for every $j = 1,\dots,n$. Then $B\mathbf{q} = -\mathbf{z}/\sum_{k=1}^n y_k \le \mathbf{0}$. Hence $v_2(\mathbf{q}) \le 0$, and therefore $v_2(B) \le 0$.

Suppose instead that (ii) holds. Then there is an $\mathbf{x} \in \mathbb{R}^m$ with $\mathbf{x} > \mathbf{0}$ and $\mathbf{x}B > \mathbf{0}$. Define $\mathbf{p} \in \Delta^m$ by $\mathbf{p} = \mathbf{x}/\sum_{i=1}^m x_i$, then $v_1(\mathbf{p}) > 0$ and therefore $v_1(B) > 0$.

We conclude that, for any matrix game B, it is not possible to have $v_1(B) \leq 0 < v_2(B)$.

Let now B be the matrix game arising by subtracting the number $v_1(A)$ from all entries of A. Then, clearly, $v_1(B) = v_1(A) - v_1(A) = 0$ and $v_2(B) = v_2(A) - v_1(A) > 0$. Hence, $v_1(B) \leq 0 < v_2(B)$, which is the desired contradiction. ∎

In view of Theorem 12.3 we can define the *value* of the game A by $v(A) = v_1(A) = v_2(A)$. An *optimal strategy* of player 1 is a strategy $\mathbf{p}$ such that $v_1(\mathbf{p}) \geq v(A)$. Similarly, an *optimal strategy* of player 2 is a strategy $\mathbf{q}$ such that $v_2(\mathbf{q}) \leq v(A)$. Theorem 12.3 implies that $v_1(\mathbf{p}) = v_2(\mathbf{q}) = v(A)$ for such optimal strategies. Thus, if $\mathbf{p}$ is an optimal strategy for player 1, then $v_1(\mathbf{p}) = \max_{\mathbf{p}' \in \Delta^m} v_1(\mathbf{p}')$, so that $\mathbf{p}$ is a maximin strategy (cf. Definition 2.3). Conversely, every maximin strategy is an optimal strategy for player 1. Similarly, the optimal strategies for player 2 are exactly the minimax strategies.

For computation of optimal strategies and the value of matrix games in some special cases, see Chap. 2 and Problems 12.3 and 12.4. In general, matrix games can be solved by linear programming. This is demonstrated in the next section.

12.2 A Linear Programming Formulation

Let A be an $m \times n$ matrix game:

$$A = \begin{pmatrix} a_{11} & \cdots & a_{1n} \\ \vdots & \ddots & \vdots \\ a_{m1} & \cdots & a_{mn} \end{pmatrix}.$$

Adding the same number to all entries of A changes the value by that same number but not the optimal strategies of the players. So we may assume without loss of generality that all entries of A are positive. We define the $(m+1) \times (n+1)$ matrix B as follows:

$$B = \begin{pmatrix} a_{11} & \cdots & a_{1n} & -1 \\ \vdots & \ddots & \vdots & \vdots \\ a_{m1} & \cdots & a_{mn} & -1 \\ -1 & \cdots & -1 & 0 \end{pmatrix}.$$

Let $\mathbf{b} = (0, \ldots, 0, -1) \in \mathbb{R}^{n+1}$ and $\mathbf{c} = (0, \ldots, 0, -1) \in \mathbb{R}^{m+1}$. Define $V := \{\mathbf{x} \in \mathbb{R}^{m+1} \mid \mathbf{x}B \geq \mathbf{b}, \mathbf{x} \geq \mathbf{0}\}$ and $W := \{\mathbf{y} \in \mathbb{R}^{n+1} \mid B\mathbf{y} \leq \mathbf{c}, \mathbf{y} \geq \mathbf{0}\}$. It is easy to check that $V, W \neq \emptyset$. The Duality Theorem of Linear Programming (Theorem 22.6) therefore implies:

Corollary 12.4 $\min\{\mathbf{x} \cdot \mathbf{c} \mid \mathbf{x} \in V\} = \max\{\mathbf{b} \cdot \mathbf{y} \mid \mathbf{y} \in W\}$.

The minimum and maximum problems in this corollary are so-called linear programming (LP) problems. If we call the minimization problem the *primal* problem, then the maximization problem is the *dual* problem—or *vice versa*. The common minimum/maximum is called the *value* of the LP, and $\mathbf{x}$ and $\mathbf{y}$ achieving the value are called *optimal solutions*. Denote the sets of optimal solutions by $O_{\min}$ and $O_{\max}$, respectively.

We have the following result. The proof uses Lemma 22.9, which states that if $\hat{\mathbf{x}} \in V$ and $\hat{\mathbf{y}} \in W$ satisfy $\hat{\mathbf{x}} \cdot \mathbf{c} = \mathbf{b} \cdot \hat{\mathbf{y}}$, then $\hat{\mathbf{x}}$ and $\hat{\mathbf{y}}$ are optimal solutions.

Theorem 12.5 *Let A be an $m \times n$ matrix game with all entries positive.*

(i) *If $\mathbf{p} \in \Delta^m$ is an optimal strategy for player 1 and $\mathbf{q} \in \Delta^n$ is an optimal strategy for player 2 in A, then $(\mathbf{p}, v(A)) \in O_{\min}$ and $(\mathbf{q}, v(A)) \in O_{\max}$. The value of the LP is $-v(A)$.*

(ii) *If $\mathbf{x} = (x_1, \ldots, x_m, x_{m+1}) \in O_{\min}$ and $\mathbf{y} = (y_1, \ldots, y_n, y_{n+1}) \in O_{\max}$, then $(x_1, \ldots, x_m)$ is an optimal strategy for player 1 in A, $(y_1, \ldots, y_n)$ is an optimal strategy for player 2 in A, and $v(A) = x_{m+1} = y_{n+1}$.*

Proof

(i) Let $\mathbf{p} \in \Delta^m$ and $\mathbf{q} \in \Delta^n$ be optimal strategies in the matrix game A. Then $\mathbf{p}A\mathbf{e}^j \geq v(A)$ and $\mathbf{e}^i A\mathbf{q} \leq v(A)$ for all $i = 1, \ldots, m$ and $j = 1, \ldots, n$. Since all entries of A are positive and therefore $v(A) > 0$, this implies $(\mathbf{p}, v(A)) \in V$ and $(\mathbf{q}, v(A)) \in W$. Since $(\mathbf{p}, v(A)) \cdot \mathbf{c} = -v(A)$ and $(\mathbf{q}, v(A)) \cdot \mathbf{b} = -v(A)$, Lemma 22.9 implies that the value of the LP is $-v(A)$, $(\mathbf{p}, v(A)) \in O_{\min}$ and $(\mathbf{q}, v(A)) \in O_{\max}$.

(ii) Let $\mathbf{x} = (x_1, \ldots, x_{m+1}) \in O_{\min}$. Since $\mathbf{x} \cdot \mathbf{c} = -v(A)$ by (i), we have $x_{m+1} = v(A)$. Since $\mathbf{x}B \geq \mathbf{b}$, we have $(x_1, \ldots, x_m)A\mathbf{e}^j \geq v(A)$ for all $j = 1, \ldots, n$, $x_i \geq 0$ for all $i = 1, \ldots, m$, and $\sum_{i=1}^{m} x_i \leq 1$. Suppose that $\sum_{i=1}^{m} x_i < 1$. Obviously, $\sum_{i=1}^{m} x_i > 0$, otherwise $\mathbf{x} = (0, \ldots, 0, v(A)) \notin V$ since $v(A) > 0$. Then, letting $t = (\sum_{i=1}^{m} x_i)^{-1} > 1$, we have $t\mathbf{x} \in V$ and $t\mathbf{x} \cdot \mathbf{c} = -t\,v(A) < -v(A)$, contradicting $\mathbf{x} \in O_{\min}$. Hence, $\sum_{i=1}^{m} x_i = 1$, and $(x_1, \ldots, x_m)$ is an optimal strategy of player 1 in A.

The proof of the second part of (ii) is analogous.   ∎

The interest of this theorem derives from the fact that solving linear programming problems is a well established area. Thus, one can apply any (computer) method for solving LPs to find the value and optimal strategies of a matrix game.

By slightly modifying part (ii) of the proof of Theorem 12.5, we can in fact derive the Minimax Theorem from the Duality Theorem (Problem 12.5). Conversely, with each LP we can associate a matrix game and thereby derive the Duality

Theorem from the Minimax Theorem. This confirms the close relationship between linear programming (Duality Theorem) and the theory of matrix games (Minimax Theorem).

12.3 Problems

12.1. *Solving a Matrix Game*
Consider the matrix game

$$A = \begin{pmatrix} 6 & 4 & 2 & 1 \\ 5 & 3 & 3 & 2 \\ 1 & 0 & 3 & 4 \\ 2 & -3 & 2 & 3 \end{pmatrix}.$$

(a) Reduce the game by iterated elimination of strictly dominated strategies. Describe exactly which pure strategy you eliminate each time, and by which pure or mixed strategy the strategy to be eliminated is strictly dominated. [See Chap. 2, also for the following questions.]

Denote the reduced game derived in (a) by B.

(b) Solve B graphically. Explicitly compute $v_1(\mathbf{p})$ and $v_2(\mathbf{q})$ for strategies $\mathbf{p}$ of player 1 and $\mathbf{q}$ of player 2. Determine the value of B and the optimal strategy or strategies of players 1 and 2 in B.

(c) Determine the value of A and the optimal strategy or strategies of players 1 and 2 in A.

Now change the entry a_{11} from 6 to $y \in \mathbb{R}$, so that we obtain the matrix game

$$A_y = \begin{pmatrix} y & 4 & 2 & 1 \\ 5 & 3 & 3 & 2 \\ 1 & 0 & 3 & 4 \\ 2 & -3 & 2 & 3 \end{pmatrix}.$$

(d) Compute $v(A_y)$ and the optimal strategies in A_y for every $y \in \mathbb{R}$.

12.2. *Proof of Lemma 12.2*
Prove Lemma 12.2.

12.3. 2×2 *Games*

Consider the 2×2 matrix game

$$A = \begin{pmatrix} a_{11} & a_{12} \\ a_{21} & a_{22} \end{pmatrix}.$$

Assume that A has no saddlepoints. [A saddlepoint is an entry (i, j) such that a_{ij} is maximal in column j and minimal in row i, cf. Definition 2.4.]

(a) Assume that $a_{11} > a_{12}$. Show that

$$a_{12} < a_{22}, \; a_{21} < a_{22}, \; a_{11} > a_{21} \; .$$

(b) Show that the unique optimal strategies **p** and **q** and the value of the game are given by:

$$\mathbf{p} = \frac{\mathbf{J}A^*}{\mathbf{J}A^*\mathbf{J}^T}, \quad \mathbf{q} = \frac{A^*\mathbf{J}^T}{\mathbf{J}A^*\mathbf{J}^T}, \quad v(A) = \frac{|A|}{\mathbf{J}A^*\mathbf{J}^T} \; ,$$

where A^* is the adjoint matrix of A, i.e.,

$$A^* = \begin{pmatrix} a_{22} & -a_{12} \\ -a_{21} & a_{11} \end{pmatrix} ,$$

$|A|$ is the determinant of A, and $\mathbf{J} := (1, 1)$.[1]

12.4. *Symmetric Games*

An $m \times n$ matrix game $A = (a_{ij})$ is called *symmetric* if $m = n$ and $a_{ij} = -a_{ji}$ for all $i, j = 1, \ldots, m$. Prove that the value of a symmetric game is zero and that the sets of optimal strategies of players 1 and 2 coincide.

12.5. *The Duality Theorem Implies the Minimax Theorem*

Modify the proof of part (ii) of Theorem 12.5 in order to derive the Minimax Theorem from the Duality Theorem. [Hint: first show that the value of the LP must be negative.]

12.6. *Infinite Matrix Games*

Consider the following two-player game. Each player mentions a natural number. The player with the higher number receives one Euro from the player with the lower number. If the numbers are equal then no player receives anything.

---

[1] **J** denotes the row vector and $\mathbf{J}^T$ the transpose, i.e., the column vector. In general, we omit the transpose notation if confusion is unlikely.

(a) Write this game in the form of an infinite matrix game A.

(b) Compute $\sup_p \inf_q pAq$ and $\inf_q \sup_p pAq$, where p and q are probability distributions over the rows and the columns of A, respectively. (Conclude that this game has no 'value'.)

12.7. *Equalizer Theorem*

Let v be the value of the $m \times n$-matrix game A, and suppose that $pAe^n = v$ for every optimal strategy p of player 1. Show that player 2 has an optimal strategy q with $q_n > 0$.

12.4 Notes

The Minimax Theorem was first proved in von Neumann (1928). The simplex algorithm was developed by George Dantzig in 1947; see Dantzig (1963) or any textbook on linear programming or operations research.

With each linear programming problem we can associate a matrix game and thereby derive the Duality Theorem from the Minimax Theorem: see, e.g., Owen (1995).

For Problem 12.7 see also Raghavan (1994).

References

Dantzig, G. B. (1963). *Linear programming and extensions*. Princeton: Princeton University Press.
Owen, G. (1995). *Game theory* (3rd ed.). San Diego: Academic.
Raghavan, T. E. S. (1994). *Zero-sum two-person games*, Chap. 20. In R. J. Aumann, & S. Hart (Eds.), *Handbook of game theory with economic applications* (Vol. 2). Amsterdam: North-Holland.
von Neumann, J. (1928). Zur Theorie der Gesellschaftsspiele. *Mathematische Annalen, 100*, 295–320.

Finite Games

13

This chapter builds on Chap. 3, where we studied finite two person games—bimatrix games. (Re)reading Chap. 3 may serve as a good preparation for the present chapter, which offers a more rigorous treatment of finite games, i.e., games with finitely many players—often two—who have finitely many pure strategies over which they can randomize. We only discuss games with complete information. In the terminology of Chap. 5, each player has only one type.

In Sect. 13.1 a proof of Nash's existence theorem is provided. Section 13.2 goes deeper into bimatrix games. In Sect. 13.3 the notion of iterated dominance is studied, and its relation with rationalizability indicated. Sections 13.4–13.6 present some basics about refinements of Nash equilibrium. Section 13.7 is on correlated equilibrium in bimatrix games, and Sect. 13.8 concludes with an axiomatic characterization of Nash equilibrium based on a reduced game (consistency) condition.

13.1 Existence of Nash Equilibrium

We start with a general definition of a finite game. Matrix and bimatrix games are special cases.

A *finite game* is a $2n + 1$-tuple

$$G = (N, S_1, \ldots, S_n, u_1, \ldots, u_n) ,$$

where

- $N = \{1, \ldots, n\}$, with $n \in \mathbb{N}$, is the set of *players*;
- for every $i \in N$, S_i is the finite *pure strategy set* of player i;
- for every $i \in N$, $u_i : S = S_1 \times \ldots \times S_n \to \mathbb{R}$ is the *payoff function* of player i; i.e., for every pure strategy combination $(s_1, \ldots, s_n) \in S$ where $s_1 \in S_1, \ldots, s_n \in S_n$, $u_i(s_1, \ldots, s_n) \in \mathbb{R}$ is player i's payoff.

© Springer-Verlag Berlin Heidelberg 2015
H. Peters, *Game Theory*, Springer Texts in Business and Economics,
DOI 10.1007/978-3-662-46950-7_13

This definition is identical to the definition of an n-person game in Chap. 6, except that the pure strategy sets are now finite. The elements of S_i are the pure strategies of player i. A (mixed) *strategy* of player i is a probability distribution over S_i. The set of (mixed) strategies of player i is denoted by $\Delta(S_i)$. Observe that, whenever we talk about a strategy, we mean a mixed strategy (which may of course be pure).

Let $(\sigma_1, \ldots, \sigma_n) \in \Delta(S_1) \times \ldots \times \Delta(S_n)$ be a strategy combination. Player i's *payoff* from this strategy combination is defined to be his expected payoff. With some abuse of notation this is also denoted by $u_i(\sigma_1, \ldots, \sigma_n)$. Formally,

$$u_i(\sigma_1, \ldots, \sigma_n) = \sum_{(s_1, \ldots, s_n) \in S} \left(\prod_{j \in N} \sigma_j(s_j) \right) u_i(s_1, \ldots, s_n) \, .$$

For a strategy combination σ and a player $i \in N$ we denote by (σ_i', σ_{-i}) the strategy combination in which player i plays $\sigma_i' \in \Delta(S_i)$ and each player $j \neq i$ plays σ_j.

A *best reply* of player i to the strategy combination σ_{-i} of the other players is a strategy $\sigma_i \in \Delta(S_i)$ such that $u_i(\sigma_i, \sigma_{-i}) \geq u_i(\sigma_i', \sigma_{-i})$ for all $\sigma_i' \in \Delta(S_i)$.

A *Nash equilibrium* of G is a strategy combination $\sigma^* \in \prod_{i \in N} \Delta(S_i)$ such that for each player i, σ_i^* is a best reply to σ_{-i}^*.

As in Chaps. 3 and 6, β_i denotes player i's *best reply correspondence*. That is, $\beta_i : \prod_{j \in N, j \neq i} \Delta(S_j) \to \Delta(S_i)$ assigns to each strategy combination of the other players the set of all best replies of player i.

Theorem 13.1 (Existence of Nash Equilibrium) *Every finite game* $G = (N, S_1, \ldots, S_n, u_1, \ldots, u_n)$ *has a Nash equilibrium.*

The proof of this theorem below is based on the Kakutani Fixed Point Theorem (Sect. 22.5).

Proof of Theorem 13.1 Consider the correspondence

$$\beta : \prod_{i \in N} \Delta(S_i) \to \prod_{i \in N} \Delta(S_i), \; (\sigma_1, \ldots, \sigma_n) \mapsto \prod_{i \in N} \beta_i(\sigma_1, \ldots, \sigma_{i-1}, \sigma_{i+1}, \ldots, \sigma_n) \, .$$

This correspondence is convex-valued and upper semi-continuous (Problem 13.2). By the Kakutani Fixed Point Theorem (Theorem 22.11) it has a fixed point σ^*. By definition of β, any fixed point is a Nash equilibrium of G. ∎

An alternative proof is obtained by using the Brouwer Fixed Point Theorem (Theorem 22.10). See Problem 13.1.

13.2 Bimatrix Games

Two-person finite games—bimatrix games—were studied in Chap. 3. Here we present some extensions. In Sect. 13.2.1 we give some formal relations between pure and mixed strategies in a Nash equilibrium. In Sect. 13.2.2 we extend the graphical method for computing Nash equilibria (cf. Sect. 3.2.2). In Sect. 13.2.3 a general mathematical programming method is described by which equilibria of bimatrix games can be found. Section 13.2.4 reconsiders matrix games as a special kind of bimatrix games. Section 13.2.5 is about Zermelo's theorem on the game of chess.

13.2.1 Pure and Mixed Strategies in Nash Equilibrium

Let (A, B) be an $m \times n$ bimatrix game (Definition 3.1). The first lemma implies that to determine whether a strategy pair is a Nash equilibrium it is sufficient to compare the expected payoff of a (mixed) strategy with the payoffs of pure strategies.

Lemma 13.2 *Let* $\mathbf{p} \in \Delta^m$ *and* $\mathbf{q} \in \Delta^n$. *Then* $\mathbf{p} \in \beta_1(\mathbf{q})$ *if and only if* $\mathbf{p}A\mathbf{q} \geq \mathbf{e}^i A\mathbf{q}$ *for all* $i = 1, \ldots, m$; *and* $\mathbf{q} \in \beta_2(\mathbf{p})$ *if and only if* $\mathbf{p}B\mathbf{q} \geq \mathbf{p}B\mathbf{e}^j$ *for all* $j = 1, \ldots, n$.

Proof Problem 13.3. ∎

The next lemma says that a player always has a pure best reply against any strategy of the opponent.

Lemma 13.3 *Let* $\mathbf{p} \in \Delta^m$ *and* $\mathbf{q} \in \Delta^n$. *Then there is an* $i \in \{1, \ldots, m\}$ *with* $\mathbf{e}^i \in \beta_1(\mathbf{q})$ *and a* $j \in \{1, \ldots, n\}$ *with* $\mathbf{e}^j \in \beta_2(\mathbf{p})$.

Proof Problem 13.4. ∎

In light of these lemmas it makes sense to introduce the pure best reply correspondences.

Definition 13.4 Let (A, B) be an $m \times n$ bimatrix game and let $\mathbf{p} \in \Delta^m$ and $\mathbf{q} \in \Delta^n$. Then

$$PB_1(\mathbf{q}) = \{i \in \{1, \ldots, m\} \mid \mathbf{e}^i A\mathbf{q} = \max_k \mathbf{e}^k A\mathbf{q}\}$$

is the set of *pure best replies* of player 1 to $\mathbf{q}$ and

$$PB_2(\mathbf{p}) = \{j \in \{1, \ldots, n\} \mid \mathbf{p}B\mathbf{e}^j = \max_k \mathbf{p}B\mathbf{e}^k\}$$

is the set of *pure best replies* of player 2 to $\mathbf{p}$. ☐

Observe that, with some abuse of notation, the pure best replies in this definition are labelled by the row and column numbers.

The *carrier* $C(\mathbf{p})$ of a mixed strategy $\mathbf{p} \in \Delta^k$, where $k \in \mathbb{N}$, is the set of coordinates that are positive, i.e.,

$$C(\mathbf{p}) = \{i \in \{1, \ldots, k\} \mid p_i > 0\} \, .$$

The next lemma formalizes the observation used already in Chap. 3, namely that in a best reply a player puts positive probability only on those pure strategies that maximize his expected payoff (cf. Problem 3.8).

Lemma 13.5 *Let* (A, B) *be an* $m \times n$ *bimatrix game,* $\mathbf{p} \in \Delta^m$ *and* $\mathbf{q} \in \Delta^n$. *Then*

$$\mathbf{p} \in \beta_1(\mathbf{q}) \Leftrightarrow C(\mathbf{p}) \subseteq PB_1(\mathbf{q})$$

and

$$\mathbf{q} \in \beta_2(\mathbf{p}) \Leftrightarrow C(\mathbf{q}) \subseteq PB_2(\mathbf{p}) \, .$$

Proof We only show the first equivalence.

First let $\mathbf{p} \in \beta_1(\mathbf{q})$, and assume $i \in C(\mathbf{p})$ and, contrary to what we want to prove, that $\mathbf{e}^i A\mathbf{q} < \max_k \mathbf{e}^k A\mathbf{q}$. Then

$$\mathbf{p}A\mathbf{q} = \max_k \mathbf{e}^k A\mathbf{q} = \sum_{k'=1}^{m} p_{k'} \max_k \mathbf{e}^k A\mathbf{q} > \sum_{k=1}^{m} p_k \mathbf{e}^k A\mathbf{q} = \mathbf{p}A\mathbf{q} \, ,$$

where the first equality follows from Lemma 13.3. This is a contradiction, hence $\mathbf{e}^i A\mathbf{q} = \max_k \mathbf{e}^k A\mathbf{q}$ and $i \in PB_1(\mathbf{q})$.

Next, assume that $C(\mathbf{p}) \subseteq PB_1(\mathbf{q})$. Then

$$\mathbf{p}A\mathbf{q} = \sum_{i=1}^{m} p_i \mathbf{e}^i A\mathbf{q} = \sum_{i \in C(\mathbf{p})} p_i \mathbf{e}^i A\mathbf{q} = \sum_{i \in C(\mathbf{p})} p_i \max_k \mathbf{e}^k A\mathbf{q} = \max_k \mathbf{e}^k A\mathbf{q} \, .$$

So $\mathbf{p}A\mathbf{q} \geq \mathbf{e}^i A\mathbf{q}$ for all $i = 1, \ldots, m$, which by Lemma 13.2 implies $\mathbf{p} \in \beta_1(\mathbf{q})$. ∎

The following corollary is an immediate consequence of Lemma 13.5. It is, in principle, helpful to find Nash equilibria or to determine whether a given strategy combination is a Nash equilibrium. See Example 13.7.

Corollary 13.6 *A strategy pair* $(\mathbf{p}, \mathbf{q})$ *is a Nash equilibrium in a bimatrix game* (A, B) *if and only if* $C(\mathbf{p}) \subseteq PB_1(\mathbf{q})$ *and* $C(\mathbf{q}) \subseteq PB_2(\mathbf{p})$.

Example 13.7 Consider the bimatrix game

$$(A, B) = \begin{pmatrix} 1,1 & 0,1 & 0,1 & 0,1 \\ 1,1 & 1,1 & 0,1 & 0,1 \\ 1,1 & 1,1 & 1,1 & 0,1 \\ 1,1 & 1,1 & 1,1 & 1,1 \end{pmatrix}$$

and the strategies $\mathbf{p} = (0, \frac{1}{3}, \frac{1}{3}, \frac{1}{3})$ and $\mathbf{q} = (\frac{1}{2}, \frac{1}{2}, 0, 0)$. Since

$$A\mathbf{q} = \begin{pmatrix} \frac{1}{2} \\ 1 \\ 1 \\ 1 \end{pmatrix} \quad \text{and} \quad \mathbf{p}B = \begin{pmatrix} 1 & 1 & 1 & 1 \end{pmatrix},$$

$PB_1(\mathbf{q}) = \{2, 3, 4\}$ and $PB_2(\mathbf{p}) = \{1, 2, 3, 4\}$. Since $C(\mathbf{p}) = \{2, 3, 4\}$ and $C(\mathbf{q}) = \{1, 2\}$, we have $C(\mathbf{p}) \subseteq PB_1(\mathbf{q})$ and $C(\mathbf{q}) \subseteq PB_2(\mathbf{p})$. So Corollary 13.6 implies that $(\mathbf{p}, \mathbf{q})$ is a Nash equilibrium.

13.2.2 Extension of the Graphical Method

In Sect. 3.2.2 we learnt how to solve 2×2 bimatrix games graphically. We now extend this method to 2×3 and 3×2 games. For larger games it becomes impractical or impossible to use this graphical method.

As an example consider the 2×3 bimatrix game

$$(A, B) = \begin{pmatrix} 2,1 & 1,0 & 1,1 \\ 2,0 & 1,1 & 0,0 \end{pmatrix}.$$

The Nash equilibria of this game are elements of the set $\Delta^2 \times \Delta^3$ of all possible strategy combinations. This set can be represented as in Fig. 13.1.

Here player 2 chooses a point in the triangle with vertices $\mathbf{e}_1, \mathbf{e}_2$ and $\mathbf{e}_3$, while player 1 chooses a point of the horizontal line segment with vertices $\mathbf{e}_1$ and $\mathbf{e}_2$.

In order to determine the best replies of player 1 note that

$$A\mathbf{q} = \begin{pmatrix} 2q_1 + q_2 + q_3 \\ 2q_1 + q_2 \end{pmatrix}.$$

Fig. 13.1 The set $\Delta^2 \times \Delta^3$

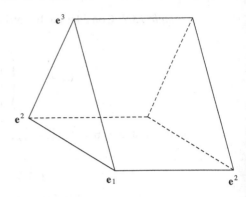

Fig. 13.2 The best reply correspondence of player 1 (*shaded*)

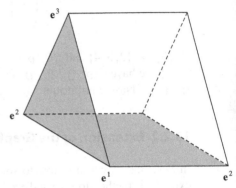

As $e_1Aq = e_2Aq \Leftrightarrow q_3 = 0$, it follows that

$$\beta_1(\mathbf{q}) = \begin{cases} \{e^1\} & \text{if } q_3 > 0 \\ \Delta^2 & \text{if } q_3 = 0 . \end{cases}$$

This yields the best reply correspondence represented in Fig. 13.2.
    Similarly,

$$\mathbf{p}B = \left(p_1 \ p_2 \ p_1 \right)$$

implies

$$\beta_2(\mathbf{p}) = \begin{cases} \{e^2\} & \text{if } p_1 < p_2 \\ \Delta^3 & \text{if } p_1 = p_2 \\ \{\mathbf{q} \in \Delta^3 \mid q_2 = 0\} & \text{if } p_1 > p_2. \end{cases}$$

This yields the best reply correspondence represented in Fig. 13.3.
    Figure 13.4 represents the intersection of the two best reply correspondences and, thus, the set of Nash equilibria.

Fig. 13.3 The best reply correspondence of player 2 (*shaded/thick*)

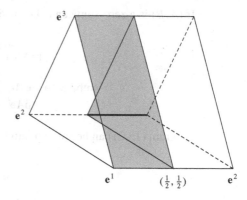

Fig. 13.4 The set of Nash equilibria:
$\{((1,0),\mathbf{q}) \mid q_2 = 0\} \cup$
$\{(\mathbf{p},(1,0,0)) \mid 1 \geq p_1 \geq \frac{1}{2}\}$
$\cup \{((\frac{1}{2},\frac{1}{2}),\mathbf{q}) \mid q_3 = 0\} \cup$
$\{(\mathbf{p},(0,1,0)) \mid \frac{1}{2} \geq p_1 \geq 0\}$

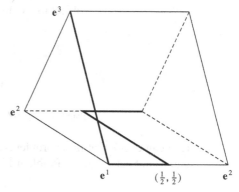

13.2.3 A Mathematical Programming Approach

In Sect. 12.2 we have seen that matrix games can be solved by linear programming. Nash equilibria of an $m \times n$ bimatrix game (A, B) can be found by considering the following *quadratic* programming problem:

$$\max_{\mathbf{p}\in\Delta^m,\ \mathbf{q}\in\Delta^n,\ a,b\in\mathbb{R}} \quad f(\mathbf{p},\mathbf{q},a,b) := \mathbf{p}A\mathbf{q} + \mathbf{p}B\mathbf{q} - a - b$$

$$\text{subject to} \quad \mathbf{e}^i A\mathbf{q} \leq a \ \text{ for all } i = 1,2,\ldots,m$$
$$\mathbf{p}B\mathbf{e}^j \leq b \ \text{ for all } j = 1,2,\ldots,n. \tag{13.1}$$

Theorem 13.8 *The following two statements are equivalent:*

(1) $(\mathbf{p},\mathbf{q},a,b)$ *is a solution of* (13.1)
(2) $(\mathbf{p},\mathbf{q})$ *is a Nash equilibrium of* (A, B), $a = \mathbf{p}A\mathbf{q}$, $b = \mathbf{p}B\mathbf{q}$.

Proof Problem 13.9.   ■

If (A, B) is a zero-sum game, i.e., if $B = -A$, then (13.1) reduces to

$$\max_{\mathbf{p}\in\Delta^m,\ \mathbf{q}\in\Delta^n,\ a,b\in\mathbb{R}} -a - b$$

$$\text{subject to}\quad \mathbf{e}^i A\mathbf{q} \le a \ \text{ for all } i = 1, 2, \ldots, m$$
$$-\mathbf{p}A\mathbf{e}^j \le b \ \text{ for all } j = 1, 2, \ldots, n . \tag{13.2}$$

Program (13.2) can be split up into two independent programs

$$\max_{\mathbf{q}\in\Delta^n,\ a\in\mathbb{R}} -a$$

$$\text{subject to}\quad \mathbf{e}^i A\mathbf{q} \le a \ \text{ for all } i = 1, 2, \ldots, m \tag{13.3}$$

and

$$\min_{\mathbf{p}\in\Delta^m,\ b\in\mathbb{R}} b$$

$$\text{subject to}\quad \mathbf{p}A\mathbf{e}^j \ge -b \ \text{ for all } j = 1, 2, \ldots, n . \tag{13.4}$$

One can check that these problems are equivalent to the LP and its dual for matrix games in Sect. 12.2, see Problem 13.10.

13.2.4 Matrix Games

Since matrix games are also bimatrix games, everything that we know about bimatrix games is also true for matrix games. In fact, the Minimax Theorem (Theorem 12.3) can be derived directly from the existence theorem for Nash equilibrium (Theorem 13.1). Moreover, each Nash equilibrium in a matrix game consists of a pair of optimal (maximin and minimax) strategies, and each such pair is a Nash equilibrium. As a consequence, in a matrix game, Nash equilibrium strategies are exchangeable—there is no coordination problem, and all Nash equilibria result in the same payoffs.

All these facts are collected in the following theorem. For terminology concerning matrix games see Chap. 12. The 'new' contribution of this theorem is part (2), part (1) is just added to provide an alternative proof of the Minimax Theorem.

Theorem 13.9 *Let A be an $m \times n$ matrix game. Then:*

(1) $v_1(A) = v_2(A)$.
(2) *A pair $(\mathbf{p}^*, \mathbf{q}^*) \in \Delta^m \times \Delta^n$ is a Nash equilibrium of $(A, -A)$ if and only if $\mathbf{p}^*$ is an optimal strategy for player 1 in A and $\mathbf{q}^*$ is an optimal strategy for player 2 in A.*

Proof

(1) In view of Lemma 12.2 it is sufficient to prove that $v_1(A) \geq v_2(A)$. Choose $(\mathbf{p}^*, \mathbf{q}^*) \in \Delta^m \times \Delta^n$ to be a Nash equilibrium of $(A, -A)$—this is possible by Theorem 13.1. Then

$$\mathbf{p}A\mathbf{q}^* \leq \mathbf{p}^*A\mathbf{q}^* \leq \mathbf{p}^*A\mathbf{q} \text{ for all } \mathbf{p} \in \Delta^m, \mathbf{q} \in \Delta^n .$$

This implies $\max_{\mathbf{p}} \mathbf{p}A\mathbf{q}^* \leq \mathbf{p}^*A\mathbf{q}$ for all $\mathbf{q}$, hence $v_2(A) = \min_{\mathbf{q}'} \max_{\mathbf{p}} \mathbf{p}A\mathbf{q}' \leq \mathbf{p}^*A\mathbf{q}$ for all $\mathbf{q}$. So

$$v_2(A) \leq \min_{\mathbf{q}} \mathbf{p}^*A\mathbf{q} \leq \max_{\mathbf{p}} \min_{\mathbf{q}} \mathbf{p}A\mathbf{q} = v_1(A) .$$

(2) First, suppose that $(\mathbf{p}^*, \mathbf{q}^*) \in \Delta^m \times \Delta^n$ is a Nash equilibrium of $(A, -A)$. Then

$$\mathbf{p}^*A\mathbf{q}^* = \max_{\mathbf{p}} \mathbf{p}A\mathbf{q}^* = v_2(\mathbf{q}^*) \geq \min_{\mathbf{q}} v_2(\mathbf{q}) = v_2(A) = v(A) .$$

If $\mathbf{p}^*$ were not optimal, then $\mathbf{p}^*A\mathbf{q} < v(A)$ for some $\mathbf{q} \in \Delta^n$, so $\mathbf{p}^*A\mathbf{q}^* \leq \mathbf{p}^*A\mathbf{q} < v(A)$, a contradiction. Similarly, $\mathbf{q}^*$ must be optimal.

Conversely, suppose that $\mathbf{p}^*$ and $\mathbf{q}^*$ are optimal strategies. Since $\mathbf{p}A\mathbf{q}^* \leq v(A)$ for all $\mathbf{p} \in \Delta^m$ and $\mathbf{p}^*A\mathbf{q} \geq v(A)$ for all $\mathbf{q} \in \Delta^n$, it follows that $\mathbf{p}^*$ and $\mathbf{q}^*$ are mutual best replies and, thus, $(\mathbf{p}^*, \mathbf{q}^*)$ is a Nash equilibrium in $(A, -A)$. ∎

13.2.5 The Game of Chess: Zermelo's Theorem

One of the earliest formal results in game theory is Zermelo's Theorem on the game of chess. In this subsection we provide a simple proof of this theorem, based on Theorem 13.9.

The game of chess is a classical example of a zero-sum game. There are three possible outcomes: a win for White, a win for Black, and a draw. Identifying player 1 with White and player 2 with Black, we can associate with these outcomes the payoffs $(1, -1)$, $(-1, 1)$, and $(0, 0)$, respectively. In order to guarantee that the (extensive form) game stops after finitely many moves, we assume the following stopping rule: if the same configuration on the chess board has occurred more than twice, the game ends in a draw. Since there are only finitely many configurations on the chess board, the game must stop after finitely many moves. Note that the chess game is a finite extensive form game of perfect information and therefore it has a Nash equilibrium in pure strategies—see Sect. 4.3. To be precise, this is a pure strategy Nash equilibrium in the associated matrix game, where mixed strategies are allowed as well.

Theorem 13.10 (Zermelo's Theorem) *In the game of chess, either White has a pure strategy that guarantees a win, or Black has a pure strategy that guarantees a win, or both players have pure strategies that guarantee at least a draw.*

Proof Let $A = (a_{ij})$ denote the associated matrix game, and let row i^* and column j^* constitute a pure strategy Nash equilibrium. We distinguish three cases.

Case 1. $a_{i^*j^*} = 1$, i.e., White wins. By Theorem 13.9, $v(A) = 1$. Hence, White has a pure strategy that guarantees a win, namely to play row i^*.

Case 2. $a_{i^*j^*} = -1$, i.e., Black wins. By Theorem 13.9, $v(A) = -1$. Hence, Black has a pure strategy that guarantees a win, namely to play column j^*.

Case 3. $a_{i^*j^*} = 0$, i.e., the game ends in a draw. By Theorem 13.9, $v(A) = 0$. Hence, both White and Black can guarantee at least a draw by playing row i^* and column j^*, respectively. ∎

13.3   Iterated Dominance and Best Reply

A pure strategy of a player in a finite game is strictly dominated if there is another (mixed or pure) strategy that yields always—whatever the other players do—a strictly higher payoff. Such a strategy is not played in a Nash equilibrium, and can therefore be eliminated. In the smaller game there may be another pure strategy of the same or of another player that is strictly dominated and again may be eliminated. This way a game may be reduced to a smaller game for which it is easier to compute the Nash equilibria. If the procedure results in a unique surviving strategy combination then the game is called *dominance solvable*, but this is a rare exception.

We applied these ideas before, in Chaps. 2 and 3. In this section we show, formally, that by this procedure of iterated elimination of strictly dominated strategies no Nash equilibria of the original game are lost, and no Nash equilibria are added.

For iterated elimination of weakly dominated strategies the situation is different: Nash equilibria may be lost, and the final result may depend on the order of elimination. See Problem 3.6.

We start with repeating the definition of a strictly dominated strategy for an arbitrary finite game.

Definition 13.11 Let $G = (N, S_1, \ldots, S_n, u_1, \ldots, u_n)$ be a finite game, $i \in N$, $s_i \in S_i$. Strategy s_i is *strictly dominated by strategy* $\sigma_i \in \Delta(S_i)$ if $u_i(\sigma_i, \sigma_{-i}) > u_i(s_i, \sigma_{-i})$ for all $\sigma_{-i} \in \prod_{j \neq i} \Delta(S_j)$. Strategy $s_i \in S_i$ is *strictly dominated* if it is strictly dominated by some strategy $\sigma_i \in \Delta(S_i)$. □

The fact that iterated elimination of strictly dominated strategies does not essentially change the set of Nash equilibria of a game is a straightforward consequence of the following lemma.

Lemma 13.12 *Let $G = (N, S_1, \ldots, S_n, u_1, \ldots, u_n)$ be a finite game, $i \in N$, and let $s_i \in S_i$ be strictly dominated. Let $G' = (N, S_1, \ldots, S_{i-1}, S_i \setminus \{s_i\}, S_{i+1}, \ldots, S_n, u_1', \ldots, u_n')$ be the game arising from G be eliminating s_i from S_i and restricting the payoff functions accordingly. Then:*

(1) *If σ is a Nash equilibrium in G, then $\sigma_i(s_i) = 0$ (where $\sigma_i(s_i)$ is the probability assigned by σ_i to pure strategy $s_i \in S_i$) and σ' is a Nash equilibrium in G', where $\sigma_j' = \sigma_j$ for each $j \in N \setminus \{i\}$ and σ_i' is the restriction of σ_i to $S_i \setminus \{s_i\}$.*

(2) *If σ' is a Nash equilibrium in G', then σ is a Nash equilibrium in G, where $\sigma_j = \sigma_j'$ for each $j \in N \setminus \{i\}$ and $\sigma_i(t_i) = \sigma_i'(t_i)$ for all $t_i \in S_i \setminus \{s_i\}$.*

Proof

(1) Let σ be a Nash equilibrium in G, and let $\tau_i \in \Delta(S_i)$ strictly dominate s_i. If $\sigma_i(s_i) > 0$, then

$$u_i(\hat{\sigma}_i + \sigma_i(s_i)\tau_i, \sigma_{-i}) > u_i(\sigma_i, \sigma_{-i}) \,,$$

where $\hat{\sigma}_i : S_i \to \mathbb{R}$ is defined by $\hat{\sigma}_i(t_i) = \sigma_i(t_i)$ for all $t_i \in S_i \setminus \{s_i\}$ and $\hat{\sigma}_i(s_i) = 0$. This contradicts the assumption that σ is a Nash equilibrium in G. Therefore, $\sigma_i(s_i) = 0$. With σ' and G' as above, we have

$$u_i'(\sigma_1', \ldots, \sigma_{i-1}', \tau_i', \sigma_{i+1}', \ldots, \sigma_n') = u_i(\sigma_1, \ldots, \sigma_{i-1}, \bar{\tau}_i', \sigma_{i+1}, \ldots, \sigma_n) \,,$$

for every $\tau_i' \in \Delta(S_i \setminus \{s_i\})$, where $\bar{\tau}_i' \in \Delta(S_i)$ assigns 0 to s_i and is equal to τ_i' otherwise. From this it follows that σ_i' is still a best reply to σ_{-i}'. It is straightforward that also for each $j \neq i$, σ_j' is still a best reply to σ_{-j}'. Hence σ' is a Nash equilibrium in G'.

(2) Let σ' and σ be as in (2) of the lemma. Obviously, for every player $j \neq i$, σ_j is still a best reply in σ since $\sigma_i(s_i) = 0$, i.e., player i puts zero probability on the new pure strategy s_i. For player i, σ_i is certainly a best reply among all strategies that put zero probability on s_i. But then, σ_i is a best reply among all strategies, since strategies that put nonzero probability on s_i can never be best replies by the first argument in the proof of (1). Hence, σ is a Nash equilibrium in G. ∎

Obviously, a strictly dominated pure strategy is not only never played in a Nash equilibrium, but, a fortiori, is never (part of) a best reply. Formally, we say that a pure strategy s_i of player i in the finite game $G = (N, S_1, \ldots, S_n, u_1, \ldots, u_n)$ is *never a best reply* if for all $(\sigma_j)_{j \neq i}$ and all $\sigma_i \in \beta_i((\sigma_j)_{j \neq i})$, we have $\sigma_i(s_i) = 0$. The following result shows that for two-player games also the converse holds.

Theorem 13.13 *In a finite two-person game every pure strategy that is never a best reply, is strictly dominated.*

Proof Let (A, B) be an $m \times n$ bimatrix game and suppose without loss of generality that pure strategy $\mathbf{e}^1 \in \Delta^m$ of player 1 is never a best reply. Let $\mathbf{b} = (-1, -1, \ldots, -1) \in \mathbb{R}^n$.

Let $\tilde{A}$ be the $(m-1) \times n$ matrix with i-th row equal to the first row of A minus the $i+1$-th row of A, i.e., $\tilde{a}_{ij} = a_{1j} - a_{i+1,j}$ for every $i = 1, \ldots, m-1$ and $j = 1, \ldots, n$. Thus,

$$\tilde{A} = \begin{pmatrix} a_{11} - a_{21} & \cdots & a_{1n} - a_{2n} \\ a_{11} - a_{31} & \cdots & a_{1n} - a_{3n} \\ \vdots & \ddots & \vdots \\ a_{11} - a_{n1} & \cdots & a_{1n} - a_{nn} \end{pmatrix}.$$

The assumption that the pure strategy $\mathbf{e}^1$ of player 1 is never a best reply is equivalent to the statement that the system

$$\tilde{A}\mathbf{q} = \begin{pmatrix} \mathbf{e}^1 A\mathbf{q} - \mathbf{e}^2 A\mathbf{q} \\ \vdots \\ \mathbf{e}^1 A\mathbf{q} - \mathbf{e}^n A\mathbf{q} \end{pmatrix} \geq \mathbf{0}, \ \mathbf{q} \in \Delta^n$$

has no solution. This, in turn, is equivalent to the statement that the system

$$\tilde{A}\mathbf{q} \geq \mathbf{0}, \ \mathbf{q} \geq \mathbf{0}, \ \mathbf{q} \cdot \mathbf{b} < 0$$

has no solution. This means that the system in (2) of Lemma 22.7 (with $\tilde{A}$ instead of A there) has no solution. Hence, this lemma implies that the system

$$\mathbf{x} \in \mathbb{R}^{m-1}, \ \mathbf{x}\tilde{A} \leq \mathbf{b}, \ \mathbf{x} \geq \mathbf{0}$$

has a solution. By definition of $\mathbf{b}$ and $\tilde{A}$ we have for such a solution $\mathbf{x} = (x_2, \ldots, x_m)$:

$$\mathbf{x} \geq \mathbf{0} \text{ and } \sum_{i=2}^{m} x_i \, \mathbf{e}^i A \geq \sum_{i=2}^{m} x_i \, \mathbf{e}^1 A + (1, \ldots, 1).$$

This implies that $\mathbf{x} \neq \mathbf{0}$ and therefore that $\mathbf{e}^1$ is strictly dominated by the strategy

$$\left(0, x_2 / \sum_{i=2}^{m} x_i, \ldots, x_m / \sum_{i=2}^{m} x_i \right) \in \Delta^m.$$

Hence, $\mathbf{e}^1$ is strictly dominated.   ∎

For games with more than two players Theorem 13.13 does not hold, see Problem 13.13 for a counterexample.

The concept of 'never a best reply' is closely related to the concept of *rationalizability*. Roughly, rationalizable strategies are strategies that survive a process of iterated elimination of strategies that are never a best reply. Just like the strategies surviving iterated elimination of strictly dominated strategies, rationalizable strategies constitute a set-valued solution concept. The above theorem implies that for two-player games the two solution concepts coincide. In general, the set of rationalizable strategies is a subset of the set of strategies that survive iterated elimination of strictly dominated strategies.

The implicit assumption justifying iterated elimination of strategies that are dominated or never a best reply is quite demanding. Not only should a player believe that some other player will not play a such a strategy, but he should also believe that the other player believes that he (the first player) believes this and, in turn, will not use such a strategy in the reduced game, and so on and so forth. See the notes at the end of the chapter for references.

13.4 Perfect Equilibrium

Since a game may have many, quite different Nash equilibria, the literature has focused since a long time on *refinements* of Nash equilibrium. We have seen examples of this in extensive form games, such as subgame perfect equilibrium and perfect Bayesian equilibrium (Chaps. 4 and 5). One of the earliest and best known refinements of Nash equilibrium in strategic form games is the concept of 'trembling hand perfection'. This refinement excludes Nash equilibria that are not robust against 'trembles' in the players' strategies.

Formally, let $G = (N, S_1, \ldots, S_n, u_1, \ldots, u_n)$ be a finite game and let μ be an *error function*, assigning a number $\mu_{ih} \in (0, 1)$ to every $i \in N$ and $h \in S_i$, such that $\sum_{h \in S_i} \mu_{ih} < 1$ for every player i. The number μ_{ih} is the minimum probability with which player i is going to play pure strategy h, perhaps by 'mistake' ('trembling hand'). Let, for each $i \in N$, $\Delta(S_i, \mu) = \{\sigma_i \in \Delta(S_i) \mid \sigma_i(h) \geq \mu_{ih}$ for all $h \in S_i\}$, and let $G(\mu)$ denote the game derived from G by assuming that each player i may only choose strategies from $\Delta(S_i, \mu)$. The game $G(\mu)$ is called the μ-perturbed game. Denote the set of Nash equilibria of G by $NE(G)$ and of $G(\mu)$ by $NE(G(\mu))$.

Lemma 13.14 *For every error function μ, $NE(G(\mu)) \neq \emptyset$.*

Proof Analogous to the proof of Theorem 13.1.   ∎

A perfect equilibrium is a strategy combination that is the limit of *some* sequence of Nash equilibria of perturbed games. Formally:

Definition 13.15 A strategy combination σ is a *perfect equilibrium* if there is a sequence $G(\mu^t)$, $t \in \mathbb{N}$, of perturbed games with $\mu^t \to \mathbf{0}$ for $t \to \infty$ and a sequence of Nash equilibria $\sigma^t \in G(\mu^t)$ such that $\sigma^t \to \sigma$ for $t \to \infty$.   □

As will follow from Theorem 13.17 below, a perfect equilibrium is a Nash equilibrium. So the expressions *perfect equilibrium* and *perfect Nash equilibrium* are equivalent.

Call a strategy combination σ in G *completely mixed* if $\sigma_i(h) > 0$ for all $i \in N$ and $h \in S_i$.

Lemma 13.16 *A completely mixed Nash equilibrium of G is a perfect equilibrium.*

Proof Problem 13.14.   ∎

Also if a game has no completely mixed Nash equilibrium, it still has a perfect equilibrium.

Theorem 13.17 *Every finite game $G = (N, S_1, \ldots, S_n, u_1, \ldots, u_n)$ has a perfect equilibrium. Every perfect equilibrium is a Nash equilibrium.*

Proof Take any sequence $(G(\mu^t))_{\mu^t \to 0}$ of perturbed games and $\sigma^t \in NE(G(\mu^t))$ for each $t \in \mathbb{N}$. Since $\prod_{i \in N} \Delta(S_i)$ is a compact set we may assume without loss of generality that the sequence $(\sigma^t)_{t \in \mathbb{N}}$ converges to some $\sigma \in \prod_{i \in N} \Delta(S_i)$. Then σ is perfect. It is easy to verify that $\sigma \in NE(G)$.   ∎

Example 13.18 Consider the bimatrix game

$$G = \begin{array}{c} \\ U \\ M \\ D \end{array} \begin{array}{ccc} L & C & R \\ \begin{pmatrix} 1,1 & 0,0 & 2,0 \\ 1,2 & 1,2 & 1,1 \\ 0,0 & 1,1 & 1,1 \end{pmatrix} \end{array}.$$

(Cf. Problem 3.6.) The set of Nash equilibria in this game is the union of the following sets (Problem 13.8):

(i) $\{((p, 0, 1-p), (0, \frac{1}{2}, \frac{1}{2})) \mid 0 \le p \le \frac{1}{2}\}$,
(ii) $\{((0, 0, 1), (0, q, 1-q)) \mid \frac{1}{2} \le q \le 1\}$,
(iii) $\{((p, 1-p, 0), (1, 0, 0)) \mid 0 \le p \le 1\}$,
(iv) $\{((0, 1, 0), (q, 1-q, 0)) \mid 0 \le q \le 1\}$,
(v) $\{((0, p, 1-p), (0, 1, 0)) \mid 0 \le p \le 1\}$.

We consider these collections one by one. The Nash equilibria in the first collection are not perfect, for the following reason. In any perturbed game, player 1 plays each pure strategy with positive probability. As a consequence, for player 2, R always gives a strictly lower expected payoff than C: if player 1 plays $\mathbf{p} = (p_1, p_2, p_3) > \mathbf{0}$, then the payoff from C is $2p_2 + p_3$, and hence strictly larger than $p_2 + p_3$, which is the payoff from R. Thus, any best reply of player 2 in a perturbed game puts minimal probability on R, so that the strategy $(0, \frac{1}{2}, \frac{1}{2})$ can never be the limit of strategies of player 2 in Nash equilibria of perturbed games. In fact, this argument preludes on Theorem 13.21, which states that weakly dominated strategies (the third column in this case) are never played with positive probability in a perfect equilibrium. For a similar reason, the equilibria in the second collection are not perfect: D gives player 1 a strictly lower payoff than M if player 2 plays each column with positive probability, as is the case in a perturbed game. Hence, player 1's strategy $(0, 0, 1)$ cannot occur as the limit of Nash equilibrium strategies in perturbed games.

Now consider the collection in (iii). We will show that all Nash equilibria in this collection are perfect. For every $\varepsilon > 0$ let $G(\varepsilon)$ be the perturbed game with $\mu_{ij} = \varepsilon$ for each player $i = 1, 2$ and each row/column $j = 1, 2, 3$. First, suppose $p = 1$ and consider the strategy combination $((1 - 2\varepsilon, \varepsilon, \varepsilon), (1 - 2\varepsilon, \varepsilon, \varepsilon))$ in $G(\varepsilon)$.[1] Given the strategy of player 2, the three pure strategies (rows) of player 1 result in the payoffs 1, 1, and 2ε, respectively, so that player 1's strategy $(1 - 2\varepsilon, \varepsilon, \varepsilon)$ is a best reply: the inferior row D is only played with the minimally required probability. Similarly, given this strategy of player 1, the three pure strategies (columns) of player 2 yield 1, 3ε, and 2ε, so that player 2's strategy $(1 - 2\varepsilon, \varepsilon, \varepsilon)$ is a best reply: it puts only the minimally required probability on the inferior columns C and R. Hence, $((1 - 2\varepsilon, \varepsilon, \varepsilon), (1 - 2\varepsilon, \varepsilon, \varepsilon))$ is a Nash equilibrium in $G(\varepsilon)$. In fact, with these arguments we are preluding on Lemma 13.19, in particular part (2). Since $((1 - 2\varepsilon, \varepsilon, \varepsilon), (1 - 2\varepsilon, \varepsilon, \varepsilon)) \to ((1, 0, 0), (1, 0, 0))$ for $\varepsilon \to 0$, we conclude that $((1, 0, 0), (1, 0, 0))$ is a perfect equilibrium.

For the other cases in (iii) the arguments are similar. For $p = 0$, take the strategy combination $((\varepsilon, 1 - 2\varepsilon, \varepsilon), (1 - 2\varepsilon, \varepsilon, \varepsilon))$ in $G(\varepsilon)$. Given the strategy of player 2 the three rows yield 1, 1, and 2ε as before, so that player 1's strategy is a best reply. Given the strategy of player 1 the three columns now yield $2 - 3\varepsilon$, $2 - 3\varepsilon$, and $1 - \varepsilon$, so that player 2 still has a best reply, putting only the minimally required probability on the inferior R. Thus, $((\varepsilon, 1 - 2\varepsilon, \varepsilon), (1 - 2\varepsilon, \varepsilon, \varepsilon))$ is a Nash equilibrium in $G(\varepsilon)$, converging to $((0, 1, 0), (1, 0, 0))$ as ε goes to zero. Hence, $((0, 1, 0), (1, 0, 0))$ is a perfect equilibrium. Finally, for $0 < p < 1$, consider the strategy combination $((p - \frac{\varepsilon}{2}, 1 - p - \frac{\varepsilon}{2}, \varepsilon), (1 - 2\varepsilon, \varepsilon, \varepsilon))$ in $G(\varepsilon)$. Player 1 still plays a best reply, as before. For player 2, the three columns yield $2 - p - \frac{3\varepsilon}{2}$, $2 - 2p$, and $1 - p + \frac{\varepsilon}{2}$. Since player 2's strategy puts only the minimally required probability on C and R, it is a best reply. Thus, the strategy combination $((p - \frac{\varepsilon}{2}, 1 - p - \frac{\varepsilon}{2}, \varepsilon), (1 - 2\varepsilon, \varepsilon, \varepsilon))$ is a Nash equilibrium in $G(\varepsilon)$, implying that its limit $((p, 1 - p, 0), (1, 0, 0))$ is perfect.

Also the Nash equilibria in (iv) are perfect: the arguments are analogous to those for the collection in (iii) (Problem 13.8). In collection (v), finally, if $1 - p > 0$, then the Nash equilibrium cannot be perfect since D always gives a strictly lower payoff than M in any perturbed game, and therefore should be played with minimal probability, hence zero in the limit. The Nash equilibrium $((0, 1, 0), (0, 1, 0))$ is also an element of the collection in (iv), and thus perfect. □

---

[1] Whenever needed we assume that ε is sufficiently small, which is without loss of generality since we consider the limit for $\varepsilon \to 0$. To comply with our definitions we may take values $\varepsilon = \varepsilon^t = 1/t$, $t \in \mathbb{N}$, but this is not essential.

The following lemma formalizes some of the arguments already used in Example 13.18. It relates the Nash equilibria of a perturbed game to pure best replies in such a game. The first part says that, in a Nash equilibrium of a perturbed game, if a player puts more than the minimally required probability on some pure strategy, then that pure strategy must be a best reply. The second part says that, in some strategy combination, if all players only put more than the minimally required probabilities on pure best replies, then that combination must be a Nash equilibrium.

Lemma 13.19 *Let $G(\mu)$ be a perturbed game and let σ a strategy combination in $G(\mu)$.*

(1) *If $\sigma \in NE(G(\mu))$, $i \in N$, $h \in S_i$, and $\sigma_i(h) > \mu_{ih}$, then $h \in \beta_i(\sigma_{-i})$.*
(2) *If for all $i \in N$ and $h \in S_i$, $\sigma_i(h) > \mu_{ih}$ implies $h \in \beta_i(\sigma_{-i})$, then $\sigma \in NE(G(\mu))$.*

Proof

(1) Let $\sigma \in NE(G(\mu))$, $i \in N$, $h \in S_i$, and $\sigma_i(h) > \mu_{ih}$. Suppose, contrary to what we wish to prove, that $h \notin \beta_i(\sigma_{-i})$. Take $h' \in S_i$ with $h' \in \beta_i(\sigma_{-i})$. (Such an h' exists by an argument similar to the proof of Lemma 13.3.) Consider the strategy σ_i' defined by $\sigma_i'(h) = \mu_{ih}$, $\sigma_i'(h') = \sigma_i(h') + \sigma_i(h) - \mu_{ih}$, and $\sigma_i'(k) = \sigma_i(k)$ for all $k \in S_i \setminus \{h, h'\}$. Then $\sigma_i' \in \Delta(S_i, \mu)$ and $u_i(\sigma_i', \sigma_{-i}) > u_i(\sigma)$, contradicting the assumption $\sigma \in NE(G(\mu))$.
(2) Let $i \in N$. The condition in (2) implies that, if $h \in S_i$ and $h \notin \beta_i(\sigma_{-i})$, then $\sigma_i(h) = \mu_{ih}$. This implies that σ_i is a best reply to $(\sigma_j)_{j \neq i}$. Thus, $\sigma \in NE(G(\mu))$. ∎

Below we present two characterizations of perfect Nash equilibrium that both avoid sequences of perturbed games. The first one is based on the notion of ε-perfect equilibrium, defined as follows. Let $\varepsilon > 0$. A strategy combination $\sigma \in \prod_{i \in N} \Delta(S_i)$ is an *ε-perfect equilibrium* of G if it is completely mixed and $\sigma_i(h) \le \varepsilon$ for all $i \in N$ and all $h \in S_i$ with $h \notin \beta_i(\sigma_{-i})$.

An ε-perfect equilibrium of G need not be a Nash equilibrium of G, but it puts probabilities of at most ε on pure strategies that are not best replies.

The announced characterizations are collected in the following theorem. The theorem says that a perfect equilibrium is a limit of ε-perfect equilibria. Also, a perfect equilibrium is a limit of completely mixed strategy combinations such that the strategy of each player in the perfect equilibrium under consideration is a best reply to the strategy combinations of the other players in those completely mixed strategy combinations.

Theorem 13.20 *Let $G = (N, S_1, \ldots, S_n, u_1, \ldots, u_n)$ and $\sigma \in \prod_{i \in N} \Delta(S_i)$. The following statements are equivalent:*

(1) *σ is a perfect equilibrium of G;*
(2) *σ is a limit of a sequence of ε-perfect equilibria σ^ε of G for $\varepsilon \to 0$;*

(3) σ *is a limit of a sequence of completely mixed strategy combinations σ^ε for $\varepsilon \to 0$, where $\sigma_i \in \beta_i(\sigma_{-i}^\varepsilon)$ for each $i \in N$ and each σ^ε in this sequence.*

Proof (1) $\Rightarrow$ (2): Take a sequence of perturbed games $G(\mu^t)$, $t \in \mathbb{N}$ with $\mu^t \to \mathbf{0}$ and a sequence $\sigma^t \in NE(G(\mu^t))$ with $\sigma^t \to \sigma$. For each t define $\varepsilon^t \in \mathbb{R}$ by $\varepsilon^t = \max\{\mu_{ih}^t \mid i \in N,\ h \in S_i\}$. Then, by Lemma 13.19(1), σ^t is an ε^t-perfect equilibrium for every t. So (2) follows.

(2) $\Rightarrow$ (3): Take a sequence of ε-perfect equilibria σ^ε as in (2) converging to σ for $\varepsilon \to 0$. Let $i \in N$. By the definition of ε-perfect equilibrium, if $\sigma_i(h) > 0$ for some $h \in S_i$, then for ε sufficiently small we have $h \in \beta_i^\varepsilon(\sigma_{-i})$. This implies $\sigma_i \in \beta_i(\sigma_{-i}^\varepsilon)$, and (3) follows.

(3) $\Rightarrow$ (1): Let σ^{ε^t}, $t \in \mathbb{N}$, be a sequence as in (3) with $\varepsilon^t \to 0$ and $\sigma^{\varepsilon^t} \to \sigma$ as $t \to \infty$. For each $t \in \mathbb{N}$, $i \in N$ and $h \in S_i$ define $\mu_{ih}^t = \sigma_i^{\varepsilon^t}(h)$ if $\sigma_i(h) = 0$ and $\mu_{ih}^t = \varepsilon^t$ otherwise. Then, for t sufficiently large, μ^t is an error function, $G(\mu^t)$ is a perturbed game, and σ^{ε^t} is a strategy combination in $G(\mu^t)$. By Lemma 13.19(2), $\sigma^{\varepsilon^t} \in NE(G(\mu^t))$. So σ is a perfect Nash equilibrium of G. ∎

There is a close relation between the concept of domination and the concept of perfection. We first extend the concept of (weak) domination to mixed strategies. In the game $G = (N, S_1, \ldots, S_n, u_1, \ldots, u_n)$, call a strategy $\sigma_i \in \Delta(S_i)$ (weakly) dominated by $\sigma_i' \in \Delta(S_i)$ if $u_i(\sigma_i, \sigma_{-i}) \le u_i(\sigma_i', \sigma_{-i})$ for all $\sigma_{-i} \in \prod_{j \ne i} \Delta(S_j)$, with at least one inequality strict. (Observe that it is actually sufficient to check this for combinations $s_{-i} \in \prod_{j \ne i} S_i$.) Call σ_i undominated if there is no σ_i' by which it is dominated, and call a strategy combination σ undominated if σ_i is undominated for every $i \in N$. We now have:

Theorem 13.21 *Every perfect Nash equilibrium in G is undominated.*

Proof Let σ be a perfect Nash equilibrium and suppose that (say) σ_1 is dominated. Then there is a $\sigma_1' \in \Delta(S_1)$ such that $u_1(\sigma_1, s_{-1}) \le u_1(\sigma_1', s_{-1})$ for all $s_{-1} \in \prod_{i=2}^n S_i$, with at least one inequality strict. Take a sequence $(\sigma^t)_{t \in \mathbb{N}}$ of strategy combinations as in (3) of Theorem 13.20, converging to σ. Then, since every σ^t is completely mixed, we have $u_1(\sigma_1, \sigma_{-1}^t) < u_1(\sigma_1', \sigma_{-1}^t)$ for every t. This contradicts the fact that σ_1 is a best reply to σ_{-1}^t. ∎

The converse of Theorem 13.21 is only true for two-person games. For a counterexample involving three players, see Problem 13.15.

For proving the converse of the theorem for bimatrix games, we use the following auxiliary lemma. In this lemma, for a matrix game $\tilde{A}$, $C_2(\tilde{A})$ denotes the set of all columns of $\tilde{A}$ that are in the carrier of some optimal strategy of player 2 in $\tilde{A}$.

Lemma 13.22 *Let $G = (A, B)$ be an $m \times n$ bimatrix game and let $\mathbf{p} \in \Delta^m$. Define the $m \times n$ matrix $\tilde{A} = (\tilde{a}_{ij})$ by $\tilde{a}_{ij} = a_{ij} - \mathbf{p}Ae^j$ for all $i = 1, \ldots, m$ and $j = 1, \ldots, n$. Then $\mathbf{p}$ is undominated in G if and only if $v(\tilde{A}) = 0$ and $C_2(\tilde{A}) = \{1, \ldots, n\}$.*

Proof First note that $\mathbf{p}\tilde{A} = \mathbf{0}$ and therefore $v(\tilde{A}) \geq 0$.

For the if-direction, suppose that $\mathbf{p}$ is dominated in G, say by $\mathbf{p}'$. Then $\mathbf{p}'A \gneqq \mathbf{p}A$, hence $\mathbf{p}'\tilde{A} \gneqq \mathbf{0}$. Therefore, if $v(\tilde{A}) = 0$, then $\mathbf{p}'$ is an optimal strategy in $\tilde{A}$; take a column j with $\mathbf{p}'\tilde{A}e^j > 0$, then we have $j \notin C_2(\tilde{A})$ and, thus, $C_2(\tilde{A}) \neq \{1,\ldots,n\}$. This proves the if-direction.

For the only-if direction suppose that $\mathbf{p}$ is undominated in G. Suppose we had $v(\tilde{A}) > 0$. Then take an optimal strategy $\tilde{\mathbf{p}}$ of player 1 in $\tilde{A}$, so that $\tilde{\mathbf{p}}\tilde{A} > \mathbf{0}$, hence $\tilde{\mathbf{p}}A > \mathbf{p}A$, a contradiction. Thus, $v(\tilde{A}) = 0$. Suppose there is a column j that is not an element of $C_2(\tilde{A})$. By Problem 12.7 there must be an optimal strategy $\mathbf{p}'$ of player 1 in $\tilde{A}$ such that $\mathbf{p}'\tilde{A}e^j > 0$, so that $\mathbf{p}'\tilde{A} \gneqq \mathbf{0}$, hence $\mathbf{p}'A \gneqq \mathbf{p}A$. So $\mathbf{p}'$ dominates $\mathbf{p}$ in G, a contradiction. This proves the only-if direction. ∎

Theorem 13.23 *Let* $G = (A, B)$ *be a bimatrix game, and let* $(\mathbf{p}, \mathbf{q})$ *be an undominated Nash equilibrium. Then* $(\mathbf{p}, \mathbf{q})$ *is perfect.*

Proof Let $\tilde{A}$ as in Lemma 13.22, then $\mathbf{p}$ is an optimal strategy for player 1 in $\tilde{A}$ since $\mathbf{p}\tilde{A} = \mathbf{0}$ and $v(\tilde{A}) = 0$. By Lemma 13.22 we can find a completely mixed optimal strategy $\mathbf{q}'$ for player 2 in $\tilde{A}$. So $\mathbf{p}$ is a best reply to $\mathbf{q}'$ in $\tilde{A}$, i.e., $\mathbf{p}\tilde{A}\mathbf{q}' \geq \tilde{\mathbf{p}}\tilde{A}\mathbf{q}'$ for all $\tilde{\mathbf{p}}$, and thus $\tilde{\mathbf{p}}A\mathbf{q}' - \mathbf{p}A\mathbf{q}' \leq 0$ for all $\tilde{\mathbf{p}}$. So $\mathbf{p}$ is also a best reply to $\mathbf{q}'$ in G. For $1 > \varepsilon > 0$ define $\mathbf{q}^\varepsilon = (1 - \varepsilon)\mathbf{q} + \varepsilon\mathbf{q}'$. Then $\mathbf{q}^\varepsilon$ is completely mixed, $\mathbf{p}$ is a best reply to $\mathbf{q}^\varepsilon$, and $\mathbf{q}^\varepsilon \to \mathbf{q}$ for $\varepsilon \to 0$. In the same way we can construct a sequence $\mathbf{p}^\varepsilon$ with analogous properties, converging to $\mathbf{p}$. Then implication (3) $\Rightarrow$ (1) in Theorem 13.20 implies that $(\mathbf{p}, \mathbf{q})$ is perfect. ∎

Example 13.24 Consider again the game G from Example 13.18. Row D is weakly dominated by M, and R by C. There are no other weakly dominated strategies in this game. Thus, as established earlier, but now using Theorems 13.21 and 13.23, the set of perfect equilibria is the set $\{((p, 1 - p, 0), (1, 0, 0)) \mid 0 \leq p \leq 1\} \cup \{((0, 1, 0), (q, 1 - q, 0)) \mid 0 \leq q \leq 1\}$. □

The following example shows an advantage but at the same time a drawback of perfect equilibrium: a perfect equilibrium may be payoff-dominated by another Nash equilibrium.

Example 13.25 Consider the bimatrix game

$$
\begin{array}{c}
\begin{array}{cc} L & \quad R \end{array} \\
\begin{array}{c} U \\ D \end{array}
\begin{pmatrix} 1,1 & 10,0 \\ 0,10 & 10,10 \end{pmatrix},
\end{array}
$$

which has two Nash equilibria, both pure, namely (U, L) and (D, R). Only (U, L) is perfect, as can be seen by direct inspection or by applying Theorems 13.21 and 13.23. At the equilibrium (D, R), each player has an incentive to deviate to the other pure strategy since the opponent may deviate by mistake. This equilibrium is excluded by perfection. On the other hand, the unique perfect equilibrium (U, L) is payoff-dominated by the equilibrium (D, R).   □

Another drawback of perfect equilibrium is the fact that adding strictly dominated strategies may result in adding perfect Nash equilibria, as the following example shows.

Example 13.26 In the game

$$
\begin{array}{c} \\ U \\ M \\ D \end{array}
\begin{array}{ccc}
L & C & R \\
\left(\begin{array}{ccc}
1, 1 & 0, 0 & -1, -2 \\
0, 0 & 0, 0 & 0, -2 \\
-2, -1 & -2, 0 & -2, -2
\end{array}\right),
\end{array}
$$

there are two perfect Nash equilibria, namely (U, L) and (M, C). If we reduce the game by deleting the strictly dominated pure strategies D and R, the only perfect equilibrium that remains is (U, L).   □

This motivated the introduction of a further refinement called *proper Nash equilibrium*. See the next section.

13.5 Proper Equilibrium

A perfect equilibrium is required to be robust only against *some* 'trembles' and, moreover, there are no further conditions on these trembles. We now propose the additional restriction that trembles be less probable if they are more 'costly'.

Given some $\varepsilon > 0$, call a strategy combination σ in the game $G = (N, S_1, \ldots, S_n, u_1, \ldots, u_n)$ an ε-*proper equilibrium* if σ is completely mixed and for all $i \in N$ and $h, k \in S_i$ we have

$$
u_i(h, \sigma_{-i}) < u_i(k, \sigma_{-i}) \Rightarrow \sigma_i(h) \leq \varepsilon \sigma_i(k) .
$$

Observe that an ε-proper equilibrium does not have to be a Nash equilibrium.

Definition 13.27 A strategy combination σ in G is *proper* if, for some sequence $\varepsilon^t \to 0, t \in \mathbb{N}$, there exist ε^t-proper equilibria $\sigma(\varepsilon^t)$ such that $\sigma(\varepsilon^t) \to \sigma$.   □

Since, in a proper strategy combination σ, a pure strategy h of a player i that is not a best reply to σ_{-i} is played with probability 0, it follows that a proper strategy combination is a Nash equilibrium. Moreover, since it is straightforward by the definitions that an ε-proper equilibrium is also an ε-perfect equilibrium, it follows from Theorem 13.20 that a proper equilibrium is perfect. Hence, properness is a refinement of perfection.

Remark 13.28 By replacing the word 'proper' by 'perfect' in Definition 13.27, we obtain an alternative definition of perfect equilibrium. This follows from Theorem 13.20.   □

Example 13.26 shows that properness is a strict refinement of perfection: the Nash equilibrium (M, C) is perfect but not proper. To see this, for $\varepsilon^t > 0$ with $\varepsilon^t \to 0$ as $t \to \infty$, let $(\mathbf{p}(\varepsilon^t), \mathbf{q}(\varepsilon^t))$ be a converging sequence of ε^t-proper equilibria. Then for each t we must have $q_3(\varepsilon^t) \leq \varepsilon^t q_1(\varepsilon^t)$. In turn, this implies that U is the unique best reply to $\mathbf{q}(\varepsilon^t)$, hence $p_2(\varepsilon^t) \leq \varepsilon^t p_1(\varepsilon^t)$ for each t. But then $\mathbf{p}(\varepsilon^t)$ cannot converge to $(0, 1, 0)$.

A proper Nash equilibrium always exists:

Theorem 13.29 *Let $G = (N, S_1, \ldots, S_n, u_1, \ldots, u_n)$ be a finite game. Then G has a proper Nash equilibrium.*

Proof It is sufficient to show that for $\varepsilon > 0$ close to 0 there exists an ε-proper equilibrium of G. Let $0 < \varepsilon < 1$ and define the error function μ by $\mu_{ik} = \varepsilon^{|S_i|}/|S_i|$ for all $i \in N$ and $k \in S_i$. For every $i \in N$ and $\sigma \in \prod_{j \in N} \Delta(S_j, \mu)$ define

$$F_i(\sigma) = \{\tau_i \in \Delta(S_i, \mu) \mid \forall k, l \in S_i \, [u_i(k, \sigma_{-i}) < u_i(l, \sigma_{-i}) \Rightarrow \tau_i(k) \leq \varepsilon\tau_i(l)]\} \, .$$

Then $F_i(\sigma) \neq \emptyset$, as can be seen as follows. Define

$$v_i(\sigma, k) = |\{l \in S_i \mid u_i(k, \sigma_{-i}) < u_i(l, \sigma_{-i})\}| \, ,$$

and define $\tau_i(k) = \varepsilon^{v_i(\sigma, k)}/\sum_{l \in S_i} \varepsilon^{v_i(\sigma, l)}$, for all $k \in S_i$. Then $\tau_i \in F_i(\sigma)$. Consider the correspondence

$$F : \prod_{j \in N} \Delta(S_j, \mu) \to \prod_{j \in N} \Delta(S_j, \mu), \quad \sigma \mapsto \prod_{i \in N} F_i(\sigma) \, .$$

Then F satisfies the conditions of the Kakutani Fixed Point Theorem (Theorem 22.11)—see Problem 13.16. Hence, F has a fixed point, and each fixed point of F is an ε-proper equilibrium of G.   ■

In spite of the original motivation for introducing properness, this concept suffers from the same deficit as perfect equilibrium: adding strictly dominated strategies may enlarge the set of proper Nash equilibria. See Problem 13.17 for an example of this.

Example 13.30 We consider again the game G from Example 13.18:

$$
G = \begin{array}{c} \\ U \\ M \\ D \end{array} \begin{pmatrix} \overset{L}{1,1} & \overset{C}{0,0} & \overset{R}{2,0} \\ 1,2 & 1,2 & 1,1 \\ 0,0 & 1,1 & 1,1 \end{pmatrix}.
$$

The perfect equilibria are:

(a) $\{((p, 1-p, 0), (1, 0, 0)) \mid 0 \le p \le 1\}$,

(b) $\{((0, 1, 0), (q, 1-q, 0)) \mid 0 \le q \le 1\}$.

First consider the equilibria in (a). Which ones are proper? In any ε-proper equilibrium $(\mathbf{p}^\varepsilon, \mathbf{q}^\varepsilon)$, player 1 plays a completely mixed strategy. This implies that C gives player 2 a higher payoff than R, so that $q_3^\varepsilon \le \varepsilon q_2^\varepsilon$. For player 1, the payoff of U is $q_1^\varepsilon + 2q_3^\varepsilon$ and the payoff of M is $q_1^\varepsilon + q_2^\varepsilon + q_3^\varepsilon > q_1^\varepsilon + 2q_3^\varepsilon$ for ε small, so that we obtain $p_1^\varepsilon \le \varepsilon p_2^\varepsilon$. Hence, for $p > 0$ the equilibria in (a) are not proper. The equilibrium $((0, 1, 0), (1, 0, 0))$ is proper: it is the limit of the ε-proper equilibria $((\varepsilon, 1, \varepsilon^2)/\eta, (1, \varepsilon, \varepsilon^2)/\eta)$, where $\eta = 1 + \varepsilon + \varepsilon^2$.

Next consider the equilibria in (b). As before, in any ε-proper equilibrium $(\mathbf{p}^\varepsilon, \mathbf{q}^\varepsilon)$ we must have $q_3^\varepsilon \le \varepsilon q_2^\varepsilon$. This implies that for player 1, the payoff from U is higher than the payoff from D, so that we must have $p_3^\varepsilon \le \varepsilon p_1^\varepsilon$. In turn, this implies that for player 2 the payoff from L is higher than the payoff from C, so that $q_2^\varepsilon \le \varepsilon q_1^\varepsilon$. Thus, the only candidate in (b) for a proper equilibrium is $((0, 1, 0), (1, 0, 0))$, and we have already established that this combination is proper indeed. Hence, it is the unique proper equilibrium of G. $\qquad\qquad\square$

13.6   Strictly Perfect Equilibrium

Another refinement of perfect equilibrium is obtained by requiring robustness of a Nash equilibrium with respect to *all* 'trembles'. This results in the concept of strictly perfect equilibrium.

Definition 13.31 A strategy combination σ in the game $G = (N, S_1, \ldots, S_n, u_1, \ldots, u_n)$ is a *strictly perfect equilibrium* if, for every sequence $\{G(\mu^t)\}, t \in \mathbb{N}$, of perturbed games with $\mu^t \to 0$ for $t \to \infty$, there exist profiles $\sigma^t \in NE(G(\mu^t))$ such that $\sigma^t \to \sigma$.

Clearly, a strictly perfect equilibrium is a perfect equilibrium. For some further observations concerning strictly perfect equilibrium see Problem 13.18.

A clear drawback of strictly perfect equilibrium is the fact that it does not have to exist, as the following (continued) example shows.

Example 13.32 We consider again the game G from Example 13.18:

$$G = \begin{array}{c} \\ U \\ M \\ D \end{array} \begin{pmatrix} \overset{L}{1,1} & \overset{C}{0,0} & \overset{R}{2,0} \\ 1,2 & 1,2 & 1,1 \\ 0,0 & 1,1 & 1,1 \end{pmatrix},$$

with perfect equilibria:

(a) $\{((p, 1-p, 0), (1, 0, 0)) \mid 0 \le p \le 1\}$,

(b) $\{((0, 1, 0), (q, 1-q, 0)) \mid 0 \le q \le 1\}$.

Consider the equilibria in (a) and suppose $p > 0$. Let $G(\mu)$ be the perturbed game as in Sect. 13.4. We may assume that, since $p > 0$, L must yield player 2 a higher payoff than C in a Nash equilibrium of $G(\mu)$. In turn this implies that player 2's equilibrium strategy in any Nash equilibrium of $G(\mu)$ is equal to $(1 - \mu_{22} - \mu_{23}, \mu_{22}, \mu_{23})$ (we already saw that player 2 must always put minimal probability on R). Now choose $\mu_{23} < \mu_{22}$, then the payoff $1 - \mu_{22} + \mu_{23}$ to player 1 from playing U is smaller than the payoff 1 from M. This implies that player 1 plays M in the limit. Thus, a perfect equilibrium of the form $((p, 1-p, 0), (1, 0, 0))$ with $p > 0$ is not strictly perfect.

Next, consider the equilibrium $((0, 1, 0), (1, 0, 0))$, which is both in (a) and in (b). Assume again that $\mu_{23} < \mu_{22}$. If player 2 plays (q_1, q_2, q_3) in $G(\mu)$, then $q_3 = \mu_{23}$ and the payoff to player 1 from U is $q_1 + 2\mu_{23}$, whereas the payoff from M is $q_1 + q_2 + \mu_{23}$. Since $q_2 \ge \mu_{22} > \mu_{23}$, this implies that M yields player 1 a higher payoff than U. In turn, this implies that player 1's strategy in any Nash equilibrium of $G(\mu)$ is equal to $(\mu_{11}, 1 - \mu_{11} - \mu_{13}, \mu_{13})$ (recall that player 1 puts minimal probability on D). Now suppose that $\mu_{11} < \mu_{13}$, then the payoff to player 2 from C is higher than the payoff from L, so that player 2 must play C in the limit. Thus, also the perfect equilibrium $((0, 1, 0), (1, 0, 0))$ with $p > 0$ is not strictly perfect.

For the equilibria in (b) with $q < 1$, the argument is almost identical to the one in the preceding paragraph. In the end, choose $\mu_{11} > \mu_{13}$, then the payoff to player 2 from L is higher than the payoff from C, so that player 2 must play L in the limit. Hence, no equilibrium of the form $\{((0, 1, 0), (q, 1-q, 0))$ with $q < 1$ is strictly perfect. We conclude that this game has no strictly perfect equilibria.   □

13.7 Correlated Equilibrium

In the preceding sections we studied several refinements of Nash equilibrium. In this section the set of Nash equilibria is extended in a way to become clear below. It is, however, not the intention to enlarge the set of Nash equilibria but rather to enable the players to reach better payoffs by allowing some communication device. This will result in the concept of *correlated equilibrium*. Attention in this section is restricted to bimatrix games.

In order to fix ideas, consider the situation where two car drivers approach a road crossing. Each driver has two pure strategies: 'stop' (s) or 'cross' (c). The preferences for the resulting combinations are as expressed by the following table:

$$
(A, B) = \begin{array}{c} \\ c \\ s \end{array} \begin{array}{cc} c & s \\ \left(\begin{array}{cc} -10, -10 & 5, 0 \\ 0, 5 & -1, -1 \end{array} \right). \end{array}
$$

This bimatrix game has two asymmetric and seemingly unfair pure Nash equilibria, and one symmetric mixed Nash equilibrium $((3/8, 5/8), (3/8, 5/8))$, resulting in an expected payoff of $-5/8$ for both, and therefore also not quite satisfying.

Now suppose that traffic lights are installed that indicate c ('green') or s ('red') according to the probabilities in the following table:

$$
\begin{array}{c} \\ c \\ s \end{array} \begin{array}{cc} c & s \\ \left(\begin{array}{cc} 0.00 & 0.55 \\ 0.40 & 0.05 \end{array} \right). \end{array}
$$

For example, with probability 0.55 (55 % of the time) the light is green for driver 1 and red for driver 2. Assume that the players (drivers) are not forced to obey the traffic lights but know the probabilities as given in the table. We argue that it is in each player's own interest to obey the lights if the other player does so.

If the light is green for player 1 then player 1 knows with certainty that the light is red for player 2. So if player 2 obeys the lights and stops, it is indeed optimal for player 1 to cross. If the light is red for player 1, then the conditional probability that player 2 crosses (if he obeys the lights) is equal to $0.4/0.45 \approx 0.89$ and the conditional probability that player 2 stops is $0.05/0.45 \approx 0.11$. So if player 1 stops, his expected payoff is $0.89 \cdot 0 + 0.11 \cdot -1 = -0.11$, and if he crosses his expected payoff is $0.89 \cdot -10 + 0.11 \cdot 5 = -8.35$. Clearly, it is optimal for player 1 to obey the light and stop.

For player 2 the argument is similar. If the light is green for player 2 then he knows with certainty that player 1 has red light. Thus, if player 1 obeys the red light and stops, it is optimal for player 2 to cross. If player 2 has red light then he knows that the conditional probabilities are $0.55/0.60 \approx 0.92$ and $0.05/0.60 \approx 0.08$ that player 1 has green light and red light, respectively. For player 2, assuming that player 1 obeys the traffic lights, stopping yields an expected payoff of $0.08 \cdot -1$ and crossing an expected payoff of $0.92 \cdot -10 + 0.08 \cdot 5$: clearly, stopping is optimal.

Thus, we can indeed talk of an equilibrium: such an equilibrium is called a *correlated equilibrium*. Note that there is no mixed strategy combination in the game (A, B) that induces these probabilities. If $\mathbf{p} = (p_1, p_2)$ and $\mathbf{q} = (q_1, q_2)$ are mixed strategies of players 1 and 2, respectively, then we would need $p_1 \neq 0$ since $p_1 q_2 = 0.55$ and we would need $q_1 \neq 0$ since $p_2 q_1 = 0.40$, but then $p_1 q_1 \neq 0$, hence the combination (c, c) would have positive probability. In terms of the situation in the example, this particular equilibrium cannot be reached without traffic lights serving as a communication device between the players. The overall expected payoffs of the players are $0.55 \cdot 5 + 0.05 \cdot -1 = 2.7$ for player 1 and $0.40 \cdot 5 + 0.05 \cdot -1 = 1.95$ for player 2, which is considerably better for both than the payoffs in the mixed Nash equilibrium.

In general, let (A, B) be an $m \times n$ bimatrix game. A *correlated strategy* is an $m \times n$ matrix $P = (p_{ij})$ with $\sum_{i=1}^{m} \sum_{j=1}^{n} p_{ij} = 1$ and $p_{ij} \geq 0$ for all $i = 1, \ldots, m$, $j = 1, \ldots, n$. A correlated strategy P can be thought of as a communication device: the pair (i, j) is chosen with probability p_{ij}, and if that happens, player 1 receives the signal i and player 2 the signal j. Suppose player 2 obeys the signal. If player 1 receives signal i and indeed plays i, his expected payoff is

$$\sum_{j=1}^{n} p_{ij} a_{ij} \bigg/ \sum_{j=1}^{n} p_{ij} \; ,$$

and if he plays row k instead, his expected payoff is

$$\sum_{j=1}^{n} p_{ij} a_{kj} \bigg/ \sum_{j=1}^{n} p_{ij} \; .$$

So to keep player 1 from disobeying the received signal, we should have

$$\sum_{j=1}^{n} (a_{ij} - a_{kj}) p_{ij} \geq 0 \text{ for all } i, k = 1, \ldots, m \; . \tag{13.5}$$

The analogous condition for player 2 is:

$$\sum_{i=1}^{m} (b_{ij} - b_{il}) p_{ij} \geq 0 \text{ for all } j, l = 1, \ldots, n \; . \tag{13.6}$$

Definition 13.33 A *correlated equilibrium* in the bimatrix game (A, B) is a correlated strategy $P = (p_{ij})$ satisfying (13.5) and (13.6).   □

For the two-driver example conditions (13.5) and (13.6) result in four inequalities, which are not difficult to solve (Problem 13.19). In general, any Nash equilibrium of a bimatrix game results in a correlated equilibrium (Problem 13.20), so existence of a correlated equilibrium is not an issue.

The set of correlated equilibria is convex (Problem 13.21), so the convex hull of all payoff pairs corresponding to the Nash equilibria of a bimatrix game consists of payoff pairs attainable in correlated equilibria. Problem 13.22 presents an example of a game in which some payoff pairs can be reached in correlated equilibria but not as convex combinations of payoff pairs of Nash equilibria.

In general, correlated equilibria can be computed using linear programming. Specifically, let (A, B) be an $m \times n$ bimatrix game. We associate with (A, B) an $mn \times (m(m-1) + n(n-1))$ matrix C as follows. For each pair (i, j) of a row and a column in (A, B) we have a row in C, and for each pair (h, k) of two different rows in (A, B) or two different columns in (A, B) we have a column in C. We define

$$c_{(i,j)(h,k)} = \begin{cases} a_{ij} - a_{kj} & \text{if } i = h \in \{1, \ldots, m\} \text{ and } k \in \{1, \ldots, m\} \\ b_{ij} - b_{ik} & \text{if } j = h \in \{1, \ldots, n\} \text{ and } k \in \{1, \ldots, n\} \\ 0 & \text{otherwise.} \end{cases}$$

Let $P = (p_{ij})$ be a correlated strategy in (A, B). Then P can be seen as a vector $\mathbf{p} \in \mathbb{R}^{mn}$. Let $c_{(h,k)}$ be a column in C. If h and k are rows of (A, B) we have

$$\mathbf{p} \cdot c_{(h,k)} = \sum_{j=1}^{n} p_{hj}(a_{hj} - a_{kj})$$

and if h and k are columns of (A, B) we have

$$\mathbf{p} \cdot c_{(h,k)} = \sum_{i=1}^{m} p_{ih}(b_{ih} - b_{ik}) .$$

Hence, by (13.5) and (13.6), P is a correlated equilibrium of (A, B) if and only if $\mathbf{p}C \geq 0$. If we consider C as a matrix game and $\mathbf{p}$ as a strategy of player 1 in C—not to be confused with player 1 in (A, B)—then the existence of a correlated equilibrium implies that $v(C)$, the value of the matrix game C, is nonnegative. In particular, this implies that any optimal strategy of player 1 in C is a correlated equilibrium in (A, B). If $v(C) = 0$ then any correlated equilibrium in (A, B) is an optimal strategy of player 1 in C, but if $v(C) > 0$ then there may be correlated equilibria in (A, B) that are not optimal strategies for player 1 in C—they may only guarantee zero. The latter is the case in the two-drivers example (Problem 13.23).

Matrix games can be solved by linear programming, see Sect. 12.2. We conclude with an example in which the described technique is applied.

Example 13.34 Consider the bimatrix game

$$(A, B) = \begin{array}{c} \\ 1 \\ 2 \end{array} \begin{pmatrix} 1' & 2' & 3' \\ 3, 1 & 2, 5 & 6, 0 \\ 1, 4 & 3, 3 & 2, 6 \end{pmatrix} .$$

The associated matrix game C is as follows.

$$
\begin{array}{c c c c c c c c c}
 & (1,2) & (2,1) & (1',2') & (1',3') & (2',1') & (2',3') & (3',1') & (3',2') \\
(1,1') & 2 & 0 & -4 & 1 & 0 & 0 & 0 & 0 \\
(1,2') & -1 & 0 & 0 & 0 & 4 & 5 & 0 & 0 \\
(1,3') & 4 & 0 & 0 & 0 & 0 & 0 & -1 & -5 \\
(2,1') & 0 & -2 & 1 & -2 & 0 & 0 & 0 & 0 \\
(2,2') & 0 & 1 & 0 & 0 & -1 & -3 & 0 & 0 \\
(2,3') & 0 & -4 & 0 & 0 & 0 & 0 & 2 & 3
\end{array}
$$

It can be checked that this game has value 0, by using the optimal strategies

$$
P = \begin{array}{c} 1 \\ 2 \end{array} \begin{pmatrix} 1' & 2' & 3' \\ 0 & 0.3 & 0.075 \\ 0 & 0.5 & 0.125 \end{pmatrix}
$$

for player 1 in C and $(0,0,1/2,1/2,0,0,0,0)$ for player 2 in C. The optimal strategy for player 1 in C is unique, and since $v(C) = 0$ this implies that the game (A,B) has a unique correlated equilibrium. Consequently, this must correspond to the unique Nash equilibrium of the game. Indeed, $((3/8,5/8),(0,4/5,1/5))$ is the unique Nash equilibrium of (A,B) and it results in the probabilities given by P. $\square$

13.8   A Characterization of Nash Equilibrium

The concept of Nash equilibrium requires strong behavioral assumptions about the players. Each player should be able to guess what other players will do, assume that other players know this and make similar conjectures, and so on, and all this should be in equilibrium. The basic difficulty is that Nash equilibrium is a circular concept: a player plays a best reply against the conjectured strategies of the opponents but, in turn, this best reply should be conjectured by the opponents and they should play best replies as well. Not surprisingly, theories of repeated play or learning or, more generally, dynamic models that aim to explain how players in a game come to play a Nash equilibrium, have in common that they change the strategic decision into a collection of single-player decision problems.

In this section we review a different approach, which is axiomatic in nature. The Nash equilibrium concept is viewed as a solution concept: a correspondence which assigns to any finite game a set of strategy combinations. One of the conditions (axioms) put on this correspondence is a condition of consistency with respect to changes in the number of players: if a player leaves the game, leaving his strategy as an input behind, then the other players should not want to change their strategies. This is certainly true for Nash equilibrium, and it can be imposed as a condition on a solution correspondence. By assuming that players in single-player games—hence, in 'simple' maximization problems—behave rationally, and by adding a converse consistency condition, it follows that the solution correspondence must be the Nash equilibrium correspondence. We proceed with a formal treatment of this axiomatic characterization.

Let Γ be a collection of finite games of the form $G = (N, S_1, \ldots, S_n, u_1, \ldots, u_n)$. (It is implicit that also the set of players N may vary in Γ.) A *solution* on Γ is a function φ that assigns to each $G \in \Gamma$ a set of strategy combinations $\varphi(G) \subseteq \prod_{i \in N} \Delta(S_i)$. A particular solution is the Nash correspondence NE, assigning to each $G \in \Gamma$ the set $NE(G)$ of all Nash equilibria of G.

Definition 13.35 The solution φ satisfies *one-person rationality* (OPR) if

$$\varphi(G) = \{\sigma_i \in \Delta(S_i) \mid u_i(\sigma_i) \geq u_i(\tau_i) \text{ for all } \tau_i \in \Delta(S_i)\}$$

for every one-person game $G = (\{i\}, S_i, u_i)$ in Γ.   □

The interpretation of OPR is clear and needs no further comment.

Let $G = (N, S_1, \ldots, S_n, u_1, \ldots, u_n)$ be a game, $\emptyset \neq M \subseteq N$, and let σ be a strategy combination in G. The *reduced game* of G with respect to M and σ is the game $G^{M,\sigma} = (M, (S_i)_{i \in M}, (u_i^\sigma)_{i \in M})$, where $u_i^\sigma(\tau) = u_i(\tau, \sigma_{N \setminus M})$ for all $\tau \in \prod_{j \in M} \Delta(S_j)$.[2] The interpretation of such a reduced game is straightforward: if the players of $N \setminus M$ leave the game, leaving their strategy combination $\sigma_{N \setminus M}$ behind, then the remaining players are faced with the game $G^{M,\sigma}$. Alternatively, if it is common knowledge among the players in M that the players outside M play according to σ, then they are faced with the game $G^{M,\sigma}$. Call a collection of games Γ *closed* if it is closed under taking reduced games.

Definition 13.36 Let Γ be a closed collection of games and let φ be a solution on Γ. Then φ is *consistent* (CONS) if for every game $G = (N, S_1, \ldots, S_n, u_1, \ldots, u_n)$, every $\emptyset \neq M \subseteq N$, and every strategy combination $\sigma \in \varphi(G)$, we have $\sigma_M \in \varphi(G^{M,\sigma})$.   □

The interpretation of consistency is as follows. If the players outside M have left the game while leaving the strategy combination $\sigma_{N \setminus M}$ behind, then there should be no need for the remaining players to revise their strategies.

The consequence of imposing OPR and CONS on a solution correspondence is that it can contain only Nash equilibria:

Proposition 13.37 *Let Γ be a closed collection of games and let φ be a solution on Γ satisfying OPR and CONS. Then $\varphi(G) \subseteq NE(G)$ for every $G \in \Gamma$.*

Proof Let $G = (N, S_1, \ldots, S_n, u_1, \ldots, u_n) \in \Gamma$ and $\sigma \in \varphi(G)$. By CONS, $\sigma_i \in \varphi(G^{\{i\},\sigma})$ for every $i \in N$. By OPR, $u_i^\sigma(\sigma_i) \geq u_i^\sigma(\tau_i)$ for every $\tau_i \in \Delta(S_i)$ and $i \in N$.

---

[2]For a subset $T \subseteq N$, we denote $(\sigma_j)_{j \in T}$ by σ_T.

Hence

$$u_i(\sigma_i, \sigma_{N\setminus\{i\}}) \geq u_i(\tau_i, \sigma_{N\setminus\{i\}}) \text{ for every } \tau_i \in \Delta(S_i) \text{ and } i \in N.$$

Thus, $\sigma \in NE(G)$.   ∎

Proposition 13.37 says that NE is the maximal solution (with respect to set-inclusion) satisfying OPR and CONS. (It it trivial to see that NE satisfies these conditions.) To derive a similar minimal set-inclusion result we use another condition.

Let Γ be a closed collection of games and let φ be a solution on Γ. For a game $G = (N, S_1, \ldots, S_n, u_1, \ldots, u_n) \in \Gamma$ with $|N| \geq 2$ denote

$$\tilde{\varphi}(G) = \{\sigma \in \prod_{i \in N} \Delta(S_i) \mid \text{ for all } \emptyset \neq M \subsetneq N, \sigma_M \in \varphi(G^{M,\sigma})\}.$$

Definition 13.38 A solution φ on a closed collection of games satisfies *converse consistency* (COCONS) if for every game G with at least two players, $\tilde{\varphi}(G) \subseteq \varphi(G)$.   □

Converse consistency says that strategy combinations of which the restrictions belong to the solution in smaller reduced games, should also belong to the solution of the game itself. Note that consistency can be defined by the converse inclusion $\varphi(G) \subseteq \tilde{\varphi}(G)$ for every $G \in \Gamma$, which explains the expression 'converse consistency'. Obviously, the Nash equilibrium correspondence satisfies COCONS.

Proposition 13.39 *Let Γ be a closed collection of games and let φ be a solution on Γ satisfying OPR and COCONS. Then $\varphi(G) \supseteq NE(G)$ for every $G \in \Gamma$.*

Proof The proof is by induction on the number of players. For one-person games the inclusion follows (with equality) from OPR. Assume that $NE(G) \subseteq \varphi(G)$ for all t-person games in Γ, where $t \leq k$ and $k \geq 1$. Let G_0 be a $k + 1$-person game in Γ. Note that $NE(G_0) \subseteq \widetilde{NE}(G_0)$ by CONS of NE. By the induction hypothesis, $\widetilde{NE}(G_0) \subseteq \tilde{\varphi}(G_0)$ and by COCONS, $\tilde{\varphi}(G_0) \subseteq \varphi(G_0)$. Thus, $NE(G_0) \subseteq \varphi(G_0)$.   ∎

Corollary 13.40 *Let Γ be a closed collection of games. The Nash equilibrium correspondence is the unique solution on Γ satisfying OPR, CONS, and COCONS.*

It can be shown that the axioms in Corollary 13.40 are independent (Problem 13.26).

In general, the consistency approach fails when applied to refinements of Nash equilibrium. For instance, Problem 13.27 shows that the correspondence of perfect equilibria is not consistent.

13.9 Problems

13.1. *Existence of Nash Equilibrium Using Brouwer*
Let $G = (N, S_1, \ldots, S_n, u_1, \ldots, u_n)$ be a finite game, as defined in Sect. 13.1. Define
the function $f : \prod_{i \in N} \Delta(S_i) \to \prod_{i \in N} \Delta(S_i)$ by

$$f_{i,s_i}(\sigma) = \frac{\sigma_i(s_i) + \max\{0, u_i(s_i, \sigma_{-i}) - u_i(\sigma)\}}{1 + \sum_{s_i' \in S_i} \max\{0, u_i(s_i', \sigma_{-i}) - u_i(\sigma)\}}$$

for all $i \in N$ and $s_i \in S_i$, where $\sigma_{-i} = (\sigma_1, \ldots, \sigma_{i-1}, \sigma_{i+1}, \ldots, \sigma_n)$.

(a) Show that f is well-defined, i.e., that $f(\sigma) \in \prod_{i \in N} \Delta(S_i)$ for every $\sigma \in \prod_{i \in N} \Delta(S_i)$.

(b) Argue that f has a fixed point, i.e., there is $\sigma^* \in \prod_{i \in N} \Delta(S_i)$ with $f(\sigma^*) = \sigma^*$, by
using the Brouwer Fixed Point Theorem (Theorem 22.10).

(c) Show that $\sigma^* \in \prod_{i \in N} \Delta(S_i)$ is a fixed point of f if and only if it is a Nash
equilibrium of G.

13.2. *Existence of Nash Equilibrium Using Kakutani*
Prove that the correspondence β in the proof of Theorem 13.1 is upper semi-continuous and convex-valued. Also check that every fixed point of β is a Nash equilibrium of G.

13.3. *Lemma 13.2*
Prove Lemma 13.2.

13.4. *Lemma 13.3*
Prove Lemma 13.3.

13.5. *Dominated Strategies*
Let (A, B) be an $m \times n$ bimatrix game. Suppose there exists a $\mathbf{q} \in \Delta^n$ such that
$q_n = 0$ and $B\mathbf{q} > Be^n$ (i.e., there exists a mixture of the first $n-1$ columns of B that
is strictly better than playing the n-th column).

(a) Prove that $q_n^* = 0$ for every Nash equilibrium $(\mathbf{p}^*, \mathbf{q}^*)$.
Let (A', B') be the bimatrix game obtained from (A, B) be deleting the last column.

(b) Prove that $(\mathbf{p}^*, \mathbf{q}')$ is a Nash equilibrium of (A', B') if and only if $(\mathbf{p}^*, \mathbf{q}^*)$ is a
Nash equilibrium of (A, B), where $\mathbf{q}'$ is the strategy obtained from $\mathbf{q}^*$ by deleting
the last coordinate.

13.6. *A* 3 × 3 *Bimatrix Game*
Consider the 3 × 3 bimatrix game

$$(A, B) = \begin{pmatrix} 0,4 & 4,0 & 5,3 \\ 4,0 & 0,4 & 5,3 \\ 3,5 & 3,5 & 6,6 \end{pmatrix}.$$

Let $(\mathbf{p}, \mathbf{q})$ be a Nash equilibrium in (A, B).

(a) Prove that $\{1, 2\} \not\subseteq C(\mathbf{p})$.
(b) Prove that $C(\mathbf{p}) \neq \{2, 3\}$.
(c) Find all Nash equilibria of this game.

13.7. *A* 3 × 2 *Bimatrix Game*
Use the graphical method to compute the Nash equilibria of the bimatrix game

$$(A, B) = \begin{pmatrix} 0,0 & 2,1 \\ 2,2 & 0,2 \\ 2,2 & 0,2 \end{pmatrix}.$$

13.8. *The Nash Equilibria in Example 13.18*

(a) Compute the Nash equilibria of the game in Example 13.18.
(b) Show that the Nash equilibria in the set $\{((0, 1, 0), (q, 1 - q, 0)) \mid 0 \leq q \leq 1\}$
    are perfect by using the definition of perfection.

13.9. *Proof of Theorem 13.8*
Prove Theorem 13.8.

13.10. *Matrix Games*
Show that the pair of linear programs (13.3) and (13.4) is equivalent to the LP and
its dual in Sect. 12.2 for solving matrix games.

13.11. *Tic-Tac-Toe*
The two-player game of Tic-Tac-Toe is played on a 3 × 3 board. Player 1 starts by
putting a cross on one of the nine fields. Next, player 2 puts a circle on one of the
eight remaining fields. Then player 1 puts a cross on one of the remaining seven
fields, etc. If player 1 achieves three crosses or player 2 achieves three circles in
a row (either vertically or horizontally or diagonally) then that player wins. If this
does not happen and the board is full, then the game ends in a draw.

(a) Design a pure maximin strategy for player 1. Show that this maximin strategy guarantees at least a draw to him.
(b) Show that player 1 cannot guarantee a win.
(c) What is the value of this game?

13.12. *Iterated Elimination in a Three-Player Game*
Solve the following three-player game, where player 1 chooses rows, player 2 columns, and player 3 one of the two games L and R:

$$L: \begin{array}{c} U \\ D \end{array} \begin{array}{cc} l & r \\ \left(\begin{array}{cc} 14,24,32 & 8,30,27 \\ 30,16,24 & 13,12,50 \end{array}\right) \end{array} \quad R: \begin{array}{c} U \\ D \end{array} \begin{array}{cc} l & r \\ \left(\begin{array}{cc} 16,24,30 & 30,16,24 \\ 30,23,14 & 14,24,32 \end{array}\right) \end{array}.$$

13.13. *Never a Best Reply and Domination*
In the following game player 1 chooses rows, player 2 chooses columns, and player 3 chooses matrices. The diagram gives the payoffs of player 3. Show that Y is never a best reply for player 3, and that Y is not strictly (and not even weakly) dominated.

$$V: \begin{array}{c} U \\ D \end{array} \begin{array}{cc} L & R \\ \left(\begin{array}{cc} 9 & 0 \\ 0 & 0 \end{array}\right) \end{array} \quad W: \begin{array}{c} U \\ D \end{array} \begin{array}{cc} L & R \\ \left(\begin{array}{cc} 0 & 9 \\ 9 & 0 \end{array}\right) \end{array}$$

$$X: \begin{array}{c} U \\ D \end{array} \begin{array}{cc} L & R \\ \left(\begin{array}{cc} 0 & 0 \\ 0 & 9 \end{array}\right) \end{array} \quad Y: \begin{array}{c} U \\ D \end{array} \begin{array}{cc} L & R \\ \left(\begin{array}{cc} 6 & 0 \\ 0 & 6 \end{array}\right) \end{array}$$

13.14. *Completely Mixed Nash Equilibria Are Perfect*
Prove Lemma 13.16.

13.15. *A 3-Player Game with an Undominated But Not Perfect Equilibrium*
Consider the following 3-player game, where player 1 chooses rows, player 2 columns, and player 3 matrices:

$$L: \begin{array}{c} U \\ D \end{array} \begin{array}{cc} l & r \\ \left(\begin{array}{cc} 1,1,1 & 1,0,1 \\ 1,1,1 & 0,0,1 \end{array}\right) \end{array} \quad R: \begin{array}{c} U \\ D \end{array} \begin{array}{cc} l & r \\ \left(\begin{array}{cc} 1,1,0 & 0,0,0 \\ 0,1,0 & 1,0,0 \end{array}\right) \end{array}.$$

(a) Show that (U,l,L) is the only perfect Nash equilibrium of this game.
(b) Show that (D,l,L) is an undominated Nash equilibrium.

13.16. *Existence of Proper Equilibrium*
Prove that the correspondence F in the proof of Theorem 13.29 satisfies the conditions of the Kakutani Fixed Point Theorem.

13.17. *Strictly Dominated Strategies and Proper Equilibrium*
Consider the 3-person game

$$L: \begin{array}{c} \\ U \\ D \end{array}\begin{array}{cc} l & r \\ \left(\begin{array}{cc} 1,1,1 & 0,0,1 \\ 0,0,1 & 0,0,1 \end{array}\right) \end{array} \quad R: \begin{array}{c} \\ U \\ D \end{array}\begin{array}{cc} l & r \\ \left(\begin{array}{cc} 0,0,0 & 0,0,0 \\ 0,0,0 & 1,1,0 \end{array}\right) \end{array},$$

where player 1 chooses rows, player 2 chooses columns, and player 3 chooses matrices.

(a) First assume that player 3 is a dummy and has only one strategy, namely L. Compute the perfect and proper Nash equilibrium or equilibria of the game.
(b) Now suppose that player 3 has two pure strategies. Compute the perfect and proper Nash equilibrium or equilibria of the game. Conclude that adding a strictly dominated strategy (namely, R) has resulted in an additional proper equilibrium.

13.18. *Strictly Perfect Equilibrium*

(a) Show that a completely mixed Nash equilibrium in a finite game G is strictly perfect.
(b) Show that a strict Nash equilibrium in a game G is strictly perfect. (A Nash equilibrium is *strict* if any unilateral deviation of a player leads to a strictly lower payoff for that player.)
(c) Compute all Nash equilibria, perfect equilibria, proper equilibria, and strictly perfect equilibria in the following game, where $\alpha, \beta > 0$. (Conclude that strictly perfect equilibria may fail to exist.)

$$(A, B) = \begin{array}{c} \\ U \\ D \end{array}\begin{array}{ccc} L & M & R \\ \left(\begin{array}{ccc} 0, \beta & \alpha, 0 & 0,0 \\ 0, \beta & 0,0 & \alpha, 0 \end{array}\right) \end{array}.$$

13.19. *Correlated Equilibria in the Two-Driver Example (1)*
Compute all correlated equilibria in the game

$$(A, B) = \begin{array}{c} \\ c \\ s \end{array}\begin{array}{cc} c & s \\ \left(\begin{array}{cc} -10,-10 & 5,0 \\ 0,5 & -1,-1 \end{array}\right) \end{array},$$

by using the definition of correlated equilibrium.

13.20. *Nash Equilibria Are Correlated*
Let $(\mathbf{p}, \mathbf{q})$ be a Nash equilibrium in the $m \times n$ bimatrix game (A, B). Let $P = (p_{ij})$ be the $m \times n$ matrix defined by $p_{ij} = p_i q_j$ for all $i = 1, \ldots, m$ and $j = 1, \ldots, n$. Show that P is a correlated equilibrium.

13.21. *The Set of Correlated Equilibria Is Convex*

Show that the set of correlated equilibria in a bimatrix game (A, B) is convex.

13.22. *Correlated vs. Nash Equilibrium*

Consider the bimatrix game

$$(A, B) = \begin{pmatrix} 6,6 & 2,7 \\ 7,2 & 0,0 \end{pmatrix}$$

and the correlated strategy

$$P = \begin{pmatrix} \frac{1}{3} & \frac{1}{3} \\ \frac{1}{3} & 0 \end{pmatrix} .$$

(a) Compute all Nash equilibria of (A, B).
(b) Show that P is a correlated equilibrium and that the associated payoffs fall outside the convex hull of the payoff pairs associated with the Nash equilibria of (A, B).

13.23. *Correlated Equilibria in the Two-Driver Example (2)*

Consider again the game of Problem 13.19 and set up the associated matrix C as in Sect. 13.7. Show that the value of the matrix game C is equal to 3, and that player 1 in C has a unique optimal strategy. (Hence, this method gives one particular correlated equilibrium.)

13.24. *Finding Correlated Equilibria*

Compute (the) correlated equilibria in the following game directly, and by using the associated matrix game.

$$\begin{array}{cc} & \begin{array}{cc} 1' & 2' \end{array} \\ (A, B) = \begin{array}{c} 1 \\ 2 \end{array} & \begin{pmatrix} 5,2 & 1,3 \\ 2,3 & 4,1 \end{pmatrix} . \end{array}$$

13.25. *Nash, Perfect, Proper, Strictly Perfect, and Correlated Equilibria*

Consider the bimatrix game $(A, B) = \begin{pmatrix} 0,6 & 0,4 & 6,0 \\ 4,0 & 0,0 & 4,0 \\ 6,0 & 0,4 & 0,6 \end{pmatrix} .$

Let $(\mathbf{p}, \mathbf{q}) \in \Delta^3 \times \Delta^3$ be a Nash equilibrium in (A, B).

(a) Show that, if $p_3 = 0$, then $\mathbf{p} = (0, 1, 0)$.
(b) Show that, if $p_1 > 0$ and $p_3 > 0$, then $\mathbf{q} = (0, 1, 0)$.
(c) Show that in each Nash equilibrium at least one player has payoff 0.
(d) Compute all Nash equilibria of (A, B).

(e) Which Nash equilibria are perfect?

(f) Which Nash equilibria are proper? Strictly perfect?

Consider the correlated strategy $P = \begin{pmatrix} 0 & \alpha & 0 \\ \beta & 0 & \gamma \\ 0 & \delta & 0 \end{pmatrix}$,

with $\alpha, \beta, \gamma, \delta \geq 0$ and $\alpha + \beta + \gamma + \delta = 1$.

(g) Under which conditions is P a correlated equilibrium?

(h) Find a correlated equilibrium with payoff 3 for player 1 and 1 for player 2.

13.26. *Independence of the Axioms in Corollary 13.40*

Show that the three conditions in Corollary 13.40 are independent: for each pair of conditions, exhibit a solution that satisfies these two conditions but not the third one.

13.27. *Inconsistency of Perfect Equilibria*

Consider the 3-person game G_0

$$
D: \begin{matrix} & L & R \\ T & \\ B \end{matrix} \begin{pmatrix} 1,1,1 & 1,0,1 \\ 1,1,1 & 0,0,1 \end{pmatrix} \quad U: \begin{matrix} & L & R \\ T & \\ B \end{matrix} \begin{pmatrix} 0,1,0 & 0,0,0 \\ 1,1,0 & 0,0,0 \end{pmatrix}
$$

where player 1 chooses rows, player 2 columns, and player 3 matrices. Let Γ consist of this game and all its reduced games. Use this collection to show that the perfect Nash equilibrium correspondence is not consistent.

13.10 Notes

The result that every finite game has a Nash equilibrium in mixed strategies is due to Nash (1951). For the quadratic programming problem in Sect. 13.2.3, see Mangasarian and Stone (1964). Lemke and Howson (1964) provide an algorithm to find at least one Nash equilibrium of a bimatrix game. See von Stengel (2002) for an overview.

For the first proof of Theorem 13.10 see Zermelo (1913). Theorem 13.13 is due to Pearce (1984). For the concept of rationalizability, see Bernheim (1984) and Pearce (1984). For a study of the assumptions underlying the procedure of iterated elimination of strictly dominated strategies see Tan and Werlang (1988) or Perea (2001). Perea (2012) presents a recent overview on epistemic game theory.

The concept of perfect equilibrium (trembling hand perfection) is due to Selten (1975). The notions of ε-perfect and (ε-)proper equilibrium were introduced in Myerson (1978). Theorem 13.20 is based on Selten (1975) and Myerson (1978) and appears as Theorem 2.2.5 in van Damme (1991). The notion of strictly perfect equilibrium was introduced in Okada (1981).

Correlated equilibria were introduced in Aumann (1974). Our treatment of the topic closely follows the presentation in Owen (1995); Example 13.34 corresponds to Example VIII.4.4 in that book.

For theories of strategic learning, see Young (2004). Section 13.8 is based on Peleg and Tijs (1996); the presentation there is for more general games. See Norde et al. (1996) for a study of the notion of consistency for refinements of Nash equilibrium.

Problem 13.12 is from Watson (2002). Problem 13.13 is from Fudenberg and Tirole (1991). Problems 13.15 and 13.17 are based on van Damme (1991). Problem 13.22 is based on Aumann (1974).

References

Aumann, R. J. (1974). Subjectivity and correlation in randomized strategies. *Journal of Mathematical Economics, 1*, 67–96.

Bernheim, B. (1984). Rationalizable strategic behavior. *Econometrica, 52*, 1007–1028.

Fudenberg, D., & Tirole, J. (1991). *Game theory*. Cambridge: MIT Press.

Lemke, C. E., & Howson, J. T. (1964). Equilibrium points of bimatrix games. *Journal of the Society for Industrial and Applied Mathematics, 12*, 413–423.

Mangasarian, O. L., & Stone, H. (1964). Two-person nonzero-sum games and quadratic programming. *Journal of Mathematical Analysis and Applications, 9*, 348–355.

Myerson, R. B. (1978). Refinements of the Nash equilibrium concept. *International Journal of Game Theory, 7*, 73–80.

Nash, J. F. (1951). Non-cooperative games. *Annals of Mathematics, 54*, 286–295.

Norde, H., Potters, J., Reijnierse, H., & Vermeulen, D. (1996). Equilibrium selection and consistency. *Games and Economic Behavior, 12*, 219–225.

Okada, A. (1981). On stability of perfect equilibrium points. *International Journal of Game Theory, 10*, 67–73.

Owen, G. (1995). *Game theory* (3rd ed.). San Diego: Academic.

Pearce, D. (1984). Rationalizable strategic behavior and the problem of perfection. *Econometrica, 52*, 1029–1050.

Peleg, B., & Tijs, S. (1996). The consistency principle for games in strategic form. *International Journal of Game Theory, 25*, 13–34.

Perea, A. (2001). *Rationality in extensive form games*. Boston: Kluwer Academic.

Perea, A. (2012). *Epistemic game theory: Reasoning and choice*. Cambridge: Cambridge University Press.

Selten, R. (1975). Reexamination of the perfectness concept for equilibrium points in extensive games. *International Journal of Game Theory, 4*, 25–55.

Tan, T., & Werlang, S. R. C. (1988). The Bayesian foundations of solution concepts of games. *Journal of Economic Theory, 45*, 370–391.

van Damme, E. C. (1991). *Stability and perfection of Nash equilibria*. Berlin: Springer.

von Stengel, B. (2002). Computing equilibria for two-person games. In: R. Aumann & S. Hart (Eds.), *Handbook of game theory with economic applications* (Vol. 3). Amsterdam: North-Holland.

Watson, J. (2002). *Strategy, an introduction to game theory*. New York: Norton.

Young, H. P. (2004). *Strategic learning and its limits*. Oxford: Oxford University Press.

Zermelo, E. (1913). Über eine Anwendung der Mengenlehre auf die Theorie des Schachspiels. In *Proceedings Fifth International Congress of Mathematicians* (Vol. 2, pp. 501–504).

Extensive Form Games

14

A game in extensive form specifies when each player in the game has to move, what his information is about the sequence of previous moves, which chance moves occur, and what the final payoffs are. Such games are discussed in Chaps. 4 and 5, and also occur in Chaps. 6 and 7. The present chapter extends the material introduced in Chaps. 4 and 5, and it may be useful to (re)read these chapters before continuing.

Section 14.1 formally introduces extensive form structures and games, and Sect. 14.2 introduces behavioral strategies and studies the relation between behavioral and mixed strategies. Section 14.3 is on Nash equilibrium and its main refinements, namely subgame perfect equilibrium and sequential equilibrium. The latter concept generalizes perfect Bayesian equilibrium.

14.1 Extensive Form Structures and Games

An extensive form game is based on a directed rooted tree. A *directed rooted tree* is a pair $T = (X, E)$, where:

- X is a finite set with $|X| \geq 2$. The elements of X are called *nodes*.
- E is a subset of $X \times X$. The elements of E are called *edges*. An edge $e = (x, y) \in E$ is called an *outgoing* edge of x and an *ingoing* edge of y.
- There is an $x_0 \in X$, called the *root*, such that for each $x \in X \setminus \{x_0\}$ there is a unique path from x_0 to x. Here, a *path* from x_0 to x is a series of edges $(x_0, x_1), (x_1, x_2), \ldots, (x_{k-1}, x_k), (x_k, x)$ for some $k \geq 0$.
- x_0 has no ingoing edges.

These conditions imply that each node which is not the root, has exactly one ingoing edge. Moreover, there are nodes which have no outgoing edges. These nodes are called *end nodes*. The set of end nodes is denoted by $Z \ (\subseteq X)$.

© Springer-Verlag Berlin Heidelberg 2015
H. Peters, *Game Theory*, Springer Texts in Business and Economics,
DOI 10.1007/978-3-662-46950-7_14

An *extensive form structure* is a tuple $S = (T, N, P, \mathcal{H}, \mathcal{A}, \tau)$, where:

- $T = (X, E)$ is a directed rooted tree with root x_0 and set of end nodes Z.
- $N = \{1, \ldots, n\}$ with $n \geq 1$ is the set of *players*.
- $P : X \setminus Z \to N \cup \{C\}$ is a function assigning to each non-end node either a player or Chance C. If $P(x)$ is a player, then node x is a *decision node* of player $P(x)$, otherwise x is a *chance node*.
- $\mathcal{H} = (H_i)_{i \in N}$ where for each $i \in N$, H_i is a partition of the set $P^{-1}(i)$ of decision nodes of player i. The sets $h \in H_i$ are called *information sets* of player i. Each $h \in H_i$ is assumed to satisfy (i) every path in T intersects h at most once and (ii) every node in h has the same number of outgoing edges.
- $\mathcal{A} = (A(h))_{h \in H}$, where $H = \cup_{i \in N} H_i$, and for each $h \in H$, $A(h)$ is a partition of the set of edges outgoing from nodes $x \in h$. The partition $A(h)$ is such that for each $x \in h$ and each $a \in A(h)$, a contains exactly one edge outgoing from x. Every set $a \in A(h)$ is called an *action* at h. It is assumed that $|A(h)| \geq 2$ for each $h \in H$.
- τ assigns to each chance node a probability distribution over the set of outgoing edges, where it is assumed that all these probabilities are positive.

In Fig. 14.1—which is a partial reproduction of Fig. 4.1—these concepts are illustrated.

In this extensive form structure, the directed rooted tree has 14 nodes $x_0, x_1, \ldots, x_{13}$ and, consequently, 13 edges. The set of end nodes is $Z = \{x_6, \ldots, x_{13}\}$ and the player set is $N = \{1, 2\}$. The function $P : \{x_0, \ldots, x_5\} \to \{1, 2\} \cup \{C\}$ is defined by $P(x) = 1$ for $x \in \{x_0, x_5\}$, $P(x) = 2$ for $x \in \{x_2, x_3, x_4\}$, and $P(x_1) = C$. The information sets are $\{x_0\}, \{x_5\} \in H_1$ and $\{x_2, x_3\}, \{x_4\} \in H_2$. The actions for player 1 are: $\{(x_0, x_2)\}, \{(x_0, x_3)\}, \{(x_0, x_1)\} \in A(\{x_0\})$ and $\{(x_5, x_{12})\}, \{(x_5, x_{13})\} \in A(\{x_5\})$. The actions for player 2 are: $\{(x_2, x_6), (x_3, x_8)\}, \{(x_2, x_7), (x_3, x_9)\} \in A(\{x_2, x_3\})$ and $\{(x_4, x_{10})\}, \{(x_4, x_{11})\} \in A(\{x_4\})$. Finally, $\tau(x_1) = (1/4, 3/4)$, where $1/4$ is the probability of (x_1, x_4) and $3/4$ is the probability of (x_1, x_5).

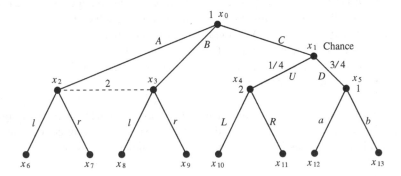

Fig. 14.1 An extensive form structure

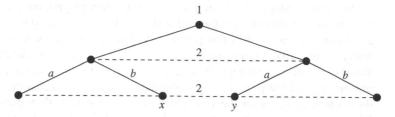

Fig. 14.2 Part of an extensive form structure without perfect recall

Clearly, this formal notation is quite cumbersome and we try to avoid it as much as possible. It is only needed to give precise definitions and proofs.

It is usually assumed that an extensive form structure S satisfies *perfect recall*: this means that each player always remembers what he did in the past. The formal definition is as follows.

Definition 14.1 An extensive form structure S satisfies *perfect recall for player $i \in N$* if for every information set $h \in H_i$ and each pair of nodes $x, y \in h$, player i's outgoing edges on the path from the root to x belong to the same player i actions as player i's outgoing edges on the path from the root to y. In other words, an action of player i is on the path to x if and only if it is on the path to y. We say that S satisfies *perfect recall* if it satisfies perfect recall for every player. □

Figure 14.2 shows part of an extensive form structure *without* perfect recall.

The condition of perfect recall plays an important role for the relation between mixed and behavioral strategies (see Sect. 14.2).

We also repeat the definitions of perfect and imperfect information (cf. Chap. 4).

Definition 14.2 An extensive form structure S has *perfect information* if for every $i \in N$ and $h \in H_i$, $|h| = 1$. Otherwise, S has *imperfect information*. □

We conclude this section with the formal definition of an extensive form game. An *extensive form game* Γ is an $n + 1$ tuple $\Gamma = (S, u_1, \ldots, u_n)$, where S is an extensive form structure and for each player $i \in N$, $u_i : Z \to \mathbb{R}$. The function u_i is player i's *payoff function*.

A game $\Gamma = (S, u_1, \ldots, u_n)$ has *(im)perfect information* if S has (im)perfect information.

14.2 Pure, Mixed and Behavioral Strategies

Let $S = (T, N, P, \mathcal{H}, \mathcal{A}, \tau)$ be an extensive form structure. A *pure strategy* s_i of player $i \in N$ is a map assigning an action $a \in A(h)$ to every information set $h \in H_i$. By S_i we denote the (finite) set of pure strategies of player i.

Any pure strategy combination $(s_1, \ldots, s_n)$, when played, results in a (unique) probability distribution over the end nodes of the game tree. (Of course, if there are no chance moves then this probability distribution is degenerate.) Therefore, with each extensive form game $\Gamma = (\mathcal{S}, u_1, \ldots, u_n)$ we can associate (with a slight abuse of notation) a strategic form game $G(\Gamma) = (S_1, \ldots, S_n, u_1, \ldots, u_n)$ in the obvious way: if the pure strategy combination $(s_1, \ldots, s_n)$ generates a probability distribution $(q_z)_{z \in Z}$ over the end nodes of the game tree, then player i receives $\sum_{z \in Z} q_z u_i(z)$. A (mixed) *strategy* of player $i \in N$ is an element of $\Delta(S_i)$, i.e., a probability distribution over the elements of S_i.

When considering an extensive form game (structure) it seems more natural to consider, instead of mixed strategies, so-called *behavioral strategies*. A behavioral strategy of a player assigns to each information set of that player a probability distribution over the actions at that information set. Formally, we have the following definition.

Definition 14.3 Let $\mathcal{S} = (T, N, P, \mathcal{H}, \mathcal{A}, \tau)$ be an extensive form structure. A *behavioral strategy* of player $i \in N$ is a map b_i assigning to each information set $h \in H_i$ a probability distribution over the set of actions $A(h)$. $\qquad\square$

Given a behavioral strategy there is an obvious way to define an associated mixed strategy: for each pure strategy, simply multiply all probabilities assigned by the behavioral strategy to the actions occurring in the pure strategy. Consider for instance the extensive form structure in Fig. 14.3. In this diagram a behavioral strategy b_1 of player 1 is indicated. The associated mixed strategy σ_1 assigns the probabilities $\sigma_1(A, l) = \frac{1}{2} \cdot \frac{1}{3} = \frac{1}{6}$, $\sigma_1(A, r) = \frac{1}{2} \cdot \frac{2}{3} = \frac{1}{3}$, $\sigma_1(B, l) = \frac{1}{2} \cdot \frac{1}{3} = \frac{1}{6}$, and $\sigma_1(B, r) = \frac{1}{2} \cdot \frac{2}{3} = \frac{1}{3}$. Strategy σ_1 is the 'right' mixed strategy associated with b_1 in the following sense. Suppose player 2 plays the mixed or behavioral strategy— there is no difference in this case—which puts probability α on L and $1 - \alpha$ on R. If player 1 plays the behavioral strategy b_1 then the probability distribution generated over the end nodes of the game is $x_5 \mapsto \frac{1}{6} \cdot \alpha$, $x_6 \mapsto \frac{1}{3} \cdot \alpha$, $x_7 \mapsto \frac{1}{6} \cdot (1 - \alpha)$, $x_8 \mapsto \frac{1}{3} \cdot (1 - \alpha)$. The same distribution is generated by the mixed strategy σ_1. For example, the probability that x_5 is reached equals $\sigma_1(A, l) \cdot \alpha = \frac{1}{6} \cdot \alpha$, etc. We call b_1

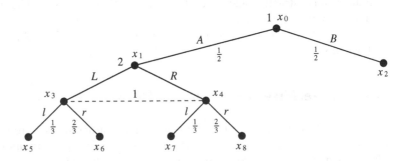

Fig. 14.3 From behavioral to mixed strategies

and σ_1 *outcome equivalent*. Obviously, it would have been sufficient to check this for the two pure strategies L and R for player 2.

We summarize these considerations in a definition and a proposition, which is presented without a formal proof.

Definition 14.4 Two (behavioral or mixed) strategies of player i in S are *outcome equivalent* if for each pure strategy combination s_{-i} of the other players the probability distributions generated by the two strategies over the end nodes are equal. □

Proposition 14.5 *Let b_i be a behavioral strategy of player i in S. Then there is a mixed strategy σ_i of player i that is outcome equivalent to b_i. Such a strategy σ_i is obtained by assigning to each pure strategy s_i of player i the product of the probabilities assigned by b_i to the actions chosen by s_i.*

It should be noted that there is not necessarily a unique mixed strategy that is outcome equivalent to a given behavioral strategy. For instance, in the example above, if we change the behavioral strategy of player 1 such that it assigns probability 0 to action A and probability 1 to action B at information set $\{x_0\}$, then all mixed strategies which put zero probability on (A, l) and (A, r) are outcome equivalent, resulting in each end node other than x_2 with zero probability.

Also for the converse question of how to associate an outcome equivalent behavioral strategy with a mixed strategy, it is not hard to figure out a procedure. Suppose that σ_1 is a mixed strategy of player i, h is an information set of player i, and a is an action in h. First, let $S_i(h)$ denote the set of pure strategies of player i such that the play of the game possibly reaches h, in other words, such that there exists a path through h containing the actions prescribed by the pure strategy under consideration. Then $\sigma_i(S_i(h))$ is the total probability assigned by σ_i to this set of pure strategies. Within this set, consider those pure strategies that assign a to h and divide their total probability by $\sigma_i(S_i(h))$ if $\sigma_i(S_i(h)) > 0$: the result is defined to be $b_i(h)(a)$. Thus, $b_i(h)(a)$ is the probability of a being played conditional on the set h being reached—it is a *conditional probability*. If $\sigma_i(S_i(h)) = 0$ then we can choose $b_i(h)$ arbitrary. This way, we construct a behavioral strategy b_i that is outcome equivalent to the mixed strategy σ_i.

As an illustration consider the extensive form structure in Fig. 14.4, which is the same as the one in Fig. 14.3.

Consider the mixed strategy σ_1 of player 1 defined by: $(A, l) \mapsto \frac{1}{5}$, $(A, r) \mapsto \frac{1}{10}$, $(B, l) \mapsto \frac{2}{5}$, $(B, r) \mapsto \frac{3}{10}$. Following the above procedure we obtain

$$b_1(A) = \frac{\sigma_1(A,l) + \sigma_1(A,r)}{1} = \frac{1}{5} + \frac{1}{10} = \frac{3}{10}$$

$$b_1(B) = \frac{\sigma_1(B,l) + \sigma_1(B,r)}{1} = \frac{2}{5} + \frac{3}{10} = \frac{7}{10}$$

$$b_1(l) = \frac{\sigma_1(A,l)}{\sigma_1(A,l) + \sigma_1(A,r)} = \frac{1/5}{1/5 + 1/10} = \frac{2}{3}$$

$$b_1(r) = \frac{\sigma_1(A,r)}{\sigma_1(A,l) + \sigma_1(A,r)} = \frac{1/10}{1/5 + 1/10} = \frac{1}{3} .$$

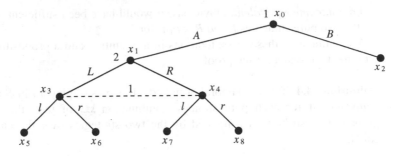

Fig. 14.4 From mixed to behavioral strategies

It is straightforward to verify that b_1 and σ_1 are outcome equivalent.

Outcome equivalence is not guaranteed without perfect recall: see Problem 14.2 for an example. With perfect recall, we have the following theorem.

Theorem 14.6 (Kuhn) *Let the extensive form structure $\mathcal{S}$ satisfy perfect recall. Then, for every player i and every mixed strategy σ_i there is a behavioral strategy b_i that is outcome equivalent to σ_i.*

The behavioral strategy b_i in Theorem 14.6 can be constructed as described in the text.

14.3   Nash Equilibrium and Refinements

Let $\Gamma = (\mathcal{S}, (u_i)_{i \in N})$ be an extensive form game with associated strategic form game $G(\Gamma) = ((S_i)_{i \in N}, (u_i)_{i \in N})$. We assume that $\mathcal{S}$ satisfies perfect recall.

A *pure strategy Nash equilibrium* of Γ is defined to be a pure strategy Nash equilibrium of $G(\Gamma)$. Note that, if Γ has perfect information, then a pure strategy Nash equilibrium exists (cf. Chap. 4).

A *mixed strategy Nash equilibrium* of Γ is defined to be a (mixed strategy) Nash equilibrium of $G(\Gamma)$. By Theorem 13.1 such an equilibrium always exists.

Consider, now, a behavioral strategy combination $b = (b_i)_{i \in N}$ in Γ. Such a strategy combination generates a probability distribution over the end nodes and, thus, an expected payoff for each player. We call b_i a *best reply* of player $i \in N$ to the strategy combination b_{-i} if there is no other behavioral strategy b_i' of player i such that (b_i', b_{-i}) generates a higher expected payoff for player i. We call b a *Nash equilibrium* (in behavioral strategies) of Γ if b_i is a best reply to b_{-i} for every player $i \in N$. (Thus, observe that a Nash equilibrium of an extensive form game is by definition a Nash equilibrium in behavioral strategies.)

Let σ be a mixed strategy Nash equilibrium of Γ. By Theorem 14.6 there is a behavioral strategy combination b that is outcome equivalent to σ. We claim that b is a Nash equilibrium of Γ. Suppose not, then there is a player $i \in N$ and a behavioral strategy b_i' that gives player i a higher expected payoff against b_{-i}. By Proposition 14.5 there is a mixed strategy σ_i' that is outcome equivalent to b_i'. Consequently, σ_i' gives player i a higher expected payoff against σ_{-i}, a contradiction. We have thus proved:

Theorem 14.7 *Every extensive form game has a Nash equilibrium.*

In fact, a similar argument as the one leading to this theorem can be applied to show that every Nash equilibrium (in behavioral strategies) results in a Nash equilibrium in mixed strategies. Hence, one way to find the (behavioral strategy) Nash equilibria of an extensive form game is to determine all (mixed strategy) Nash equilibria of the associated strategic form game. Which way is most convenient depends on the game at hand. In particular for refinements it is often easier to compute behavioral equilibrium strategies directly, without first computing the mixed strategy Nash equilibria. Before discussing these refinements we first consider an example.

Example 14.8 Consider the extensive form game Γ_1 in Fig. 14.5.
This game is based on the extensive form structure of Fig. 14.4, in which the symbols for the end nodes are replaced by payoffs for player 1 (upper number) and player 2 (lower number). The associated strategic form of this game is given in Table 14.1. (Note that in $G(\Gamma_1)$ there is no essential difference between BC and BD.)

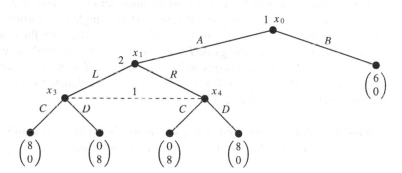

Fig. 14.5 The game Γ_1

Table 14.1 The strategic form $G(\Gamma_1)$ of Γ_1

	L	R
AC	$8,0$	$0,8$
AD	$0,8$	$8,0$
BC	$6,0$	$6,0$
BD	$6,0$	$6,0$

To find the Nash equilibria in $G(\Gamma_1)$, first note that player 2 will never play pure in a Nash equilibrium. Suppose, in equilibrium, that player 2 plays $(\alpha, 1 - \alpha)$ with $0 < \alpha < 1$, and player 1 plays $\mathbf{p} = (p_1, p_2, p_3, p_4)$, where p_1 is the probability of AC, p_2 the probability of AD, p_3 the probability of BC, and p_4 the probability of BD. Since $0 < \alpha < 1$, player 2 is indifferent between L and R, which implies that $p_1 = p_2$. Suppose $p_1 = p_2 > 0$. Then we must have $8\alpha \geq 6$ and $8(1 - \alpha) \geq 6$, which is impossible. Hence $p_3 + p_4 = 1$ and both $6 \geq 8\alpha$ and $6 \geq 8(1 - \alpha)$, so $1/4 \leq \alpha \leq 3/4$. This implies that $b = (b_1, b_2)$ is a (behavioral strategy) Nash equilibrium of Γ if and only if

$$b_1(A) = 0, \ b_1(B) = 1, \ 1/4 \leq b_2(L) \leq 3/4 .$$

So $b_1(C)$ may take any arbitrary value. □

In the remainder of this section we consider refinements of Nash equilibrium.

14.3.1 Subgame Perfect Equilibrium

Let x be a non-end node in an extensive form structure S and let $T^x = (V^x, E^x)$ be the subtree starting from x—i.e., V^x is the subset of V consisting of $\{x\}$ and all nodes of V that can be reached by a path starting from x, and E^x is the subset of E of all edges between nodes in V^x. If every information set of S is contained either in V^x or in $V \setminus V^x$ (this implies, in particular, that $\{x\}$ is a singleton information set), then we call S^x, the restriction of S to T^x, a *substructure*. Then, for the extensive form game $\Gamma = (S, (u_i)_{i \in N})$, the game $\Gamma^x = (S^x, (u_i^x)_{i \in N})$ is defined by restricting the payoff functions to the end nodes still available in V^x. We call Γ^x a *subgame* of Γ. For a behavioral strategy combination $b = (b_i)_{i \in N}$ we denote by $b^x = (b_i^x)_{i \in N}$ the restriction to the substructure S^x.

Definition 14.9 A behavioral strategy combination b in Γ is a *subgame perfect equilibrium* if b^x is a Nash equilibrium for every subgame Γ^x. □

Clearly, this definition extends the definition of subgame perfection for pure strategy combinations given in Chap. 4.

Since the whole game Γ is a subgame ($\Gamma = \Gamma^{x_0}$, where x_0 is the root of the game tree), every subgame perfect equilibrium is a Nash equilibrium. By carrying out a backward induction procedure as in Sect. 4.3, it can be seen that a subgame perfect equilibrium exists in any extensive form game.

Subgame perfection often implies a considerable reduction of the set of Nash equilibria. The following example is the continuation of Example 14.8.

Example 14.10 To find the subgame perfect equilibria in Γ_1, we only have to analyze the subgame $\Gamma_1^{x_1}$. It is easy to see that this subgame has a unique Nash equilibrium, namely player 2 playing L and R each with probability $1/2$, and player

1 playing C and D each with probability $1/2$. This results in a unique subgame perfect equilibrium $b = (b_1, b_2)$ given by

$$b_1(B) = 1, \; b_1(C) = 1/2, \; b_2(L) = 1/2 \; .$$

To reach this conclusion we have used that in any Nash equilibrium player 1 plays B, as established in Example 14.8. This, however, can also be seen from the fact that the only Nash equilibrium in the subgame starting at x_1 gives player 1 a payoff of 4, which is smaller than the payoff of 6 which player 1 obtains from playing B. In fact, this argument is a kind of 'backward induction', where in the first step we analyze the whole game $\Gamma_1^{x_1}$. $\square$

14.3.2 Perfect Bayesian and Sequential Equilibrium

In games without proper subgames and in games of imperfect information the subgame perfection requirement may not have much bite (see the examples in Chaps. 4 and 5). The concepts of perfect Bayesian and sequential equilibrium allow to distinguish between Nash equilibria by considering beliefs of players on information sets.

Consider an extensive form structure $S = (T, N, P, \mathcal{H}, \mathcal{A}, \tau)$. A *belief system* β assigns to every information set $h \in H = \cup_{i \in N} H_i$ a probability distribution β_h over the nodes in h. An *assessment* is a pair (b, β) of a behavioral strategy combination $b = (b_i)_{i \in N}$ and a belief system β.

The first requirement we consider is that of *sequential rationality*, which requires a player at an information set to choose only those actions that yield maximal expected payoff, given that player's beliefs. Consider an extensive form game $\Gamma = (S, (u_i)_{i \in N})$. Let (b, β) be an assessment. Let $i \in N$, $h \in H_i$, $a \in A(h)$, and $x \in h$. Suppose player i is at node x and takes action a. This corresponds to an edge (x, y) in the game tree. Then each end node on a path starting from x and passing though y is reached with a probability that is equal to the product of the probabilities of all edges on this path following y, given by b and the probabilities of eventual chance nodes on this path. This way, we can compute the expected payoff to player i from playing a, conditional on being at node x: denote this payoff by $u_i(a|b, x)$. Player i's expected payoff from action a, given information set h, is then equal to $\sum_{x \in h} \beta(x) u_i(a|b, x)$.

Definition 14.11 An assessment (b, β) in $\Gamma = (S, (u_i)_{i \in N})$ is *sequentially rational* if for every $i \in N$, $h \in H_i$, and $a \in A(h)$ we have:

$$b_i(h)(a) > 0 \Rightarrow \sum_{x \in h} \beta(x) \, u_i(a|b, x) = \max_{a' \in A(h)} \sum_{x \in h} \beta(x) \, u_i(a'|b, x) \; .$$

$\square$

Thus, indeed, sequential rationality of an assessment (b, β) means that a player puts only positive probability on those actions at an information set h that maximize his expected payoff, given h and his belief $\{\beta(x) \mid x \in h\}$.

Sequential rationality is defined with respect to a given belief system. We now formulate conditions on such belief systems which, in turn, are defined with respect to a given profile of behavioral strategies.

Let x be a node in the game tree T such that $\mathcal{S}^x$ is a substructure. For any node x' in the subtree T^x, let $\mathbb{P}_b^x(x')$ denote the probability that x' is reached in T^x given b, that is, $\mathbb{P}_b^x(x')$ is the product of the probabilities of all edges on the unique path from x to x': these probabilities are given by b, or by τ in case of a chance node on the path. For every information set h of $\mathcal{S}^x$, $\mathbb{P}_b^x(h) = \sum_{x' \in h} \mathbb{P}_b^x(x')$ is the probability that, within $\mathcal{S}^x$, the information set h is reached, given b.

The next definition is the formal version of the consistency requirement introduced already in Chap. 4.

Definition 14.12 An assessment (b, β) in $\mathcal{S}$ is *Bayesian consistent* if $\beta_h(x') = \mathbb{P}_b^x(x')/\mathbb{P}_b^x(h)$ for all $x' \in h$, for each substructure $\mathcal{S}^x$ and each information set h in $\mathcal{S}^x$ for which $\mathbb{P}_b^x(h) > 0$.   □

It is not difficult to see that, if h is an information set within a substructure $\mathcal{S}^x$ and $\mathcal{S}^x$ is in turn a substructure of another substructure $\mathcal{S}^y$, such that $\mathbb{P}_b^y(h) > 0$, then $\mathbb{P}_b^x(h) > 0$ and $\mathbb{P}_b^x(x')/\mathbb{P}_b^x(h) = \mathbb{P}_b^y(x')/\mathbb{P}_b^y(h)$ for all $x' \in h$. In other words, if beliefs are restricted by Bayesian consistency, this restriction does not depend on the particular substructure (or subgame) considered.

By Bayesian consistency the players' beliefs are determined, as conditional probabilities, by the behavioral strategies on all information sets that are reached with positive probability in some subgame. This requirement can be quite weak; for instance, it does not even imply that the beliefs of one and the same player have to be internally consistent, as is illustrated by the following example. Consider the extensive form structure in Fig. 14.6. Suppose that player 1 plays a and player 2 plays e. Then the beliefs of player 2 at the information sets $\{x_1, x_2\}$ and $\{x_3, x_4\}$ are not restricted by Bayesian consistency: the only substructure is the whole structure, within which the information sets of player 2 are reached with zero probability.

Fig. 14.6 If player 1 plays a, then player 2's beliefs at $\{x_1, x_2\}$ and $\{x_3, x_4\}$ are independent under Bayesian consistency

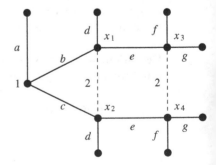

Moreover, a belief $\beta(x_1) = \beta(x_4) = 1$ is allowed, which means that player 2's beliefs are not internally consistent.

In many applications this drawback does not occur, and Bayesian consistency is strong enough. For instance, in signaling games (see Chap. 5) Bayesian consistency implies the stronger version of consistency in Definition 14.13 below. See Problem 14.4.

Call a behavioral strategy b_i of player i *completely mixed* if $b_i(h)(x) > 0$ for each $h \in H_i$ and $x \in h$. A behavioral strategy combination $b = (b_i)_{i \in N}$ is *completely mixed* if b_i is completely mixed for every $i \in N$. Observe that, if b is completely mixed and the assessment (b, β) is Bayesian consistent, then β is uniquely determined by b: all information sets in all substructures are reached with positive probability. The announced stronger version of consistency, simply called *consistency*, is defined as follows.

Definition 14.13 An assessment (b, β) in S is *consistent* if there exists a sequence $(b^m, \beta^m)_{m \in \mathbb{N}}$ of Bayesian consistent assessments with each b^m completely mixed and $\lim_{m \to \infty}(b^m, \beta^m) = (b, \beta)$. □

Consistency implies Bayesian consistency (Problem 14.3). Consistency is clearly stronger than Bayesian consistency. For instance, in the extensive form structure of Fig. 14.6, it is easily seen that consistency requires player 2 to have identical beliefs on his two information sets, i.e., $\beta(x_1) = \beta(x_3)$. This is true even if on his right information set player 2 is replaced by some other player 3, as in Fig. 14.7.

We can now define the announced equilibrium refinements.

Definition 14.14 An assessment (b, β) in $\Gamma = (S, (u_i)_{i \in N})$ is a *perfect Bayesian equilibrium* if it is sequentially rational and Bayesian consistent; it is a *sequential equilibrium* if it is sequentially rational and consistent. □

The following theorem, stated without a proof, collects some facts about perfect Bayesian and sequential equilibrium.[1]

Fig. 14.7 Under consistency, the beliefs of players 2 and 3 are identical: $\beta(x_1) = \beta(x_3)$

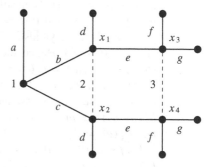

Fig. 14.8 The game of
Example 14.16

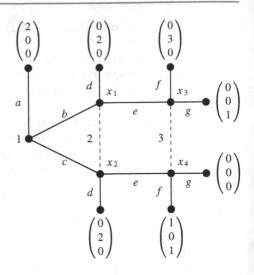

Theorem 14.15 *Every extensive form game has a sequential equilibrium. Every sequential equilibrium is perfect Bayesian. Every perfect Bayesian equilibrium is subgame perfect and, in particular, a Nash equilibrium.*

The following example shows that a perfect Bayesian equilibrium does not have to be sequential.

Example 14.16 We consider again the extensive form structure of Fig. 14.7, supplemented with payoffs: see Fig. 14.8. An example of a perfect Bayesian equilibrium in this game is the assessment (b, β) with $\beta(x_1) = \beta(x_4) = 1$ and $b_1(a) = b_2(e) = b_3(f) = 1$. This equilibrium is not sequential, since any consistent assessment should attach equal belief probabilities to x_1 and x_3. In fact, there is no sequential equilibrium in which $b_1(a) = b_2(e) = b_3(f) = 1$. To see this, let $\alpha = \beta(x_1) = \beta(x_3)$. Then still $b_1(a) = 1$. If $\alpha < 1/2$, then $b_3(f) = 1$, if $\alpha > 1/2$, then $b_3(f) = 0$, and if $\alpha = 1/2$, then $b_3(f)$ is arbitrary. For player 2, playing d yields 2. Playing e yields 3α if $\alpha < 1/2$, 0 if $\alpha > 1/2$, and at most $3/2$ if $\alpha = 1/2$. Hence, $b_2(d) = 1$. Hence, the sequential equilibria in this game are the assessments (b, β) with $b_1(a) = b_2(d) = 1$, $\beta(x_1) = \beta(x_3) = \alpha \in [0, 1]$, and if $\alpha < 1/2$, then $b_3(f) = 1$, if $\alpha > 1/2$, then $b_3(f) = 0$, and if $\alpha = 1/2$, then $b_3(f)$ is arbitrary. For $0 < \alpha < 1$, such an assessment is obtained as the limit of assessments (b^m, β^m) with: $b_1^m(a) = 1 - \alpha/m - (1-\alpha)/m$, $b_1^m(b) = \alpha/m$, and $b_1^m(c) = (1-\alpha)/m$; $b_2^m(e) = 1/m$; and $b_3^m(g) = 1/m$ if $\alpha < 1/2$, $b_3^m(f) = 1/m$ if $\alpha > 1/2$, and $b_3^m(f) = b_3(f)$ if $b_3(f) \in (0, 1)$. If $\alpha = 1$, then $b_1^m(a) = 1 - 1/m - 1/m^2$, $b_1^m(b) = 1/m$, $b_1^m(c) = 1/m^2$; if $\alpha = 0$, then $b_1^m(a) = 1 - 1/m - 1/m^2$, $b_1^m(b) = 1/m^2$, $b_1^m(c) = 1/m$; in both these cases the behavioral strategies b_2^m and b_3^m can remain unchanged. The belief systems β^m are determined by Bayesian consistency. □

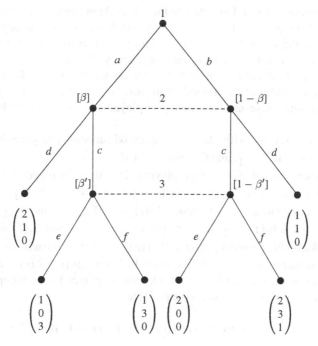

Fig. 14.9 Sequential equilibrium analysis in a three-player game

The advantage of perfect Bayesian equilibrium is that it avoids the condition involving limiting assessments in the definition of consistency. The idea behind perfect Bayesian equilibrium is that beliefs of players should be obtained by the method of Bayesian updating 'as much as possible'. In the preceding examples, however, it has become clear that the formal definition of Bayesian consistency and perfect Bayesian equilibrium does not capture this completely. (See also the Notes to this chapter.) Therefore, we stick to sequential equilibrium but keep in mind that, in order to compute sequential equilibria for a given game, the method of Bayesian updating 'as much as possible' is a good heuristic. There is hardly any general method available to compute sequential equilibria: it depends very much on the game at hand what the best way is. We conclude with two additional examples and refer to the problem section for other examples.

Example 14.17 Consider the three-player game in Fig. 14.9. The extensive form structure is similar to the one in Example 14.16, but without the 'outside option' of player 1. Note that the only subgame is the whole game—the notation $\mathbb{P}(\cdot)$ refers to the probabilities in this trivial subgame. To find the sequential equilibria of this game, first observe that consistency requires the beliefs of players 2 and 3 to be the same, so $\beta = \beta'$. For completeness we spell out the argument. Denote the behavioral strategies of the players by $b_1(a)$ and $b_1(b) = 1 - b_1(a)$, $b_2(c)$ and $b_2(d) = 1 - b_2(c)$, and $b_3(e)$ and $b_3(f) = 1 - b_3(e)$. Suppose b_1, b_2, and b_3 are completely mixed. Let x denote the left node in player 2's information set h_2. Then $\mathbb{P}_b(x) = b_1(a)$ and $\mathbb{P}_b(h_2) = 1$, so by Bayesian consistency $\beta = b_1(a)$. Let y denote the left node in player 3's information set h_3. Then $\mathbb{P}_b(y) = b_1(a)b_2(c)$ and $\mathbb{P}_b(h_3) = b_2(c)$, so by Bayesian consistency $\beta' = b_1(a)b_2(c)/b_2(c) = b_1(a)$. Thus, $\beta = \beta'$ if the behavioral strategies are completely mixed. By consistency, $\beta = \beta'$ for all profiles of behavioral strategies. Note that this is not necessarily the case under only Bayesian consistency: if player 2 plays d with probability 1, then β' is 'free'.

Now, starting with player 3, sequential rationality requires $b_3(e) = 1$ if $\beta > \frac{1}{4}$, $b_3(e) = 0$ if $\beta < \frac{1}{4}$, and $0 \le b_3(e) \le 1$ if $\beta = \frac{1}{4}$.

Using this, if $\beta > \frac{1}{4}$ then playing c yields 0 for player 2 and playing d yields 1. Therefore $b_2(c) = 0$. Similarly, $b_2(c) = 1$ if $\beta < \frac{1}{4}$. If $\beta = \frac{1}{4}$ then d yields 1 whereas c yields $3b_3(f)$. Hence, $b_2(c) = 0$ if $\beta = \frac{1}{4}$ and $b_3(f) < \frac{1}{3}$; $b_2(c) = 1$ if $\beta = \frac{1}{4}$ and $b_3(f) > \frac{1}{3}$; and $0 \le b_2(c) \le 1$ if $\beta = \frac{1}{4}$ and $b_3(f) = \frac{1}{3}$.

We finally consider player 1. If $b_1(a) > \frac{1}{4}$ then consistency requires $\beta = b_1(a) > \frac{1}{4}$ and therefore $b_2(c) = 0$. So player 1 obtains $2b_1(a) + 1(1 - b_1(a)) = 1 + b_1(a)$, which is maximal for $b_1(a) = 1$. Obviously, player 1 cannot improve on this. So we have the following sequential equilibrium:

$$b_1(a) = 1, \ b_2(c) = 0, \ b_3(e) = 1, \ \beta = \beta' = 1 \,.$$

If $b_1(a) < \frac{1}{4}$ then $\beta = b_1(a) < \frac{1}{4}$ and therefore $b_2(c) = 1$ and $b_3(e) = 0$. So player 1 obtains $1b_1(a) + 2(1 - b_1(a)) = 2 - b_1(a)$, which is maximal for $b_1(a) = 0$. Obviously again, player 1 cannot improve on this. So we have a second sequential equilibrium:

$$b_1(a) = 0, \ b_2(c) = 1, \ b_3(e) = 0, \ \beta = \beta' = 0 \,.$$

If $b_1(a) = \frac{1}{4}$ then player 1 must be indifferent between a and b. This implies that the expected payoff from a should be equal to the expected payoff from b, hence that $2(1 - b_2(c)) + 1b_2(c) = 2b_2(c) + 1(1 - b_2(c))$ which is true for $b_2(c) = \frac{1}{2}$. The preceding analysis for player 2 shows that for player 2 to play completely mixed we need $b_3(e) = \frac{2}{3}$. So we have a third sequential equilibrium

$$b_1(a) = \frac{1}{4}, \ b_2(c) = \frac{1}{2}, \ b_3(e) = \frac{2}{3}, \ \beta = \beta' = \frac{1}{4} \,.$$

Also in this game there are perfect Bayesian equilibria which are not sequential, e.g., $b_1(a) = b_2(d) = 1$, $b_3(e) = 2/3$, $\beta = 1$, and $\beta' = 1/4$.   □

Example 14.18 Consider the signaling game in Fig. 14.10 (this is the same game as in Fig. 5.4). To find the sequential equilibria in this game, we start with the beliefs of player 2. Clearly, and independently of α, sequential rationality requires player 2

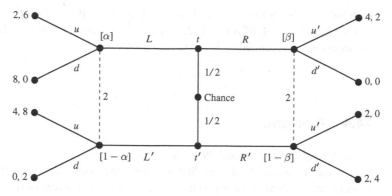

Fig. 14.10 The game of Example 14.18

to play u at his left information set. At his right information set, u' is at least as good as d' if $2\beta \geq 4(1 - \beta)$, hence if $\beta \geq 2/3$. Therefore, we distinguish three cases:

- $\beta > 2/3$. Then player 2 plays the pure strategy uu', hence player 1 plays RL', implying $\beta = 1$ and $\alpha = 0$. This is a sequential equilibrium (in fact, a pure strategy Nash equilibrium). In behavioral strategies: $b_1(R) = b_1(L') = 1$, $b_2(u) = b_2(u') = 1$.
- $\beta < 2/3$. Then player 2 plays ud', player 1 plays LL', $\alpha = 1/2$, and $\beta < 2/3$. In behavioral strategies: $b_1(L) = b_1(L') = 1$, $b_2(u) = b_2(d') = 1$.
- $\beta = 2/3$. Again $b_2(u) = 1$ implies $b_1(L') = 1$, but then we must have $b_1(L) = 1$ since otherwise we would have $\beta = 1$ (by Bayesian consistency). We then must have $b_2(u') \leq 1/2$ to keep type t from deviating to R. Further, $\alpha = 1/2$ completes the description of this collection of sequential equilibria.

The analysis of this game is facilitated by the fact that player 2 always prefers u over d. In general, analysis of even these relatively simple signaling games can be quite cumbersome. □

Remark 14.19 With the exception of Example 14.16 we did not really check that the sequential equilibria, computed in the examples, are indeed consistent. In general, however, this is rather straightforward. For instance, for the sequential equilibria in Example 14.17 we can proceed as follows, for each of the three equilibria:

- $b_1(a) = 1$, $b_2(c) = 0$, $b_3(e) = 1$, $\beta = \beta' = 1$. For each $m \in \mathbb{N}$ take $b_1^m(b) = b_2^m(c) = b_3^m(f) = 1/m$. The associated beliefs are determined by Bayesian consistency.
- $b_1(a) = 0$, $b_2(c) = 1$, $b_3(e) = 0$, $\beta = \beta' = 0$. For each $m \in \mathbb{N}$ take $b_1^m(a) = b_2^m(d) = b_3^m(e) = 1/m$. The associated beliefs are determined by Bayesian consistency.

- $b_1(a) = 1/4, b_2(c) = 1/2, b_3(e) = 2/3, \beta = \beta' = 1/4$. For each $m \in \mathbb{N}$ take $b_1^m(a) = 1/4, b_2^m(c) = 1/2, b_3(e) = 2/3$ (hence, constant). The associated beliefs are determined by Bayesian consistency.

□

14.4   Problems

14.1. *Mixed and Behavioral Strategies*
Determine all mixed strategies that are outcome equivalent with the behavioral strategy represented in the following one-player extensive form structure.

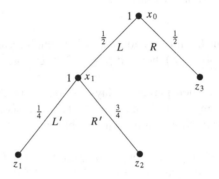

14.2. *An Extensive Form Structure Without Perfect Recall*
Consider the following extensive form structure:

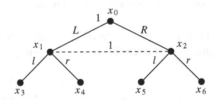

(a) Show that this one-player extensive form structure has no perfect recall.
(b) Consider the mixed strategy σ_1 that assigns probability $1/2$ to both (L, l) and (R, r). Show that there is no behavioral strategy that generates the same probability distribution over the end nodes as σ_1 does.

14.3. *Consistency Implies Bayesian Consistency*
Let (b, β) be a consistent assessment in an extensive form structure $\mathcal{S}$. Show that (b, β) is Bayesian consistent.

14.4. *(Bayesian) Consistency in Signaling Games*

Prove that Bayesian consistency implies consistency in a signaling game.
[A general definition of a signaling game (cf. also Chap. 5) is as follows. The set of players is $N = \{1, 2\}$ and for the extensive form structure $\mathcal{S}$ we have:

1. The directed rooted tree is $T = (X, E)$ with root x_0,

$$X = \{x_0, x_1, \ldots, x_k, x_{i1}, \ldots, x_{il}, x_{ij1}, \ldots, x_{ijm} \mid i = 1, \ldots, k, \ j = 1, \ldots, l\},$$

where $k, l, m \geq 2$,

$$E = \{(x_0, x_i), (x_i, x_{ij}), (x_{ij}, x_{ijj'}) \mid i = 1, \ldots, k, \ j = 1, \ldots, l, \ j' = 1, \ldots, m\}.$$

Hence

$$Z = \{x_{ijj'} \mid i = 1, \ldots, k, \ j = 1, \ldots, l, \ j' = 1, \ldots, m\}.$$

2. The chance and player assignment P is defined by $P(x_0) = C$, $P(x_i) = 1$ for all $i = 1, \ldots, k$, $P(x_{ij}) = 2$ for all $i = 1, \ldots, k, j = 1, \ldots, l$.
3. The information sets are

$$H_1 = \{\{x_1\}, \ldots, \{x_k\}\}, \ H_2 = \{\{x_{1j}, \ldots, x_{kj}\} \mid j = 1, \ldots, l\}.$$

4. The action sets are

$$A(\{x_i\}) = \{\{(x_i, x_{ij})\} \mid j = 1, \ldots, l\} \text{ for every } i = 1, \ldots, k$$

for player 1 and

$$A(\{x_{1j}, \ldots, x_{kj}\}) = \{\{(x_{1j}, x_{1jj'}), \ldots, (x_{kj}, x_{kjj'})\} \mid j' = 1, \ldots, m\}$$

for every $j = 1, \ldots, l$, for player 2.
5. The map τ assigns a positive probability to each edge in the set $\{(x_0, x_1), \ldots, (x_0, x_k)\}$. (Player 1 has k 'types'.)

Finally, the players have payoff functions $u_1, u_2 : Z \to \mathbb{R}$.]

14.5. *Sequential Equilibria in a Signaling Game*
Compute the sequential equilibria in the signaling game in Fig. 14.11 (this is the game from Fig. 5.8).

14.6. *Computation of Sequential Equilibrium (1)*
Compute the sequential equilibrium or equilibria in the game Γ_1 in Fig. 14.5.

14.7. *Computation of Sequential Equilibrium (2)*
Consider the following extensive form game below.

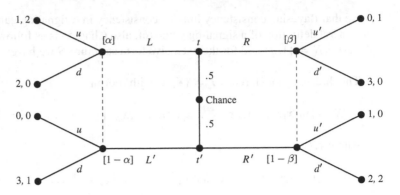

Fig. 14.11 The signaling game of Problem 14.5

(a) Determine the strategic form of this game.
(b) Determine all Nash and subgame perfect Nash equilibria of this game.
(c) Determine all sequential equilibria of this game.

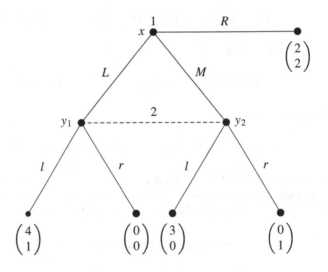

14.8. *Computation of Sequential Equilibrium (3)*
Consider the following extensive form game below.

(a) Determine the strategic form of this game.
(b) Compute the Nash equilibria and subgame perfect Nash equilibria.
(c) Compute the sequential equilibria.

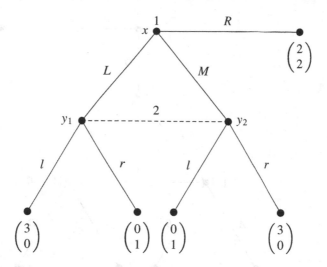

14.9. *Computation of Sequential Equilibrium (4)*
Consider the following extensive form game.

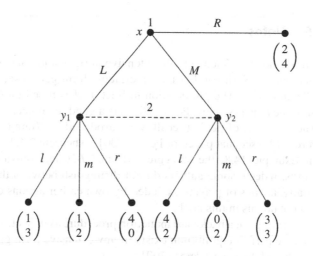

Compute all sequential equilibria of this game.

14.10. *Computation of Sequential Equilibrium (5)*
Compute all Nash, subgame perfect and sequential equilibria in the following game.

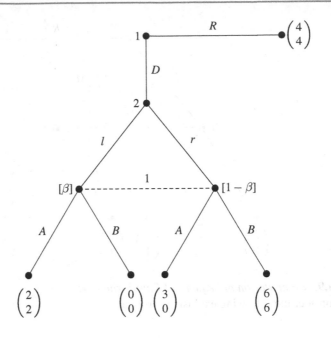

14.5   Notes

For more about refinements in extensive form games and some relations with refinements of Nash equilibrium in strategic form games see van Damme (1991) and Perea (2001). The presentation in Sect. 14.1 is based on the latter source. For most of the remarks below Perea (2001) is a good reference.

The condition of perfect recall was introduced by Kuhn (1953). For a proof of Theorem 14.6 see that paper, or Perea (2001), Theorem 2.4.4.

In Example 14.8, the strategies BC and BD are equivalent for all practical purposes. Indeed, some authors do not distinguish between these strategies. More generally, actions of players excluded by own earlier actions could be left out, but we do not do this in this book.

By carrying out a backward induction procedure as in Sect. 4.3, it can be seen that a subgame perfect equilibrium exists in any extensive form game. This is intuitive but not trivial. See, e.g., Perea (2001).

In the literature the condition of *updating consistency* has been proposed to remedy the defect in a game like the one in Fig. 14.6: with player 1 playing a and player 2 playing e, updating consistency would imply that $\beta(x_3) = \beta(x_1)$, as seems natural. Consistency of beliefs based on limits of completely mixed strategies was introduced in Kreps and Wilson (1982). This consistency condition is stronger than Bayesian consistency combined with 'updating consistency', since the latter condition would only require a player to 'update' his own earlier beliefs. Also sequential equilibria were introduced in Kreps and Wilson (1982). The definition of sequential rationality in Definition 14.11 is actually called 'local' sequential rationality. Together with consistency, it implies the sequential rationality condition in Kreps and Wilson. It is, however, easier to apply. See also Perea (2001), Chap. 4, or van Damme (1991), Chap. 6. See the same sources for a proof that sequential equilibria always exist.

There is no unified definition of perfect Bayesian equilibrium in the literature, since it is hard to capture all instances of Bayesian updating for general extensive for games. For instance, Fudenberg and Tirole (1991b) provide two definitions of perfect Bayesian equilibrium for different classes of games and show that their concepts coincide with sequential equilibrium.

References

Fudenberg, D., & Tirole, J. (1991b) Perfect Bayesian equilibrium and sequential equilibrium. *Journal of Economic Theory, 53*, 236–260.

Kreps, D.M., & Wilson, R.B. (1982). Sequential equilibria. *Econometrica, 50*, 863–894.

Kuhn, H. W. (1953). Extensive games and the problem of information. In H. W. Kuhn & A. W. Tucker (Eds.), *Contributions to the theory of games II (Annals of Mathematics Studies 28)* (pp. 193–216). Princeton: Princeton University Press.

Perea, A. (2001). *Rationality in extensive form games*. Boston: Kluwer Academic.

van Damme, E. C. (1991). *Stability and perfection of Nash equilibria*. Berlin: Springer.

Evolutionary Games

15

In this chapter we go deeper into evolutionary game theory. The concepts of evolutionary stable strategy and replicator dynamics, introduced in Chap. 8, are further explored. It may be helpful to study Chap. 8 first, although the present chapter is largely self-contained.

In Sect. 15.1 we briefly review symmetric two-player games. Section 15.2 discusses evolutionary stable strategies and Sect. 15.3 replicator dynamics.

15.1 Symmetric Two-Player Games

Much of evolutionary game theory is concerned with symmetric two-player games. A (finite) symmetric two-player game is a pair of $m \times m$ payoff matrices (A, B) such that $B = A^T$, i.e., $B = (b_{ij})_{i,j=1}^m$ is the transpose of $A = (a_{ij})_{i,j=1}^m$. In other words, for all $i, j \in \{1, 2, \ldots, m\}$, we have $b_{ij} = a_{ji}$.

In such a game we are particularly interested in symmetric (pure and mixed strategy) Nash equilibria. A Nash equilibrium (σ_1, σ_2) is *symmetric* if $\sigma_1 = \sigma_2$. We denote by $NE(A, A^T)$ the set of all Nash equilibria of (A, A^T) and by

$$NE(A) = \{\mathbf{x} \in \Delta^m \mid (\mathbf{x}, \mathbf{x}) \in NE(A, A^T)\}$$

the set of all strategies that occur in a symmetric Nash equilibrium. By a standard application of the Kakutani Fixed Point Theorem we prove that this set is nonempty.

Proposition 15.1 *For any $m \times m$-matrix A, $NE(A) \neq \emptyset$.*

Proof For each $\mathbf{x} \in \Delta^m$, viewed as a strategy of player 2 in (A, A^T), let $\beta_1(\mathbf{x})$ be the set of best replies of player 1 in (A, A^T). Then the correspondence $\mathbf{x} \mapsto \beta_1(\mathbf{x})$ is upper semi-continuous and convex-valued (check this), so that by the Kakutani Fixed Point Theorem 22.11 there is an $\mathbf{x}^* \in \Delta^m$ with $\mathbf{x}^* \in \beta_1(\mathbf{x}^*)$. Since player

© Springer-Verlag Berlin Heidelberg 2015
H. Peters, *Game Theory*, Springer Texts in Business and Economics,
DOI 10.1007/978-3-662-46950-7_15

2's payoff matrix is the transpose of A, it also follows that $\mathbf{x}^* \in \beta_2(\mathbf{x}^*)$. Hence, $(\mathbf{x}^*, \mathbf{x}^*) \in NE(A, A^T)$, so $\mathbf{x}^* \in NE(A)$. ∎

15.1.1 Symmetric 2×2-Games

For later reference it is convenient to have a classification of symmetric 2×2-games with respect to their symmetric Nash equilibria. Such a game is described by the payoff matrix

$$A = \begin{pmatrix} a_{11} & a_{12} \\ a_{21} & a_{22} \end{pmatrix}.$$

For the purpose of Nash equilibrium analysis, we may consider without loss of generality the matrix

$$A' = \begin{pmatrix} a_{11} - a_{21} & a_{12} - a_{12} \\ a_{21} - a_{21} & a_{22} - a_{12} \end{pmatrix} = \begin{pmatrix} a_{11} - a_{21} & 0 \\ 0 & a_{22} - a_{12} \end{pmatrix} = \begin{pmatrix} a_1 & 0 \\ 0 & a_2 \end{pmatrix},$$

where $a_1 := a_{11} - a_{21}$ and $a_2 := a_{22} - a_{12}$. Indeed, it is straightforward to verify that $(\mathbf{p}, \mathbf{q}) \in \Delta^2 \times \Delta^2$ is a Nash equilibrium of (A, A^T) if and only if it is a Nash equilibrium of (A', A'^T).

For a generic matrix A, implying $a_1, a_2 \neq 0$, there are essentially three different cases:

(1) $a_1 < 0, a_2 > 0$. In this case, $NE(A') = \{\mathbf{e}^2\}$, i.e., each player playing the second strategy is the unique symmetric Nash equilibrium.
(2) $a_1, a_2 > 0$. In this case, $NE(A') = \{\mathbf{e}^1, \mathbf{e}^2, \hat{\mathbf{x}}\}$, where $\hat{\mathbf{x}} = (a_2/(a_1+a_2), a_1/(a_1+a_2))$.
(3) $a_1, a_2 < 0$. In this case, $NE(A') = \{\hat{\mathbf{x}}\}$ with $\hat{\mathbf{x}}$ as in (2).

15.2   Evolutionary Stability

15.2.1 Evolutionary Stable Strategies

In evolutionary game theory the interpretation of a symmetric two-person game is that players in a possibly large population randomly meet in pairs. Let such a game be described by A, then a mixed strategy $\mathbf{x} \in \Delta^m$ is interpreted as a vector of population shares: for each k, x_k is the share of the population that 'plays' pure strategy k. Such a strategy is called *evolutionary stable* if it performs better against a small 'mutation' than that mutation performs against itself. Formally, we have the following definition.

Definition 15.2 A strategy $x \in \Delta^m$ is an *evolutionary stable strategy (ESS)* in A if for every strategy $y \in \Delta^m$, $y \neq x$, there exists some $0 < \varepsilon_y < 1$ such that for all $0 < \varepsilon < \varepsilon_y$ we have

$$xA(\varepsilon y + (1 - \varepsilon)x) > yA(\varepsilon y + (1 - \varepsilon)x). \tag{15.1}$$

The set of all ESS is denoted by *ESS(A)*. □

Again, the interpretation of an ESS x is as follows. Consider any small *mutation* $\varepsilon y + (1 - \varepsilon)x$ of x. Condition (15.1) then says that against such a small mutation, the original strategy x is better than the *mutant* strategy y. In other words, if the population x is *invaded* by a small part of the *mutant* population y, then x survives since it fares better against this small mutation than the mutant y itself does.

Evolutionary stable strategies can be characterized as follows. In fact, this characterization was used as the definition of an evolutionary stable strategy in Chap. 8 (Definition 8.4).

Theorem 15.3 *Let A be a symmetric $m \times m$ game. Then*

$$ESS(A) = \{x \in NE(A) \mid \forall y \in \Delta^m, y \neq x \, [xAx = yAx \Rightarrow xAy > yAy]\} \, .$$

This theorem follows from Propositions 15.4 and 15.5.

Proposition 15.4 *Let A be an $m \times m$-matrix and let $x \in ESS(A)$. Then $x \in NE(A)$.*

Proof Let $y \in \Delta^m$, then it is sufficient to show $xAx \geq yAx$. Let ε_y be as in Definition 15.2, then

$$xA(\varepsilon y + (1 - \varepsilon)x) > yA(\varepsilon y + (1 - \varepsilon)x)$$

for all $0 < \varepsilon < \varepsilon_y$ by (15.1). By letting ε go to zero, this implies $xAx \geq yAx$. ∎

Proposition 15.5 *Let A be an $m \times m$-matrix. If $x \in ESS(A)$, then, for all $y \in \Delta^m$ with $y \neq x$ we have:*

$$xAx = yAx \Rightarrow xAy > yAy \, . \tag{15.2}$$

Conversely, if $x \in NE(A)$ and (15.2) holds, then $x \in ESS(A)$.

Proof Let $x \in ESS(A)$. Let $y \in \Delta^m$ with $y \neq x$ and $xAx = yAx$. Suppose that $yAy \geq xAy$. Then, for any $\varepsilon \in [0, 1]$, $yA(\varepsilon y + (1 - \varepsilon)x) \geq xA(\varepsilon y + (1 - \varepsilon)x)$, contradicting (15.1).

Conversely, let $x \in NE(A)$ and let (15.2) hold for x. If $xAx > yAx$, then also $xA(\varepsilon y + (1 - \varepsilon)x) > yA(\varepsilon y + (1 - \varepsilon)x)$ for small enough ε. If $xAx = yAx$, then $xAy > yAy$, hence (15.1) holds for any $\varepsilon \in (0, 1]$. ∎

Clearly, Theorem 15.3 follows from the preceding propositions.

From Problems 15.1–15.3 it follows that the concept of ESS does not 'solve' the prisoners' dilemma nor the 'coordination problem'. Also, an ESS may be completely mixed (Hawk-Dove), or fail to exist (Rock-Paper-Scissors).

15.2.2 The Structure of the Set $ESS(A)$

Let $\mathbf{x}$ be an ESS and let $\mathbf{y} \in \Delta^m$, $\mathbf{y} \neq \mathbf{x}$ such that the carrier[1] of $\mathbf{y}$ is contained in the carrier of $\mathbf{x}$, i.e., $C(\mathbf{y}) \subseteq C(\mathbf{x})$. Since $\mathbf{x} \in NE(A)$ by Theorem 15.3, this implies $\mathbf{x}A\mathbf{x} = \mathbf{y}A\mathbf{x}$ and hence, again by Theorem 15.3, $\mathbf{x}A\mathbf{y} > \mathbf{y}A\mathbf{y}$. We have established:

Proposition 15.6 *If* $\mathbf{x} \in ESS(A)$ *and* $\mathbf{y} \in \Delta^m$ *with* $\mathbf{y} \neq \mathbf{x}$ *and* $C(\mathbf{y}) \subseteq C(\mathbf{x})$*, then* $\mathbf{y} \notin NE(A)$.

This implies the following corollary (check!):

Corollary 15.7 *The set* $ESS(A)$ *is finite. If* $\mathbf{x} \in ESS(A)$ *is completely mixed, then* $ESS(A) = \{\mathbf{x}\}$.

15.2.3 Relations with Other Refinements

If $\mathbf{x} \in NE(A)$ is weakly dominated by $\mathbf{y} \in \Delta^m$, then $\mathbf{x}A\mathbf{x} = \mathbf{y}A\mathbf{x}$ and $\mathbf{y}A\mathbf{y} \geq \mathbf{x}A\mathbf{y}$; so by Theorem 15.3, $\mathbf{x} \notin ESS(A)$. Therefore, if $\mathbf{x} \in ESS(A)$, then $(\mathbf{x}, \mathbf{x})$ is an undominated equilibrium and hence perfect by Theorem 13.23. It can even be shown that $(\mathbf{x}, \mathbf{x})$ is proper. The next proposition summarizes these facts.

Proposition 15.8 *If* $\mathbf{x} \in ESS(A)$*, then* $(\mathbf{x}, \mathbf{x}) \in NE(A, A^T)$ *is undominated, perfect and proper.*

The unique (symmetric) equilibrium in the Rock-Paper-Scissors game in Problem 15.3 is proper (why?), but the associated equilibrium strategy is not ESS, so the converse of Proposition 15.8 does not hold.

15.2.4 Other Characterizations of ESS

15.2.4.1 Uniform Invasion Barriers
The number $\varepsilon_\mathbf{y}$ in the definition of an ESS can be interpreted as an 'invasion barrier': if the share of the mutant strategy $\mathbf{y}$ is smaller than $\varepsilon_\mathbf{y}$, then the 'incumbent' strategy $\mathbf{x}$ fares better against the mutated population than the mutant $\mathbf{y}$ itself does, so that

---

[1]Recall—see Chap. 13—that the carrier of $\mathbf{y}$, $C(\mathbf{y})$, is the set $\{i \in \{1, \ldots, m\} \mid y_i > 0\}$.

the mutant strategy becomes extinct. In a large but finite population, it would not make sense if this invasion barrier could become arbitrarily small since then the 'mutant' population would sometimes have to consist of less than one individual to guarantee survival of the strategy $\mathbf{x}$ under consideration. This gives rise to the following definition.

Definition 15.9 A strategy $\mathbf{x} \in \Delta^m$ has a *uniform invasion barrier* if there exists an $\bar{\varepsilon} \in (0, 1)$ such that (15.1) holds for all strategies $\mathbf{y} \neq \mathbf{x}$ and every $\varepsilon \in (0, \bar{\varepsilon})$. $\qquad\square$

It turns out that possessing a uniform invasion barrier characterizes an evolutionary stable strategy.

Proposition 15.10 *For each* $\mathbf{x} \in \Delta^m$, $\mathbf{x} \in ESS(A)$ *if and only if* $\mathbf{x}$ *has a uniform invasion barrier.*

Proof Let $\mathbf{x} \in \Delta^m$. If $\mathbf{x}$ has a uniform invasion barrier $\bar{\varepsilon}$, then clearly $\mathbf{x}$ is an ESS by choosing, in (15.1), $\varepsilon_{\mathbf{y}} = \bar{\varepsilon}$ for each $\mathbf{y} \in \Delta^m$.

Conversely, let $\mathbf{x}$ be an ESS. Define the function $b : \Delta^m \setminus \{\mathbf{x}\} \to [0, 1]$ by

$$b(\mathbf{y}) = \sup\{\delta \in [0, 1] \mid \forall \varepsilon \in (0, \delta) \; [(\mathbf{x} - \mathbf{y})A(\varepsilon\mathbf{y} + (1 - \varepsilon)\mathbf{x}) > 0]\}$$

for all $\mathbf{y} \in \Delta^m \setminus \{\mathbf{x}\}$. We first consider the function b on the compact set $Z = \{\mathbf{z} \in \Delta^m \mid z_i = 0 \text{ for some } i \in C(\mathbf{x})\}$. Consider $\mathbf{y} \in Z$. Since $\mathbf{x}$ is an ESS, we have that $(\mathbf{x} - \mathbf{y})A(\varepsilon\mathbf{y} + (1 - \varepsilon)\mathbf{x})$ is positive for small positive values of ε. Since this expression depends linearly on ε, this implies that there can be at most one value of ε, which we denote by $\varepsilon'_{\mathbf{y}}$, such that $(\mathbf{x} - \mathbf{y})A(\varepsilon'_{\mathbf{y}}\mathbf{y} + (1 - \varepsilon'_{\mathbf{y}})\mathbf{x}) = 0$. If $\varepsilon'_{\mathbf{y}} \in (0, 1)$, then $(\mathbf{x} - \mathbf{y})A(\varepsilon'_{\mathbf{y}}\mathbf{y} + (1 - \varepsilon'_{\mathbf{y}})\mathbf{x}) = 0$ implies that $(\mathbf{x} - \mathbf{y})A(\mathbf{x} - \mathbf{y}) \neq 0$. To see this, suppose that $(\mathbf{x} - \mathbf{y})A(\mathbf{x} - \mathbf{y}) = 0$. Then

$$0 = (\mathbf{x} - \mathbf{y})A(\varepsilon'_{\mathbf{y}}\mathbf{y} + (1 - \varepsilon'_{\mathbf{y}})\mathbf{x}) = (\mathbf{x} - \mathbf{y})A\mathbf{x} \,,$$

hence also $(\mathbf{x} - \mathbf{y})A\mathbf{y} = 0$ and, thus, $(\mathbf{x} - \mathbf{y})A(\varepsilon\mathbf{y} + (1 - \varepsilon)\mathbf{x}) = 0$ for all ε, a contradiction. Hence, in that case, $b(\mathbf{y}) = \varepsilon'_{\mathbf{y}} = (\mathbf{x} - \mathbf{y})A\mathbf{x}/(\mathbf{x} - \mathbf{y})A(\mathbf{x} - \mathbf{y})$; otherwise, $b(\mathbf{y}) = 1$. Clearly, b is a continuous function. Since b is positive and Z is compact, $\min_{\mathbf{y} \in Z} b(\mathbf{y}) > 0$. Hence, $\mathbf{x}$ has a uniform invasion barrier, namely this minimum value, on the set Z.

Now suppose that $\mathbf{y} \in \Delta^m$, $\mathbf{y} \neq \mathbf{x}$. We claim that there is a $\lambda \in (0, 1]$ such that $\mathbf{y} = \lambda\mathbf{z} + (1 - \lambda)\mathbf{x}$ for some $\mathbf{z} \in Z$. To see this, first note that we can take $\lambda = 1$ and $\mathbf{z} = \mathbf{y}$ if $\mathbf{y} \in Z$. If $\mathbf{y} \notin Z$ then consider, for each $\mu \geq 0$, the point $\mathbf{z}(\mu) = (1 - \mu)\mathbf{x} + \mu\mathbf{y}$, and let $\hat{\mu} \geq 1$ be the largest value of μ such that $\mathbf{z}(\mu) \in \Delta^m$. Then there is a coordinate $i \in \{1, \ldots, m\}$ with $z_i(\hat{\mu}) = 0$, $z_i(\mu) > 0$ for all $\mu < \hat{\mu}$, and $z_i(\mu) < 0$ for all $\mu > \hat{\mu}$. Clearly, this implies $x_i > y_i$, hence $i \in C(\mathbf{x})$, and thus $\mathbf{z}(\hat{\mu}) \in Z$. Then, for $\mathbf{z} = \mathbf{z}(\hat{\mu})$ and $\lambda = 1/\hat{\mu}$, we have $\mathbf{y} = \lambda\mathbf{z} + (1 - \lambda)\mathbf{x}$.

By straightforward computation we have

$$(\mathbf{x} - \mathbf{y})A(\varepsilon \mathbf{y} + (1 - \varepsilon)\mathbf{x}) = \lambda(\mathbf{x} - \mathbf{z})A(\varepsilon \lambda \mathbf{z} + (1 - \varepsilon \lambda)\mathbf{x})$$

for each $\varepsilon > 0$, so that $b(\mathbf{y}) = \min\{b(\mathbf{z})/\lambda, 1\} \geq b(\mathbf{z})$.

We conclude that $\bar{\varepsilon} = \min_{\mathbf{y} \in Z} b(\mathbf{y})$ is a uniform invasion barrier for $\mathbf{x}$. ∎

15.2.4.2 Local Superiority
By Theorem 15.3, a completely mixed ESS earns a higher payoff against any mutant than such a mutant earns against itself. This global superiority property can be generalized to the following local version.

Definition 15.11 The strategy $\mathbf{x} \in \Delta^m$ is *locally superior* if it has an open neighborhood U such that $\mathbf{x}A\mathbf{y} > \mathbf{y}A\mathbf{y}$ for all $\mathbf{y} \in U \setminus \{\mathbf{x}\}$. □

The local superiority condition provides another characterization of ESS.

Proposition 15.12 *For each* $\mathbf{x} \in \Delta^m$, $\mathbf{x} \in ESS(A)$ *if and only if* $\mathbf{x}$ *is locally superior.*

Proof Let $\mathbf{x} \in \Delta^m$.

First suppose that $\mathbf{x}$ is locally superior, and let U be as in Definition 15.11. Let $\mathbf{z} \in \Delta^m \setminus \{\mathbf{x}\}$ and define for each $0 < \varepsilon < 1$ the point $\mathbf{w}(\varepsilon)$ by $\mathbf{w}(\varepsilon) = \varepsilon \mathbf{z} + (1 - \varepsilon)\mathbf{x}$. Then there is $\varepsilon_\mathbf{z} > 0$ such that $\mathbf{w}(\varepsilon) \in U$, hence $\mathbf{x}A\mathbf{w}(\varepsilon) > \mathbf{w}(\varepsilon)A\mathbf{w}(\varepsilon)$, for all $\varepsilon \in (0, \varepsilon_\mathbf{z})$. By slight rewriting, this implies $\mathbf{x}A\mathbf{w}(\varepsilon) > \mathbf{z}A\mathbf{w}(\varepsilon)$ for all $\varepsilon \in (0, \varepsilon_\mathbf{z})$. In particular, we have

$$\varepsilon \mathbf{x}A\mathbf{z} + (1 - \varepsilon)\mathbf{x}A\mathbf{x} > \varepsilon \mathbf{z}A\mathbf{z} + (1 - \varepsilon)\mathbf{z}A\mathbf{x}$$

for all $\varepsilon \in (0, \varepsilon_\mathbf{z})$, hence $\mathbf{x}A\mathbf{x} \geq \mathbf{z}A\mathbf{x}$. So $\mathbf{x} \in NE(A)$. Suppose now that $\mathbf{z}A\mathbf{x} = \mathbf{x}A\mathbf{x}$. Then, for $\varepsilon \in (0, \varepsilon_\mathbf{z})$,

$$\varepsilon \mathbf{x}A\mathbf{z} = \mathbf{x}A\mathbf{w}(\varepsilon) - (1 - \varepsilon)\mathbf{x}A\mathbf{x}$$
$$> \mathbf{z}A\mathbf{w}(\varepsilon) - (1 - \varepsilon)\mathbf{x}A\mathbf{x}$$
$$= \varepsilon \mathbf{z}A\mathbf{z} + (1 - \varepsilon)\mathbf{z}A\mathbf{x} - (1 - \varepsilon)\mathbf{x}A\mathbf{x}$$
$$= \varepsilon \mathbf{z}A\mathbf{z},$$

so that $\mathbf{x}$ is an ESS.

Conversely, let $\mathbf{x}$ be an ESS with uniform invasion barrier (cf. Proposition 15.10) $\bar{\varepsilon} \in (0, 1)$, and let Z be as in the proof of Proposition 15.10. Let

$$V = \{\mathbf{y} \in \Delta^m \mid \mathbf{y} = \varepsilon \mathbf{z} + (1 - \varepsilon)\mathbf{x} \text{ for some } \mathbf{z} \in Z \text{ and } \varepsilon \in [0, \bar{\varepsilon})\}.$$

Since Z is closed and $\mathbf{x} \notin Z$, there is an open neighborhood U of $\mathbf{x}$ such that $U \cap \Delta^m \subseteq V$. Suppose that $\mathbf{y} \neq \mathbf{x}$, $\mathbf{y} \in U \cap \Delta^m$. Then $\mathbf{y} \in V$, and by Proposition 15.10, $\mathbf{z}A\mathbf{y} = \mathbf{z}A(\varepsilon \mathbf{z} + (1-\varepsilon)\mathbf{x}) < \mathbf{x}A(\varepsilon \mathbf{z} + (1-\varepsilon)\mathbf{x}) = \mathbf{x}A\mathbf{y}$, with $\mathbf{z}$ as in the definition of V. This implies $\mathbf{y}A\mathbf{y} = \varepsilon \mathbf{z}A\mathbf{y} + (1-\varepsilon)\mathbf{x}A\mathbf{y} < \mathbf{x}A\mathbf{y}$. ∎

15.2.4.3 Local Strict Efficiency

Consider the special case of a symmetric game (A, B) with $A^T = A$, hence A is itself symmetric and $B = A$. Call such a game *doubly symmetric*.

Definition 15.13 A strategy $\mathbf{x} \in \Delta^m$ is *locally strictly efficient* if it has an open neighborhood U such that $\mathbf{x}A\mathbf{x} > \mathbf{y}A\mathbf{y}$ for all $\mathbf{y} \in U \setminus \{\mathbf{x}\}$. □

For doubly symmetric games, local strict efficiency characterizes ESS.

Proposition 15.14 *Let $A = A^T$. Then $\mathbf{x} \in ESS(A)$ if and only if $\mathbf{x}$ is locally strictly efficient.*

Proof Let $\mathbf{x} \in \Delta^m$. For any $\mathbf{y} \neq \mathbf{x}$ and $\mathbf{z} = \frac{1}{2}\mathbf{x} + \frac{1}{2}\mathbf{y}$, we have

$$\mathbf{y}A\mathbf{y} = \mathbf{x}A\mathbf{x} - 2\mathbf{x}A\mathbf{z} - 2\mathbf{z}A\mathbf{x} + 4\mathbf{z}A\mathbf{z} .$$

Hence, using the symmetry of A,

$$\mathbf{x}A\mathbf{x} - \mathbf{y}A\mathbf{y} = 4 \left[\mathbf{x}A\mathbf{z} - \mathbf{z}A\mathbf{z} \right] .$$

If $\mathbf{x}$ is locally strictly efficient, then this identity implies that $\mathbf{x}$ is locally superior, and conversely. By Proposition 15.12, it follows that $\mathbf{x}$ is an ESS if and only if $\mathbf{x}$ is locally strictly efficient. ∎

15.3 Replicator Dynamics and ESS

The concept of an evolutionary stable strategy is based on the idea of *mutation*. Incorporation of the evolutionary concept of *selection* calls for a more explicitly dynamic approach.

15.3.1 Replicator Dynamics

As before, consider a symmetric game described by the $m \times m$ matrix A. A mixed strategy $\mathbf{x} \in \Delta^m$ can be interpreted as a vector of population shares (a *state*) over the pure strategies, evolving over time. To express time dependence, we write $\mathbf{x} = \mathbf{x}(t)$. For each pure strategy i, the expected payoff of playing i when the population is in state $\mathbf{x}$ is equal to $\mathbf{e}^i A\mathbf{x}$, hence the average population payoff is equal to $\sum_{i=1}^m x_i \mathbf{e}^i A\mathbf{x} = \mathbf{x}A\mathbf{x}$. In the *replicator dynamics* it is assumed that population shares develop according to the differential equation

$$\dot{x}_i = dx_i(t)/dt = \left[e^i A\mathbf{x} - \mathbf{x}A\mathbf{x} \right] x_i \qquad (15.3)$$

for each pure strategy $i = 1, 2, \ldots, m$, where dependence on t is (partly) suppressed from the notation. In other words, the share of the population playing strategy i changes with rate proportional to the difference between the expected payoff of i (individual fitness) and the average population payoff (average fitness).

To study the replicator dynamics in (15.3) one needs to apply the theory of differential equations and dynamical systems. For a first analysis we can restrict attention to a few basic concepts and facts.

For each *initial state* $\mathbf{x}(0) = \mathbf{x}^0 \in \Delta^m$, the system (15.3) induces a *solution* or *trajectory* $\xi(t, \mathbf{x}^0)$ in Δ^m. Call state $\mathbf{x}$ a *stationary point* of the dynamics (15.3) if $\dot{\mathbf{x}} = (\dot{x}_1, \ldots, \dot{x}_m) = (0, \ldots, 0)$. If $m = 2$ then $\dot{x}_1 = 0$ or $\dot{x}_2 = 0$ is sufficient for $\mathbf{x}$ to be a stationary point, since (15.3) implies the natural condition $\sum_{i=1}^m \dot{x}_i = 0$. Note that any e^i is a stationary point—this is a more or less artificial property of the replicator dynamics. A state $\mathbf{x}$ is *Lyapunov stable* if every open neighborhood B of $\mathbf{x}$ contains an open neighborhood B^0 of $\mathbf{x}$ such that $\xi(t, \mathbf{x}^0) \in B$ for all $\mathbf{x}^0 \in B^0$ and $t \geq 0$. A state $\mathbf{x}$ *asymptotically stable* if it is Lyapunov stable and it has an open neighborhood B^* such that $\lim_{t \to \infty} \xi(t, \mathbf{x}^0) = \mathbf{x}$ for all $\mathbf{x}^0 \in B^*$. It is easy to see that Lyapunov stability implies stationarity.

Before studying the replicator dynamics in more detail, we state the following useful fact without proof—see the Notes to this chapter for a reference. By Δ_0^m we denote the (relative) interior of the set Δ^m, i.e., $\Delta_0^m = \{ \mathbf{x} \in \Delta^m \mid \mathbf{x} > \mathbf{0} \}$ is the set of completely mixed strategies or states.

Proposition 15.15 *The replicator dynamics (15.3) has a unique solution* $\xi(t, \mathbf{x}^0)$, $t \in \mathbb{R}$, *through any initial state* $\mathbf{x}^0 \in \Delta^m$. *The solution mapping* $\xi : \mathbb{R} \times \Delta^m \to \Delta^m$ *is continuous, and continuously differentiable with respect to time. Both* Δ_0^m *and* $\Delta^m \setminus \Delta_0^m$ *are invariant, that is,* $\xi(t, \mathbf{x}^0) \in \Delta_0^m$ *for all* t *whenever* $\mathbf{x}^0 \in \Delta_0^m$ *and* $\xi(t, \mathbf{x}^0) \in \Delta^m \setminus \Delta_0^m$ *for all* t *whenever* $\mathbf{x}^0 \in \Delta^m \setminus \Delta_0^m$.

15.3.2 Symmetric 2×2 Games

In order to analyze the replicator dynamics for symmetric 2×2 games corresponding to A, we can without loss of generality restrict attention again to the normalized game

$$A' = \begin{pmatrix} a_1 & 0 \\ 0 & a_2 \end{pmatrix}.$$

Now (15.3) reduces to

$$\dot{x}_1 = [a_1 x_1 - a_2 x_2] x_1 x_2 \qquad (15.4)$$

(and $\dot{x}_2 = -\dot{x}_1$). For case (1) in Sect. 15.1.1, $a_1 < 0$ and $a_2 > 0$, the stationary points of the dynamics are $\mathbf{x} = \mathbf{e}^1$ and $\mathbf{x} = \mathbf{e}^2$. For all other $\mathbf{x}$, $\dot{x}_1 < 0$, which implies that the system then converges to $\mathbf{e}^2$, the unique ESS. Hence, the (unique) ESS is also the (unique) asymptotically stable state.

From the answers to Problems 15.5 and 15.6 we have:

Proposition 15.16 *Let A be a generic* 2×2 *matrix and let* $\mathbf{x} \in \Delta^2$. *Then* $\mathbf{x} \in ESS(A)$ *if and only if* $\mathbf{x}$ *is an asymptotically stable state of the replicator dynamics.*

Note that this proposition implies Proposition 8.5(b). Part (a) of Proposition 8.5 follows from Problem 15.2.

15.3.3 Dominated Strategies

Does the replicator dynamics discard of dominated strategies? One answer to this question is provided by the following proposition, which states that if we start from a completely mixed strategy eventually all strictly dominated pure strategies vanish, i.e., their population shares converge to zero.

Proposition 15.17 *Let* $\mathbf{x}^0 \in \Delta^m$ *be completely mixed and let pure strategy i be strictly dominated. Then* $\lim_{t\to\infty} \xi_i(t, \mathbf{x}^0) = 0$.

Proof Let i be strictly dominated by $\mathbf{y} \in \Delta^m$ and let

$$\varepsilon = \min_{\mathbf{x}\in\Delta^m} \mathbf{y}A\mathbf{x} - \mathbf{e}^i A\mathbf{x} .$$

By continuity of the expected payoff function and compactness of Δ^m, $\varepsilon > 0$. Define $v_i : \Delta_0^m \to \mathbb{R}$ by $v_i(\mathbf{x}) = \ln x_i - \sum_{j=1}^m y_j \ln(x_j)$. The function v_i is differentiable, with time derivative at any point $\mathbf{x} = \xi(t, \mathbf{x}^0)$ equal to

$$\dot{v}_i(\mathbf{x}) = \left[\frac{dv_i(\xi(t, \mathbf{x}^0))}{dt} \right]_{\xi(t,\mathbf{x}^0)=\mathbf{x}}$$

$$= \sum_{j=1}^m \frac{\partial v_i(\mathbf{x})}{\partial x_j} \dot{x}_j$$

$$= \frac{\dot{x}_i}{x_i} - \sum_{j=1}^m \frac{y_j \dot{x}_j}{x_j}$$

$$= (\mathbf{e}^i - \mathbf{x})A\mathbf{x} - \sum_{j=1}^m y_j(\mathbf{e}^j - \mathbf{x})A\mathbf{x}$$

$$= (\mathbf{e}^i - \mathbf{y})A\mathbf{x} \le -\varepsilon < 0 .$$

(Cf. Proposition 15.15.) Hence, $v_i(\xi(t, \mathbf{x}^0))$ decreases to minus infinity as $t \to \infty$. This implies $\ln(\xi(t, \mathbf{x}^0))$ decreases to minus infinity, so that $\xi_i(t, \mathbf{x}^0) \to 0$. ∎

Proposition 15.17 remains true for pure strategies i that are iteratively strictly dominated. For weakly dominated strategies several things may happen, see Problem 15.7.

15.3.4 Nash Equilibrium Strategies

Consider again the finite symmetric two-player game with payoff matrix A. What is the relation between the replicator dynamics and Nash equilibrium strategies? The answer is given by the following proposition, where $ST(A)$ denotes the set of stationary states, hence (check!):

$$ST(A) = \{\mathbf{x} \in \Delta^m \mid \forall i \in C(\mathbf{x}) \ [\mathbf{e}^i A \mathbf{x} = \mathbf{x} A \mathbf{x}]\} . \tag{15.5}$$

Proposition 15.18 *For any finite symmetric two-player game with payoff matrix A we have:*

(a) $\{\mathbf{e}^1, \ldots, \mathbf{e}^m\} \cup NE(A) \subseteq ST(A)$,
(b) $ST(A) \cap \Delta_0^m = NE(A) \cap \Delta_0^m$,
(c) $ST(A) \cap \Delta_0^m$ *is a convex set and if* $\mathbf{z} \in \Delta^m$ *is a linear combination of states in this set, then* $\mathbf{z} \in NE(A)$.

Proof It is straightforward from (15.3) that $\mathbf{e}^i \in ST(A)$ for every pure strategy i. If $\mathbf{x} \in NE(A)$, then every $i \in C(\mathbf{x})$ is a pure best reply, hence $\mathbf{e}^i A \mathbf{x} = \mathbf{x} A \mathbf{x}$. Hence, $\mathbf{x} \in ST(A)$. This proves (a). As to (b), part (a) implies that $ST(A) \cap \Delta_0^m \supseteq NE(A) \cap \Delta_0^m$. Further, $\mathbf{e}^i A \mathbf{x} = \mathbf{x} A \mathbf{x}$ for every $\mathbf{x} \in ST(A) \cap \Delta_0^m$ and every $i \in \Delta^m$, so that $\mathbf{x} \in NE(A) \cap \Delta_0^m$. This proves (b).

It remains to prove the last claim. Let $\mathbf{x}$ and $\mathbf{y}$ be completely mixed stationary points, and let $\alpha, \beta \in \mathbb{R}$ and $\mathbf{z} = \alpha\mathbf{x} + \beta\mathbf{y} \in \Delta^m$. For any pure strategy i we have

$$\mathbf{e}^i A \mathbf{z} = \alpha\mathbf{e}^i A \mathbf{x} + \beta\mathbf{e}^i A \mathbf{y} = \alpha\mathbf{x} A \mathbf{x} + \beta\mathbf{y} A \mathbf{y}$$

since $\mathbf{x}, \mathbf{y} \in ST(A) \cap \Delta_0^m$. This implies that actually $\mathbf{e}^i A \mathbf{z} = \mathbf{z} A \mathbf{z}$ for all pure strategies i, hence $\mathbf{z}$ is stationary. If $\mathbf{z}$ is completely mixed, then we are done by part (b). Otherwise, $\mathbf{z}$ is a boundary point of $ST(A) \cap \Delta_0^m$ and hence of $NE(A) \cap \Delta_0^m$, so $\mathbf{z} \in NE(A)$ since $NE(A)$ is a closed set. Finally, since Δ^m is convex and $\mathbf{z} \in ST(A) \cap \Delta_0^m$ for all $\alpha, \beta \geq 0$ with $\alpha + \beta = 1$, $ST(A) \cap \Delta_0^m$ is a convex set. ∎

Proposition 15.18 implies that every (symmetric) Nash equilibrium is stationary. The weakest form of dynamical stability, Lyapunov stability, leads to a refinement of Nash equilibrium, as the next result shows.

Proposition 15.19 *Let* $\mathbf{x} \in \Delta^m$ *be a Lyapunov stable stationary state. Then* $\mathbf{x} \in NE(A)$.

Proof Suppose $\mathbf{x} \notin NE(A)$. Then $\mathbf{e}^i A\mathbf{x} - \mathbf{x}A\mathbf{x} > 0$ for some $i \in N$. By continuity, there is a $\delta > 0$ and an open neighborhood U of $\mathbf{x}$ such that $\mathbf{e}^i A\mathbf{y} - \mathbf{y}A\mathbf{y} \geq \delta$ for all $\mathbf{y} \in U \cap \Delta^m$. Let $\mathbf{x}^0 \in U \cap \Delta^m$ with $x_i^0 > 0$. Then $\xi_i(t, \mathbf{x}^0) \geq x_i^0 \exp(\delta t)$ for all $t \geq 0$; this follows from the fact that the system $\dot{\eta} = \delta\eta$ has solution $\eta(t) = \eta(0)\exp(\delta t)$. So $\xi_i(t, \mathbf{x}^0)$ increases exponentially from any $\mathbf{x}^0 \in U \cap \Delta_0^m$ with $x_i^0 > 0$. This contradicts Lyapunov stability of $\mathbf{x}$. ■

The final result in this subsection says that if a trajectory of the replicator dynamics starts from an interior (completely mixed) state and converges, then the limit state is a Nash equilibrium strategy.

Proposition 15.20 *Let* $\mathbf{x}^0 \in \Delta_0^m$ *and* $\mathbf{x} \in \Delta^m$ *such that* $\mathbf{x} = \lim_{t\to\infty}\xi(t, \mathbf{x}^0)$. *Then* $\mathbf{x} \in NE(A)$.

Proof Suppose that $\mathbf{x} \notin NE(A)$. Then there is a pure strategy i and an $\varepsilon > 0$ such that $\mathbf{e}^i A\mathbf{x} - \mathbf{x}A\mathbf{x} = \varepsilon$. Hence, there is a $T \in \mathbb{R}$ such that $\mathbf{e}^i A\xi(t, \mathbf{x}^0) - \xi(t, \mathbf{x}^0)A\xi(t, \mathbf{x}^0) > \varepsilon/2$ for all $t \geq T$. By (15.3), $\dot{x}_i > x_i\varepsilon/2$ for all $t \geq T$, and hence (as in the proof of Proposition 15.19) $\xi_i(t, \mathbf{x}^0) > \xi_i(T, \mathbf{x}^0)\exp(\varepsilon(t - T)/2)$ for all $t \geq T$. Since $\xi_i(T, \mathbf{x}^0) > 0$, this implies $\xi_i(t, \mathbf{x}^0) \to \infty$ as $t \to \infty$, a contradiction. ■

15.3.5 Perfect Equilibrium Strategies

In the preceding subsection we have seen that Lyapunov stability implies Nash equilibrium. What are the implications of asymptotic stability?

First, asymptotic stability implies Lyapunov stability and therefore also Nash equilibrium. Since Nash equilibrium implies stationarity, however, it must be the case that an asymptotically stable Nash equilibrium strategy is *isolated*, meaning that it has an open neighborhood in which there are no other Nash equilibrium strategies. If not, there would be arbitrarily close stationary states, which conflicts with asymptotic stability.

Second, asymptotic stability also implies perfection.

Proposition 15.21 *Let* $\mathbf{x} \in \Delta^m$ *be asymptotically stable. Then* $\mathbf{x} \in NE(A)$ *and* $\mathbf{x}$ *is isolated. Moreover,* $(\mathbf{x}, \mathbf{x})$ *is a perfect equilibrium in* (A, A^T).

Proof We still have to prove that $(\mathbf{x}, \mathbf{x})$ is a perfect equilibrium in (A, A^T). Suppose not. Then $\mathbf{x}$ is weakly dominated by some $\mathbf{y} \in \Delta^m \setminus \{\mathbf{x}\}$, see Theorem 13.23. Hence

$\mathbf{y}A\mathbf{z} \geq \mathbf{x}A\mathbf{z}$ for all $\mathbf{z} \in \Delta^m$. Define $v : \Delta^m \to \mathbb{R}$ by

$$v(\mathbf{z}) = \sum_{i \in C(\mathbf{z})} (y_i - x_i) \ln(z_i)$$

for all $\mathbf{z} \in \Delta^m$. Similarly as in the proof of Proposition 15.17, we obtain that v is nondecreasing along all interior solution trajectories of (15.3), i.e., at any $\mathbf{z} \in \Delta_0^m$ (so $\mathbf{z}$ completely mixed),

$$\dot{v}(\mathbf{z}) = \sum_{i \in C(\mathbf{z})} (y_i - x_i)\frac{\dot{z}_i}{z_i} = \sum_{i=1}^m (y_i - x_i)[e^i A\mathbf{z} - \mathbf{z}A\mathbf{z}] = (\mathbf{y} - \mathbf{x})A\mathbf{z} \geq 0 \ .$$

Since $\mathbf{x}$ is asymptotically stable, it has an open neighborhood U such that $\xi(t, \mathbf{x}^0) \to \mathbf{x}$ for all $\mathbf{x}^0 \in U \cap \Delta^m$. By nondecreasingness of v along all interior solution trajectories this implies $v(\mathbf{x}) \geq v(\mathbf{z})$ for all $\mathbf{z} \in U \cap \Delta_0^m$. We will construct, however, a $\mathbf{z}$ in $U \cap \Delta_0^m$ with $v(\mathbf{z}) > v(\mathbf{x})$. This is a contradiction and, hence, $\mathbf{x}$ must be perfect.

To construct such a $\mathbf{z}$, define for $\delta \in (0, 1)$, $\mathbf{w} \in \Delta_0^m$, and $\varepsilon \in [0, 1]$,

$$\mathbf{z} = (1 - \varepsilon)[(1 - \delta)\mathbf{x} + \delta\mathbf{y}] + \varepsilon\mathbf{w} \ .$$

For ε sufficiently small, we have $y_i > x_i \Rightarrow z_i > x_i$ and $y_i < x_i \Rightarrow z_i < x_i$. Moreover, $\mathbf{z} \in \Delta_0^m$ and

$$v(\mathbf{z}) - v(\mathbf{x}) = \sum_{i=1}^m (y_i - x_i) \ln(z_i) - \sum_{i \in C(\mathbf{x})} (y_i - x_i) \ln(x_i)$$

$$= \sum_{i \in C(\mathbf{x})} (y_i - x_i)[\ln(z_i) - \ln(x_i)] + \sum_{i \notin C(\mathbf{x})} y_i \ln(z_i) \ .$$

Note that the first term after the second inequality sign is positive. We will show that the second term is zero, which completes the proof of the proposition, since then $v(\mathbf{z}) > v(\mathbf{x})$. To show that $\sum_{i \notin C(\mathbf{x})} y_i \ln(z_i) = 0$, it is sufficient to show that $C(\mathbf{y}) \subseteq C(\mathbf{x})$. Suppose that $j \in C(\mathbf{y})$ and $j \notin C(\mathbf{x})$. By asymptotic stability of $\mathbf{x}$, $\xi(t, \mathbf{x}^0) \to \mathbf{x}$ for all $\mathbf{x}^0 \in U \cap \Delta_0^m$. Write

$$v(\xi(t, \mathbf{x}^0)) = \sum_{i : \xi_i(t,\mathbf{x}^0) > 0, i \in C(\mathbf{x})} (y_i - x_i) \ln(\xi_i(t, \mathbf{x}^0))$$

$$+ \sum_{i : \xi_i(t,\mathbf{x}^0) > 0, i \notin C(\mathbf{x})} y_i \ln(\xi_i(t, \mathbf{x}^0)) \ .$$

The first term after the inequality sign is bounded by some constant γ, since $\xi_i(t, \mathbf{x}^0) \to x_i > 0$. The second term converges to $-\infty$ since $\xi_j(t, \mathbf{x}^0) \to x_j = 0$ and $y_j > 0$. But $v(\xi(t, \mathbf{x}^0)) \to -\infty$ contradicts the nondecreasingness of v along the trajectory $\xi(t, \mathbf{x}^0)$. ∎

15.4 Problems

15.1. *Computing ESS in 2×2 Games (1)*
Compute $ESS(A)$ for the following payoff matrices A.

(a) $A = \begin{pmatrix} 4 & 0 \\ 5 & 3 \end{pmatrix}$ (Prisoners' Dilemma)

(b) $A = \begin{pmatrix} 2 & 0 \\ 0 & 1 \end{pmatrix}$ (Coordination game)

(c) $A = \begin{pmatrix} -1 & 4 \\ 0 & 2 \end{pmatrix}$ (Hawk-Dove game)

15.2. *Computing ESS in 2×2 Games (2)*
Compute $ESS(A')$ for each of the cases (1), (2), and (3) in Sect. 15.1.1. Compare with your answers to Problem 15.1.

15.3. *Rock-Paper-Scissors (1)*
Show that the Rock-Paper-Scissors game

$$A = \begin{pmatrix} 1 & 2 & 0 \\ 0 & 1 & 2 \\ 2 & 0 & 1 \end{pmatrix}$$

has no ESS.

15.4. *Uniform Invasion Barriers*
Find the maximal value of the uniform invasion barrier for the ESS's in each of the cases (1), (2), and (3) in Sect. 15.1.1.

15.5. *Replicator Dynamics in Normalized Game (1)*
Show that A and A' (see Sect. 15.3.2) result in the same replicator dynamics.

15.6. *Replicator Dynamics in Normalized Game (2)*

(a) Simplify the dynamics (15.4) for case (1) in Sect. 15.1.1 by substituting $x_2 = 1 - x_1$ and plot $\dot{x}_1$ as a function of $x_1 \in [0, 1]$.
(b) Carry out this analysis also for cases (2) and (3). What is your conclusion?

15.7. *Weakly Dominated Strategies and Replicator Dynamics*

(a) Consider the matrix

$$A = \begin{pmatrix} 0 & 1 \\ 0 & 0 \end{pmatrix}.$$

Investigate the trajectories of the replicator dynamics.

(b) Consider the matrix

$$A = \begin{pmatrix} 1 & 1 & 1 \\ 1 & 1 & 0 \\ 0 & 0 & 0 \end{pmatrix}.$$

Investigate the trajectories of the replicator dynamics.
[Cf. Proposition 15.17.]

15.8. *Stationary Points and Nash Equilibria (1)*
Consider the two-person symmetric game with payoff matrix

$$A = \begin{pmatrix} 0 & 2 & 0 \\ 2 & 0 & 0 \\ 1 & 1 & 0 \end{pmatrix}.$$

(a) Compute $NE(A)$.
(b) Compute $ST(A)$.

15.9. *Stationary Points and Nash Equilibria (2)*
Consider the two-person symmetric game with payoff matrix

$$A = \begin{pmatrix} 3 & 3 & 1 \\ 4 & 4 & 0 \\ 0 & 2 & 4 \end{pmatrix}.$$

(a) Compute $NE(A)$.
(b) Compute $ST(A)$.
(c) Which stationary states are asymptotically stable, which are Lyapunov stable, and which are not Lyapunov stable?
(d) Show directly from the definition of local superiority that the strategy $x = (1/2, 0, 1/2)$ is not locally superior. [Hint: consider the strategy $z = (1/2, \lambda, 1/2 - \lambda)$ for $0 < \lambda < 1/2$.]

15.10. *Lyapunov Stable States in* 2×2 *Games*
Consider the normalized two-player symmetric 2×2 game A'. Compute the Lyapunov stable states for cases (1), (2), and (3).

15.11. *Nash Equilibrium and Lyapunov Stability*
Consider the symmetric game with payoff matrix

$$A = \begin{pmatrix} 0 & 1 & 0 \\ 0 & 0 & 2 \\ 0 & 0 & 1 \end{pmatrix}.$$

Compute $NE(A)$. Show that the unique element in this set is not Lyapunov stable.

15.12. *Rock-Paper-Scissors (2)*
Consider the generalized Rock-Paper-Scissors game with payoff matrix

$$A = \begin{pmatrix} 1 & 2+a & 0 \\ 0 & 1 & 2+a \\ 2+a & 0 & 1 \end{pmatrix}$$

where $a \in \mathbb{R}$.

(a) Write down the three equations of the replicator dynamics.
(b) Define $h(\mathbf{x}) = \ln(x_1 x_2 x_3)$ for $\mathbf{x}$ positive and show that $\dot{h}(\mathbf{x}) = 3 + a - 3\mathbf{x}A\mathbf{x}$.
(c) Show that the average payoff is equal to

$$\mathbf{x}A\mathbf{x} = 1 + \frac{a}{2}(1 - ||\mathbf{x}||^2)$$

for each $\mathbf{x} \in \Delta^3$, where $||\mathbf{x}||$ is the Euclidean norm of $\mathbf{x}$. Conclude that $\dot{h}(\mathbf{x}) = \frac{a}{2}(3||\mathbf{x}||^2 - 1)$.
(d) Show that $\dot{h}(\frac{1}{3}, \frac{1}{3}, \frac{1}{3}) = 0$ and $\dot{h}(\mathbf{x})$ has the same sign as a for other $\mathbf{x} \in \Delta^m$.
(e) Show that the unique Nash equilibrium in this game is (i) asymptotically stable for $a > 0$; (ii) Lyapunov but not asymptotically stable for $a = 0$; (iii) not Lyapunov stable for $a < 0$.

15.5 Notes

This chapter is based mainly on Weibull (1995). The concept of an evolutionary stable strategy was introduced by Maynard Smith and Price (1973). See Selten (1980, 1983) for early applications of evolutionary stable strategies in game theory.

For properness of $(\mathbf{x}, \mathbf{x})$, where $\mathbf{x}$ is an ESS (Proposition 15.8), see van Damme (1991).

Uniform invasion barriers were introduced by Vickers and Cannings (1987), and local superiority by Hofbauer et al. (1979). Proposition 15.14 is due to Hofbauer and Sigmund (1988).

Replicator dynamics were introduced by Taylor and Jonker (1978). Proposition 15.17 remains true for pure strategies i that are iteratively strictly dominated: see Samuelson and Zhang (1992). For Proposition 15.21, in particular the result that asymptotic stability implies perfection, see Bomze (1986). For more on weakly dominated strategies see Weibull (1995), p. 83 ff. See Sect. 3.3.2 in the same book for more on the relation between Lyapunov stability and Nash equilibrium. For a proof of Proposition 15.15, see Weibull (1995), Proposition 3.20.

A general reference to the theory of differential equations and dynamical systems is Hirsch and Smale (1974).

References

Bomze, I. (1986). Non-cooperative two-person games in biology: a classification. *International Journal of Game Theory, 15*, 31–57.
Hirsch, M., & Smale, S. (1974). *Differential equations, dynamical systems, and linear algebra*. San Diego: Academic.
Hofbauer, J., & Sigmund, K. (1988). The theory of evolution and dynamical systems. Cambridge: Cambridge University Press.
Hofbauer, J., Schuster, P., & Sigmund, K. (1979). A note on evolutionary stable strategies and game dynamics. *Journal of Theoretical Biology, 81*, 609–612.
Maynard Smith, J., & Price, G. R. (1973). The logic of animal conflict. *Nature, 246*, 15–18.
Samuelson, L., & Zhang, J. (1992). Evolutionary stability in asymmetric games. *Journal of Economic Theory, 57*, 363–391.
Selten, R. (1980). A note on evolutionary stable strategies in asymmetric animal conflicts. *Journal of Theoretical Biology, 84*, 93–101.
Selten, R. (1983). Evolutionary stability in extensive-form two-person games. *Mathematical Social Sciences, 5*, 269–363.
Taylor, P., & Jonker, L. (1978). Evolutionary stable strategies and game dynamics. *Mathematical Biosciences, 40*, 145–156.
van Damme, E. C. (1991). *Stability and perfection of Nash equilibria*. Berlin: Springer.
Vickers, G., & Cannings, C. (1987). On the definition of an evolutionarily stable strategy. *Journal of Theoretical Biology, 132*, 387–408.
Weibull, J. W. (1995). *Evolutionary game theory*. Cambridge: MIT Press.

Part III

Cooperative Games

TU-Games: Domination, Stable Sets, and the Core

16

In a game with transferable utility (TU-game) each coalition (subset of players) is characterized by its worth, i.e., a real number representing the payoff or utility that the coalition can achieve if it forms. It is assumed that this payoff can be freely distributed among the members of the coalition in any way desired.

For some examples the reader is referred to Chap. 1. Chapter 9 presents a first acquaintance with transferable utility games. Although the present chapter and the following ones are self-contained, it may be helpful to study the relevant parts of Chaps. 1 and 9 first.

In this chapter the focus is on the core of a transferable utility game. Section 16.1 starts with a weaker concept, the imputation set, and introduces the concept of domination. Section 16.2 introduces the domination core and the core. Section 16.3 studies these solution concepts for a special class of TU-games called simple games. In Sect. 16.4 we briefly review von Neumann and Morgenstern's stable sets, which are also based on the concept of domination. Section 16.5, finally, presents a characterization of games with non-empty cores in terms of balancedness.

16.1 Imputations and Domination

We start with repeating the definition of a game with transferable utility (cf. Definition 9.1).

Definition 16.1 A *cooperative game with transferable utility* or *TU-game* is a pair (N, v), where $N = \{1, \ldots, n\}$ with $n \in \mathbb{N}$ is the set of *players*, and v is a function assigning to each *coalition* S, i.e., to each subset $S \subseteq N$ a real number $v(S)$, such that $v(\emptyset) = 0$. The function v is called the *characteristic function* and $v(S)$ is called the *worth* of S. The coalition N is called the *grand coalition*. A *payoff distribution for* coalition S is a vector of real numbers $(x_i)_{i \in S}$. □

© Springer-Verlag Berlin Heidelberg 2015
H. Peters, *Game Theory*, Springer Texts in Business and Economics,
DOI 10.1007/978-3-662-46950-7_16

The set of coalitions is also denoted by 2^N, so that a TU-game is a pair (N, v) with $v : 2^N \to \mathbb{R}$ such that $v(\emptyset) = 0$. The game (N, v) is often denoted by v if no confusion about the set of players is likely to arise. Also, for a coalition $\{i, j, \ldots, k\}$ we sometimes write $i, j, \ldots, k$ or $ij \ldots k$ instead of $\{i, j, \ldots, k\}$. By $|S|$ we denote the cardinality of a coalition S. By $\mathcal{G}^N$ the set of all TU-games with player set N is denoted.

We frequently use the notation $x(S) := \sum_{i \in S} x_i$ for a payoff distribution $\mathbf{x} = (x_1, \ldots, x_n) \in \mathbb{R}^N$ and a coalition $S \subseteq N$.

Let (N, v) be a TU-game. A vector $\mathbf{x} \in \mathbb{R}^N$ is called an *imputation* if

(a) $\mathbf{x}$ is *individually rational* i.e.

$$x_i \geq v(i) \text{ for all } i \in N ,$$

(b) $\mathbf{x}$ is *efficient* i.e.

$$x(N) = v(N) .$$

The set of imputations of (N, v) is denoted by $I(v)$. An element $\mathbf{x} \in I(v)$ is a payoff distribution of the worth $v(N)$ of the grand coalition N which gives each player i a payoff x_i which is at least as much as he can obtain when he operates alone.

Example 16.2 A game v is called *additive* if $v(S \cup T) = v(S) + v(T)$ for all disjoint coalitions S and T. Such a game is completely determined by the worths of the one-person coalitions $v(i)$ $(i \in N)$, since $v(S) = \sum_{i \in S} v(i)$ for every coalition S. For an additive game v, $I(v)$ consists of one point: $I(v) = \{(v(1), v(2), \ldots, v(n))\}$. $\square$

Note that for a game v

$$I(v) \neq \emptyset \text{ if and only if } v(N) \geq \sum_{i=1}^{n} v(i) .$$

For an *essential* game v, that is, a game with $v(N) \geq \sum_{i=1}^{n} v(i)$, $I(v)$ is the convex hull of the points: $\mathbf{f}^1, \mathbf{f}^2, \ldots, \mathbf{f}^n$ where $f_k^i := v(k)$ if $k \neq i$ and $f_i^i := v(N) - \sum_{k \in N \setminus \{i\}} v(k)$. (See Problem 16.1.)

Example 16.3 Let (N, v) be a three-person game with $v(1) = v(3) = 0, v(2) = 3,$ $v(1, 2, 3) = 5$. Then $I(v)$ is the triangle with vertices $\mathbf{f}^1 = (2, 3, 0), \mathbf{f}^2 = (0, 5, 0)$ and $\mathbf{f}^3 = (0, 3, 2)$. (See Fig. 16.1.) $\square$

Fig. 16.1 Example 16.3

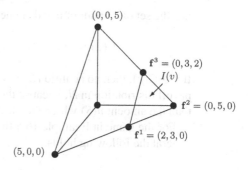

Definition 16.4 Let (N, v) be a game. Let $\mathbf{y}, \mathbf{z} \in I(v)$, $S \in 2^N \setminus \{\emptyset\}$. Then $\mathbf{y}$ *dominates* $\mathbf{z}$ *via coalition* S, denoted by $\mathbf{y} \operatorname{dom}_S \mathbf{z}$, if

(1) $y_i > z_i$ for all $i \in S$,
(2) $y(S) \leq v(S)$.

For $\mathbf{y}, \mathbf{z} \in I(v)$, $\mathbf{y}$ is said to *dominate* $\mathbf{z}$ (notation: $\mathbf{y} \operatorname{dom} \mathbf{z}$) if there is an $S \in 2^N \setminus \{\emptyset\}$ such that $\mathbf{y} \operatorname{dom}_S \mathbf{z}$.   □

Thus, imputation $\mathbf{y}$ dominates imputation $\mathbf{z}$ via coalition S if $\mathbf{y}$ is better than $\mathbf{z}$ for all members $i \in S$—this is condition (1)—and the payoffs $(y_i)_{i \in S}$ are attainable for the members of S by cooperation—this is condition (2). Against each $\mathbf{z}$ in

$$D(S) := \{\mathbf{z} \in I(v) \mid \text{ there exists } \mathbf{y} \in I(v) \text{ with } \mathbf{y} \operatorname{dom}_S \mathbf{z}\}$$

the players of S can protest successfully. The set $D(S)$ consists of the imputations which are dominated via S. Note that always $D(N) = \emptyset$ (see Problem 16.3). We call $\mathbf{x} \in I(v)$ *undominated* if $\mathbf{x} \in I(v) \setminus \bigcup_{S \in 2^N \setminus \{\emptyset\}} D(S)$.

Example 16.5 Let (N, v) be the three-person game with $v(1, 2) = 2$, $v(N) = 1$ and $v(S) = 0$ if $S \neq \{1, 2\}$, N. Then $D(S) = \emptyset$ if $S \neq \{1, 2\}$ and $D(\{1, 2\}) = \{\mathbf{x} \in I(v) \mid x_3 > 0\}$. The elements $\mathbf{x}$ in $I(v)$ which are undominated are those that satisfy $x_3 = 0$.   □

16.2 The Core and the Domination-Core

The concept of domination defined in the preceding section gives rise to the following definition.

Definition 16.6 The *domination core* (*D-core*) of a game (N, v) is the set

$$DC(v) := I(v) \setminus \bigcup_{S \in 2^N \setminus \{\emptyset\}} D(S),$$

i.e., the set of all undominated elements in $I(v)$. The *core* of a game (N, v) is the set

$$C(v) := \{\mathbf{x} \in I(v) \mid x(S) \geq v(S) \text{ for all } S \in 2^N \setminus \{\emptyset\}\}. \qquad \square$$

If $\mathbf{x} \in C(v)$, then no coalition $S \neq N$ has an incentive to split off if $\mathbf{x}$ is the proposed payoff distribution in N, because the total amount $x(S)$ allocated to S is not smaller than the amount $v(S)$ which the players in S can obtain by forming the coalition S.

For the game in Example 16.5 the D-core is nonempty and the core is empty. In general the following holds.

Theorem 16.7 *The core is a subset of the D-core for each TU-game.*

Proof Let (N, v) be a game and $\mathbf{x} \in I(v)$, $\mathbf{x} \notin DC(v)$. Then there is a $\mathbf{y} \in I(v)$ and a coalition $S \neq \emptyset$ such that $\mathbf{y} \, \mathrm{dom}_S \, \mathbf{x}$. Thus, $v(S) \geq y(S) > x(S)$, which implies that $\mathbf{x} \notin C(v)$. ∎

Elements of $C(v)$ can easily be obtained because the core is defined with the aid of linear inequalities. The core is a polytope. Also the D-core is a convex set: see Problem 16.2.

A natural question that arises is: for which games is the core equal to the D-core? Consider the following condition on a game (N, v):

$$v(N) \geq v(S) + \sum_{i \in N \setminus S} v(i) \text{ for all } S \in 2^N \setminus \{\emptyset\}. \qquad (16.1)$$

It turns out that this condition is sufficient for the equality of core and D-core.

Theorem 16.8 *Let (N, v) be a game satisfying* (16.1). *Then $DC(v) = C(v)$.*

Proof In view of Theorem 16.7 it is sufficient to show that $DC(v) \subseteq C(v)$.

Claim Let $\mathbf{x} \in I(v)$ with $x(S) < v(S)$ for some S, then there is a $\mathbf{y} \in I(v)$ such that $\mathbf{y} \, \mathrm{dom}_S \, \mathbf{x}$.

To prove this claim, define $\mathbf{y}$ as follows. If $i \in S$, then $y_i := x_i + |S|^{-1}(v(S) - x(S))$. If $i \notin S$, then $y_i := v(i) + (v(N) - v(S) - \sum_{i \in N \setminus S} v(i))|N \setminus S|^{-1}$. Then $\mathbf{y} \in I(v)$, where $y_i \geq v(i)$ for $i \in N \setminus S$ follows from (16.1). Furthermore, $\mathbf{y} \, \mathrm{dom}_S \, \mathbf{x}$. This proves the claim.

To prove $DC(v) \subseteq C(v)$, suppose $\mathbf{x} \in DC(v)$. Then there is no $\mathbf{y} \in I(v)$ with $\mathbf{y} \, \mathrm{dom} \, \mathbf{x}$. In view of the Claim it follows that $x(S) \geq v(S)$ for all $S \in 2^N \setminus \{\emptyset\}$. Hence, $\mathbf{x} \in C(v)$. ∎

Remark 16.9 Condition (16.1) is satisfied if the game v has a non-empty core. So in that case, $C(v) = DC(v)$. See also Problem 16.11. □

Many games v derived from practical situations have the following property:

$$v(S \cup T) \geq v(S) + v(T) \text{ for all disjoint } S, T \subseteq N. \tag{16.2}$$

A game satisfying (16.2) is called *super-additive*. Observe that (16.2) implies (16.1), so that Theorem 16.8 holds for super-additive games in particular.

16.3 Simple Games

In this section we study the core and D-core of simple games. Simple games arise in particular in political situations, see for instance the United Nations Security Council example in Chap. 1.

Definition 16.10 A *simple game* (N, v) is a game where every coalition has either worth 0 or worth 1, and the grand coalition N has worth 1. Coalitions with worth 1 are called *winning*, the other coalitions are called *losing*. A *minimal* winning coalition is a winning coalition for which every proper subset is losing. A player i is called a *dictator* in a simple game (N, v) if $v(S) = 1$ if and only if $i \in S$. A player i is called a *veto player* in a simple game (N, v) if i belongs to all winning coalitions. The *set of veto players* of v is denoted by veto(v). Hence,

$$\text{veto}(v) = \bigcap \{S \in 2^N \mid v(S) = 1\}. \qquad \square$$

The next example suggests that non-emptiness of the core has something to do with the existence of veto players.

For each $i \in N$ let $\mathbf{e}^i \in \mathbb{R}^n$ denote the vector with i-th coordinate equal to 1 and all other coordinates equal to 0.

Example 16.11 (1) Let $i \in N$. For the *dictator game* δ_i, which is the simple game with $\delta_i(S) = 1$ if and only if $i \in S$ one has $I(\delta_i) = \{\mathbf{e}^i\}$, veto$(\delta_i) = \{i\}$ and $C(\delta_i) = DC(\delta_i) = \{\mathbf{e}^i\}$.

(2) For the three-person *majority game* with $v(S) = 1$ if $|S| \in \{2, 3\}$ and $v(S) = 0$ if $|S| \in \{0, 1\}$ one has:

$$\{1, 2\} \cap \{1, 3\} \cap \{2, 3\} \cap \{1, 2, 3\} = \emptyset = \text{veto}(v)$$

and

$$C(v) = DC(v) = \emptyset .$$

(3) Let T be a nonempty coalition. For the *T-unanimity game* u_T, which is the simple game with $u_T(S) = 1$ if and only if $T \subseteq S$, veto$(u_T) = T$ and

$$C(u_T) = DC(u_T) = \text{conv}\{\mathbf{e}^i \mid i \in T\} . \qquad \square$$

The following theorem shows that the core of a simple game is nonempty if and only if the game has veto players. Furthermore, core elements divide the total amount $v(N) = 1$ of the grand coalition among the veto players. The D-core is equal to the core for simple games except in one case where there is exactly one $k \in N$ with $v(k) = 1$ and k is not a veto player. See also Example 16.13 below.

Theorem 16.12 *Let (N, v) be a simple game. Then:*

(1) $C(v) = conv\{e^i \in \mathbb{R}^n \mid i \in veto(v)\}$.
(2) *If $veto(v) = \emptyset$ and $\{i \in N \mid v(i) = 1\} = \{k\}$, then $C(v) = \emptyset$ and $DC(v) = \{e^k\}$. Otherwise, $DC(v) = C(v)$.*

Proof

(a) Suppose $i \in veto(v)$. Let $S \in 2^N \setminus \{\emptyset\}$. If $i \in S$ then $e^i(S) = 1 \geq v(S)$, otherwise $e^i(S) = 0 = v(S)$. Obviously, $e^i(N) = 1 = v(N)$. So $e^i \in C(v)$. This proves the inclusion $\supseteq$ in (1) because $C(v)$ is a convex set.

(b) To prove the inclusion $\subseteq$ in (1), let $\mathbf{x} \in C(v)$. It is sufficient to prove: $i \notin veto(v) \Rightarrow x_i = 0$. Suppose, to the contrary, that $x_i > 0$ for some non-veto player i. Take S with $v(S) = 1$ and $i \notin S$ (such an S exists otherwise i would be a veto player). Then $x(S) = x(N) - x(N \setminus S) \leq 1 - x_i < 1$, contradicting the fact that $\mathbf{x}$ is a core element. This concludes the proof of (1).

(c) If $veto(v) = \emptyset$ and k is the only player in the set $\{i \in N \mid v(i) = 1\}$, then $C(v) = \emptyset$ by part (1), whereas $I(v) = \{e^k\}$, hence $DC(v) = \{e^k\}$. If $veto(v) = \emptyset$ and $\{i \in N \mid v(i) = 1\} = \emptyset$ then (16.1) is satisfied, so that core and D-core are equal by Theorem 16.8. If $veto(v) = \emptyset$ and $|\{i \in N \mid v(i) = 1\}| \geq 2$ then $I(v) = \emptyset$ so that $C(v) = DC(v) = \emptyset$.

(d) To complete the proof of (2), suppose $veto(v) \neq \emptyset$. Then $C(v) \neq \emptyset$ by part (1). Hence $C(v) = DC(v)$ by Remark 16.9.   ∎

Example 16.13 Let $N = \{1, 2, 3\}$, $v(1) = v(2, 3) = v(1, 2, 3) = 1$ and $v(S) = 0$ for the other coalitions. Then $veto(v) = \emptyset$, $C(v) = \emptyset$, $DC(v) = \{e^1\}$. Note that this simple game is not super-additive, and does not satisfy (16.1).   □

16.4 Stable Sets

The definition of a stable set is again based on the concept of domination. By way of example, let v be the three-person game with all worths equal to 1 except for the one-person coalitions, which have worth equal to 0. Observe that the three vectors $(\frac{1}{2}, \frac{1}{2}, 0)$, $(\frac{1}{2}, 0, \frac{1}{2})$, and $(0, \frac{1}{2}, \frac{1}{2})$ are imputations that do not dominate each other. Moreover, each imputation other than one of these three is dominated by one of these three (see Problem 16.5). For this reason, von Neumann and Morgenstern called the set of these three imputations a 'solution' of the game.

Definition 16.14 Let v be a game and let $A \subseteq I(v)$. The set A is called a *stable set* if

(1) if $\mathbf{x}, \mathbf{y} \in A$ then $\mathbf{x}$ does not dominate $\mathbf{y}$,
(2) if $\mathbf{x} \in I(v) \setminus A$ then there is a $\mathbf{y} \in A$ that dominates $\mathbf{x}$. □

The first property in Definition 16.14 is called *internal stability* and the second one *external stability*.

The three-person game described above has many stable sets: see Problem 16.5. But even if a game has only one stable set then still a selection would have to be made, for practical purposes; stability, however, is a property of sets, not of single payoff distributions. The core does not suffer from this problem and, moreover, in that case there exist some plausible choices (like the nucleolus, see Chap. 19). Moreover, games with non-empty cores have been exactly characterized (see Sect. 16.5), whereas the problem of existence of stable sets is only partially solved.

Some partial existence results are given now. First, essential simple games always have stable sets:

Theorem 16.15 *Let v be a simple game and let S be a minimal winning coalition. Let Δ^S be the set of those imputations $\mathbf{x}$ with $x_i = 0$ for every $i \notin S$. Then, if $\Delta^S \neq \emptyset$, it is a stable set.*

Proof Problem 16.8. ■

A game (N, v) is called a *zero-one game* if all one-person coalitions have worth 0 and the grand coalition N has worth 1. In the following example symmetric three-person zero-one games are considered.

Example 16.16 Let (N, v) be a game with $N = \{1, 2, 3\}$ and $v(i) = 0$ for all $i \in N$, $v(N) = 1$, and $v(S) = \alpha$ for every two-person coalition S, where $0 \leq \alpha \leq 1$. Then:

(a) Let $\alpha \geq \frac{2}{3}$. Then

$$\{(x, x, 1 - 2x), (x, 1 - 2x, x), (1 - 2x, x, x) \mid \frac{\alpha}{2} \leq x \leq \frac{1}{2}\} \qquad (16.3)$$

is a stable set.
(b) For $\alpha < \frac{2}{3}$, the set in (16.3) is internally but not externally stable. The union of this set with the core of the game is a stable set.
(c) For $\alpha \leq \frac{1}{2}$ the core is a (the unique) stable set.

For the proofs of these statements see Problem 16.9. □

The next theorem gives the relation between the domination core and stable sets.

Theorem 16.17 *Let* (N, v) *be a game. Then:*

(a) *The D-core of* v *is a subset of any stable set.*
(b) *If* A *and* B *are stable sets and* $A \neq B$, *then* $A \not\subseteq B$.
(c) *Suppose the D-core of* v *is a stable set. Then it is the unique stable set of the game.*

Proof Problem 16.10.   ∎

16.5 Balanced Games and the Core

In this section we derive the Bondareva-Shapley Theorem, which characterizes games with non-empty cores in terms of balancedness. First, the concepts of balanced maps, collections, and games are introduced.

Let $N = \{1, 2, \ldots, n\}$. A map $\lambda : 2^N \setminus \{\emptyset\} \to \mathbb{R}_+ := \{t \in \mathbb{R} \mid t \geq 0\}$ is called a *balanced map* if

$$\sum_{S \in 2^N \setminus \{\emptyset\}} \lambda(S) \mathbf{e}^S = \mathbf{e}^N .$$

Here $\mathbf{e}^S \in \mathbb{R}^N$ is the *characteristic vector* for coalition S with

$$e_i^S = 1 \text{ if } i \in S \text{ and } e_i^S = 0 \text{ if } i \in N \setminus S .$$

A collection B of nonempty coalitions is called a *balanced collection* if there is a balanced map λ such that

$$B = \{S \in 2^N \mid \lambda(S) > 0\} .$$

Example 16.18

(1) Let the nonempty coalitions $N_1, N_2, \ldots, N_k$ form a partition of N, i.e., $N = \bigcup_{r=1}^{k} N_r$ and $N_s \cap N_t = \emptyset$ if $s \neq t$. Then $\{N_1, N_2, \ldots, N_k\}$ is a balanced collection, corresponding to the balanced map λ with $\lambda(S) = 1$ if $S \in \{N_1, N_2, \ldots, N_k\}$ and $\lambda(S) = 0$ otherwise.
(2) For $N = \{1, 2, 3\}$ the set $B = \{\{1, 2\}, \{1, 3\}, \{2, 3\}\}$ is balanced and corresponds to the balanced map λ with $\lambda(S) = 0$ if $|S| \in \{1, 3\}$ and $\lambda(S) = \frac{1}{2}$ if $|S| = 2$. □

In order to have an interpretation of a balanced map, one can think of each player having one unit of time (or energy, labor, …) to spend. Each player can distribute his time over the various coalitions of which he is a member. Such a distribution is 'balanced' if it corresponds to a balanced map λ, where $\lambda(S)$ is interpreted as the length of time that the coalition S exists ('cooperates'); balancedness of λ means that each player spends exactly his one unit of time over the various coalitions.

Definition 16.19 A game (N, v) is called a *balanced game* if for each balanced map $\lambda : 2^N \setminus \{\emptyset\} \to \mathbb{R}_+$ we have

$$\sum_S \lambda(S)v(S) \leq v(N) . \tag{16.4}$$

$\square$

Extending the interpretation of a balanced map in terms of a distribution of time to a game, balancedness of a game could be interpreted as saying that it is at least as productive to have the grand coalition operate during one unit of time as to have a balanced distribution of time over various smaller coalitions—worths of coalitions being interpreted as productivities. Thus, in a balanced game, it seems advantageous to form the grand coalition. Indeed, technically the importance of the notion of balancedness follows from Theorem 16.22. This theorem characterizes games with a nonempty core. Its proof is based on the following duality theorem.

For $\mathbf{x}, \mathbf{y} \in \mathbb{R}^n$, $\mathbf{x} \cdot \mathbf{y}$ denotes the usual inner product: $\mathbf{x} \cdot \mathbf{y} = \sum_{i=1}^{n} x_i y_i$.

Theorem 16.20 *Let A be an $n \times p$-matrix, $\mathbf{b} \in \mathbb{R}^p$ and $\mathbf{c} \in \mathbb{R}^n$, and let $\{\mathbf{x} \in \mathbb{R}^n \mid \mathbf{x}A \geq \mathbf{b}\} \neq \emptyset$ and $\{\mathbf{y} \in \mathbb{R}^p \mid A\mathbf{y} = \mathbf{c}, \mathbf{y} \geq 0\} \neq \emptyset$. Then*

$$\min\{\mathbf{x} \cdot \mathbf{c} \mid \mathbf{x}A \geq \mathbf{b}\} = \max\{\mathbf{b} \cdot \mathbf{y} \mid A\mathbf{y} = \mathbf{c}, \mathbf{y} \geq 0\} .$$

Proof Problem 16.13.   ■

Remark 16.21 In Theorem 16.20 also the following holds: if one of the programs is infeasible (i.e., one of the two sets in the theorem is empty), then both programs do not have an optimal solution (i.e., neither the minimum nor the maximum are attained). See Problem 16.14 for a proof. $\square$

Theorem 16.22 *Let (N, v) be a TU-game. Then the following two assertions are equivalent:*

(1) $C(v) \neq \emptyset$,
(2) (N, v) *is a balanced game.*

Proof First note that $C(v) \neq \emptyset$ if and only if

$$v(N) = \min\{\sum_{i=1}^{n} x_i \mid \mathbf{x} \in \mathbb{R}^N, \ x(S) \geq v(S) \text{ for all } S \in 2^N \setminus \{\emptyset\}\} . \tag{16.5}$$

By the duality theorem, Theorem 16.20, equality (16.5) holds if and only if

$$v(N) = \max\{\sum \lambda(S)v(S) \mid \sum \lambda(S)e^S = e^N, \lambda \geq 0\} . \tag{16.6}$$

(Take for A the matrix with the characteristic vectors $\mathbf{e}^S$ as columns, let $\mathbf{c} := \mathbf{e}^N$ and let $\mathbf{b}$ be the vector of coalitional worths. Obviously, the non-emptiness conditions in Theorem 16.20 are satisfied.) Now (16.6) holds if and only if (16.4) holds. Hence (1) and (2) are equivalent.   ∎

An alternative proof of Theorem 16.22 can be based directly on Farkas' Lemma (Lemma 22.5): see Problem 16.15.

16.6   Problems

16.1. *Imputation Set of an Essential Game*
Prove that for an essential game v, $I(v)$ is the convex hull of the points $\mathbf{f}^1, \mathbf{f}^2, \ldots, \mathbf{f}^n$, as claimed in Sect. 16.1.

16.2. *Convexity of the Domination Core*
Prove that for each game the domination core is a convex set.

16.3. *Dominated Sets of Imputations*

(a) Prove that for each game (N, v), $D(S) = \emptyset$ if $|S| \in \{1, n\}$.
(b) Determine for each S the set $D(S)$ for the cost savings game (three communities game) in Chap. 1. Answer the same question for the glove game in Chap. 1.

16.4. *The Domination Relation*

(a) Prove that dom and dom_S are irreflexive relations and that dom_S is transitive and antisymmetric.[1]
(b) Construct a game (N, v) and imputations $\mathbf{x}$ and $\mathbf{y}$ such that $\mathbf{x} \operatorname{dom} \mathbf{y}$ and $\mathbf{y} \operatorname{dom} \mathbf{x}$.
(c) Construct a game (N, v) and $\mathbf{x}, \mathbf{y}, \mathbf{z} \in I(v)$ with $\mathbf{x} \operatorname{dom} \mathbf{y}$ and $\mathbf{y} \operatorname{dom} \mathbf{z}$ and not $\mathbf{x} \operatorname{dom} \mathbf{z}$.

16.5. *Stable Sets in a Three-Person Game*
Let $(\{1, 2, 3\}, v)$ be the game with all worths equal to 1 except for the one-person and the empty coalitions, which have worth equal to 0.

(a) Prove that each element of the imputation set of this game is dominated by another element.
(b) Prove that in this game each $\mathbf{x} \in I(v) \setminus A$ is dominated by an element of $A := \{(\frac{1}{2}, \frac{1}{2}, 0), (\frac{1}{2}, 0, \frac{1}{2}), (0, \frac{1}{2}, \frac{1}{2})\}$.
(c) If $c \in [0, \frac{1}{2})$ and $B := \{\mathbf{x} \in I(v) \mid x_3 = c\}$, then each element of $I(v) \setminus B$ is dominated by an element of B. Show this.

16.6. *Singleton Stable Set*

---

[1] See Sect. 11.1 for definitions.

Prove that if a game (N, v) has a one-element stable set then $v(N) = \sum_{i \in N} v(i)$.

16.7. *A Glove Game*

Consider the three-person simple game v defined by

$$v(S) := \begin{cases} 1 & \text{if } S = \{1, 2\} \text{ or } \{2, 3\} \text{ or } \{1, 2, 3\} \\ 0 & \text{otherwise.} \end{cases}$$

(a) Show that any imputation (x_1, x_2, x_3) that is not equal to $\mathbf{e}^2$ is dominated by another imputation.
(b) Compute the core and the domination core.
(c) Show that the domination core is not a stable set.
(d) Show that

$$B := \{(\lambda, 1 - 2\lambda, \lambda) \mid 0 \le \lambda \le \frac{1}{2}\}$$

is a stable set.

16.8. *Proof of Theorem 16.15*

Prove Theorem 16.15.

16.9. *Example 16.16*

Prove the statements in Example 16.16.

16.10. *Proof of Theorem 16.17*

Prove Theorem 16.17. Does this theorem also hold for the core instead of the D-core?

16.11. *Core and D-Core*

Is (16.1) also a necessary condition for equality of the core and the D-core? (Cf. Theorem 16.8.)

16.12. *Strategic Equivalence*

Let (N, w) be *strategically equivalent* to (N, v), that is, there are $k \in \mathbb{R}$, $k > 0$ and $\mathbf{a} \in \mathbb{R}^N$ such that for each coalition S: $w(S) = kv(S) + a(S)$. Show that

(i) $C(w) = kC(v) + \mathbf{a}$ $(:= \{\mathbf{x} \in \mathbb{R}^N \mid \mathbf{x} = k\mathbf{y} + \mathbf{a} \text{ for some } \mathbf{y} \in C(v)\})$
(ii) $DC(w) = kDC(v) + \mathbf{a}$.

[The equalities (i) and (ii) express that the core and the D-core are covariant w.r.t. strategic equivalence.]

16.13. *Proof of Theorem 16.20*

Prove Theorem 16.20. [Hint: use Theorem 22.6.]

16.14. *Infeasible Programs in Theorem 16.20*

Prove the claim made in Remark 16.21. [Hint: Suppose, say, that there is no $\mathbf{y} \ge \mathbf{0}$ with $A\mathbf{y} = \mathbf{c}$. Then, certainly, the max-program does not have an optimal solution.

Use Farkas' Lemma (Lemma 22.5) to conclude that there exists a vector $\mathbf{z}$ with $\mathbf{z}A \geq \mathbf{0}$ and $\mathbf{z} \cdot \mathbf{c} < 0$. Suppose the min-program is feasible, i.e., there is an $\mathbf{x}$ with $\mathbf{x}A \geq \mathbf{b}$. Then, show that the min-program does not have an optimal solution by considering the vectors $\mathbf{x} + t\mathbf{z}$ for $t \in \mathbb{R}, t > 0$.]

16.15. *Proof of Theorem 16.22 Using Lemma 22.5*
Prove Theorem 16.22 with the aid of Lemma 22.5. [Hint: List the nonempty coalitions $S \subseteq N$ as $S_1, \ldots, S_p$ ($p = 2^n - 1$) with $S_p = N$. Define the $(n + n + p) \times p$ matrix A as follows. Column $k < p$ is $(\mathbf{e}^{S_k}, -\mathbf{e}^{S_k}, -\mathbf{e}^k)$ where: $\mathbf{e}^{S_k} \in \mathbb{R}^n$, $e_i^{S_k} = 1$ if $i \in S_k$, $e_i^{S_k} = 0$ if $i \notin S_k$. Column p is $(\mathbf{e}^N, -\mathbf{e}^N, \mathbf{0})$. Then $C(N, v) \neq \emptyset$ iff there exists $(\mathbf{z}, \mathbf{z}', \mathbf{w}) \in \mathbb{R}^n \times \mathbb{R}^n \times \mathbb{R}^p$ with $(\mathbf{z}, \mathbf{z}', \mathbf{w}) \geq \mathbf{0}$ and $(\mathbf{z}, \mathbf{z}', \mathbf{w})A = \mathbf{b}$, where $\mathbf{b} = (v(S_k))_{k=1}^p$. This has the form as in (a) of Lemma 22.5.]

16.16. *Balanced Maps and Collections*

(a) Show that for any balanced map λ one has $\sum_S \lambda(S) \geq 1$, with equality if and only if the corresponding balanced collection equals $\{N\}$.
(b) If B is a balanced collection unequal to $\{N\}$, then

$$B^c := \{S \in 2^N \setminus \{\emptyset\} \mid N \setminus S \in B\}$$

is also a balanced collection. Give the corresponding balanced map.
(c) Let $S \in 2^N \setminus \{\emptyset, N\}$. Prove that $\{S, (N \setminus \{i\})_{i \in S}\}$ is balanced collection.
(d) Prove that the balanced maps form a convex set Λ^n.

16.17. *Minimum of Balanced Games*
Show that the minimum of two balanced games is again balanced.

16.18. *Balanced Simple Games*
A simple game has a non-empty core if and only if it has veto players, cf. Theorem 16.12(1). Derive this result from Theorem 16.22.

16.7 Notes

The concepts of domination, imputation, and stable set were introduced by von Neumann and Morgenstern (1944/1947). The core was introduced by Gillies (1953).

Lucas (1969) gives an example of a(n essential) game that does not have a stable set; see also Owen (1995), p. 253. Theorem 16.22 is due to Bondareva (1962) and Shapley (1967).

Problem 16.6 is taken from Morris (1994).

References

Bondareva, O. N. (1962). Theory of the core in the *n*-person game. *Vestnik Leningradskii Universitet, 13*, 141–142 (in Russian).

Gillies, D. B. (1953). *Some theorems on n-person games*. Ph.D. Thesis, Princeton University Press, Princeton, New Jersey.

Lucas, W. F. (1969). The proof that a game may not have a solution. *Transactions of the American Mathematical Society, 136*, 219–229.

Morris, P. (1994). *Introduction to game theory*. New York: Springer.

Owen, G. (1995). *Game theory* (3rd ed.). San Diego: Academic.

Shapley, L. S. (1967). On balanced sets and cores. *Naval Research Logistics Quarterly, 14*, 453–460.

von Neumann, J., & Morgenstern, O. (1944/1947). *Theory of games and economic behavior*. Princeton: Princeton University Press.

The Shapley Value

17

In Chap. 16 set-valued solution concepts for games with transferable utilities were studied: the imputation set, core, domination core, and stable sets. In this chapter, a one-point (single-valued) solution concept is discussed: the Shapley value. It may again be helpful to first study the relevant parts of Chaps. 1 and 9.

Section 17.1 introduces the Shapley value by several formulas and presents (a variation on) Shapley's axiomatic characterization using additivity. In Sect. 17.2 we present three other characterizations of the Shapley value: a description in terms of Harsanyi dividends; an axiomatic characterization of Young based on strong monotonicity; and Owen's formula for the Shapley value based on a multilinear extension of games. Section 11.3 discusses Hart and Mas-Colell's approach to the Shapley value based on potential and reduced games.

17.1 Definition and Shapley's Characterization

Let (N, v) be a TU-game and let $\sigma : N \to N$ be a permutation of the player set. Imagine that the players enter a room one by one in the ordering $\sigma(1)$, $\sigma(2)$, ..., $\sigma(n)$ and give each player the marginal contribution he creates in the game. To be more specific, let the *set of predecessors of i in* σ be the coalition

$$P_\sigma(i) := \{r \in N \mid \sigma^{-1}(r) < \sigma^{-1}(i)\} .$$

For example, if $N = \{1, 2, 3, 4, 5\}$ and $\sigma(1) = 2$, $\sigma(2) = 5$, $\sigma(3) = 4$, $\sigma(4) = 1$, and $\sigma(5) = 3$, player 2 enters first, next players 5, 4, 1, and 3. So $P_\sigma(1) = \{2, 5, 4\}$.

Define the *marginal vector* m^σ by

$$m_i^\sigma = v(P_\sigma(i) \cup \{i\}) - v(P_\sigma(i)) . \tag{17.1}$$

© Springer-Verlag Berlin Heidelberg 2015
H. Peters, *Game Theory*, Springer Texts in Business and Economics,
DOI 10.1007/978-3-662-46950-7_17

Thus, the marginal vector m^σ gives each player his marginal contribution to the coalition formed by his entrance, according to the ordering σ.[1]

Definition 17.1 The *Shapley value* $\Phi(v)$ of a game (N, v) is the average of the marginal vectors of the game, i.e.

$$\Phi(v) := \frac{1}{n!} \sum_{\sigma \in \Pi(N)} m^\sigma \ . \tag{17.2}$$

Here, $\Pi(N)$ denotes the set of permutations of N.   □

Example 17.2

(1) For a two-person game (N, v) the Shapley value is

$$\Phi(v) = \left(v(1) + \frac{v(N) - v(1) - v(2)}{2}, v(2) + \frac{v(N) - v(1) - v(2)}{2} \right) .$$

(2) Let (N, v) be the three-person game with $v(1) = v(2) = v(3) = 0$, $v(1, 2) = 4$, $v(1, 3) = 7$, $v(2, 3) = 15$, $v(1, 2, 3) = 20$. Then the marginal vectors are given in Table 17.1. The Shapley value of this game is equal to $\frac{1}{6}(21, 45, 54)$, as one easily obtains from this table.

(3) The Shapley value $\Phi(v)$ for an additive game is equal to $(v(1), v(2), \dots, v(n))$.   □

Based on (17.2), a probabilistic interpretation of the Shapley value is as follows. Suppose we draw from an urn, containing the elements of $\Pi(N)$, a permutation σ (probability $(n!)^{-1}$). Then let the players enter a room one by one in the order σ and give each player the marginal contribution created by him. Then the i-th coordinate $\Phi_i(v)$ of $\Phi(v)$ is the expected payoff to player i according to this random procedure.

Table 17.1
Example 17.2(2)

$(\sigma(1), \sigma(2), \sigma(3))$	m_1^σ	m_2^σ	m_3^σ
$(1, 2, 3)$	0	4	16
$(1, 3, 2)$	0	13	7
$(2, 1, 3)$	4	0	16
$(2, 3, 1)$	5	0	15
$(3, 1, 2)$	7	13	0
$(3, 2, 1)$	5	15	0
Σ	21	45	54

---

[1] Of course, m^σ depends on the game v.

Using (17.1) formula (17.2) can be rewritten as

$$\Phi_i(v) = \frac{1}{n!} \sum_{\sigma \in \Pi(N)} v(P_\sigma(i) \cup \{i\}) - v(P_\sigma(i)) .$$ (17.3)

The terms at the right hand side of the summation sign are of the form $v(S \cup \{i\}) - v(S)$, where S is a subset of N not containing i. For how many orderings does one have $P_\sigma(i) = S$? The answer is $|S|!(n - 1 - |S|)!$, where the first factor $|S|!$ corresponds to the number of orderings of S and the second factor $(n - 1 - |S|)!$ to the number of orderings of $N \setminus (S \cup \{i\})$. Hence, (17.3) can be rewritten to obtain

$$\Phi_i(v) = \sum_{S:i \notin S} \frac{|S|!(n - 1 - |S|)!}{n!} (v(S \cup \{i\}) - v(S)) .$$ (17.4)

Note that

$$\frac{|S|!(n - 1 - |S|)!}{n!} = \frac{1}{n} \binom{n-1}{|S|}^{-1} .$$

This gives rise to a second probabilistic interpretation of the Shapley value. Construct a subset S to which i does not belong, as follows. First, draw at random a number from the urn containing the numbers (possible sizes) $0, 1, 2, \ldots, n - 1$, where each number has probability n^{-1} to be drawn. If size s is chosen, draw a set from the urn containing the subsets of $N \setminus \{i\}$ of size s, where each set has the same probability $\binom{n-1}{s}^{-1}$ to be drawn. If S is drawn with $|S| = s$, then pay player i the amount $v(S \cup \{i\}) - v(S)$. Then, in view of (17.4), the expected payoff for player i in this random procedure is the Shapley value for player i of the game (N, v).

We next give an axiomatic characterization of the Shapley value. That is, we formulate a number of properties that a one-point solution should (or might) have and then show that the Shapley value is the only solution with these properties.

Definition 17.3 A *value* on $\mathcal{G}^N$ is a map $\psi : \mathcal{G}^N \to \mathbb{R}^N$.[2] □

The following axioms for a value $\psi : \mathcal{G}^N \to \mathbb{R}^N$ are used in the announced characterization of the Shapley value.

Efficiency (EFF): $\sum_{i=1}^n \psi_i(v) = v(N)$ for all $v \in \mathcal{G}^N$.

The efficiency (sometimes called Pareto optimality or Pareto efficiency) axiom needs no further explanation.

Call a player i in a game (N, v) a *null-player* if $v(S \cup i) - v(S) = 0$ for every coalition $S \in 2^N$. Such a player does not contribute anything to any coalition, in

---

[2]Occasionally, also the word *solution* will be used.

particular also $v(i) = 0$. So it seems reasonable that such a player obtains zero according to the value. This is what the following axiom requires.

Null-player Property (NP): $\psi_i(v) = 0$ for all $v \in \mathcal{G}^N$ and all null-players i in v.

Call players i and j *symmetric* in the game (N, v) if $v(S \cup i) = v(S \cup j)$ for every coalition $S \subseteq N \setminus \{i,j\}$. Symmetric players have the same contribution to any coalition, and therefore it seems reasonable that they should obtain the same payoff according to the value. That is the content of the following axiom.

Symmetry (SYM): $\psi_i(v) = \psi_j(v)$ for all $v \in \mathcal{G}^N$ and all symmetric players i and j in v.

In order to formulate the last axiom in the announced characterization, we define the sum $(N, v + w)$ of two games (N, v) and (N, w) by $(v + w)(S) = v(S) + w(S)$ for every $S \in 2^N$.

This last axiom can be interpreted as follows. Suppose the game (N, v) is played today and the game (N, w) tomorrow. If the value ψ is applied then player i obtains in total: $\psi_i(N, v) + \psi_i(N, w)$. One may also argue that, in total, the game $(N, v + w)$ is played and that, accordingly, player i should obtain $\psi_i(N, v + w)$. The following axiom expresses the possible point of view that these two evaluations should not make a difference.

Additivity (ADD): $\psi(v + w) = \psi(v) + \psi(w)$ for all $v, w \in \mathcal{G}^N$.

The announced characterization is the following theorem.

Theorem 17.4 *Let $\psi : \mathcal{G}^N \to \mathbb{R}^N$ be a value. Then ψ satisfies EFF, NP, SYM, and ADD, if, and only if, ψ is the Shapley value Φ.*

The proof of Theorem 17.4 uses, through the additivity axiom, the fact that $\mathcal{G}^N$ is a linear space, with addition $v + w$ and scalar multiplication $(\alpha v)(S) = \alpha v(S)$ for all $v, w \in \mathcal{G}^N$, $S \in 2^N$, and $\alpha \in \mathbb{R}$. An obvious basis for $\mathcal{G}^N$ is the set $\{1_T \in \mathcal{G}^N \mid T \in 2^N \setminus \{\emptyset\}\}$, where 1_T is the game defined by $1_T(T) = 1$ and $1_T(S) = 0$ for all $S \neq T$ (cf. Problem 17.1). This basis is not very well suited for the present purpose because the Shapley value $\Phi(1_T)$ $(T \in 2^N \setminus \{\emptyset\})$ cannot easily be determined from the axioms in Theorem 17.4: in particular, there is no null-player in the game 1_T (cf. Problem 17.1).

Another basis is the collection of unanimity games $\{u_T \in \mathcal{G}^N \mid T \in 2^N \setminus \{\emptyset\}\}$, see Example 16.11(3) for the definition, and Problem 17.2. This basis is used in the following proof.

Proof of Theorem 17.4 That the Shapley value satisfies the four axioms in the theorem is the subject of Problem 17.3.

Conversely, suppose ψ satisfies the four axioms. It has to be proved that $\psi = \Phi$. Take $v \in \mathcal{G}^N$. Then there are unique numbers c_T $(T \neq \emptyset)$ such that $v = \sum_{T \neq \emptyset} c_T u_T$. (Cf. Problem 17.2.) By ADD of ψ and Φ it follows that

$$\psi(v) = \sum_{T \neq \emptyset} \psi(c_T u_T) \, , \quad \Phi(v) = \sum_{T \neq \emptyset} \Phi(c_T u_T) \, .$$

So it is sufficient to show that for all $T \neq \emptyset$ and $c \in \mathbb{R}$:

$$\psi(c u_T) = \Phi(c u_T) \, . \tag{17.5}$$

Take $T \neq \emptyset$ and $c \in \mathbb{R}$. Note first that for all $i \in N \setminus T$:

$$c u_T(S \cup \{i\}) - c u_T(S) = 0 \text{ for all } S \, ,$$

implying that i is a null-player in $c u_T$. So, by NP of ψ and Φ:

$$\psi_i(c u_T) = \Phi_i(c u_T) = 0 \text{ for all } i \in N \setminus T \, . \tag{17.6}$$

Now suppose that $i, j \in T$, $i \neq j$. Then, for every coalition $S \subseteq N \setminus \{i, j\}$, $c u_T(S \cup i) = c u_T(S \cup j) = 0$, which implies that i and j are symmetric in $c u_T$. Hence, by SYM of ψ and Φ:

$$\Phi_i(c u_T) = \Phi_j(c u_T) \text{ for all } i, j \in T \tag{17.7}$$

and

$$\psi_i(c u_T) = \psi_j(c u_T) \text{ for all } i, j \in T \, . \tag{17.8}$$

Then EFF and (17.6)–(17.8) imply that

$$\psi_i(c u_T) = \Phi_i(c u_T) = |T|^{-1} c \text{ for all } i \in T \, . \tag{17.9}$$

Now (17.6) and (17.9) imply (17.5).   ∎

The following two axioms (for a value ψ) are stronger versions of the null-player property and symmetry, respectively. Call a player i in a game (N, v) a *dummy player* if $v(S \cup i) - v(S) = v(i)$ for all $S \subseteq N \setminus \{i\}$.

Dummy player Property (DUM): $\psi_i(v) = v(i)$ for all $v \in \mathcal{G}^N$ and all dummy players i in v.

A dummy player only contributes his own worth to every coalition, and that is what he should be payed according to the dummy property.

For a permutation $\sigma \in \Pi(N)$ define the game v^σ by

$$v^\sigma(S) := v(\sigma^{-1}(S)) \text{ for all } S \in 2^N$$

and define $\sigma^* : \mathbb{R}^N \to \mathbb{R}^N$ by $(\sigma^*(x))_{\sigma(k)} := x_k$ for all $x \in \mathbb{R}^N$ and $k \in N$.

Anonymity (AN): $\psi(v^\sigma) = \sigma^*(\psi(v))$ for all $v \in \mathcal{G}^N$ and all $\sigma \in \Pi(N)$.

Anonymity implies that a value does not discriminate between the players solely on the basis of their 'names' (i.e., numbers).

The dummy player property implies the null-player property, and anonymity implies symmetry. The Shapley value has the dummy player property. See Problem 17.4 for these claims.

The Shapley value is also anonymous.

Lemma 17.5 *The Shapley value Φ is anonymous.*

Proof

(1) First we show that

$$\rho^*(m^\sigma(v)) = m^{\rho\sigma}(v^\rho) \text{ for all } v \in \mathcal{G}^N, \rho, \sigma \in \pi(N) .$$

This follows because for all $i \in N$:

$$m^{\rho\sigma}(v^\rho)_{\rho\sigma(i)} = v^\rho(\{\rho\sigma(1), \ldots, \rho\sigma(i)\} - v^\rho(\{\rho\sigma(1), \ldots, \rho\sigma(i-1)\})$$
$$= v(\{\sigma(1), \ldots, \sigma(i)\}) - v(\{\sigma(1), \ldots, \sigma(i-1)\})$$
$$= (m^\sigma(v))_{\sigma(i)} = (\rho^*(m^\sigma(v)))_{\rho\sigma(i)} .$$

(2) Take $v \in \mathcal{G}^N$ and $\rho \in \Pi(N)$. Then (1), the fact that $\rho \mapsto \rho\sigma$ is a surjection on $\Pi(N)$ and the linearity of ρ^* imply:

$$\Phi(v^\rho) = \frac{1}{n!} \sum_{\sigma \in \Pi(N)} m^\sigma(v^\rho) = \frac{1}{n!} \sum_\sigma m^{\rho\sigma}(v^\rho) =$$

$$= \frac{1}{n!} \sum_\sigma \rho^*(m^\sigma(v)) = \rho^*(\frac{1}{n!} \sum_\sigma m^\sigma) = \rho^*(\Phi(v)) .$$

This proves the anonymity of Φ.   ∎

We have seen so far that the Shapley value has many appealing properties. In the following sections even more properties and characterizations of the Shapley value are considered. However, it also has some drawbacks. In a balanced game the Shapley value does not necessarily assign a core element. Also, it does not have to be individually rational (cf. Problem 17.5).

Variations of the Shapley value, obtained by omitting the symmetry (or anonymity) requirement in particular, are discussed in Chap. 18.

17.2   Other Characterizations

In this section three other characterizations of the Shapley value are discussed: the dividend approach, an axiomatization based on a monotonicity condition, and the so-called multilinear approach.

17.2.1   Dividends

With each game we can associate another game by computing the dividends of coalitions.

Definition 17.6 Let (N, v) be game. For each coalition T the *dividend* $\Delta_v(T)$ is defined, recursively, as follows.

$$\Delta_v(\emptyset) := 0$$

$$\Delta_v(T) := v(T) - \sum_{S:S \subsetneq T} \Delta_v(S) \text{ if } |T| \geq 1 \, .$$

$\square$

The relation between dividends and the Shapley value is described in the next theorem. The Shapley value of a player in a game turns out to be the sum of all equally distributed dividends of coalitions to which the player belongs.

Theorem 17.7 *Let* $v = \sum_{T \in 2^N \setminus \{\emptyset\}} c_T u_T$ *(as in Problem 17.2). Then:*

(a) $\Delta_v(T) = c_T$ *for all* $T \neq \emptyset$.
(b) *The Shapley value* $\Phi_i(v)$ *for player* i *is equal to the sum of the equally distributed dividends of the coalitions to which player i belongs i.e.,*

$$\Phi_i(v) = \sum_{T:i \in T} \frac{\Delta_v(T)}{|T|} \, .$$

Proof In the proof of Theorem 17.4 it was shown that $\Phi(c_T u_T) = |T|^{-1} c_T \mathbf{e}^T$ for each T, so by ADD, $\Phi(v) = \sum_{T \neq \emptyset} c_T |T|^{-1} \mathbf{e}^T$. Hence, $\Phi_i(v) = \sum_{T:i \in T} c_T |T|^{-1}$. The only thing left to show is that

$$c_T = \Delta_v(T) \text{ for all } T \neq \emptyset \, . \tag{17.10}$$

The proof of this is done by induction. If $|T| = 1$, say $T = \{i\}$, then $c_T = v(i) = \Delta_v(T)$. Suppose (17.10) holds for all $S \subsetneq T$. Then $\Delta_v(T) = v(T) - \sum_{S \subsetneq T} \Delta_v(S) = v(T) - \sum_{S \subsetneq T} c_S = c_T$ because $v(T) = \sum_{S \subseteq T} c_S$. ∎

17.2.2 Strong Monotonicity

The Shapley value obviously has the property that if a player contributes at least as much to any coalition in a game v than in a game w, then his payoff from the Shapley value in v is at least as large as that in w. Formally, the Shapley value satisfies the following axiom for a value $\psi : \mathcal{G}^N \to \mathbb{R}^N$; a proof is immediate from (17.4).

Strong Monotonicity (SMON): $\psi_i(v) \geq \psi_i(w)$ for all $v, w \in \mathcal{G}^N$ that satisfy

$$v(S \cup \{i\}) - v(S) \geq w(S \cup \{i\}) - w(S) \text{ for all } S \in 2^N.$$

We now show that together with efficiency and symmetry this axiom characterizes the Shapley value. For $T \subseteq N$ denote by $\mathbf{e}^T \in \mathbb{R}^N$ the vector with $\mathbf{e}_i^T = 1$ if $i \in T$ and $\mathbf{e}_i^T = 0$ if $i \notin T$.

Theorem 17.8 *Let* $\psi : \mathcal{G}^N \to \mathbb{R}^N$ *be a value. Then* ψ *satisfies EFF, SYM, and SMON, if and only if* ψ *is the Shapley value* Φ.

Proof Obviously, Φ satisfies the three axioms. Conversely, suppose ψ satisfies the three axioms.

(1) Let z be the game that is identically zero. In this game, all players are symmetric, so SYM and EFF together imply $\psi(z) = 0$.
(2) Let i be a null-player in a game v. Then the condition in SMON applies to z and v with all inequalities being equalities. So SMON yields $\psi_i(v) \geq \psi_i(z)$ and $\psi_i(z) \geq \psi_i(v)$. Hence by (1), $\psi_i(v) = \mathbf{0}$.
(3) Let $c \in \mathbb{R}$ and $T \in 2^N \setminus \{\emptyset\}$. Then (2) implies $\psi_i(cu_T) = 0$ for every $i \in N \setminus T$. This, SYM, and EFF imply $\psi_i(cu_T) = c|T|^{-1}$ for every $i \in T$. Hence, $\psi(cu_T) = c|T|^{-1}\mathbf{e}^T$.
(4) Each $v \in \mathcal{G}^N$ can be written in a unique way as a linear combination of $\{u_T \mid T \in 2^N \setminus \{\emptyset\}\}$ (see Problem 17.2). So v is of the form $\sum c_T u_T$. The proof of $\psi(v) = \Phi(v)$ will be completed by induction on the number $\alpha(v)$ of terms in $\sum c_T u_T$ with $c_T \neq 0$.

From (1), $\psi(v) = \Phi(v) = \mathbf{0}$ if $\alpha(v) = 0$, and from (3), $\psi(v) = \Phi(v)$ if $\alpha(v) = 1$ because $\Phi(cu_T) = c|T|^{-1}\mathbf{e}^T$. Suppose $\psi(w) = \Phi(w)$ for all $w \in \mathcal{G}^N$ with $\alpha(w) < k$, where $k \geq 2$. Let v be a game with $\alpha(v) = k$. Then there are coalitions $T_1, T_2, \ldots, T_k$ and real numbers $c_1, c_2, \ldots, c_k$, unequal to zero, such that $v = \sum_{r=1}^{k} c_r u_{T_r}$. Let $D := \cap_{r=1}^{k} T_r$.

For $i \in N \setminus D$, define $w^i := \sum_{r:i \in T_r} c_r u_{T_r}$. Because $\alpha(w^i) < k$, the induction hypothesis implies: $\psi_i(w^i) = \Phi_i(w^i)$. Further, for every $S \in 2^N$:

$$v(S \cup i) - v(S) = \sum_{r=1}^{k} c_r u_{T_r}(S \cup i) - \sum_{r=1}^{k} c_r u_{T_r}(S)$$

$$= \sum_{r:i \in T_r} c_r u_{T_r}(S \cup i) - \sum_{r:i \in T_r} c_r u_{T_r}(S)$$

$$= w^i(S \cup i) - w^i(S) ,$$

so that, by SMON of ψ and Φ, it follows that $\psi_i(v) = \psi_i(w^i) = \Phi_i(w^i) = \Phi_i(v)$. So

$$\psi_i(v) = \Phi_i(v) \text{ for all } i \in N \setminus D . \tag{17.11}$$

Equation (17.11) and EFF for ψ and Φ yield

$$\sum_{i \in D} \psi_i(v) = \sum_{i \in D} \Phi_i(v) . \tag{17.12}$$

Let $i, j \in D$, then for every $S \subseteq N \setminus \{i, j\}$:

$$(0 =) v(S \cup i) = \sum_{r=1}^{k} c_r u_{T_r}(S \cup i) = \sum_{r=1}^{k} c_r u_{T_r}(S \cup j) = v(S \cup j) ,$$

so i and j are symmetric. Hence, by SYM of ψ and Φ:

$$\psi_i(v) = \psi_j(v), \quad \Phi_i(v) = \Phi_j(v) . \tag{17.13}$$

Now $\psi(v) = \Phi(v)$ follows from (17.11)–(17.13). ∎

17.2.3 Multilinear Extension

The Shapley value of a game may also be described by means of the multilinear extension of a game. Let (N, v) be game. Consider the function $f : [0, 1]^N \to \mathbb{R}$ on the hypercube $[0, 1]^N$, defined by

$$f(x_1, x_2, \ldots, x_n) = \sum_{S \in 2^N} \left(\prod_{i \in S} x_i \prod_{i \in N \setminus S} (1 - x_i) \right) v(S) . \tag{17.14}$$

Observe that the set of extreme points[3] of $[0, 1]^N$, $\text{ext}([0, 1]^N)$, is equal to $\{e^S \mid S \in 2^N\}$. Clearly,

$$f(e^S) = v(S) \text{ for each } S \in 2^N .\tag{17.15}$$

So f can be seen as an extension of $\tilde{v}$: $\text{ext}([0, 1]^N) \to \mathbb{R}$ with $\tilde{v}(e^S) := v(S)$. In view of Problem 17.8, f is called the *multilinear extension* of ($\tilde{v}$ or) v.

One can give a probabilistic interpretation of $f(\mathbf{x})$. Suppose that each of the players $i \in N$, independently, decides whether to cooperate (probability x_i) or not (probability $1-x_i$). So with probability $\prod_{i \in S} x_i \prod_{i \in N \setminus S} (1-x_i)$ the coalition S forms, which has worth $v(S)$. Then $f(\mathbf{x})$ as given in (17.14) can be seen as the expectation of the worth of the formed coalition.

Another interpretation is to see $\mathbf{x} \in [0, 1]^N$ as a fuzzy set, where x_i is the intensity of availability of player i and to see f as a characteristic function, defined for fuzzy coalitions in N.

Denote by $D_k f(\mathbf{x})$ the derivative of f w.r.t. the k-th coordinate in $\mathbf{x}$. The following result provides another description of the Shapley value, namely as the integral along the main diagonal of $[0, 1]^N$ of $D_k f$.

Theorem 17.9 $\Phi_k(v) = \int_0^1 (D_k f)(t, t, \ldots, t)dt$ for each $k \in N$.

Proof

$$D_k f(\mathbf{x}) = \sum_{T: k \in T} \left[\prod_{i \in T \setminus \{k\}} x_i \prod_{i \in N \setminus T} (1-x_i) \right] v(T)$$

$$- \sum_{S: k \notin S} \left[\prod_{i \in S} x_i \prod_{i \in N \setminus (S \cup \{k\})} (1-x_i) \right] v(S)$$

$$= \sum_{S: k \notin S} \left[\prod_{i \in S} x_i \prod_{i \in N \setminus (S \cup \{k\})} (1-x_i) \right] (v(S \cup \{k\}) - v(S)) .$$

Hence, $\int_0^1 (D_k f)(t, t, \ldots, t)dt = \sum_{S: k \notin S} (\int_0^1 t^{|S|}(1-t)^{n-|S|-1}dt) (v(S \cup \{k\}) - v(S))$. Using the well-known (beta-)integral formula (cf. Problem 17.9)

$$\int_0^1 t^{|S|}(1-t)^{n-|S|-1}dt = \frac{|S|!(n-|S|-1)!}{n!}$$

<hr>

[3]Cf. Sect. 22.6.

it follows that $\int_0^1 (D_k f)(t, t, \ldots, t) dt = \sum_{S:k \notin S} \frac{|S|!(n-|S|-1)!}{n!} (v(S \cup \{k\}) - v(S)) = \Phi_k(v)$ by (17.4). ∎

Example 17.10 Let (N, v) be the three-person game with $v(1) = v(2) = v(3) = v(1,2) = 0$, $v(1,3) = 1$, $v(2,3) = 2$, $v(N) = v(1,2,3) = 4$. Then $f(x_1, x_2, x_3) = x_1(1 - x_2)x_3 + 2(1 - x_1) x_2 x_3 + 4x_1 x_2 x_3 = x_1 x_3 + 2x_2 x_3 + x_1 x_2 x_3$ for all $x \in [0, 1]^N$.

So $D_1 f(\mathbf{x}) = x_3 + x_2 x_3$, $D_2 f(\mathbf{x}) = 2x_3 + x_1 x_3$, $D_3 f(\mathbf{x}) = x_1 + 2x_2 + x_1 x_2$.
Theorem 17.9 implies:

$$\Phi_1(v) = \int_0^1 D_1 f(t, t, t) dt = \int_0^1 (t + t^2) dt = \frac{5}{6},$$

$$\Phi_2(v) = \int_0^1 (2t + t^2) dt = 1\frac{1}{3}, \quad \Phi_3(v) = \int_0^1 (3t + t^2) dt = 1\frac{5}{6}.$$

□

17.3 Potential and Reduced Game

This section starts with discussing the potential approach to the Shapley value. The potential is, in a sense, dual to the concept of dividends. Next, reduced games are considered, which leads to another axiomatic characterization of the Shapley value.

17.3.1 The Potential Approach to the Shapley Value

Denote by $\mathcal{G}$ the family of all games (N, v) with an arbitrary finite (player) set N (not necessarily the set of the first n natural numbers). It is convenient to include also the game $(\emptyset, v)$, with empty player set. Thus,

$$\mathcal{G} = \bigcup_{N \subseteq \mathbb{N},\ |N| < \infty} \mathcal{G}^N.$$

Definition 17.11 A *potential* is a function $P : \mathcal{G} \to \mathbb{R}$ satisfying

$$P(\emptyset, v) = 0 \tag{17.16}$$

$$\sum_{i \in N} D_i P(N, v) = v(N) \text{ for all } (N, v) \in \mathcal{G} . \tag{17.17}$$

Here $D_i P(N, v) := P(N, v) - P(N \setminus \{i\}, v)$, where v in the last expression is the restriction to $N \setminus \{i\}$. □

If P is a potential then (17.17) says that the *gradient* grad $P(N, v) := (D_i P(N, v))_{i \in N}$ is an efficient payoff vector for the game (N, v).

Note that

$$P(\{i\}, v) = v(\{i\})$$

$$P(\{i, j\}, v) = \frac{1}{2}(v(\{i, j\}) + v(\{i\}) + v(\{j\}))$$

if P is a potential. More generally, it follows from (17.17) that

$$P(N, v) = |N|^{-1}(v(N) + \sum_{i \in N} P(N \setminus \{i\}, v)) . \tag{17.18}$$

So the potential of (N, v) is uniquely determined by the potential of subgames of (N, v), which implies with (17.16) the first assertion in Theorem 17.12 below. The second assertion in this theorem connects the potential of a game to the dividends (see Sect. 17.2.1) of a game. Further, the gradient of P in (N, v) is equal to the Shapley value of (N, v). The theorem also provides an algorithm to calculate the Shapley value, by calculating with (17.16) and (17.18) the potentials of the game and its subgames and then using Theorem 17.12(c).

Theorem 17.12

(a) *There is a unique potential* $P : \mathcal{G} \to \mathbb{R}$.
(b) $P(N, v) = \sum_{\emptyset \neq T \subseteq N} c_T |T|^{-1}$ *for* $v = \sum_{\emptyset \neq T \subseteq N} c_T u_T$.
(c) grad $P(N, v) = \Phi(N, v)$.

Proof Part (a) follows immediately from (17.16) and (17.18). For (b) and (c), let $Q : \mathcal{G} \to \mathbb{R}$ be defined by $Q(\emptyset, v) := 0$ and $Q(N, v) := \sum_{T \in 2^N \setminus \{\emptyset\}} c_T |T|^{-1}$ for all (N, v), where $v = \sum c_T u_T$, if $N \neq \emptyset$. Further, for each (N, v) and $i \in N$

$$D_i Q(N, v) = \sum_{\emptyset \neq T \subseteq N} c_T |T|^{-1} - \sum_{\emptyset \neq T \subseteq N \setminus \{i\}} c_T' |T|^{-1} \tag{17.19}$$

if $v = \sum_{\emptyset \neq T \subseteq N} c_T u_T$ and $v' = \sum_{\emptyset \neq T \subseteq N \setminus \{i\}} c_T' u_T$ where v' is the restriction of v to $2^{N \setminus \{i\}}$. Since, for each $S \subseteq N \setminus \{i\}$, we have

$$\sum_{\emptyset \neq T \subseteq N \setminus \{i\}} c_T u_T(S) = v(S) = v'(S) = \sum_{\emptyset \neq T \subseteq N \setminus \{i\}} c_T' u_T(S)$$

we obtain by recursion that $c_T = c'_T$ for all $\emptyset \neq T \subseteq N \setminus \{i\}$. But then (17.19) and Theorem 17.7 imply

$$D_i Q(N, v) = \sum_{T: i \in T} c_T |T|^{-1} = \Phi_i(N, v) .$$

Further, by efficiency (EFF) of Φ:

$$\sum_{i \in N} D_i Q(N, v) = \sum_{i \in N} \Phi_i(N, v) = v(N) .$$

So Q is a potential. From (a) it follows that $Q = P$ and then (c) holds. ∎

The potential of a game can also be expressed directly in terms of the coalitional worths, as in the following theorem.

Theorem 17.13 *For each game $(N, v) \in \mathcal{G}$:*

$$P(N, v) = \sum_{S \subseteq N} \frac{(|S| - 1)!(|N| - |S|)!}{|N|!} v(S) .$$

Proof Let for each $(N, v) \in \mathcal{G}$,

$$Q(N, v) := \sum_{S \subseteq N} \frac{(|S| - 1)!(|N| - |S|)!}{|N|!} v(S) .$$

Then $Q(\emptyset, v) = 0$. It is sufficient for the proof of $Q(N, v) = P(N, v)$, to show that $D_i Q(N, v) - \Phi_i(N, v)$ for all $i \in N$. Now

$$D_i Q(N, v) = Q(N, v) - Q(N \setminus \{i\}, v)$$

$$= \sum_{T \subseteq N} \frac{(|T| - 1)!(|N| - |T|)!}{|N|!} v(T)$$

$$- \sum_{S \subseteq N \setminus \{i\}} \frac{(|S| - 1)!(|N| - 1 - |S|)!}{(|N| - 1)!} v(S)$$

$$= \sum_{S \subseteq N \setminus \{i\}} \frac{|S|!(|N| - |S| - 1)!}{|N|!} v(S \cup \{i\})$$

$$+ \sum_{S \subseteq N \setminus \{i\}} \frac{(|S| - 1)!(|N| - |S|)!}{|N|!} v(S)$$

$$- \sum_{S \subseteq N \setminus \{i\}} \frac{(|S| - 1)!(|N| - 1 - |S|)!}{(|N| - 1)!} v(S)$$

$$= \sum_{S \subseteq N \setminus \{i\}} \frac{|S|!(|N| - |S| - 1)!}{|N|!} (v(S \cup \{i\}) - v(S)) = \Phi_i(N, v)$$

where the last equality follows from (17.4).   ∎

A probabilistic interpretation of Theorem 17.13 is the following. The number $|N|^{-1}P(N, v)$ is the expectation of the normalized worth $|S|^{-1}v(S)$ of the formed coalition $S \subseteq N$ if the probability that S forms is $|N|^{-1}\binom{|N|}{|S|}^{-1}$ (corresponding to drawing first a size $s \in \{1, 2, \ldots, |N|\}$ and then a set S with $|S| = s$).

Theorem 17.12 can be used to calculate the Shapley value, as the following example shows.

Example 17.14 Consider the three-person game (N, v) given in Table 17.2. The dividends of the subcoalitions and the potential of the subgames are given in lines 3 and 4 of this table, respectively. It follows that

$$\Phi(N, v) = (D_1 P(N, v), D_2 P(N, v), D_3 P(N, v))$$

$$= (10\frac{1}{3} - 7, 10\frac{1}{3} - 5, 10\frac{1}{3} - 4) = (3\frac{1}{3}, 5\frac{1}{3}, 6\frac{1}{3})$$

$$\Phi(\{2, 3\}, v) = (7 - 3, 7 - 2) = (4, 5) .$$

□

17.3.2 Reduced Games

To introduce the concept of a reduced game, consider the game in Example 17.14. Suppose that the players agree on using the Shapley value, and consider the coalition $\{1, 2\}$. If players 1 and 2 pool their Shapley value payoffs then together they have $3\frac{1}{3} + 5\frac{1}{3} = 8\frac{2}{3}$. Another way to obtain this amount is to take the worth of the grand coalition, 15, and to subtract player 3's payoff, $6\frac{1}{3}$. Consider $\{1\}$ as a subcoalition of $\{1, 2\}$. Player 1 could form a coalition with player 3 and obtain the worth 6, but

Table 17.2 Example 17.14

S	∅	{1}	{2}	{3}	{1, 2}	{1, 3}	{2, 3}	{1, 2, 3}
$v(S)$	0	1	2	3	5	6	9	15
$\Delta_v(S)$	0	1	2	3	2	2	4	1
$\Delta_v(S)/|S|$	0	1	2	3	1	1	2	$\frac{1}{3}$
$P(S, v)$	0	1	2	3	4	5	7	$10\frac{1}{3}$

$$\Phi(N, v) = (3\frac{1}{3}, 5\frac{1}{3}, 6\frac{1}{3})$$

he would have to pay player 3 according to the Shapley value of the two-player game $(\{1, 3\}, v)$, which is the vector $(2, 4)$; recall that the players agree on using the Shapley value. So player 1 is left with $6 - 4 = 2$. Similarly, player 2 could form a coalition with player 3 and obtain $v(2, 3) = 9$ minus the Shapley value payoff for player 3 in the game $(\{2, 3\}, v)$, which is 5. So player 2 is left with $9 - 5 = 4$. So a 'reduced' game $(\{1, 2\}, \tilde{v})$ has been constructed with $\tilde{v}(1) = 2$, $\tilde{v}(2) = 4$, and $\tilde{v}(1, 2) = 8\frac{2}{3}$. The Shapley value of this game is the vector $(3\frac{1}{3}, 5\frac{1}{3})$. Observe that these payoffs are equal to the Shapley value payoffs in the original game. This is not a coincidence; the particular way of constructing a reduced game as illustrated here leaves the Shapley value invariant.

A general game theoretic principle is the following. Suppose in a game a subset of the players consider the game arising among themselves; then, if they apply the same 'solution rule' as in the original game, their payoffs should not change—they should have no reason to renegotiate. Of course, the formulation of this principle leaves open many ways to define 'the game arising among themselves'. Different definitions correspond to different solution rules. Put differently, there are many ways to define reduced games, leading to many different 'reduced game properties' as specific instances of the above general game theoretic principle. Instead of 'reduced game property' also the term 'consistency' is used. This concept has been very fruitful over the past decades—in cooperative as well as noncooperative game theory. (For a reduced game approach to Nash equilibrium see Sect. 13.8.)

For the Shapley value the following reduced game turns out to be relevant. It is the reduced game applied in the example above.

Definition 17.15 Let ψ be a value, assigning to each $(N, v) \in \mathcal{G}$ a vector $\psi(N, v) \in \mathbb{R}^N$. For $(N, v) \in \mathcal{G}$ and $U \subseteq N$, $U \neq \emptyset$, the game $(N \setminus U, v_{U,\psi})$ is defined by $v_{U,\psi}(\emptyset) = 0$ and

$$v_{U,\psi}(S) := v(S \cup U) - \sum_{k \in U} \psi_k(S \cup U, v) \text{ for all } S \in 2^{N \setminus U} \setminus \{\emptyset\} .$$

$v_{U,\psi}$ is called the (U, ψ)-*reduced game* of v.   □

Thus, the worth of coalition S in the game $v_{U,\psi}$ is obtained by subtracting from the worth of $S \cup U$ in the game $(S \cup U, v)$ the payoffs of the players in U according to the value ψ. The reduced game property or consistency property for a value ψ on $\mathcal{G}$ based on this reduced game is the following.

HM-consistency (HMC)[4]: for all games (N, v) and all $U \subseteq N$, $U \neq \emptyset$

$$\psi_i(N \setminus U, v_{U,\psi}) = \psi_i(N, v) \text{ for all } i \in N \setminus U . \tag{17.20}$$

The following lemma is used in the proof of HM-consistency of the Shapley value.

---

[4]Named after Hart and Mas-Colell.

Lemma 17.16 *Let* $(N, v) \in \mathcal{G}$. *Suppose* $Q : 2^N \to \mathbb{R}$ *satisfies*

$$\sum_{i \in S} (Q(S) - Q(S \setminus \{i\})) = v(S) \text{ for each } S \in 2^N \setminus \{\emptyset\} .$$

Then for each $S \in 2^N$:

$$Q(S) = P(S, v) + Q(\emptyset) . \tag{17.21}$$

Proof The proof of (17.21) is by induction on $|S|$. Obviously, (17.21) holds if $|S| = 0$. Take T with $|T| > 0$ and suppose (17.21) holds for all S with $|S| < |T|$. Then

$$Q(T) = |T|^{-1}(v(T) + \sum_{i \in T} Q(T \setminus \{i\}))$$

$$= |T|^{-1}(v(T) + |T|Q(\emptyset) + \sum_{i \in T} P(T \setminus \{i\}, v))$$

$$= Q(\emptyset) + |T|^{-1}(v(T) + \sum_{i \in T} P(T \setminus \{i\}, v))$$

$$= Q(\emptyset) + P(T, v) ,$$

where the last equality follows from (17.18).   ∎

Lemma 17.17 *The Shapley value* Φ *is HM-consistent.*

Proof Take (N, v) in $\mathcal{G}$, and $\emptyset \neq U \subseteq N$. We have to prove that

$$\Phi_i(N \setminus U, v_{U,\Phi}) = \Phi_i(N, v) \text{ for all } i \in N \setminus U . \tag{17.22}$$

Note that, in view of the definition of $v_{U,\Phi}$, efficiency of Φ, and Theorem 17.12(c), we have for all $S \subseteq N \setminus U$:

$$v_{U,\Phi}(S) = v(S \cup U) - \sum_{i \in U} \Phi_i(S \cup U, v) = \sum_{i \in S} \Phi_i(S \cup U, v)$$

$$= \sum_{i \in S} P(S \cup U, v) - P((S \cup U) \setminus \{i\}, v) . \tag{17.23}$$

For $S \in 2^{N \setminus U}$ define $Q(S) := P(S \cup U, v)$. Then, by Lemma 17.16, with $N \setminus U$ and $v_{U,\Phi}$ in the roles of N and v, (17.23) implies:

$$Q(S) = P(S, v_{U,\Phi}) + Q(\emptyset) = P(S, v_{U,\Phi}) + P(U, v)$$

for all $S \subseteq N \setminus U$. Hence, by the definition of Q:

$$P(S \cup U, v) = P(S, v_{U,\Phi}) + P(U, v). \tag{17.24}$$

Equation (17.24) and Theorem 17.12(c) imply

$$\Phi_i(N \setminus U, v_{U,\Phi}) = P(N \setminus U, v_{U,\Phi}) - P((N \setminus U) \setminus \{i\}, v_{U,\Phi})$$
$$= P(N, v) - P(N \setminus \{i\}, v) = \Phi_i(N, v) .$$

So (17.22) holds.   ∎

Call a value ψ *standard for two-person games* if for all $(\{i,j\}, v) \in \mathcal{G}$

$$\psi_i(\{i,j\}, v) = \frac{1}{2}(v(i,j) + v(i) - v(j)),$$

$$\psi_j(\{i,j\}, v) = \frac{1}{2}(v(i,j) - v(i) + v(j)) .$$

Note that the Shapley value is standard for 2-person games. It turns out that this property, together with HM-consistency, characterizes the Shapley value.

Theorem 17.18 *Let ψ be a value on $\mathcal{G}$. Then ψ is standard for two-person games and HM-consistent if, and only if, ψ is the Shapley value.*

Proof

(a) By Lemma 17.17, Φ is HM-consistent. Further, Φ is standard for two-person games.
(b) For the proof of the converse, assume ψ has the two properties in the theorem. It will first be proved that ψ is efficient. For two-person games this is true by standardness. Let $(\{i\}, v)$ be a one-person game. To prove that $\psi_i(\{i\}, v) = v(i)$, construct a two-person game $(\{i,j\}, v^*)$ (with $j \neq i$) by defining

$$v^*(i) := v(i), \; v^*(j) := 0, \; v^*(i,j) := v(i) .$$

From the standardness of ψ it follows that

$$\psi_i((i,j), v^*) = v(i), \; \psi_j((i,j), v^*) = 0 . \tag{17.25}$$

The definition of $v^*_{j,\psi}$ and (17.25) imply: $v^*_{j,\psi}(i) = v^*(i,j) - \psi_j((i,j), v^*) = v(i) - 0 = v(i)$. So

$$(\{i\}, v^*_{j,\psi}) = (\{i\}, v) . \tag{17.26}$$

Then, by (17.26), the consistency of ψ and (17.25):

$$\psi_i(\{i\}, v) = \psi_i(\{i\}, v_{j,\psi}^*) = \psi_i(\{i,j\}, v^*) = v(i) .$$

So for one- and two-person games the rule ψ is efficient. Take a game (S, v) with $|S| \geq 3$ and suppose that for all games (T, v) with $|T| < |S|$ the rule ψ is efficient. Take $k \in S$. Then the consistency of ψ implies:

$$\sum_{i \in S} \psi_i(S, v) = \psi_k(S, v) + \sum_{i \in S \setminus \{k\}} \psi_i(S, v)$$

$$= \psi_k(S, v) + \sum_{i \in S \setminus \{k\}} \psi_i(S \setminus \{k\}, v_{k,\psi})$$

$$= \psi_k(S, v) + v_{k,\psi}(S \setminus \{k\})$$

$$= v(S) .$$

Hence, ψ is an efficient rule.
(c) If a $Q : \mathcal{G} \to \mathbb{R}$ can be constructed with $Q(\emptyset, v) = 0$ and

$$\psi_i(N, v) = Q(N, v) - Q(N \setminus \{i\}, v) \tag{17.27}$$

for all (N, v) in $\mathcal{G}$ and $i \in N$, then by (b) and Theorem 17.12(a), $Q = P$ and then $\psi_i(N, v) = \Phi_i(N, v)$ by Theorem 17.12(c).

Hence, the proof is complete if we can construct such a Q. To achieve this, start with $Q(\emptyset, v) := 0$, $Q(\{i\}, v) := v(i)$, $Q(\{i,j\}, v) := \frac{1}{2}(v(i,j) + v(i) + v(j))$ and continue in a recursive way as follows. Let $(N, v) \in \mathcal{G}$, $|N| \geq 3$ and suppose Q with property (17.27) has been defined already for games with less than $|N|$ players. Then one can define $Q(N, v) := \alpha$ if, and only if

$$\alpha - Q(N \setminus \{i\}, v) = \psi_i(N, v) \text{ for all } i \in N .$$

This implies that the proof is complete if it can be shown that

$$\psi_i(N, v) + Q(N \setminus \{i\}, v) = \psi_j(N, v) + Q(N \setminus \{j\}, v) \text{ for all } i, j \in N . \tag{17.28}$$

To prove (17.28) take $k \in N \setminus \{i,j\}$ ($|N| \geq 3$). Then

$$\psi_i(N, v) - \psi_j(N, v) = \psi_i(N \setminus \{k\}, v_{k,\psi}) - \psi_j(N \setminus \{k\}, v_{k,\psi})$$

$$= Q(N \setminus \{k\}, v_{k,\psi}) - Q(N \setminus \{k, i\}, v_{k,\psi})$$

$$- Q(N \setminus \{k\}, v_{k,\psi}) + Q(N \setminus \{k,j\}, v_{k,\psi})$$

$$= (-Q(N \setminus \{k, i\}, v_{k,\psi}) + Q(N \setminus \{i,j,k\}, v_{k,\psi}))$$

$$+ Q(N \setminus \{k,j\}, v_{k,\psi}) - Q(N \setminus \{i,j,k\}, v_{k,\psi})$$
$$= - \psi_j(N \setminus \{k,i\}, v_{k,\psi}) + \psi_i(N \setminus \{k,j\}, v_{k,\psi})$$
$$= - \psi_j(N \setminus \{i\}, v) + \psi_i(N \setminus \{j\}, v)$$
$$= - Q(N \setminus \{i\}, v) + Q(N \setminus \{i,j\}, v)$$
$$+ Q(N \setminus \{j\}, v) - Q(N \setminus \{i,j\}, v)$$
$$= Q(N \setminus \{j\}, v) - Q(N \setminus \{i\}, v) \, ,$$

where HM-consistency of ψ is used in the first and fifth equality and (17.27) for subgames in the second, fourth and sixth equality.

This proves (17.28).   ∎

17.4   Problems

17.1. *The Games* 1_T

(a) Show that $\{1_T \in \mathcal{G}^N \mid T \in 2^N \setminus \{\emptyset\}\}$ is a basis for $\mathcal{G}^N$.
(b) Show that there is no null-player in a game $(N, 1_T)$ $(T \neq \emptyset)$.
(c) Determine the Shapley value $\Phi(N, 1_T)$.

17.2. *Unanimity Games*

(a) Prove that the collection of unanimity games $\{u_T \mid T \in 2^N \setminus \{\emptyset\}\}$ is a basis for $\mathcal{G}^N$. [Hint: In view of Problem 17.1(a) it is sufficient to show linear independence.]
(b) Prove that for each game $v \in \mathcal{G}^N$:

$$v = \sum_{T \in 2^N \setminus \{\emptyset\}} c_T u_T \text{ with } c_T = \sum_{S: S \subseteq T} (-1)^{|T| - |S|} v(S).$$

17.3. *If-Part of Theorem 17.4*
Show that the Shapley value satisfies EFF, NP, SYM, and ADD.

17.4. *Dummy Player Property and Anonymity*
Show that DUM implies NP, and that AN implies SYM, but that the converses of these implications do not hold. Show that the Shapley value has the dummy player property.

17.5. *Shapley Value, Core, and Imputation Set*
Show that the Shapley value of a game does not have to be an element of the core or of the imputation set, even if these sets are non-empty. How about the case of two players?

17.6. *Shapley Value as a Projection*
The Shapley value $\Phi : \mathcal{G}^N \to \mathbb{R}^N$ can be seen as a map from $\mathcal{G}^N$ to the space A^N of additive games by identifying $\mathbb{R}^N$ with A^N. Prove that $\Phi : \mathcal{G}^N \to A^N$ is a projection i.e. $\Phi \circ \Phi = \Phi$.

17.7. *Shapley Value of Dual Game*
The dual game (N, v^*) of a game (N, v) is defined by $v^*(S) = v(N) - v(N \setminus S)$ for every $S \subseteq N$. Prove that the Shapley value of v^* is equal to the Shapley value of v. [Hint: If $v = \sum \alpha_T u_T$, then $v^* = \sum \alpha_T u_T^*$.]

17.8. *Multilinear Extension*

(a) Show that f in (17.14) is a multilinear function. [A function $g : \mathbb{R}^n \to \mathbb{R}$ is called multilinear if g is of the form $g(\mathbf{x}) = \sum_{T \subseteq N} c_T \left(\prod_{i \in T} x_i \right)$ for arbitrary real numbers c_T.]
(b) Prove that f is the unique multilinear extension of $\tilde{v}$ to $[0, 1]^N$.

17.9. *The Beta-Integral Formula*
Prove the beta-integral formula used in the proof of Theorem 17.9.

17.10. *Path Independence of Φ*
Let $(N, v) \in \mathcal{G}^N$ with $N = \{1, 2, \ldots, n\}$. Prove that for each permutation $\tau : N \to N$ we have

$$\sum_{k=1}^{n} \Phi_{\tau(k)}(\{\tau(1), \tau(2), \ldots, \tau(k)\}, v) = \sum_{k=1}^{n} \Phi_k(\{1, 2, \ldots, k\}, v) .$$

17.11. *An Alternative Characterization of the Shapley Value*
Let ψ be a value on $\mathcal{G}$. Prove: $\psi = \Phi$ if and only if ψ has the following four properties.

(a) ψ is HM-consistent.
(b) ψ is efficient for two-person games.
(c) ψ is anonymous for two-person games.
(d) ψ is relative invariant w.r.t. strategic equivalence for two-person games. [This means: $(\psi_i(\tilde{v}), \psi_j(\tilde{v})) = \alpha(\psi_i(v), \psi_j(v)) + (\beta_i, \beta_j)$ whenever $\tilde{v}(S) = \alpha v(S) + \sum_{i \in S} \beta_i$ for every coalition S, where $\alpha > 0$ and $\beta_i, \beta_j \in \mathbb{R}$.]

17.5 Notes

The Shapley value was introduced in Shapley (1953). This paper is the starting point of a large literature on this solution concept and related concepts. For overviews see Roth (1988) and Chaps. 53 and 54 in Aumann and Hart (2002).

The coalitional dividend approach is due to Harsanyi (1959). The axiomatization based on strong monotonicity (Theorem 17.8) is from Young (1985), and the multilinear extension approach is from Owen (1972), see also Owen (1995).

Section 17.3 is based on Hart and Mas-Colell (1989).

References

Aumann, R. J., & Hart, S. (Eds.), (2002). *Handbook of game theory with economic aplications* (Vol. 3). Amsterdam: North-Holland.

Harsanyi, J. C. (1959). A bargaining model for the cooperative *n*-person game. In *Contributions to the theory of games* (Vol. 4, pp. 324–356). Princeton: Princeton University Press.

Hart, S., & Mas-Colell, A. (1989). Potential, value, and consistency. *Econometrica, 57,* 589–614.

Owen, G. (1972). Multilinear extensions of games. *Management Science, 18,* 64–79.

Owen, G. (1995). *Game theory* (3rd ed.). San Diego: Academic.

Roth, A. E. (Ed.), (1988). *The Shapley value, essays in honor of Lloyd S. Shapley.* Cambridge: Cambridge University Press.

Shapley, L. S. (1953). A value for *n*-person games. In A. W. Tucker & Kuhn, H. W. (Eds.), *Contributions to the theory of games* (Vol. II, pp. 307–317). Princeton: Princeton University Press.

Young, H. P. (1985). Monotonic solutions of cooperative games. *International Journal of Game Theory, 14,* 65–72.

Core, Shapley Value, and Weber Set

18

In Chap. 17 we have seen that the Shapley value of a game does not have to be in the core of the game, nor even an imputation (Problem 17.5). In this chapter we introduce a set-valued extension of the Shapley value, the Weber set, and show that it always contains the core (Sect. 18.1). Next, we study so-called convex games and show that these are exactly those games for which the core and the Weber set coincide. Hence, for such games the Shapley value is an attractive core selection (Sect. 18.2). Finally, we study random order values (Sect. 18.3), which fill out the Weber set, and the subset of weighted Shapley values, which still cover the core (Sect. 18.4).

18.1 The Weber Set

Let $(N, v) \in \mathcal{G}^N$. Recall the definition of a marginal vector from Sect. 17.1.

Definition 18.1 The *Weber set* of a game $(N, v) \in \mathcal{G}^N$ is the convex hull of its marginal vectors:

$$W(v) := \text{conv}\{m^{\sigma}(v) \mid \sigma \in \Pi(N)\} .$$

$\square$

Example 18.2 Consider the three-person game $(\{1, 2, 3\}, v)$ defined by $v(12) = v(13) = 1$, $v(23) = -1$, $v(123) = 3$, and $v(i) = 0$ for every $i \in \{1, 2, 3\}$. The marginal vectors of this game, the core and the Weber set are given in Fig. 18.1. $\square$

We show now that the core is always a subset of the Weber set.

Theorem 18.3 *Let $(N, v) \in \mathcal{G}^N$. Then $C(v) \subseteq W(v)$.*

© Springer-Verlag Berlin Heidelberg 2015
H. Peters, *Game Theory*, Springer Texts in Business and Economics,
DOI 10.1007/978-3-662-46950-7_18

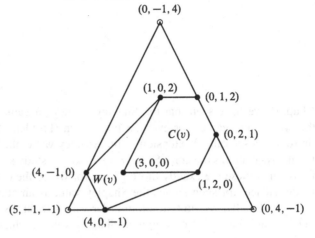

$(\sigma(1),\sigma(2),\sigma(3))$	m_1^σ	m_2^σ	m_3^σ
$(1,2,3)$	0	1	2
$(1,3,2)$	0	2	1
$(2,1,3)$	1	0	2
$(2,3,1)$	4	0	−1
$(3,1,2)$	1	2	0
$(3,2,1)$	4	−1	0

Fig. 18.1 Example 18.2. The core is the convex hull of the vectors $(3,0,0)$, $(1,2,0)$, $(0,2,1)$, $(0,1,2)$, and $(1,0,2)$. The Weber set is the convex hull of the six marginal vectors

Proof Suppose there is an $\mathbf{x} \in C(v) \setminus W(v)$. By a separation theorem (Theorem 22.1), there exists a vector $\mathbf{y} \in \mathbb{R}^N$ such that $\mathbf{w} \cdot \mathbf{y} > \mathbf{x} \cdot \mathbf{y}$ for every $\mathbf{w} \in W(v)$. In particular,

$$m^\sigma \cdot \mathbf{y} > \mathbf{x} \cdot \mathbf{y} \quad \text{for every } \sigma \in \Pi(N) . \tag{18.1}$$

Let $\pi \in \Pi(N)$ with $y_{\pi(1)} \geq y_{\pi(2)} \geq \ldots \geq y_{\pi(n)}$. Since $\mathbf{x} \in C(v)$,

$$m^\pi \cdot \mathbf{y} = \sum_{i=1}^{n} y_{\pi(i)}(v(\pi(1),\pi(2),\ldots,\pi(i)) - v(\pi(1),\pi(2),\ldots,\pi(i-1)))$$

$$= y_{\pi(n)}v(N) - y_{\pi(1)}v(\emptyset) + \sum_{i=1}^{n-1}(y_{\pi(i)} - y_{\pi(i+1)})v(\pi(1),\pi(2)\ldots,\pi(i))$$

$$\leq y_{\pi(n)}\sum_{j=1}^{n} x_{\pi(j)} + \sum_{i=1}^{n-1}(y_{\pi(i)} - y_{\pi(i+1)})\sum_{j=1}^{i} x_{\pi(j)}$$

$$= \sum_{i=1}^{n} y_{\pi(i)} \sum_{j=1}^{i} x_{\pi(j)} - \sum_{i=2}^{n} y_{\pi(i)} \sum_{j=1}^{i-1} x_{\pi(j)}$$

$$= \sum_{i=1}^{n} y_{\pi(i)} x_{\pi(i)} = \mathbf{x} \cdot \mathbf{y}$$

contradicting (18.1). ∎

18.2 Convex Games

For the coincidence of core and Weber set the following possible property of a game plays a crucial role.

Definition 18.4 A TU-game (N, v) is *convex* if the following condition holds for all $S, T \subseteq N$:

$$v(S) + v(T) \leq v(S \cup T) + v(S \cap T) . \tag{18.2}$$

$\square$

Observe that convexity of a game implies super-additivity [cf. (16.2)]: v is super-additive if (18.2) holds whenever S and T have empty intersection. The intuition is similar: larger coalitions have a relatively larger worth. This intuition is also apparent in the following condition:

For all $i \in N$ and $S \subseteq T \subseteq N \setminus \{i\}$: $v(S \cup i) - v(S) \leq v(T \cup i) - v(T) .$ (18.3)

Lemma 18.5 *A game* $(N, v) \in \mathcal{G}^N$ *is convex if and only if it satisfies* (18.3).

Proof Let $v \in \mathcal{G}^N$. Obviously, (18.3) follows from (18.2) by taking, for S and T in (18.2), $S \cup i$ and T from (18.3), respectively.

In order to derive (18.2) from (18.3), first take $S_0 \subseteq T_0 \subseteq N$ and $R \subseteq N \setminus T_0$, say $R = \{i_1, \ldots, i_k\}$. By repeated application of (18.3) one obtains

$$v(S_0 \cup i_1) - v(S_0) \leq v(T_0 \cup i_1) - v(T_0)$$

$$v(S_0 \cup i_1 i_2) - v(S_0 \cup i_1) \leq v(T_0 \cup i_1 i_2) - v(T_0 \cup i_1)$$

$$\vdots$$

$$v(S_0 \cup i_1 \cdots i_k) - v(S_0 \cup i_1 \cdots i_{k-1}) \leq v(T_0 \cup i_1 \cdots i_k) - v(T_0 \cup i_1 \cdots i_{k-1}) .$$

Adding these inequalities yields

$$v(S_0 \cup R) - v(S_0) \le v(T_0 \cup R) - v(T_0)$$

for all $R \subseteq N \setminus T_0$. Applying this inequality to arbitrary S, T by setting $S_0 = S \cap T$, $T_0 = T$, and $R = S \setminus T$, yields (18.2).   ∎

The importance of convexity of a game for the relation between the core and the Weber set follows from the following theorem.

Theorem 18.6 *Let $v \in \mathcal{G}^N$. Then v is convex if and only if $C(v) = W(v)$.*

Proof

(a) Suppose v is convex. For the 'only if' part it is, in view of Theorem 18.3 and convexity of the core, sufficient to prove that each marginal vector $m^\pi(v)$ of v is in the core. In order to show this, assume for notational simplicity that π is identity. Let $S \subseteq N$ be an arbitrary coalition, say $S = \{i_1, \dots, i_s\}$ with $i_1 \le \dots \le i_s$. Then, for $1 \le k \le s$, by (18.3):

$$v(i_1, \dots, i_k) - v(i_1, \dots, i_{k-1}) \le v(1, 2, \dots, i_k) - v(1, 2, \dots, i_k - 1) = m_{i_k}^\pi(v) .$$

Summing these inequalities from $k = 1$ to $k = s$ yields

$$v(S) = v(i_1, \dots, i_s) \le \sum_{k=1}^{s} m_{i_k}^\pi(v) = \sum_{i \in S} m_i^\pi(v) ,$$

which shows $m^\pi(v) \in C(v)$.

(b) For the converse, suppose that all marginal vectors of v are in the core. Let $S, T \subseteq N$ be arbitrary. Order the players of N as follows:

$$N = \{\underbrace{i_1, \dots, i_k}_{S \cap T}, \underbrace{i_{k+1}, \dots, i_\ell}_{T \setminus S}, \underbrace{i_{\ell+1}, \dots, i_s}_{S \setminus T}, \underbrace{i_{s+1}, \dots, i_n}_{N \setminus (S \cup T)}\} .$$

This defines an ordering or permutation π, namely by $\pi(j) = i_j$ for all $j \in N$, with corresponding marginal vector $m(v) = m^\pi(v)$. Since $m(v) \in C(v)$,

$$v(S) \le \sum_{i \in S} m_i(v)$$

$$= \sum_{j=1}^{k} m_{i_j}(v) + \sum_{j=\ell+1}^{s} m_{i_j}(v)$$

$$= v(i_1, \dots, i_k) +$$

$$[v(i_1, \ldots, i_{\ell+1}) - v(i_1, \ldots, i_\ell)] +$$
$$[v(i_1, \ldots, i_{\ell+2}) - v(i_1, \ldots, i_{\ell+1}] + \ldots$$
$$[v(i_1, \ldots, i_s) - v(i_1, \ldots, i_{s-1})]$$
$$= v(S \cap T) + v(S \cup T) - v(T) ,$$

which implies (18.2). So v is a convex game.   ∎

An immediate consequence of Theorem 18.6 and the definition of the Shapley value (Definition 17.1) is the following corollary.

Corollary 18.7 *Let $v \in \mathcal{G}^N$ be convex. Then $\Phi(v) \in C(v)$, i.e., the Shapley value is in the core.*

18.3   Random Order Values

A value $\psi : \mathcal{G}^N \to \mathbb{R}^N$ is called a *random order value* if there is a probability distribution p over the set of permutations $\Pi(N)$ of N such that

$$\psi(N, v) = \sum_{\pi \in \Pi(N)} p(\pi) m^\pi(N, v)$$

for every $(N, v) \in \mathcal{G}^N$. In that case, we denote ψ by Φ^p. Observe that Φ^p is the Shapley value Φ if $p(\pi) = 1/n!$ for every $\pi \in \Pi(N)$. Obviously,

$$W(v) = \{\mathbf{x} \in \mathbb{R}^N \mid \mathbf{x} = \Phi^p(v) \text{ for some } p\}.$$

Random order values satisfy the following two conditions.

Monotonicity (MON): $\psi(v) \geq \mathbf{0}$ for all monotonic games $v \in \mathcal{G}^N$. [The game v is *monotonic* if $S \subseteq T$ implies $v(S) \leq v(T)$ for all $S, T \subseteq N$.]

Linearity (LIN): $\psi(\alpha v + \beta w) = \alpha \psi(v) + \beta \psi(w)$ for all $v, w \in \mathcal{G}^N$, and $\alpha, \beta \in \mathbb{R}$ [where, for each S, $(\alpha v + \beta w)(S) = \alpha v(S) + \beta w(S)$].

Monotonicity says that in a monotonic game, where larger coalitions have higher worths, i.e., all marginal contributions are nonnegative, every player should receive a nonnegative payoff. Linearity is a strengthening of additivity. The main result in this section is the following characterization of random order values. (See Problem 18.7 for a strengthening of the 'only if' part of this theorem.)

Theorem 18.8 *Let $\psi : \mathcal{G}^N \to \mathbb{R}^N$ be a value. Then ψ satisfies LIN, DUM, EFF, and MON if and only if it is a random order value.*

The proof of Theorem 18.8 is based on a sequence of propositions and lemmas, which are of independent interest.

Proposition 18.9 *Let $\psi : \mathcal{G}^N \to \mathbb{R}^N$ be a linear value. Then there is a collection of constants $\{a_T^i \in \mathbb{R} \mid i \in N,\ \emptyset \neq T \subseteq N\}$ such that $\psi_i(v) = \sum_{\emptyset \neq T \subseteq N} a_T^i v(T)$ for every $v \in \mathcal{G}^N$ and $i \in N$.*

Proof Let $a_T^i := \psi_i(1_T)$ for all $i \in N$ and $\emptyset \neq T \subseteq N$ (cf. Problem 17.1). For every $v \in \mathcal{G}^N$ we have $v = \sum_{T \neq \emptyset} v(T) 1_T$. The desired conclusion follows from linearity of ψ. ∎

Proposition 18.10 *Let $\psi : \mathcal{G}^N \to \mathbb{R}^N$ be a linear value satisfying DUM. Then there is a collection of constants $\{p_T^i \in \mathbb{R} \mid i \in N,\ T \subseteq N \setminus i\}$ with $\sum_{T \subseteq N \setminus i} p_T^i = 1$ for all $i \in N$, such that for every $v \in \mathcal{G}^N$ and every $i \in N$:*

$$\psi_i(v) = \sum_{T \subseteq N \setminus i} p_T^i [v(T \cup i) - v(T)] .$$

Proof Let $v \in \mathcal{G}^N$ and $i \in N$. By Proposition 18.9 there are a_T^i such that $\psi_i(v) = \sum_{\emptyset \neq T \subseteq N} a_T^i v(T)$. Then $0 = \psi_i(u_{N \setminus i}) = a_N^i + a_{N \setminus i}^i$, where the first equality follows from DUM. Assume now as induction hypothesis that $a_{T \cup i}^i + a_T^i = 0$ for all $T \subseteq N \setminus i$ with $|T| \geq k \geq 2$ (we have just established this for $k = n - 1$), and let $S \subseteq N \setminus i$ with $|S| = k - 1$. Then

$$0 = \psi_i(u_S)$$

$$= \sum_{T : S \subseteq T} a_T^i$$

$$= \sum_{T : S \subsetneq T \subseteq N \setminus i} (a_{T \cup i}^i + a_T^i) + a_{S \cup i}^i + a_S^i$$

$$= a_{S \cup i}^i + a_S^i ,$$

where the last equality follows by induction and the first one by DUM. Hence, we have proved that $a_{T \cup i}^i + a_T^i = 0$ for all $T \subseteq N \setminus i$ with $0 < |T| \leq n - 1$. Now define, for all $i \in N$ and all such T, $p_T^i := a_{T \cup i}^i = -a_T^i$, and define $p_\emptyset^i := a_{\{i\}}^i$. Then for every $v \in \mathcal{G}^N$ and $i \in N$:

$$\psi_i(v) = \sum_{\emptyset \neq T \subseteq N} a_T^i v(T)$$

$$= \sum_{T :\, i \in T} a_T^i v(T) + \sum_{\emptyset \neq T \subseteq N \setminus i} a_T^i v(T)$$

$$= \sum_{T:\, i \notin T} a_{T \cup i}^i v(T \cup i) - \sum_{T \subseteq N \setminus i} (-a_T^i) v(T)$$

$$= \sum_{T \subseteq N \setminus i} p_T^i [v(T \cup i) - v(T)] .$$

Finally, by DUM,

$$1 = u_{\{i\}}(i) = \psi_i(u_{\{i\}}) = \sum_{T \subseteq N \setminus i} p_T^i ,$$

which completes the proof of the proposition.   ∎

Proposition 18.11 *Let* $\psi : \mathcal{G}^N \to \mathbb{R}^N$ *be a linear value satisfying DUM and MON. Then there is a collection of constants* $\{p_T^i \in \mathbb{R} \mid i \in N, \ T \subseteq N \setminus i\}$ *with* $\sum_{T \subseteq N \setminus i} p_T^i = 1$ *and* $p_S^i \geq 0$ *for all* $S \subseteq N \setminus i$ *and* $i \in N$, *such that for every* $v \in \mathcal{G}^N$ *and every* $i \in N$:

$$\psi_i(v) = \sum_{T \subseteq N \setminus i} p_T^i [v(T \cup i) - v(T)] .$$

Proof In view of Proposition 18.10 we only have to prove that the weights p_T^i are nonnegative. Let $i \in N$ and $T \subseteq N \setminus i$ and consider the game $\hat{u}_T$ assigning worth 1 to all strict supersets of T and 0 otherwise. Then $\psi_i(\hat{u}_T) = \sum_{S \subseteq N \setminus i} p_S^i [\hat{u}_T(S \cup i) - \hat{u}_T(S)]$ by Proposition 18.10, hence $\psi_i(\hat{u}_T) = p_T^i$. Since $\hat{u}_T$ is a monotonic game, this implies $p_T^i \geq 0$.   ∎

Lemma 18.12 *Let* $\psi : \mathcal{G}^N \to \mathbb{R}^N$ *be a value and* $\{p_T^i \in \mathbb{R} \mid i \in N, \ T \subseteq N \setminus i\}$ *be a collection of constants such that for every* $v \in \mathcal{G}^N$ *and every* $i \in N$:

$$\psi_i(v) = \sum_{T \subseteq N \setminus i} p_T^i [v(T \cup i) - v(T)] .$$

Then ψ *is efficient if and only if* $\sum_{i \in N} p_{N \setminus i}^i = 1$ *and* $\sum_{i \in T} p_{T \setminus i}^i = \sum_{j \in N \setminus T} p_T^j$ *for all* $\emptyset \neq T \neq N$.

Proof Let $v \in \mathcal{G}^N$. Then

$$\psi(v)(N) = \sum_{i \in N} \sum_{T \subseteq N \setminus i} p_T^i [v(T \cup i) - v(T)]$$

$$= \sum_{i \in N} \left(\sum_{T \subseteq N:\, i \in T} p_{T \setminus i}^i v(T) - \sum_{T \subseteq N:\, i \notin T} p_T^i v(T) \right)$$

$$= \sum_{T \subseteq N} v(T) \left(\sum_{i \in T} p^i_{T \setminus i} - \sum_{i \notin T} p^i_T \right)$$

$$= \sum_{T \subseteq N} v(T) \left(\sum_{i \in T} p^i_{T \setminus i} - \sum_{j \in N \setminus T} p^j_T \right) .$$

Clearly, this implies efficiency of ψ if the relations in the lemma hold. Conversely, suppose that ψ is efficient. Let $\emptyset \neq T \subseteq N$ and consider the games u_T and $\hat{u}_T$, where $\hat{u}$ was defined in the proof of Proposition 18.11. Then the preceding equation implies that

$$\psi(u_T)(N) - \psi(\hat{u}_T)(N) = \sum_{S \subseteq N} (u_T(S) - \hat{u}_T(S)) \left(\sum_{i \in S} p^i_{S \setminus i} - \sum_{j \in N \setminus S} p^j_S \right)$$

$$= \sum_{i \in T} p^i_{T \setminus i} - \sum_{j \in N \setminus T} p^j_T .$$

The relations in the lemma now follow by efficiency of ψ, since $u_T(N) - \hat{u}_T(N)$ is equal to 1 if $T = N$ and equal to 0 otherwise.   ■

We are now sufficiently equipped to prove Theorem 18.8.

Proof of Theorem 18.8 We leave it to the reader to verify that random order values satisfy the four axioms in the theorem. Conversely, let ψ satisfy these four axioms. By Proposition 18.11 there is a collection of constants $\{p^i_T \in \mathbb{R} \mid i \in N, \ T \subseteq N \setminus i\}$ with $\sum_{T \subseteq N \setminus i} p^i_T = 1$ and $p^i_S \geq 0$ for all $S \subseteq N \setminus i$ and $i \in N$, such that for every $v \in \mathcal{G}^N$ and every $i \in N$:

$$\psi_i(v) = \sum_{T \subseteq N \setminus i} p^i_T [v(T \cup i) - v(T)] .$$

For all $i \in N$ and $T \subseteq N \setminus i$ define $A(T) := \sum_{j \in N \setminus T} p^j_T$ and $A(i; T) := p^i_T / A(T)$ if $A(T) \neq 0$, $A(i; T) := 0$ if $A(T) = 0$. For any permutation $\pi \in \Pi(N)$ write $\pi = (i_1, \ldots, i_n)$ (that is, $\pi(k) = i_k$ for all $k \in N$). Define

$$p(\pi) = p^{i_1}_\emptyset A(i_2; \{i_1\}) A(i_3; \{i_1, i_2\}) \cdots A(i_n; \{i_1, \ldots, i_{n-1}\}) .$$

Then we claim that

$$\sum_{\pi \in \Pi(N)} p(\pi) = \sum_{i \in N} p^i_\emptyset . \tag{18.4}$$

In order to prove (18.4), note that when considering the sum $\sum_{\pi \in \Pi(N)} p(\pi)$ we may leave out any π with $p(\pi) = 0$. This means in particular that in what follows all expressions $A(\cdot\,;\cdot)$ are positive. Now

$$
\begin{aligned}
\sum_{\pi \in \Pi(N)} p(\pi) &= \sum_{i_1 \in N} \sum_{i_2 \in N\setminus i_1} \cdots \sum_{i_{n-1} \in N\setminus i_1 \cdots i_{n-2}} p_{\emptyset}^{i_1} A(i_2; \{i_1\}) \cdot A(i_3; \{i_1, i_2\}) \cdot \\
&\qquad \cdots A(i_{n-1}; \{i_1, \ldots, i_{n-2}\}) \cdot A(i_n; \{i_1, \ldots, i_{n-1}\}) \\
&= \sum_{i_1 \in N} \sum_{i_2 \in N\setminus i_1} \cdots \sum_{i_{n-2} \in N\setminus i_1 \cdots i_{n-3}} p_{\emptyset}^{i_1} A(i_2; \{i_1\}) \cdot A(i_3; \{i_1, i_2\}) \cdot \\
&\qquad \cdots A(i_{n-2}; \{i_1, \ldots, i_{n-3}\}) \\
&\quad \cdot \left(\frac{p_{N\setminus i_1 \cdots i_{n-2}}^{\ell}}{p_{N\setminus i_1 \cdots i_{n-2}}^{\ell} + p_{N\setminus i_1 \cdots i_{n-2}}^{k}} + \frac{p_{N\setminus i_1 \cdots i_{n-2}}^{k}}{p_{N\setminus i_1 \cdots i_{n-2}}^{\ell} + p_{N\setminus i_1 \cdots i_{n-2}}^{k}} \right) \\
&= \sum_{i_1 \in N} \sum_{i_2 \in N\setminus i_1} \cdots \sum_{i_{n-2} \in N\setminus i_1 \cdots i_{n-3}} p_{\emptyset}^{i_1} A(i_2; \{i_1\}) \cdot A(i_3; \{i_1, i_2\}) \cdot \\
&\qquad \cdots A(i_{n-2}; \{i_1, \ldots, i_{n-3}\}) \\
&= \sum_{i_1 \in N} \sum_{i_2 \in N\setminus i_1} \cdots \sum_{i_{n-3} \in N\setminus i_1 \cdots i_{n-4}} p_{\emptyset}^{i_1} A(i_2; \{i_1\}) \cdot A(i_3; \{i_1, i_2\}) \cdot \\
&\qquad \cdots A(i_{n-3}; \{i_1, \ldots, i_{n-4}\}) \\
&= \cdots \\
&= \sum_{i_1 \in N} p_{\emptyset}^{i_1} \,,
\end{aligned}
$$

where after the first equality sign, $i_n \in N \setminus \{i_1, \ldots, i_{n-1}\}$, and $A(i_n; \{i_1, \ldots, i_{n-1}\}) = 1$ by definition; after the second equality sign $\ell, k \in N \setminus \{i_1, \ldots, i_{n-2}\}$ with $\ell \neq k$; the third equality sign follows since the sum involving ℓ and k is equal to 1; the remaining equality signs follow from repetition of this argument. This concludes the proof of (18.4).

We next claim that for every $0 \leq t \leq n - 1$ we have

$$
\sum_{T: |T| = t} \sum_{i \in N \setminus T} p_T^i = 1 . \tag{18.5}
$$

To prove this, first let $t = n - 1$. Then the sum on the left-hand side of (18.5) is equal to $\sum_{i \in N} p_{N \setminus i}^i$, which is equal to 1 by Lemma 18.12. Now as induction hypothesis assume that (18.5) holds for $t + 1$. Then

$$\sum_{T: \, |T|=t} \sum_{i \in N \setminus T} p_T^i = \sum_{i \in N} \sum_{T: \, |T|=t+1, \, i \in T} p_{T \setminus i}^i$$

$$= \sum_{T: \, |T|=t+1} \sum_{i \in T} p_{T \setminus i}^i$$

$$= \sum_{T: \, |T|=t+1} \sum_{i \in N \setminus T} p_T^i$$

$$= 1 \, ,$$

where the second equality follows by Lemma 18.12 and the last equality by induction. This proves (18.5). In particular, for $t = 0$, we have $\sum_{i \in N} p_\emptyset^i = 1$. Together with (18.4) this implies that $p(\cdot)$ as defined above is a probability distribution on $\Pi(N)$.

In order to complete the proof of the theorem, it is sufficient to show that $\psi = \Phi^p$. For every game $v \in \mathcal{G}^N$ and $i \in N$ we can write

$$\Phi_i^p(v) = \sum_{\pi \in \Pi(N)} p(\pi)[v(P_\pi(i) \cup i) - v(P_\pi(i))] \, ,$$

where $P_\pi(i)$ denotes the set of predecessors of player i under the permutation π (cf. Sect. 17.1). Hence, it is sufficient to prove that for all $i \in N$ and $T \subseteq N \setminus i$ we have

$$p_T^i = \sum_{\pi \in \Pi(N): \, T = P_\pi(i)} p(\pi) \, . \tag{18.6}$$

In order to prove (18.6), first let $|T| = t$. By using a similar argument as for the proof of (18.4), we can write

$$\sum_{\pi: T = P_\pi(i)} p(\pi) = \sum_{i_1 \in T} \sum_{i_2 \in T \setminus i_1} \cdots \sum_{i_t \in T \setminus i_1 i_2 \cdots i_{t-1}} p_\emptyset^{i_1} A(i_2; \{i_1\}) \cdots A(i_t; T \setminus \{i_t\}) \cdot A(i; T) \, .$$

Hence,

$$\sum_{\pi: T = P_\pi(i)} p(\pi) = \frac{p_T^i}{\sum_{j \in N \setminus T} p_T^j}$$

$$\cdot \sum_{i_t \in T} \frac{p_{T \setminus i_t}^{i_t}}{\sum_{j \in (N \setminus T) \cup i_t} p_{T \setminus i_t}^j} \cdot \sum_{i_{t-1} \in T \setminus i_t} \frac{p_{T \setminus i_t i_{t-1}}^{i_{t-1}}}{\sum_{j \in (N \setminus T) \cup i_t i_{t-1}} p_{T \setminus i_t i_{t-1}}^j}$$

$$\cdots \cdots \sum_{i_1 \in T \setminus i_t \cdots i_2} p_\emptyset^{i_1}$$

$$= \frac{p_T^i}{\sum_{j \in T} p_{T \setminus j}^j}$$

$$\cdot \sum_{i_t \in T} \frac{p_{T \setminus i_t}^{i_t}}{\sum_{j \in T \setminus i_t} p_{(T \setminus i_t) \setminus j}^j} \cdot \sum_{i_{t-1} \in T \setminus i_t} \frac{p_{T \setminus i_t i_{t-1}}^{i_{t-1}}}{\sum_{j \in T \setminus i_t i_{t-1}} p_{(T \setminus i_t i_{t-1}) \setminus j}^j}$$

$$\cdots \cdots \sum_{i_1 \in T \setminus i_t \cdots i_2} p_\emptyset^{i_1}$$

$$= p_T^i .$$

Here, the first equality sign follows from rearranging terms and substituting the expressions for $A(\cdot; \cdot)$; the second equality sign follows from Lemma 18.12; the final equality sign follows from reading the preceding expression from right to left, noting that the remaining sum in each enumerator cancels against the preceding denominator.   ■

18.4   Weighted Shapley Values

The Shapley value is a random order value that distributes the dividend of each coalition equally among all the members of that coalition (see Theorem 17.7). In this sense, it treats players consistently over coalitions. This is not necessarily the case for every random order value. To be specific, consider Example 18.2. The payoff vector $(2\frac{1}{2}, -\frac{1}{2}, 1)$ is a point of the Weber set, namely the midpoint of the marginal vectors $(4, -1, 0)$ and $(1, 0, 2)$. Thus, it can be obtained uniquely as $\Phi^p(v)$, where the probability distribution p assigns weights $1/2$ to the permutations $(3, 2, 1)$ and $(2, 1, 3)$. In terms of dividends, the two marginal vectors can be written as

$$(\Delta_v(1) + \Delta_v(12) + \Delta_v(13) + \Delta_v(123), \Delta_v(2) + \Delta_v(23), \Lambda_v(3))$$

for the permutation $(3, 2, 1)$ and

$$(\Delta_v(1) + \Delta_v(12), \Delta_v(2), \Delta_v(3) + \Delta_v(13) + \Delta_v(23) + \Delta_v(123))$$

for the permutation $(2, 1, 3)$. (Cf. Problem 18.1, where this is generalized.) Thus, $\Delta_v(123)$ is split equally between players 1 and 3, whereas $\Delta_v(23)$ is split equally between players 2 and 3. Hence, whereas player 2 has zero power compared to player 3 in distributing $\Delta_v(123)$, they have equal power in distributing $\Delta_v(23)$. In this respect, players 2 and 3 are not treated consistently by Φ^p.

In order to formalize the idea of consistent treatment, we first define positively weighted Shapley values. Let $\omega \in \mathbb{R}^N$ with $\omega > 0$. The *positively weighted Shapley value* Φ^ω is defined as the unique linear value which assigns to each unanimity game (N, u_S):

$$\Phi_i^\omega (u_S) = \begin{cases} \omega_i/\omega(S) \text{ for } i \in S \\ 0 \qquad\quad \text{ for } i \in N \setminus S . \end{cases} \tag{18.7}$$

We will show that these positively weighted Shapley values are random order values. Define independently distributed random variables X_i ($i \in N$) on $[0, 1]$ by their cumulative distribution functions $[0, 1] \ni t \mapsto t^{\omega_i}$ (that is, $X_i \le t$ with probability t^{ω_i}). Then, define the probability distribution p^ω by

$$p^\omega (\pi) = \int_0^1 \int_0^{t_n} \int_0^{t_{n-1}} \cdots \int_0^{t_2} dt_1^{\omega_{i_1}} \dots dt_{n-2}^{\omega_{i_{n-2}}} dt_{n-1}^{\omega_{i_{n-1}}} dt_n^{\omega_{i_n}} \tag{18.8}$$

for every permutation $\pi = (i_1, i_2, \dots, i_n)$. That is, $p^\omega (\pi)$ is defined as the probability that i_1 comes before i_2, i_2 before i_3, etc., evaluated according to the independent random variables X_i. It is straightforward to check that, indeed, p^ω is a probability distribution over the set of permutations. Then we have:

Theorem 18.13 *For every* $\omega \in \mathbb{R}^N$ *with* $\omega > 0$, $\Phi^\omega = \Phi^{p^\omega}$.

Proof Let S be a nonempty coalition and $i \in S$. Since Φ^ω and Φ^{p^ω} are linear, it is sufficient to prove that $\Phi_i^{p^\omega} (u_S) = \omega_i/\omega(S)$. Note that

$$\Phi_i^{p^\omega} (u_S) = \sum_{\pi \in \Pi(N)} p^\omega (\pi) m_i^\pi (u_S) = \sum_{\pi \in \Pi(N): \, S \setminus i \subseteq P_\pi (i)} p^\omega (\pi) ,$$

and the right-hand side of this identity is equal to

$$\int_0^1 \int_0^{t_i} dt^{\omega(S \setminus i)} dt_i^{\omega_i} .$$

Hence,

$$\Phi_i^{p^\omega} (u_S) = \int_0^1 \int_0^{t_i} dt^{\omega(S \setminus i)} dt_i^{\omega_i} = \int_0^1 t_i^{\omega(S \setminus i)} dt_i^{\omega_i} = \int_0^1 \omega_i t^{\omega(S)-1} dt = \omega_i/\omega(S)$$

which concludes the proof of the theorem. ∎

Next, we extend the concept of weighted Shapley value to include zero weights. Consider for instance, the three-person random order value that puts weight $1/2$ on the permutations $(1, 2, 3)$ and $(1, 3, 2)$. Then each of the marginal vectors

$$(\Delta_v(1), \Delta_v(2) + \Delta_v(12), \Delta_v(3) + \Delta_v(13) + \Delta_v(23) + \Delta_v(123))$$

and

$$(\Delta_v(1), \Delta_v(2) + \Delta_v(12) + \Delta_v(23) + \Delta_v(123), \Delta_v(3) + \Delta_v(13))$$

gets weight $1/2$. (Cf. again Problem 18.1.) The dividend $\Delta_v(12)$ goes to player 2, the dividend $\Delta_v(13)$ to player 3, and the dividends $\Delta_v(23)$ and $\Delta_v(123)$ are split equally between players 2 and 3. Thus, this random order value treats players consistently but we cannot just formalize this by giving player 1 weight 0 since player 1 does obtain $\Delta_v(1)$.

To accommodate this kind of random order values we introduce the concept of a weight system. A *weight system* w is an ordered partition $(S_1, \ldots, S_k)$ of N together with a vector $\omega \in \mathbb{R}^N$ such that $\omega > \mathbf{0}$. The *weighted Shapley value* Φ^w is defined as the unique linear value which assigns to each unanimity game $u_S \in \mathcal{G}^N$, $S \neq \emptyset$:

$$\Phi_i^w(u_S) = \begin{cases} \omega_i/\omega(S \cap S_m) & \text{for } i \in S \cap S_m \text{ and } m = \max\{h : S_h \cap S \neq \emptyset\} \\ 0 & \text{otherwise.} \end{cases}$$

$$(18.9)$$

Hence, S_h is more powerful as h is larger; for each coalition S we consider the subset of the most powerful players $S \cap S_m$, where m is the largest index h such that the intersection of S_h with S is nonempty, and they distribute the dividend of coalition S according to their (relative) weights $\omega_i/\omega(S \cap S_m)$. Clearly, for $k = 1$ we obtain a positively weighted Shapley value as defined above.

Weighted Shapley values are again random order values. For a weight system w with ordered partition $(S_1, \ldots, S_k)$ we only assign positive probability to those permutations in which all players of S_1 enter before all players of S_2, all players of S_2 enter before all players of S_3, etc. Given such a permutation we can assign probability $p_1(\pi)$ to the order induced by π on S_1 in the same way as we did above in Eq. (18.8); similarly, we assign probabilities $p_2(\pi), \ldots, p_k(\pi)$ to the orders induced on $S_2, \ldots, S_k$, respectively. Then we define

$$p^w(\pi) = \prod_{h=1}^{k} p_h(\pi) .$$

It can be shown again that $\Phi^w = \Phi^{p^w}$.

We conclude with an axiomatic characterization of weighted Shapley values. To this end we consider the following axiom for a value $\psi : \mathcal{G}^N \to \mathbb{R}^N$.

Partnership (PA): $\quad \psi_i(\psi(u_T)(S)u_S) = \psi_i(u_T)$ for all $S \subseteq T \subseteq N$ and all $i \in S$.

Theorem 18.14 *Let* $\psi : \mathcal{G}^N \to \mathbb{R}^N$ *be a value. Then* ψ *satisfies LIN, DUM, EFF, MON, and PA, if and only if it is a weighted Shapley value.*

Proof Problem 18.8. ∎

By Theorem 18.3 we know that the core of any game is included in the Weber set and, thus, in any game any core element corresponds to at least one random order value. The following theorem, included here without a proof, states that, in fact, the core is always covered by the set of weighted Shapley values.

Theorem 18.15 *Let $v \in \mathcal{G}^N$ and $\mathbf{x} \in C(v)$. Then there is a weight system w such that $\mathbf{x} = \Phi^w(v)$.*

18.5   Problems

18.1. *Marginal Vectors and Dividends*
Let $(N, v) \in \mathcal{G}^N$.

(a) Show that

$$v(S) = \sum_{T \subseteq S} \Delta_v(T) \tag{18.10}$$

where $\Delta_v(T)$ are the dividends defined in Sect. 17.1.
(b) Express each marginal vector m^π in terms of dividends.

18.2. *Convexity and Marginal Vectors*
Prove that a game (N, v) is convex if and only if for all $T \in 2^N \setminus \{\emptyset\}$:

$$v(T) = \min_{\pi \in \Pi(N)} \sum_{i \in T} m_i^\pi(v) .$$

18.3. *Strictly Convex Games*
Call a game (N, v) *strictly convex* if all inequalities in (18.3) hold strictly. Show that in a strictly convex game all marginal vectors are different.

18.4. *Sharing Profits*
Consider the following situation with $n + 1$ players. Player 0 (the landlord) owns the land and players $1, 2, \ldots, n$ are n identical workers who own their labor only. The production $f : \{0, 1, \ldots, n\} \to \mathbb{R}$ describes how much is produced by the workers. Assume that f is nondecreasing and that $f(0) = 0$. We associate with this situation a TU-game that reflects the production possibilities of coalitions. Without agent 0 a coalition has zero worth, otherwise the worth depends on the number of workers.

More precisely,

$$v(S) := \begin{cases} 0 & \text{if } 0 \notin S \\ f(|S| - 1) & \text{if } 0 \in S \end{cases}$$

for every coalition $S \subseteq \{0, 1, \ldots, n\}$.

(a) Compute the marginal vectors and the Shapley value of this game.
(b) Compute the core of this game.
(c) Give a necessary and sufficient condition on f such that the game is convex. [So in that case, the core and the Weber set coincide and the Shapley value is in the core.]

18.5. *Sharing Costs*
Suppose that n airlines share the cost of a runway. To serve the planes of company i, the length of the runway must be c_i, which is also the cost of a runway of that length. Assume $0 \leq c_1 \leq c_2 \leq \ldots \leq c_n$. The cost of coalition S is defined as $c_S = \max_{i \in S} c_i$ for every nonempty coalition S.

(a) Model this situation as a cost savings game (cf. the three communities game in Chap. 1).
(b) Show that the resulting game is convex, and compute the marginal vectors, the Shapley value, and the core.

18.6. *Independence of the Axioms in Theorem 18.8*
Show that the axioms in Theorem 18.8 are independent.

18.7. *Null-Player in Theorem 18.8*
Show that Theorem 18.8 still holds if DUM is replaced by NP.

18.8. *Characterization of Weighted Shapley Values*
Prove Theorem 18.14. Also show that the axioms are independent.

18.9. *Core and Weighted Shapley Values in Example 18.2*
In Example 18.2, determine for each $\mathbf{x} \in C(v)$ a weight system w such that $\mathbf{x} = \Phi^w(v)$.

18.6 Notes

The Weber set was introduced by Weber (1988), who also gave a proof of Theorem 18.3 (the core is always a subset of the Weber set); our proof is due to Derks (1992). Theorem 18.6 (coincidence of the core and Weber set of convex games) is from Shapley (1971) and Ichiishi (1981). The proof of Theorem 18.13

(positively weighted Shapley values are random order values) is from Owen (1972). Theorem 18.14 (axiomatic characterization of weighted Shapley values) is from Kalai and Samet (1987), see also Derks et al. (2000). Theorem 18.15 is due to Monderer et al. (1992).

Problem 18.5 is based on Littlechild and Owen (1974).

References

Derks, J. (1992). A short proof of the inclusion of the core in the Weber set. *International Journal of Game Theory, 21,* 149–150.

Derks, J., Haller, H., & Peters, H. (2000). The Selectope for cooperative games. *International Journal of Game Theory, 29,* 23–38.

Ichiishi, T. (1981). Super-modularity: Applications to convex games and to the greedy algorithm for LP. *Journal of Economic Theory, 25,* 283–286.

Kalai, E., & Samet, D. (1987). On weighted Shapley values. *International Journal of Game Theory, 16,* 205–222.

Littlechild, S. C., & Owen, G. (1974). A simple expression for the Shapley Value in a special case. *Management Science, 20,* 370–372.

Monderer, D., Samet, D., & Shapley, L. S. (1992). Weighted values and the core. *International Journal of Game Theory, 21,* 27–39.

Owen, G. (1972). Multilinear extensions of games. *Management Science, 18,* 64–79.

Shapley, L. S. (1971). Cores of convex games. *International Journal of Game Theory, 1,* 11–26.

Weber, R. J. (1988). Probabilistic values for games. In Roth, A. E. (Ed.), *The Shapley value. Essays in honor of Lloyd S. Shapley* (pp. 101–119). Cambridge: Cambridge University Press.

The Nucleolus

19

The core of a game with transferable utility can be a large set, but it can also be empty. The Shapley value assigns to each game a unique point, which, however, does not have to be in the core. The *nucleolus* assigns to each game with a nonempty imputation set a unique element of that imputation set; moreover, this element is in the core if the core of the game is nonempty. The *pre-nucleolus* exists for every essential game (and does not have to be an imputation, even if the imputation set is nonempty), but for balanced games it coincides with the nucleolus.

In this chapter we consider both the nucleolus and the pre-nucleolus. The reader may want to read the relevant part of Chap. 9 first.

In Sect. 19.1 we start with an example illustrating the (pre-)nucleolus and Kohlberg's balancedness criterion. Section 19.2 introduces the lexicographic order, on which the definition of the (pre-)nucleolus in Sect. 19.3 is based. Section 19.4 presents the Kohlberg criterion, which is a characterization of the (pre-)nucleolus in terms of balanced collections of coalitions. Computational aspects are discussed in Sect. 19.5, while Sect. 19.6 presents Sobolev's characterization of the pre-nucleolus based on a reduced game property.

19.1 An Example

Consider the three-person TU-game given by Table 19.1. It is easy to see that this game has a nonempty core; for instance, $(8, 8, 8)$ is a core element. Let $\mathbf{x} = (x_1, x_2, x_3)$ be an arbitrary efficient payoff distribution. For a nonempty coalition S, define the *excess of S at $\mathbf{x}$* as $e(S, \mathbf{x}) := v(S) - x(S)$. For an efficient vector $\mathbf{x}$ the excess of the grand coalition N is always zero, and is therefore omitted from consideration. The idea underlying the nucleolus is as follows. For an arbitrary efficient vector consider the corresponding vector of excesses. Among all imputations [and for the pre-nucleolus: among all efficient vectors] find those where the maximal excess is minimal. If this set consists of one point, then this is the

© Springer-Verlag Berlin Heidelberg 2015
H. Peters, *Game Theory*, Springer Texts in Business and Economics,
DOI 10.1007/978-3-662-46950-7_19

Table 19.1 The example of Sect. 19.1

S	$\emptyset$	$\{1\}$	$\{2\}$	$\{3\}$	$\{1,2\}$	$\{1,3\}$	$\{2,3\}$	$\{1,2,3\}$
$v(S)$	0	4	4	4	8	12	16	24
$e(S,(8,8,8))$		-4	-4	-4	-8	-4	0	
$e(S,(6,9,9))$		-2	-5	-5	-7	-3	-2	
$e(S,(6,8,10))$		-2	-4	-6	-6	-4	-2	

(pre-)nucleolus. Otherwise, determine the coalitions for which this maximal excess cannot be further decreased, and repeat the procedure for the remaining coalitions, keeping the excesses of the other coalitions fixed. By successively minimizing maximal excesses, the (pre-)nucleolus results. Note that if a game has a nonempty core then every core element has by definition only non-positive excesses, whereas efficient payoff vectors outside the core have at least one positive excess. This implies that for balanced games the successive minimization of excesses can be restricted to core elements, and the pre-nucleolus and the nucleolus coincide.

In order to illustrate these ideas, consider again Table 19.1, where the excesses of some core vectors for this example are calculated. The highest excess for the core vector $(8,8,8)$ is equal to zero, attained for the coalition $\{2,3\}$. Obviously, this excess can be decreased by increasing the payoff for players 2 and 3 together, at the expense of player 1, who has an excess of -4. Thus, a next 'try' is the payoff vector $(6,9,9)$, which indeed has maximal excess -2 reached for coalitions $\{1\}$ and $\{2,3\}$. It is then obvious that this is indeed the minimal maximal excess, because the excess for coalition $\{1\}$ can only be decreased by increasing the excess for $\{2,3\}$, and vice versa. In particular, this implies that the excesses of these two coalitions will indeed be fixed at -2. Observe that the collection $\{\{1\},\{2,3\}\}$ is balanced (in particular, it is a partition).[1] At $(6,9,9)$ the second maximal excess is equal to -3, reached by the coalition $\{1,3\}$. Again, this might be decreased by improving the payoff for players 1 and 3 together at the expense of player 2. Because the payoff for player 1 has already been fixed at 6, this means that the payoff for player 3 has to be increased and that of player 2 has to be decreased. These observations lead to the next 'try' $(6,8,10)$, where the maximal excess is still equal to -2, and the second maximal excess equals -4, reached by the coalitions $\{2\}$ and $\{1,3\}$. It is obvious that this second maximal excess, as well as the third maximal excess of -6, cannot be decreased any further. Observe that also the collections $\{\{1\},\{2,3\},\{2\},\{1,3\}\}$ and $\{\{1\},\{2,3\},\{2\},\{1,3\},\{3\},\{1,2\}\}$ are all balanced.

It follows that $(6,8,10)$ is the (pre-)nucleolus of this game. Moreover, the excesses are closely related to balanced collections of coalitions; this will appear to be a more general phenomenon, known as the Kohlberg criterion.

---

[1] See Chap. 16.

19.2 The Lexicographic Order

The definition of the nucleolus is based on a comparison of vectors by means of the lexicographic order. We briefly discuss this order and examine some of its properties. Let $\mathbb{R}^k$ be the real vector space of dimension k. A binary relation $\succeq$ on $\mathbb{R}^k$ that satisfies

(1) *reflexivity*: $\mathbf{x} \succeq \mathbf{x}$ for all $\mathbf{x} \in \mathbb{R}^k$ and
(2) *transitivity*: $\mathbf{x} \succeq \mathbf{z}$ for all $\mathbf{x}, \mathbf{y}, \mathbf{z} \in \mathbb{R}^k$ with $\mathbf{x} \succeq \mathbf{y}$ and $\mathbf{y} \succeq \mathbf{z}$

is called a *partial order*. For a partial order $\succeq$, we write $\mathbf{x} \succ \mathbf{y}$ to indicate that $\mathbf{x} \succeq \mathbf{y}$ and $\mathbf{y} \not\succeq \mathbf{x}$. A partial order $\succeq$ is called a *weak order* if it satisfies

(3) *completeness*: $\mathbf{x} \succeq \mathbf{y}$ or $\mathbf{y} \succeq \mathbf{x}$ for all $\mathbf{x}, \mathbf{y} \in \mathbb{R}^k$ with $\mathbf{x} \neq \mathbf{y}$.

If it also satisfies

(4) *antisymmetry*: $\mathbf{x} = \mathbf{y}$ for all $\mathbf{x}, \mathbf{y} \in \mathbb{R}^k$ with $\mathbf{x} \succeq \mathbf{y}$ and $\mathbf{y} \succeq \mathbf{x}$

the relation is called a *linear order*.[2]
  On the vector space $\mathbb{R}^k$ we define the linear order $\succeq_{\text{lex}}$ as follows. For any two vectors $\mathbf{x}$ and $\mathbf{y}$ in $\mathbb{R}^k$, $\mathbf{x}$ is *lexicographically* larger than or equal to $\mathbf{y}$, notation: $\mathbf{x} \succeq_{\text{lex}} \mathbf{y}$, if either $\mathbf{x} = \mathbf{y}$, or $\mathbf{x} \neq \mathbf{y}$ and for

$$i = \min\{j \in \{1, \dots, k\} \mid x_j \neq y_j\}$$

we have that $x_i > y_i$. In other words, $\mathbf{x}$ should assign a higher value than $\mathbf{y}$ to the *first* coordinate at which $\mathbf{x}$ and $\mathbf{y}$ are different. For obvious reasons the order $\succeq_{\text{lex}}$ is called the *lexicographic order* on $\mathbb{R}^k$.
  The lexicographic order cannot be represented by a continuous utility function (Problem 19.5). In fact, it can be shown that the lexicographic order on $\mathbb{R}^k$ cannot be represented by *any* utility function (Problem 19.6).

19.3 The (Pre-)Nucleolus

Let (N, v) be a TU-game and let $X \subseteq \mathbb{R}^N$ be some set of payoff distributions. For every nonempty coalition $S \subseteq N$ and every $\mathbf{x} \in X$ the *excess of S at $\mathbf{x}$* is the number

$$e(S, \mathbf{x}, v) := v(S) - x(S) .$$

---

[2] See also Chap. 11. Here, we repeat some of the definitions for convenience.

This number can be interpreted as the dissatisfaction (complaint, regret) of the coalition S if $\mathbf{x}$ is the payoff vector. For every $\mathbf{x} \in X$ let $\theta(\mathbf{x})$ denote the vector of the excesses of all nonempty coalitions at $\mathbf{x}$, arranged in non-increasing order, hence

$$\theta(\mathbf{x}) = (e(S_1, \mathbf{x}, v), \ldots, e(S_{2^n-1}, \mathbf{x}, v))$$

such that $e(S_t, \mathbf{x}, v) \geq e(S_p, \mathbf{x}, v)$ for all $1 \leq t \leq p \leq 2^n - 1$. Let $\succeq_{\text{lex}}$ be the lexicographic order on $\mathbb{R}^{2^n-1}$, as defined in Sect. 19.2. The *generalized nucleolus of* (N, v) *with respect to* X is the set

$$\mathcal{N}(N, v, X) := \{\mathbf{x} \in X \mid \theta(\mathbf{y}) \succeq_{\text{lex}} \theta(\mathbf{x}) \text{ for all } \mathbf{y} \in X\} .$$

So the generalized nucleolus consists of all payoff vectors in X at which the excess vectors are lexicographically minimized. The motivation for this is that we first try to minimize the dissatisfaction of those coalitions for which this is maximal, next for those coalitions that have second maximal excess, etc.

We first establish conditions under which the generalized nucleolus with respect to a set X is non-empty.

Theorem 19.1 *Let* $X \subseteq \mathbb{R}^N$ *be nonempty and compact. Then* $\mathcal{N}(N, v, X)$ *is nonempty and compact for every game* v.

Proof First observe that all excess functions $e(S, \cdot, v)$ are continuous and therefore $\theta(\cdot)$ is continuous. Define $X_0 := X$ and, recursively,

$$X_t := \{\mathbf{x} \in X_{t-1} \mid \theta_t(\mathbf{y}) \geq \theta_t(\mathbf{x}) \text{ for all } \mathbf{y} \in X_{t-1}\}$$

for all $t = 1, \ldots, 2^n - 1$. Since $\theta(\cdot)$ is continuous, Weierstrass' Theorem implies that every X_t is a non-empty compact subset of X_{t-1}. This holds in particular for $t = 2^n - 1$ and, clearly, $X_{2^n-1} = \mathcal{N}(N, v, X)$. ∎

We will show that, if X is, moreover, convex, then the generalized nucleolus with respect to X consists of a single point. We start with the following lemma.

Lemma 19.2 *Let* $X \subseteq \mathbb{R}^N$ *be convex,* $\mathbf{x}, \mathbf{y} \in X$, *and* $0 \leq \alpha \leq 1$. *Then*

$$\alpha\theta(\mathbf{x}) + (1 - \alpha)\theta(\mathbf{y}) \succeq_{\text{lex}} \theta(\alpha\mathbf{x} + (1 - \alpha)\mathbf{y}) . \tag{19.1}$$

Proof Let $S_1, \ldots, S_{2^n-1}$ be an ordering of the non-empty coalitions such that

$$\theta(\alpha\mathbf{x} + (1 - \alpha)\mathbf{y}) = (e(S_1, \alpha\mathbf{x} + (1 - \alpha)\mathbf{y}, v), \ldots, e(S_{2^n-1}, \alpha\mathbf{x} + (1 - \alpha)\mathbf{y}, v)) .$$

The right hand side of this equation is equal to $\alpha\mathbf{a} + (1 - \alpha)\mathbf{b}$, where $\mathbf{a} = (e(S_1, \mathbf{x}, v), \ldots, e(S_{2^n-1}, \mathbf{x}, v))$ and $\mathbf{b} = (e(S_1, \mathbf{y}, v), \ldots, e(S_{2^n-1}, \mathbf{y}, v))$. Since

$\theta(\mathbf{x}) \succeq_{\text{lex}} \mathbf{a}$ and $\theta(\mathbf{y}) \succeq_{\text{lex}} \mathbf{b}$ it follows that

$$\alpha\theta(\mathbf{x}) + (1-\alpha)\theta(\mathbf{y}) \succeq_{\text{lex}} \alpha\mathbf{a} + (1-\alpha)\mathbf{b} = \theta(\alpha\mathbf{x} + (1-\alpha)\mathbf{y}) \, ,$$

hence (19.1) holds. ∎

Theorem 19.3 *Let $X \subseteq \mathbb{R}^N$ be nonempty, compact, and convex. Then, for every game (N, v), the generalized nucleolus with respect to X consists of a single point.*

Proof By Theorem 19.1 the generalized nucleolus is nonempty. Let $\mathbf{x}, \mathbf{y} \in \mathcal{N}(N, v, X)$ and $0 < \alpha < 1$. Then $\theta(\mathbf{x}) = \theta(\mathbf{y})$. Since on the one hand

$$\theta(\alpha\mathbf{x} + (1-\alpha)\mathbf{y}) \succeq_{\text{lex}} \theta(x) = \theta(y)$$

and on the other hand, by Lemma 19.2,

$$\theta(x) = \theta(y) = \alpha\theta(\mathbf{x}) + (1-\alpha)\theta(\mathbf{y}) \succeq_{\text{lex}} \theta(\alpha\mathbf{x} + (1-\alpha)\mathbf{y})$$

antisymmetry of the lexicographic order implies

$$\theta(\alpha\mathbf{x} + (1-\alpha)\mathbf{y}) = \theta(\mathbf{x}) = \theta(\mathbf{y}) \, .$$

Therefore $\theta(\mathbf{x}) = \theta(\mathbf{y}) = \alpha\mathbf{a} + (1-\alpha)\mathbf{b}$, with $\mathbf{a}$ and $\mathbf{b}$ as in the proof of Lemma 19.2. Since $\theta(\mathbf{x}) \succeq_{\text{lex}} \mathbf{a}$ and $\theta(\mathbf{y}) \succeq_{\text{lex}} \mathbf{b}$ it follows that $\mathbf{a} = \theta(\mathbf{x})$ and $\mathbf{b} = \theta(\mathbf{y})$. In $\mathbf{a}$ and $\mathbf{b}$ the coalitions are ordered in the same way and therefore this is also the case in $\theta(\mathbf{x})$ and $\theta(\mathbf{y})$. Hence $\mathbf{x}$ and $\mathbf{y}$ have all excesses equal, and thus $\mathbf{x} = \mathbf{y}$. ∎

Well-known choices for the set X are the imputation set $I(N, v)$ of a game (N, v) and the set of efficient payoff distributions $I^*(N, v) = \{\mathbf{x} \in \mathbb{R}^N \mid x(N) = v(N)\}$. For an essential game (N, v) the set $I(N, v)$ is nonempty, compact and convex, and therefore Theorem 19.3 implies that the generalized nucleolus with respect to $I(N, v)$ consists of a single point, called the *nucleolus* of (N, v), and denoted by $v(N, v)$. Although the set $I^*(N, v)$ is not compact, the generalized nucleolus of (N, v) with respect to this set exists and is also single-valued (Problem 19.7): its unique member is called the *pre-nucleolus* of (N, v), denoted by $v^*(N, v)$. In Problem 19.8 the reader is asked to show that both points are in the core of the game if this set is nonempty, and then coincide.

19.4 The Kohlberg Criterion

In this section we derive the so-called Kohlberg criterion for the pre-nucleolus, which characterizes this solution in terms of balanced sets (cf. Chap. 16). A similar result can be derived for the nucleolus (Problem 19.9) but the formulation for the pre-nucleolus is slightly simpler.

We start, however, with Kohlberg's characterization in terms of side-payments. A *side-payment* is a vector $\mathbf{y} \in \mathbb{R}^N$ satisfying $y(N) = 0$. Let (N, v) be a game and for every $\alpha \in \mathbb{R}$ and $\mathbf{x} \in \mathbb{R}^N$ denote by

$$\mathcal{D}(\alpha, \mathbf{x}, v) = \{S \subseteq N \setminus \{\emptyset\} \mid e(S, \mathbf{x}, v) \geq \alpha\}$$

the set of (nonempty) coalitions with excess at least α at $\mathbf{x}$.

Theorem 19.4 *Let (N, v) be a game and $\mathbf{x} \in I^*(N, v)$. Then the following two statements are equivalent.*

(a) $\mathbf{x} = v^*(N, v)$.
(b) *For every α such that $\mathcal{D}(\alpha, \mathbf{x}, v) \neq \emptyset$ and for every side-payment $\mathbf{y}$ with $y(S) \geq 0$ for every $S \in \mathcal{D}(\alpha, \mathbf{x}, v)$ we have $y(S) = 0$ for every $S \in \mathcal{D}(\alpha, \mathbf{x}, v)$.*

Proof Assume that $\mathbf{x} = v^*(N, v)$ and that the conditions in (b) are fulfilled for $\mathbf{x}$, α, and $\mathbf{y}$. Define $\mathbf{z}_\varepsilon = \mathbf{x} + \varepsilon \mathbf{y}$ for every $\varepsilon > 0$. Then $\mathbf{z}_\varepsilon \in I^*(N, v)$. Choose $\varepsilon^* > 0$ such that, for all $S \in \mathcal{D}(\alpha, \mathbf{x}, v)$ and nonempty $T \notin \mathcal{D}(\alpha, \mathbf{x}, v)$,

$$e(S, \mathbf{z}_{\varepsilon^*}, v) > e(T, \mathbf{z}_{\varepsilon^*}, v) . \tag{19.2}$$

For every $S \in \mathcal{D}(\alpha, \mathbf{x}, v)$,

$$e(S, \mathbf{z}_{\varepsilon^*}, v) = v(S) - (x(S) + \varepsilon^* y(S))$$

$$= e(S, \mathbf{x}, v) - \varepsilon^* y(S)$$

$$\leq e(S, \mathbf{x}, v) . \tag{19.3}$$

Assume, contrary to what we want to prove, that $y(S) > 0$ for some $S \in \mathcal{D}(\alpha, \mathbf{x}, v)$. Then, by (19.2) and (19.3), $\theta(\mathbf{x}) \succeq_{\text{lex}} \theta(\mathbf{z}_{\varepsilon^*})$ and $\mathbf{x} \neq \mathbf{z}_{\varepsilon^*}$, a contradiction.

Next, let $\mathbf{x} \in I^*(N, v)$ satisfy (b). Let $\mathbf{z} = v^*(N, v)$. Denote

$$\{e(S, \mathbf{x}, v) \mid S \in 2^N \setminus \{\emptyset\}\} = \{\alpha_1, \ldots, \alpha_p\}$$

with $\alpha_1 > \cdots > \alpha_p$. Define $\mathbf{y} = \mathbf{z} - \mathbf{x}$. Hence, $\mathbf{y}$ is a side-payment. Since $\theta(\mathbf{x}) \succeq_{\text{lex}} \theta(\mathbf{z})$, we have $e(S, \mathbf{x}, v) = \alpha_1 \geq e(S, \mathbf{z}, v)$ for all $S \in \mathcal{D}(\alpha_1, \mathbf{x}, v)$ and thus

$$e(S, \mathbf{x}, v) - e(S, \mathbf{z}, v) = (z - x)(S) = y(S) \geq 0 .$$

Therefore, by (b), $y(S) = 0$ for all $S \in \mathcal{D}(\alpha_1, \mathbf{x}, v)$.
Assume now that $y(S) = 0$ for all $S \in \mathcal{D}(\alpha_t, \mathbf{x}, v)$ for some $1 \leq t \leq p$. Then, since $\theta(\mathbf{x}) \succeq_{\text{lex}} \theta(\mathbf{z})$,

$$e(S, \mathbf{x}, v) = \alpha_{t+1} \geq e(S, \mathbf{z}, v) \text{ for all } S \in \mathcal{D}(\alpha_{t+1}, \mathbf{x}, v) \setminus \mathcal{D}(\alpha_t, \mathbf{x}, v) .$$

Hence $y(S) \geq 0$ and thus, by (b), $y(S) = 0$ for all $S \in \mathcal{D}(\alpha_{t+1}, \mathbf{x}, v)$. We conclude that $y(S) = 0$ for all $S \in 2^N \setminus \{\emptyset\}$, so $\mathbf{y} = \mathbf{0}$ and $\mathbf{x} = \mathbf{z}$. ∎

We can now prove Kohlberg's characterization of the pre-nucleolus by balanced collections.

Theorem 19.5 (Kohlberg) *Let* (N, v) *be a game and* $\mathbf{x} \in I^*(N, v)$. *Then the following two statements are equivalent.*

(a) $\mathbf{x} = v^*(N, v)$.
(b) *For every* α, $\mathcal{D}(\alpha, \mathbf{x}, v) \neq \emptyset$ *implies that* $\mathcal{D}(\alpha, \mathbf{x}, v)$ *is a balanced collection.*

Proof Assume that $\mathbf{x}$ satisfies (b). Let $\alpha \in \mathbb{R}$ such that $\mathcal{D}(\alpha, \mathbf{x}, v) \neq \emptyset$, and let $\mathbf{y}$ be a side-payment with $y(S) \geq 0$ for all $S \in \mathcal{D}(\alpha, \mathbf{x}, v)$. Since, by (b), $\mathcal{D}(\alpha, \mathbf{x}, v)$ is balanced there are numbers $\lambda(S) > 0$, $S \in \mathcal{D}(\alpha, \mathbf{x}, v)$, such that

$$\sum_{S \in \mathcal{D}(\alpha, \mathbf{x}, v)} \lambda(S) e^S = e^N .$$

By taking the product on both sides with $\mathbf{y}$ this implies

$$\sum_{S \in \mathcal{D}(\alpha, \mathbf{x}, v)} \lambda(S) y(S) = y(N) = 0 .$$

Therefore $y(S) = 0$ for every $S \in \mathcal{D}(\alpha, \mathbf{x}, v)$. Thus, Theorem 19.4 implies $\mathbf{x} = v^*(N, v)$.

Assume next that $\mathbf{x} = v^*(N, v)$. Let $\alpha \in \mathbb{R}$ such that $\mathcal{D}(\alpha, \mathbf{x}, v) \neq \emptyset$. Consider the linear program

$$\max \sum_{S \in \mathcal{D}(\alpha, \mathbf{x}, v)} y(S) \text{ subject to } -y(S) \leq 0, \ S \in \mathcal{D}(\alpha, \mathbf{x}, v), \text{ and } y(N) = 0 .$$

(19.4)

This program is feasible and, by Theorem 19.4, its value is 0. Hence (Problem 19.10) its dual is feasible, that is, there are $\lambda(S) \geq 0$, $S \in \mathcal{D}(\alpha, \mathbf{x}, v)$, and $\lambda(N) \in \mathbb{R}$ such that

$$-\sum_{S \in \mathcal{D}(\alpha, \mathbf{x}, v)} \lambda(S) e^S + \lambda(N) e^N = \sum_{S \in \mathcal{D}(\alpha, \mathbf{x}, v)} e^S .$$

Hence $\lambda(N) e^N = \sum_{S \in \mathcal{D}(\alpha, \mathbf{x}, v)} (1 + \lambda(S)) e^S$. Since $1 + \lambda(S) > 0$ for every $S \in \mathcal{D}(\alpha, \mathbf{x}, v)$, we have $\lambda(N) > 0$ and thus $\mathcal{D}(\alpha, \mathbf{x}, v)$ is balanced, with balancing weights $(1 + \lambda(S))/\lambda(N)$ for $S \in \mathcal{D}(\alpha, \mathbf{x}, v)$. ∎

19.5 Computation of the Nucleolus

For two-person games, the (pre-)nucleolus is easy to compute (Problem 19.13). In general, the computation of the nucleolus can be based on the subsequent determination of the sets $X_0, X_1, X_2, \ldots$ in Theorem 19.1, but this may not be easy, as the following example shows.

Example 19.6 Consider the TU-game v with player set $N = \{1, 2, 3, 4\}$ defined by[3]

$$v(S) = \begin{cases} 20 & \text{if } S = N \\ 8 & \text{if } S = \{1, 2\} \\ 8 & \text{if } S = \{3, 4\} \\ 4 & \text{if } S = \{1\} \\ 2 & \text{if } S = \{3\} \\ 0 & \text{otherwise.} \end{cases}$$

First observe that it is easy to find some imputation [e.g., $\mathbf{x} = (6, 4, 5, 5)$] such that the excesses of $\{1, 2\}$ and $\{3, 4\}$ are both equal to -2 and all other excesses are at most -2. Clearly, this must be the minimal (over all imputations) maximal excess attainable, since decreasing the excess of $\{1, 2\}$ implies increasing the excess of $\{3, 4\}$ by efficiency, and vice versa. Thus,

$$X_1 = \{\mathbf{x} \in I(v) \mid \theta(\mathbf{x})_1 = -2\},$$

and $X_2 = X_1$ since the excess of -2 is reached at the two coalitions $\{1, 2\}$ and $\{3, 4\}$. Consistently with the Kohlberg criterion, these coalitions form a balanced collection. Next, observe that the remaining excesses are always at most as large as at least one of the excesses of the four one-person coalitions. So we can find the second-highest excess by minimizing α subject to the constraints

$$\begin{aligned} 8 - x_1 - x_2 &= -2 \\ 8 - x_3 - x_4 &= -2 \\ 4 - x_1 &\leq \alpha \\ -x_2 &\leq \alpha \\ 2 - x_3 &\leq \alpha \\ -x_4 &\leq \alpha. \end{aligned}$$

---

[3]Observe that this game has a non-empty core and therefore the nucleolus and pre-nucleolus coincide and are elements of the core. Cf. Problem 19.8.

This can be rewritten as the system

$$
\begin{aligned}
x_1 + x_2 &= 10 \\
x_3 + x_4 &= 10 \\
x_1 &\geq 4 - \alpha \\
x_2 &\geq -\alpha \\
x_3 &\geq 2 - \alpha \\
x_4 &\geq -\alpha .
\end{aligned}
$$

By considering the equation and inequalities involving x_1 and x_2, we obtain $\alpha \geq -3$, and those with x_3 and x_4 yield $\alpha \geq -4$. So the obvious minimum value is $\alpha = -3$. Hence, the next two coalitions of which the excesses become fixed are $\{1\}$ and $\{2\}$, and, thus, the nucleolus allocates $x_1 = 7$ to player 1 and $x_2 = 3$ to player 2. The third step in the computation is to minimize α subject to the constraints

$$
\begin{aligned}
x_3 + x_4 &= 10 \\
x_3 &\geq 2 - \alpha \\
x_4 &\geq -\alpha .
\end{aligned}
$$

(Note that the constraints that only refer to x_1 and x_2 have become superfluous.) This linear program has the obvious solution $\alpha = -4$, which yields $x_3 = 6$ and $x_4 = 4$. Thus, the (pre-)nucleolus of this game is

$$
\nu(v) = \nu^*(v) = (7, 3, 6, 4) .
$$

It is interesting to see that, even though at first glance player 4 does not seem to have any noticeable advantage over player 2, he is still doing better in the nucleolus. This is due to the fact that early on in the process player 4 was grouped together with player 3, who has a lower individual worth than player 1, with whom player 2 becomes partnered. Thus, player 4 obtains a larger slice of the cake of size 10 that he has to share with player 3, than player 2 does in a similar situation.   □

This example raises the question how the nucleolus of a given TU-game can be computed in a systematic way. More generally, let (N, v) be a game. In order to compute the nucleolus $\mathcal{N}(N, v, X)$ for a compact polyhedral set $X \subseteq \mathbb{R}^N$, determined by a system of linear (in)equalities, we can start by solving the linear program[4]

Minimize α subject to $x(S) + \alpha \geq v(S)$, $\forall \emptyset \neq S \in 2^N$, $\mathbf{x} \in X$.

---

[4] Under appropriate restrictions this program is feasible and bounded.

Let α_1 denote the minimum of this program and let $X_1 \subseteq X$ be the set of points where the minimum is obtained. If $|X_1| = 1$ then $\mathcal{N}(N, v, X) = X_1$. Otherwise, let $\mathcal{B}_1$ be the set of coalitions S such that $e(S, \mathbf{x}, v) = \alpha_1$ for all $\mathbf{x} \in X_1$, and solve the linear program

$$\text{Minimize } \alpha \text{ subject to } x(S) + \alpha \geq v(S), \ \forall \emptyset \neq S \in 2^N \setminus \mathcal{B}_1, \mathbf{x} \in X_1.$$

Continuing in this way, we eventually reach a unique point, which is the generalized nucleolus of v with respect to X (cf. Theorem 19.3). The following example illustrates this.

Example 19.7 Let $N = \{1, 2, 3, 4\}$ and $v(N) = 100$, $v(123) = 95$, $v(124) = 85$, $v(134) = 80$, $v(234) = 55$, $v(ij) = 50$ for all $i \neq j$, $v(i) = 0$ for all i.[5] We compute the pre-nucleolus (which will turn out to be an imputation and therefore equal to the nucleolus). We start with the linear program

Minimize α subject to

$$
\begin{aligned}
x_1 + x_2 + x_3 \quad\quad + \alpha &\geq 95 \\
x_1 + x_2 \quad\quad + x_4 + \alpha &\geq 85 \\
x_1 \quad\quad + x_3 + x_4 + \alpha &\geq 80 \\
x_2 + x_3 + x_4 + \alpha &\geq 55 \\
x_i + x_j + \alpha &\geq 50 \\
x_i \quad\quad + \alpha &\geq 0 \\
x_1 + x_2 + x_3 + x_4 \quad\quad &= 100
\end{aligned}
$$

for all $i, j \in N$ with $i \neq j$.

Solving this program results in $\alpha_1 = 10$, obtained over the set X_1 given by

$$x_1 + x_2 = 60, \ x_1 \geq 30, \ x_2 \geq 25, \ x_3 = 25, \ x_4 = 15,$$

and

$$\mathcal{B}_1 = \{123, 124, 34\},$$

which is a balanced set, as was to be expected (cf. Theorem 19.5).

---

[5]We write $v(123)$ instead of $v(\{1, 2, 3\})$, etc.

The second linear program is now

Minimize α subject to

$$
\begin{aligned}
x_1 \quad + x_3 + x_4 + \alpha &\geq 80 \\
x_2 + x_3 + x_4 + \alpha &\geq 55 \\
x_1 \quad + x_3 \qquad + \alpha &\geq 50 \\
x_1 \qquad\quad + x_4 + \alpha &\geq 50 \\
x_2 + x_3 \qquad + \alpha &\geq 50 \\
x_2 \qquad + x_4 + \alpha &\geq 50 \\
x_i + \alpha &\geq 0 \\
\mathbf{x} &\in X_1
\end{aligned}
$$

for all $i \in N$.

By some simplifications this program reduces to

Minimize α subject to

$$
\begin{aligned}
x_1 + \alpha &\geq 40 \\
x_2 + \alpha &\geq 35 \\
x_1 + x_2 &= 60 \\
x_1 &\geq 30 \\
x_2 &\geq 25,
\end{aligned}
$$

with solution $\alpha_2 = 7.5$, $x_1 = 32.5$, $x_2 = 27.5$. Hence, the (pre-)nucleolus of this game is

$$
\nu(N, v) = \nu^*(N, v) = (32.5, 27.5, 25, 15),
$$

and $B_1 = \{123, 124, 34\}$, $B_2 = \{134, 234\}$. By Theorem 19.5 it can be verified that we have indeed found the pre-nucleolus (Problem 19.15).   □

19.6   A Characterization of the Pre-nucleolus

The pre-nucleolus is a single-valued solution concept, defined for any game (N, v). Hence, it is an example of a value on $\mathcal{G}$ (see Chap. 17). In this section we provide a characterization based on a reduced game property. We will not give the complete proof of this characterization but, instead, refer the reader to the literature.

Let ψ be a value on $\mathcal{G}$, and let $\mathbf{x} \in \mathbb{R}^N$. Let S be a non-empty coalition. The *Davis-Maschler reduced game for S at* $\mathbf{x}$ is the game $(S, v_{S,\mathbf{x}}) \in \mathcal{G}^S$ defined by

$$
v_{S,\mathbf{x}}(T) := \begin{cases} 0 & \text{if } T = \emptyset \\ v(N) - x(N \setminus S) & \text{if } T = S \\ \max_{Q \subseteq N \setminus S} v(T \cup Q) - x(Q) & \text{otherwise.} \end{cases}
$$

The interpretation of this reduced game is as follows. Suppose $\mathbf{x}$ is the payoff vector for the grand coalition. The coalition S could renegotiate these payoffs among themselves. Assume that the outside players are happy with $\mathbf{x}$. Hence, S has $v(N) - x(N \setminus S)$ to redistribute. Any smaller coalition T, however, could cooperate with zero or more outside players and pay them according to $\mathbf{x}$: then $v_{S,\mathbf{x}}(T)$ as defined above is the maximum they could get. Hence, the redistribution game takes the form $v_{S,\mathbf{x}}$.

The following axiom for a value ψ on $\mathcal{G}$ requires the outcome of the redistribution game for S to be equal to the original outcome.

Davis-Maschler consistency (DMC): $\psi_i(S, v_{S,\mathbf{x}}) = \psi_i(N, v)$ for every $(N, v) \in \mathcal{G}$, $\emptyset \neq S \subseteq N$, $\mathbf{x} = \psi(N, v)$ and $i \in S$.

The announced characterization is based on two other axioms, namely Anonymity (AN, see Sect. 17.1) and the following axiom.

Covariance (COV): $\psi(N, \alpha v + \mathbf{b}) = \alpha \psi(N, v) + \mathbf{b}$ for all $(N, v) \in \mathcal{G}$, every $\alpha \in \mathbb{R}$, $\alpha > 0$, and every $\mathbf{b} \in \mathbb{R}^N$, where $(\alpha v + \mathbf{b})(S) := \alpha v(S) + b(S)$ for every non-empty coalition $S \subseteq N$.

Remark 19.8 Covariance requires that the value ψ respects *strategic equivalence* of games, cf. Problem 16.12.

We first prove that the pre-nucleolus is Davis-Maschler consistent.

Lemma 19.9 *The pre-nucleolus, as a value on $\mathcal{G}$, is Davis-Maschler consistent.*

Proof Let $(N, v) \in \mathcal{G}$, $\mathbf{x} = v^*(N, v)$ and $\emptyset \neq S \subseteq N$. Let $\mathbf{x}_S \in \mathbb{R}^S$ be the restriction of $\mathbf{x}$ to S, then we have to prove that $\mathbf{x}_S = v^*(S, v_{S,\mathbf{x}})$. Let $\alpha \in \mathbb{R}$ with $\mathcal{D}(\alpha, \mathbf{x}_S, v_{S,\mathbf{x}}) \neq \emptyset$ and let $\mathbf{y}_S \in \mathbb{R}^S$ be a side-payment with $y_S(R) \geq 0$ for every $R \in \mathcal{D}(\alpha, \mathbf{x}_S, v_{S,\mathbf{x}})$ then, in view of Theorem 19.4 it is sufficient to prove that $y_S(R) = 0$ for every $R \in \mathcal{D}(\alpha, \mathbf{x}_S, v_{S,\mathbf{x}})$. We claim that

$$\{T \cap S \mid T \in \mathcal{D}(\alpha, \mathbf{x}, v), \ \emptyset \neq T \cap S \neq S\} = \mathcal{D}(\alpha, \mathbf{x}_S, v_{S,\mathbf{x}}) \setminus \{S\} . \tag{19.5}$$

To see this, first assume $T \in \mathcal{D}(\alpha, \mathbf{x}, v)$, $\emptyset \neq T \cap S \neq S$. Then

$$v_{S,\mathbf{x}}(T \cap S) - \mathbf{x}_S(T \cap S) \geq v((T \cap S) \cup (T \setminus S)) - x(T \setminus S) - \mathbf{x}(T \cap S)$$
$$= v(T) - x(T)$$
$$\geq \alpha$$

which proves one direction. For the other direction, let $S' \in \mathcal{D}(\alpha, \mathbf{x}_S, v_{S,\mathbf{x}})$, $S' \neq S$. Then

$$\max_{Q \subseteq N \setminus S} v(S' \cup Q) - x(Q) - x_S(S') \geq \alpha$$

hence

$$v(S' \cup Q') - x(Q' \cup S') \geq \alpha$$

for some $Q' \subseteq N \setminus S$; now take $T = S' \cup Q'$ to establish the converse direction.

Extend $\mathbf{y}_S$ to a vector $\mathbf{y} \in \mathbb{R}^N$ by setting $y_i = 0$ for all $i \in N \setminus S$. Then $y(N) = 0$ and, by (19.5), $y(R) \geq 0$ for all $R \in \mathcal{D}(\alpha, \mathbf{x}, v)$. By Theorem 19.4, it follows that $y(R) = 0$ for all $R \in \mathcal{D}(\alpha, \mathbf{x}, v)$. Hence, by (19.5), $y_S(R) = 0$ for all $R \in \mathcal{D}(\alpha, \mathbf{x}_S, v_{S,\mathbf{x}})$, which completes the proof. ∎

As an additional result, we prove that COV and DMC imply Efficiency (EFF).

Lemma 19.10 *Let ψ be a value on $\mathcal{G}$ satisfying COV and DMC. Then ψ satisfies EFF.*

Proof Let $(\{i\}, v)$ be a one-person game. If $v(i) = 0$ then, by COV, $\psi(\{i\}, 0) = \psi(\{i\}, 2 \cdot 0) = 2\psi(\{i\}, 0)$, hence $\psi(\{i\}, 0) = 0$. Again by COV,

$$\psi(\{i\}, v) = \psi(\{i\}, 0 + v) = \psi(\{i\}, 0) + v(i) = v(i) \,,$$

so EFF on one-person games is satisfied. Now let $(N, v) \in \mathcal{G}$ with at least two players. Let $\mathbf{x} = \psi(N, v)$ and $i \in N$. By DMC, $x_i = \psi(\{i\}, v_{\{i\},\mathbf{x}})$. Hence, $x_i = v_{\{i\},\mathbf{x}}(i) = v(N) - x(N \setminus \{i\})$, where the second equality follows by definition of the reduced game. Thus, $x(N) = v(N)$ and the proof is complete. ∎

The announced characterization of the pre-nucleolus is as follows.

Theorem 19.11 *A value ψ on $\mathcal{G}$ satisfies COV, AN, and DMC if and only if it is the pre-nucleolus.*

Proof COV and AN of the pre-nucleolus follow from Problem 19.17. DMC follows from Lemma 19.9. For the only-if part see the Notes to this chapter. ∎

19.7 Problems

19.1. *Binary Relations*
Give an example of a relation that satisfies (1)–(3), but not (4) in Sect. 19.2. Also find an example that only violates (3), and one that only violates (2). What about (1)? Give an example of a partial order that is neither antisymmetric nor complete.

19.2. *Linear Orders*
Let $\succeq$ be a linear order. Show that $\mathbf{x} \succ \mathbf{y}$ holds if and only if $\mathbf{x} \succeq \mathbf{y}$ and $\mathbf{x} \neq \mathbf{y}$.

19.3. *The Lexicographic Order (1)*
Show that $\succeq_{\text{lex}}$ is indeed a linear order.

19.4. *The Lexicographic Order (2)*
Find the set of points (x_1, x_2) in $\mathbb{R}^2$ for which $(x_1, x_2) \succeq_{\text{lex}} (3, 1)$. Draw this set in the Cartesian plane. Is this set closed?

19.5. *Representability of Lexicographic Order (1)*
Let $u: \mathbb{R}^n \to \mathbb{R}$ be a continuous function. Define $\succeq_u$ by $\mathbf{x} \succeq_u \mathbf{y}$ if and only if $u(\mathbf{x}) \geq u(\mathbf{y})$. Use Problem 19.4 to show that $\succeq_u \neq \succeq_{\text{lex}}$.

19.6. *Representability of Lexicographic Order (2)*
Show that the lexicographic order cannot be represented by any utility function. [Hint: take the lexicographic order on $\mathbb{R}^2$ and argue that representability implies that for each pair of real numbers t and s we can find rational numbers $q(t)$ and $q(s)$ such that $q(t) \neq q(s)$ whenever $t \neq s$. Hence, we have uncountably many different rational numbers, a contradiction.]

19.7. *Single-Valuedness of the Pre-nucleolus*
Prove that the nucleolus of any game (N, v) with respect to $I^*(N, v)$ is single-valued.

19.8. *(Pre-)Nucleolus and Core*
Let (N, v) be a game with $C(N, v) \neq \emptyset$. Prove that $v(N, v) = v^*(N, v) \in C(N, v)$.

19.9. *Kohlberg Criterion for the Nucleolus*
Let (N, v) be a game satisfying $I(N, v) \neq \emptyset$, and let $\mathbf{x} \in I(N, v)$. Prove that $\mathbf{x} = v(N, v)$ if and only if for every $\alpha \in \mathbb{R}$: if $\mathcal{D}(\alpha, v, \mathbf{x}) \neq \emptyset$ then there exists a set $\mathcal{E}(\alpha, \mathbf{x}, v) \subseteq \{\{j\} \mid j \in N, \ x_j = v(j)\}$ such that $\mathcal{D}(\alpha, \mathbf{x}, v) \cup \mathcal{E}(\alpha, \mathbf{x}, v)$ is balanced.

19.10. *Proof of Theorem 19.5*
In the proof of Theorem 19.5, determine the dual program and conclude that it is feasible. Hint: use Theorem 16.20 and Remark 16.21.

19.11. *Nucleolus of a Three-Person Game (1)*
Compute the nucleolus of the three-person game v defined by

S	$\{1\}$	$\{2\}$	$\{3\}$	$\{1,2\}$	$\{1,3\}$	$\{2,3\}$	$\{1,2,3\}$
$v(S)$	4	3	2	4	3	2	12

19.12. *Nucleolus of a Three-Person Game (2)*

(a) Compute the nucleolus of the three-person TU-game defined by

S	{1}	{2}	{3}	{1,2}	{1,3}	{2,3}	{1,2,3}
$v(S)$	0	0	1	7	5	3	10

(b) Make a graphical representation of the sets $X_0, X_1, \ldots$

19.13. *Nucleolus of a Two-Person Game*
Compute the pre-nucleolus of a two-person game and the nucleolus of an essential two-person game.

19.14. *Individual Rationality Restrictions for the Nucleolus*
Compute the nucleolus and the pre-nucleolus of the three-person TU-game defined by $v(12) = v(13) = 2$, $v(123) = 1$ and $v(S) = 0$ for all other coalitions.

19.15. *Example 19.7*
Verify that $(32.5, 27.5, 25, 15)$ is indeed the pre-nucleolus of the game in Example 19.7, by applying Theorem 19.5.

19.16. *(Pre-)Nucleolus of a Symmetric Game*
Let v be an essential game. Suppose that v is symmetric (meaning that there exists a function $f: \mathbb{R} \to \mathbb{R}$ such that $v(S) = f(|S|)$ for every coalition S.)

(a) Prove that the (pre-)nucleolus is symmetric, that is, $v(v)_i = v(v)_j$ and $v^*(v)_i = v^*(v)_j$ for all players $i, j \in N$. Give a formula for the (pre-)nucleolus.
(b) Suppose that $\mathcal{D}(\alpha, v(v), v) \neq \{N\}$, where α is the maximal excess at the (pre-) nucleolus. Prove that X_1 (cf. Theorem 19.1) is a singleton set.

19.17. *COV and AN of the Pre-nucleolus*
Prove that the pre-nucleolus satisfies COV and AN.

19.18. *Apex Game*
Consider the 5-person *apex game* (N, v) with $N = \{1, 2, 3, 4, 5\}$ and $v(S) = 1$ if $1 \in S$ and $|S| \geq 2$ or if $|S| \geq 4$, and $v(S) = 0$ otherwise. Compute the (pre-) nucleolus of this game.

19.19. *Landlord Game*
Consider the landlord game in Problem 18.4.

(a) Assume that for all $i = 1, \ldots, n-1$ we have $f(i) - f(i-1) \geq f(i+1) - f(i)$. Show that the (pre-)nucleolus of this game assigns $f(n) - \frac{n}{2}[f(n) - f(n-1)]$ to the landlord. Compare with the Shapley value.

(b) Assume that for all $i = 1, \ldots, n-1$ we have $f(i) - f(i-1) \leq f(i+1) - f(i)$, and that $\frac{1}{n+1}f(n) \leq \frac{1}{2}[f(n) - f(n-1)]$. Show that the (pre-)nucleolus of this game treats all players (including the landlord) equally.

19.20. *Game in Sect. 19.1*
Use the algorithm of solving successive linear programs to find the (pre-)nucleolus of the game discussed in Sect. 19.1. Use Theorem 19.5 to verify that the (pre-)nucleolus has been found.

19.21. *The Prekernel*
For a game (N, v) define the *pre-kernel* $\mathcal{K}^*(N, v) \subseteq I^*(N, v)$ by

$$\mathcal{K}^*(N, v) = \{x \in I^*(N, v) \mid \max_{S \subseteq N \setminus \{j\},\ i \in S} e(S, \mathbf{x}, v)$$

$$= \max_{S \subseteq N \setminus \{i\},\ j \in S} e(S, \mathbf{x}, v) \text{ for all } i, j \in N\} \,.$$

Prove that $v^*(N, v) \in \mathcal{K}^*(N, v)$.

19.8   Notes

The nucleolus was introduced in Schmeidler (1969). The treatment in the present chapter is partially based on the treatment of the subject in Peleg and Sudhölter (2003) and Owen (1995).

Theorem 19.5 is due to Kohlberg (1971). Example 19.7 is taken from Owen (1995).

The Davis-Maschler reduced game stems from Davis and Maschler (1965).

The characterization of the pre-nucleolus in Theorem 19.11 is due to Sobolev (1975). The reader may also consult Peleg and Sudhölter (2003) for a complete proof of the theorem. Snijders (1995) provides a characterization of the nucleolus on the class of all games with non-empty imputation set by modifying the Davis-Maschler consistency condition.

For the landlord game in Problem 19.19 see also Moulin (1988).

References

Davis, M., & Maschler, M. (1965). The kernel of a cooperative game. *Naval Research Logistics Quarterly, 12*, 223–259.
Kohlberg, E. (1971). On the nucleolus of a characteristic function game. *SIAM Journal of Applied Mathematics, 20*, 62–66.
Moulin, H. (1988). *Axioms of cooperative decision making*. Cambridge: Cambridge University Press.
Owen, G. (1995). *Game theory* (3rd ed.). San Diego: Academic.
Peleg, B., & Sudhölter, P. (2003). *Introduction to the theory of cooperative games*. Boston: Kluwer Academic.

Schmeidler, D. (1969). The nucleolus of a characteristic function game. *SIAM Journal on Applied Mathematics, 17*, 1163–1170.

Snijders, C. (1995). Axiomatization of the nucleolus. *Mathematics of Operations Research, 20*, 189–196.

Sobolev, A. I. (1975). Characterization of the principle of optimality through functional equations. *Mathematical Methods in the Social Sciences, 6*, 92–151.

Special Transferable Utility Games

20

In this chapter we consider a few classes of games with transferable utility which are derived from specific economic (or political) models or combinatorial problems. In particular, we study assignment and permutation games, flow games, and voting games.

20.1 Assignment and Permutation Games

An example of a permutation game is the 'dentist game' described in Sect. 1.3.4. An example of an assignment game is the following.

Example 20.1 Vladimir (player 1), Wanda (player 2), and Xavier (player 3) each own a house that they want to sell. Yolanda (player 4) and Zarik (player 5) each want to buy a house. Vladimir, Wanda, and Xavier value their houses at 1, 1.5, and 2, respectively (each unit is 100,000 Euros). The worths of their houses to Yolanda and Zarik, respectively, are 0.8 and 1.5 for Vladimir's house, 2 and 1.2 for Wanda's house, and 2.2 and 2.3 for Xavier's house.

This situation gives rise to a five-player TU-game, where the worth of each coalition is defined to be the maximal surplus that can be generated by buying and selling within the coalition. For instance, in the coalition $\{2, 3, 5\}$ the maximum surplus is generated if Zarik buys the house of Xavier, namely $2.3 - 2 = 0.3$, which is greater than the $1.2 - 1.5 = -0.3$ that results if Zarik buys Wanda's house. Each coalition can generate a payoff of at least 0 because it can refrain from trading at all. The complete game is described in Table 20.1, where coalitions with only buyers or only sellers are left out. A game like this is called an assignment game. □

We will examine such games in detail, starting with the basic definitions.

Let M and P be two finite, disjoint sets. For each pair $(i, j) \in M \times P$ the number $a_{ij} \geq 0$ is interpreted as the value of the matching between i and j. With this situation a cooperative game (N, v) can be associated, as follows. The player set N is the set

© Springer-Verlag Berlin Heidelberg 2015
H. Peters, *Game Theory*, Springer Texts in Business and Economics,
DOI 10.1007/978-3-662-46950-7_20

Table 20.1 Worths for the assignment game in Example 20.1

S	$v(S)$	S	$v(S)$	S	$v(S)$
14	0	125	0.5	345	0.3
15	0.5	134	0.2	1,234	0.5
24	0.5	135	0.5	1,235	0.5
25	0	145	0.5	1,245	1
34	0.2	234	0.5	1,345	0.7
35	0.3	235	0.3	2,345	0.8
124	0.5	245	0.5	12,345	1

$M \cup P$. For each coalition $S \subseteq N$ the worth $v(S)$ is the maximum that S can achieve by making pairs among its own members. Formally, if $S \subseteq M$ or $S \subseteq P$ then $v(S) = 0$, because no pairs can be formed at all. Otherwise, $v(S)$ is equal to the value of the following integer programming problem.

$$\max \sum_{i \in M} \sum_{j \in P} a_{ij} x_{ij}$$
$$\text{subject to } \sum_{j \in P} x_{ij} \le 1_S(i) \text{ for all } i \in M$$
$$\sum_{i \in M} x_{ij} \le 1_S(j) \text{ for all } j \in P \tag{20.1}$$
$$x_{ij} \in \{0, 1\} \quad \text{for all } i \in M, \ j \in P.$$

Here, $1_S(i) := 1$ if $i \in S$ and equal to zero otherwise. Games defined by (20.1) are called *assignment games*. The reader may verify that in Example 20.1 the numbers a_{ij} are given by $a_{ij} = \max\{h_{ij} - c_i, 0\}$, where h_{ij} is the value of the house of player i to player j and c_i is the value of the house of player i for himself.

As will become clear below, a more general situation is the following. For each $i \in N = \{1, 2, \ldots, n\}$ let $k_{i\pi(i)}$ be the value placed by player i on the permutation $\pi \in \Pi(N)$. (The implicit assumption is that $k_{i\pi(i)} = k_{i\sigma(i)}$ whenever $\pi(i) = \sigma(i)$.) Each coalition $S \subseteq N$ may achieve a permutation π involving only the players of S, that is, $\pi(i) = i$ for all $i \notin S$. Let $\Pi(S)$ denote the set of all such permutations. Then a game v results by defining, for each nonempty coalition S, the worth by

$$v(S) := \max_{\pi \in \Pi(S)} \sum_{i \in S} k_{i\pi(i)} . \tag{20.2}$$

The game thus obtained is called a *permutation game*. Alternatively, the worth $v(S)$ in such a game can be defined by the following integer programming problem.

$$\max \sum_{i \in N} \sum_{j \in N} k_{ij} x_{ij}$$
$$\text{subject to } \sum_{j \in N} x_{ij} = 1_S(i) \text{ for all } i \in N$$
$$\sum_{i \in N} x_{ij} = 1_S(j) \text{ for all } j \in N \tag{20.3}$$
$$x_{ij} \in \{0, 1\} \quad \text{for all } i, j \in N.$$

The two definitions are equivalent, and both can be used to verify that the 'dentist game' of Sect. 1.3.4 is indeed a permutation game (Problem 20.1).

The relation between the class of assignment games and the class of permutation games is a simple one. The former class is contained in the latter, as the following theorem shows.

Theorem 20.2 *Every assignment game is a permutation game.*

Proof Let v be an assignment game with player set $N = M \cup P$. For all $i, j \in N$ define

$$k_{ij} := \begin{cases} a_{ij} & \text{if } i \in M, \ j \in P \\ 0 & \text{otherwise.} \end{cases}$$

Let w be the permutation game defined by (20.3) with k_{ij} as above. Note that the number of variables in the integer programming problem defining $v(S)$ is $|M| \times |P|$, while the number of variables in the integer programming problem defining $w(S)$ is $(|M| + |P|)^2$. For $S \subseteq M$ or $S \subseteq P$, $w(S) = 0 = v(S)$. Let now $S \subseteq N$ with $S \not\subseteq M$ and $S \not\subseteq P$. Let $x \in \{0, 1\}^{|M| \times |P|}$ be an optimal solution for (20.1). Define $\hat{x} \in \{0, 1\}^{(|M| + |P|)^2}$ by

$$\hat{x}_{ij} := x_{ij} \quad \text{if } i \in M, j \in P$$

$$\hat{x}_{ij} := x_{ji} \quad \text{if } i \in P, j \in M$$

$$\hat{x}_{ii} := 1_S(i) - \sum_{j \in P} x_{ij} \quad \text{if } i \in M$$

$$\hat{x}_{jj} := 1_S(j) - \sum_{i \in M} x_{ij} \quad \text{if } j \in P$$

$$\hat{x}_{ij} := 0 \quad \text{in all other cases.}$$

Then $\hat{x}$ satisfies the conditions in problem (20.3). Hence, for every S,

$$w(S) \geq \sum_{i \in N} \sum_{j \in N} k_{ij} \hat{x}_{ij} = \sum_{i \in M} \sum_{j \in P} a_{ij} x_{ij} = v(S) \, .$$

On the other hand, let $z \in \{0, 1\}^{(|M| + |P|)^2}$ be an optimal solution for (20.3). Define $\hat{z} \in \{0, 1\}^{|M| \times |P|}$ by

$$\hat{z}_{ij} := z_{ij} \text{ for } i \in M, \ j \in P \, .$$

Then $\hat{z}$ satisfies the conditions in problem (20.1). Hence, for every S,

$$v(S) \geq \sum_{i \in M} \sum_{j \in P} a_{ij} \hat{z}_{ij} = \sum_{i \in M} \sum_{j \in P} k_{ij} z_{ij} = w(S) \, .$$

Consequently, $v = w$. ∎

The converse of Theorem 20.2 is not true, as the following example shows. As a matter of fact, a necessary condition for a permutation game to be an assignment game is the existence of a partition of the player set N of the permutation game into two subsets N_1 and N_2, such that the value of a coalition S is 0 whenever $S \subseteq N_1$ or $S \subseteq N_2$. The example shows that this is not a sufficient condition.

Example 20.3 Let $N = \{1, 2, 3\}$ and let v be the permutation game with the numbers k_{ij} given in the following matrix:

$$\begin{pmatrix} 0 & 2 & 1 \\ 1 & 0 & 0 \\ 2 & 0 & 0 \end{pmatrix}.$$

Then $v(i) = 0$ for every $i \in N$, $v(1, 2) = v(1, 3) = 3$, $v(2, 3) = 0$, and $v(N) = 4$. Note that this game satisfies the condition formulated above with $N_1 = \{1\}$ and $N_2 = \{2, 3\}$, but it is not an assignment game (Problem 20.2).   □

The main purpose of this section is to show that permutation games and, hence, assignment games are balanced and, in fact, totally balanced. A TU-game (N, v) is *totally balanced* if the *subgame* (M, v)—where v is the restriction to M—is balanced for every $M \subseteq N$. Balanced games are exactly those games that have a non-empty core, see Chap. 16.

Theorem 20.4 *Assignment games and permutation games are totally balanced.*

Proof In view of Theorem 20.2, it is sufficient to prove that permutation games are totally balanced. Because any subgame of a permutation game is again a permutation game (see Problem 20.3), it is sufficient to prove that any permutation game is balanced.

Let (N, v) be a permutation game, defined by (20.3). By the Birkhoff–von Neumann Theorem (Theorem 22.12) the integer restriction can be dropped so that each $v(S)$ is also defined by the following program:

$$\max \sum_{i \in N} \sum_{j \in N} k_{ij} x_{ij}$$
$$\text{subject to } \sum_{j \in N} x_{ij} = 1_S(i) \text{ for all } i \in N$$
$$\sum_{i \in N} x_{ij} = 1_S(j) \text{ for all } j \in N \qquad (20.4)$$
$$x_{ij} \geq 0 \qquad \text{for all } i, j \in N.$$

Note that this is a linear programming problem of the same format as the maximization problem in Theorem 16.20. Namely, with notations as there, take

$$\mathbf{y} = (x_{11}, \ldots, x_{1n}, x_{21}, \ldots, x_{2n}, \ldots, x_{n1}, \ldots, x_{nn})$$
$$\mathbf{b} = (k_{11}, \ldots, k_{1n}, k_{21}, \ldots, k_{2n}, \ldots, k_{n1}, \ldots, k_{nn})$$
$$\mathbf{c} = (1_S, 1_S) \,.$$

Further, let A be the $2n \times n^2$-matrix with row $\ell \in \{1, \ldots, n\}$ containing a 1 at columns $\ell, \ell + n, \ell + 2n, \ldots, \ell + (n-1)n$ and zeros otherwise; and with row $\ell + n$ ($\ell \in \{1, \ldots, n\}$) containing a 1 at columns $(\ell - 1)n + 1, \ldots, \ell n$ and zeros otherwise. The corresponding dual problem, the minimization problem in Theorem 16.20, then has the form:

$$\min \sum_{i \in N} 1_S(i) y_i + \sum_{j \in N} 1_S(j) z_j$$

$$\text{subject to} \qquad y_i + z_j \geq k_{ij} \qquad \text{for all } i, j \in N. \qquad (20.5)$$

Let $(\hat{\mathbf{y}}, \hat{\mathbf{z}})$ be an optimal solution of problem (20.5) for $S = N$. Then, by Theorem 16.20 and the fact that the maximum in problem (20.4) for $S = N$ is equal to $v(N)$ by definition, it follows that

$$\sum_{i \in N} (\hat{y}_i + \hat{z}_i) = v(N) \,.$$

Since $(\hat{\mathbf{y}}, \hat{\mathbf{z}})$ satisfies the restrictions in problem (20.5) for every $S \subseteq N$, it furthermore holds that for every $S \subseteq N$,

$$\sum_{i \in S} (\hat{y}_i + \hat{z}_i) = \sum_{i \in N} 1_S(i) \hat{y}_i + \sum_{i \in N} 1_S(i) \hat{z}_i \geq v(S) \,.$$

Therefore, $\mathbf{u} \in \mathbb{R}^N$ defined by $u_i := \hat{y}_i + \hat{z}_i$ is in the core of v. ∎

20.2 Flow Games

In this section another class of balanced games is considered. These games are derived from the following kind of situation. There is a given capacitated network, the edges of which are controlled by subsets of players. These coalitions can send a flow through the network. The flow is maximal if all players cooperate, and then the question arises how to distribute the profits. One can think of an almost literal example, where the edges represent oil pipelines, and the players are in power in different countries through which these pipelines cross. Alternatively, one can think of rail networks between cities, or information channels between different users.

Fig. 20.1 Example 20.5

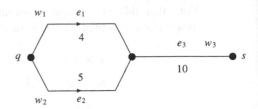

Capacitated networks are treated in Sect. 22.7, which the reader may consult before continuing.

Consider a capacitated network (V, E, k) and a set of players $N := \{1, \ldots, n\}$. Suppose that with each edge in E a simple game (cf. Sect. 16.3.) is associated. The winning coalitions in this simple game are supposed to control the corresponding edge; the capacitated network is called a *controlled* capacitated network. For any coalition $S \subseteq N$ consider the capacitated network arising from the given network by deleting the edges that are *not* controlled by S. A game can be defined by letting the worth of S be equal to the value of a maximal flow through this restricted network. The game thus arising is called a *flow game*.

Example 20.5 Consider the capacitated network in Fig. 20.1. This network has three edges denoted e_1, e_2, and e_3 with capacities 4, 5 and 10, respectively. The control games are w_1, w_2, w_3 with $N = \{1, 2, 3\}$ and

$$w_1(S) = 1 \text{ if } S \in \{\{1, 2\}, N\} \text{ and } w_1(S) = 0 \text{ otherwise}$$
$$w_2(S) = 1 \text{ if } S \in \{\{1, 3\}, N\} \text{ and } w_2(S) = 0 \text{ otherwise}$$
$$w_3(S) = 1 \text{ if, and only if, } 1 \in S.$$

The coalition $\{1, 2\}$ can only use the edges e_1 and e_3, so the maximal flow (per time unit) for $\{1, 2\}$ is 4. This results in $v(\{1, 2\}) = 4$ for the corresponding flow game (N, v). The complete game is given by $v(i) = 0$ for all $i \in N$, $v(\{1, 2\}) = 4$, $v(\{1, 3\}) = 5$, $v(\{2, 3\}) = 0$ and $v(N) = 9$.

A minimum cut in this network corresponding to the grand coalition is $(\{q\}, V \setminus \{q\})$. By the Max Flow Min Cut Theorem of Ford and Fulkerson (Theorem 22.16), the sum of the capacities of e_1 and e_2 $(4 + 5)$ is equal to $v(N)$. Divide $v(N)$ as follows. Divide 4 equally among the veto players of w_1, and 5 equally among the veto players of w_2. The result for the players is the payoff vector $(4\frac{1}{2}, 2, 2\frac{1}{2})$. Note that this vector is in $C(v)$.   □

The next theorem shows that the non-emptiness of the core of the control games is inherited by the flow game.

Theorem 20.6 *Suppose all control games in a controlled capacitated network have veto players. Then the corresponding flow game is balanced.*

Proof Take a maximal flow for the grand coalition and a minimum cut in the network for the grand coalition, consisting of the edges

$$e_1, e_2, \ldots, e_p \text{ with capacities } k_1, k_2, \ldots, k_p$$

and control games w_1, w_2, ..., w_p, respectively. Then Theorem 22.16 implies that $v(N) = \sum_{r=1}^{p} k_r$. For each r take $x^r \in C(w_r)$ and divide k_r according to the division key x^r (i.e. $k_r x_i^r$ is the amount for player i). Note that non-veto players get nothing. Then $\sum_{r=1}^{p} k_r x^r \in C(v)$. To see this, first note that

$$\sum_{i=1}^{n} \sum_{r=1}^{p} k_r x_i^r = \sum_{r=1}^{p} k_r \sum_{i=1}^{n} x_i^r = \sum_{r=1}^{p} k_r = v(N) \,.$$

Next, for each coalition S, the set

$$E_S := \{e_r : r \in \{1, \ldots p\}, w_r(S) = 1\}$$

is associated with a cut of the network, governed by the coalition S. Hence, $\sum_{i \in S}$ ($\sum_{r=1}^{p} k_r x_i^r$) $= \sum_{r=1}^{p} k_r \sum_{i \in S} x_i^r \geq \sum_{r=1}^{p} k_r w_r (S) = \sum_{e_r \in E_S} k_r =$ capacity$(E_S) \geq v(S)$, where the last inequality follows from Theorem 22.16. ∎

The next theorem is a partial converse to Theorem 20.6.

Theorem 20.7 *Each nonnegative balanced game arises from a controlled capacitated network where all control games possess veto players.*

Proof See Problem 20.5. ∎

20.3 Voting Games: The Banzhaf Value

Voting games constitute another special class of TU-games. Voting games are simple games which reflect the distribution of voting power within, for instance, political systems. There is a large body of work on voting games within the political science literature. In this section we restrict ourselves to a brief discussion of a well-known example of a power index, to so-called Banzhaf–Coleman index and the associated value, the Banzhaf value.

A *power index* is a value applied to voting (simple) games. The payoff vector assigned to a game is interpreted as reflecting power distribution—e.g., the probability of having a decisive vote—rather than utility.

We start with an illustrating example.

Example 20.8 Consider a parliament with three parties 1, 2, and 3. The numbers of votes are, respectively, 50, 30, and 20. To pass any law, a two-third majority is needed. This leads to a simple game with winning coalitions $\{1, 2\}$, $\{1, 3\}$, and $\{1, 2, 3\}$. The Shapley value[1] of this game is $(\frac{2}{3}, \frac{1}{6}, \frac{1}{6})$, as can easily be checked. By definition of the Shapley value this means that in four of the six orderings player 1 makes the coalition of his predecessors winning by joining them, whereas for players 2 and 3 this is only the case in one ordering each. The coalitions that are made winning by player 1 if he joins, are $\{2\}$, $\{3\}$, and $\{2, 3\}$. In the Shapley value the last coalition is counted double. It might be more natural to count this coalition only once. This would lead to an outcome $(\frac{3}{5}, \frac{1}{5}, \frac{1}{5})$, instead of the Shapley value. The associated value is called the *normalized Banzhaf–Coleman index*. $\square$

For a simple game (N, v) (see Sect. 16.3), the *normalized Banzhaf–Coleman index* can be defined as follows. Define a *swing* for player i as a coalition $S \subseteq N$ with $i \in S$, S wins, and $S \setminus \{i\}$ looses. Let θ_i be the number of swings for player i, and define the numbers

$$\beta_i(N, v) := \frac{\theta_i}{\sum_{j=1}^{n} \theta_j} .$$

The vector $\beta(N, v)$ is the normalized Banzhaf–Coleman index of the simple game (N, v).

For a general game (N, v) write

$$\theta_i(v) := \sum_{S \subseteq N:\ i \notin S} [v(S \cup i) - v(S)] .$$

For a simple game v this number $\theta_i(v)$ coincides with the number θ_i above.

Next, define the value $\Psi : \mathcal{G}^N \to \mathbb{R}^N$ by

$$\Psi_i(v) := \frac{\theta_i(v)}{2^{|N|-1}} = \sum_{S \subseteq N:\ i \notin S} \frac{1}{2^{|N|-1}} [v(S \cup i) - v(S)] . \qquad (20.6)$$

The value Ψ is called the *Banzhaf value*. The remainder of this section is devoted to an axiomatic characterization of this value. This characterization uses the axioms SYM (Symmetry), SMON (Strong Monotonicity), and DUM (Dummy Property), which were all introduced in Chap. 17. Besides, it uses an 'amalgamation' property, as follows.

---

[1] Also called the *Shapley-Shubik power index* in this context.

For a game (N, v) (with at least two players) and different players i, j put $p = \{i, j\}$ and define the game $((N \setminus p) \cup \{p\}, v_p)$ by

$$v_p(S) = v(S) \text{ and } v_p(S \cup \{p\}) = v(S \cup p), \text{ for any } S \subseteq N \setminus p. \tag{20.7}$$

Thus, v_p is an $(n - 1)$-person game obtained by amalgamating players i and j in v into one player p in v_p.

Let ψ be an arbitrary value (on the class $\mathcal{G}$ of all games with arbitrary player set). The announced axiom is as follows.

2-Efficiency (2-EFF): $\quad \psi_i(v) + \psi_j(v) = \psi_p(v_p)$ for all v, i, j, p, v_p as above.

The following theorem gives a characterization of the Banzhaf value.

Theorem 20.9 *The value ψ on $\mathcal{G}$ satisfies 2-EFF, SYM, DUM, and SMON, if and only if ψ is the Banzhaf value Ψ.*

Proof That the Banzhaf value satisfies the four axioms in the theorem is the subject of Problem 20.9. For the converse, let ψ be a value satisfying the four axioms. We prove that $\psi = \Psi$.

Step 1
Let u_T be a unanimity game. We first show that

$$\psi_i(u_T) = 1/2^{|T|-1} \text{ if } i \in T \text{ and } \psi_i(u_T) = 0 \text{ if } i \notin T. \tag{20.8}$$

If $|T| = 1$ then every player is a dummy, so that (20.8) follows from DUM. Suppose (20.8) holds whenever $|T| \leq k$ or $|N| \leq m$, and consider a unanimity game u_T where now the number of players is $m + 1$, and T contains $k + 1$ players. Let $i, j \in T$, put $p = \{i, j\}$ and consider the game $(u_T)_p$. Then $(u_T)_p$ is the m-person unanimity game of the coalition $T' = (T \setminus p) \cup \{p\}$, and $|T'| = k$. By the induction hypothesis

$$\psi_p((u_T)_p) = 1/2^{|T'|-1} = 1/2^{k-1}.$$

By 2-EFF this implies

$$\psi_i(u_T) + \psi_j(u_T) = 1/2^{k-1}.$$

From this and SYM it follows that

$$\psi_i(u_T) = 1/2^k = 1/2^{|T|-1}$$

for all $i \in T$, and by DUM, $\psi_j(u_T) = 0$ for all $j \notin T$. Thus, ψ is the Banzhaf value on unanimity games for any finite set of players. In the same way, one shows that this is true for any real multiple cu_T of a unanimity game.

Step 2
For an arbitrary game v write $v = \sum_{\emptyset \neq T} c_T u_T$, and let $\alpha(v)$ denote the number of nonzero coefficients in this representation. The proof will be completed by induction on the number $\alpha(v)$ and the number of players. For $\alpha(v) = 1$ Step 1 implies $\psi(v) = \Psi(v)$ independent of the number of players. Assume that $\psi(v) = \Psi(v)$ on any game v with at most n players, and also any game v with $\alpha(v) \leq k$ for some natural number k and with $n + 1$ players, and let v be a game with $n + 1$ players and with $\alpha(v) = k + 1$. There are $k + 1$ different nonempty coalitions $T_1, \ldots, T_{k+1}$ with

$$v = \sum_{r=1}^{k+1} c_{T_r} u_{T_r} ,$$

where all coefficients are nonzero. Let $T := T_1 \cap \ldots \cap T_{k+1}$. Because $k + 1 \geq 2$, it holds that $N \setminus T \neq \emptyset$. Assume $i \notin T$. Define the game w by

$$w = \sum_{r: \, i \in T_r} c_{T_r} u_{T_r} .$$

Then $\alpha(w) \leq k$ and $v(S \cup i) - v(S) = w(S \cup i) - w(S)$ for every coalition S not containing player i. By SMON and the induction hypothesis it follows that $\psi_i(v) = \psi_i(w) = \Psi_i(w) = \Psi_i(v)$. Hence,

$$\psi_i(v) = \Psi_i(v) \text{ for every } i \in N \setminus T. \tag{20.9}$$

Let $j \in T$ and $i \in N \setminus T$, put $p = \{i, j\}$, and consider the game v_p. Because the game v_p has n players the induction hypothesis implies

$$\psi_p(v_p) = \Psi_p(v_p) . \tag{20.10}$$

Applying 2-EFF to both ψ and Ψ yields

$$\psi_p(v_p) = \psi_i(v) + \psi_j(v) \text{ and } \Psi_p(v_p) = \Psi_i(v) + \Psi_j(v) . \tag{20.11}$$

Combining (20.9)–(20.11) implies $\psi_j(v) = \Psi_j(v)$ for every $j \in T$. Together with (20.9) this completes the induction argument, and therefore the proof. ∎

20.4 Problems

20.1. *The Dentist Game*
Show that (20.2) and (20.3) are equivalent, and use each of these to verify that the dentist game of Sect. 1.3.4 is a permutation game.

Fig. 20.2 The network of
Problem 20.4

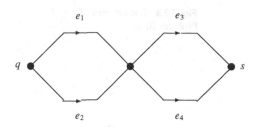

20.2. *Example 20.3*
Show that the game in Example 20.3 is not an assignment game.

20.3. *Subgames of Permutation Games*
Prove that subgames of permutation games are again permutation games. Is this also true for assignment games?

20.4. *A Flow Game*
Consider the network in Fig. 20.2. Suppose that this is a controlled capacitated network with player set $N = \{1, 2, 3, 4\}$, suppose that all edges have capacity 1 and that $w_1 = \delta_1$, $w_2 = \delta_2$, $w_3 = \delta_3$ and $w_4(S) = 1$ iff $S \in \{\{3, 4\}, N\}$. [Here, δ_i is the simple game where a coalition is winning if, and only if, it contains player i.]

(a) Calculate the corresponding flow game (N, v).
(b) Calculate $C(v)$.
(c) The proof of Theorem 20.6 describes a way to find core elements by looking at minimum cuts and dividing the capacities of edges in the minimum cut in some way among the veto players of the corresponding control game. Which elements of $C(v)$ can be obtained in this way?

20.5. *Every Nonnegative Balanced Game is a Flow Game*
Prove that every nonnegative balanced game is a flow game. [Hint: You may use the following result: every nonnegative balanced game can be written as a positive linear combination of balanced simple games.]

20.6. *On Theorem 20.6 (1)*

(a) Consider a controlled capacitated network with a minimum cut, where all control games corresponding to the edges in this minimum cut (connecting vertices between the two sets in the cut) have veto players. Prove that the corresponding flow game is balanced.
(b) Show that the flow game, corresponding to Fig. 20.3, where the winning coalitions of w_1 are $\{1, 3\}$, $\{2, 4\}$ and $N = \{1, 2, 3, 4\}$, where the winning coalitions of w_2 are $\{1, 2\}$ and N and of w_3 $\{3, 4\}$ and N and where the capacities are 1, 10, 10 respectively, has a nonempty core. Note that there is no minimum cut where all control games have veto players.

Fig. 20.3 The network of
Problem 20.6

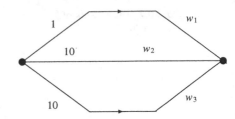

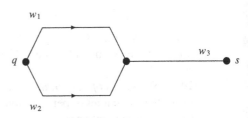

Fig. 20.4 The network of
Problem 20.7

20.7. *On Theorem 20.6 (2)*
Prove that the two-person flow game corresponding to the controlled capacitated
network of Fig. 20.4 has an empty core, where $w_1 = \delta_1$, $w_2 = \delta_2$, $w_3(S) = 1$ if
$S \neq \emptyset$, and where the capacities of the edges are equal to 1.

20.8. *Totally Balanced Flow Games*
Let (N, v) be the flow game corresponding to a controlled capacitated network
where all control games are dictatorial games (games of the form δ_i, see Problem 20.4). Prove that each subgame (S, v_S) (where v_S is the restriction of v to 2^S)
has a nonempty core, i.e., that the game (N, v) is totally balanced.

20.9. *If-Part of Theorem 20.9*
Prove that the Banzhaf value satisfies 2-EFF, SYM, DUM, and SMON. Is it possible
to weaken DUM to NP (the null-player property) in Theorem 20.9? Give an example
showing that the Banzhaf value is not efficient.

20.5   Notes

The presentation in Sect. 20.1 is mainly based on Curiel (1997, Chap. 3). Example 20.1 is from this book. Assignment games were introduced by Shapley and
Shubik (1972). Permutation games were introduced by Tijs et al. (1984).

Theorem 20.6 on flow games is due to Curiel et al. (1986).

In the literature many characterizations of power indices are available. The one
presented in Sect. 20.3 is based on Nowak (1997).

For the result in the hint to Problem 20.5 see Derks (1987). Problem 20.8 refers
to Kalai and Zemel (1982).

References

Curiel, I. (1997). *Cooperative game theory and applications: Cooperative games arising from combinatorial optimization problems.* Boston: Kluwer Academic.

Curiel, I., Derks, J., & Tijs, S. H. (1986). On balanced games and games with committee control. *Operations Research Spektrum, 11,* 83–88.

Derks, J. (1987). Decomposition of games with non-empty core into veto-controlled simple games. *Operations Research Spektrum, 9,* 81–85.

Kalai, E., & Zemel, E. (1982). Totally balanced games and games of flow. *Mathematics of Operations Research, 7,* 476–478.

Nowak, A. S. (1997). On an axiomatization of the Banzhaf value without the additivity axiom. *International Journal of Game Theory, 26,* 137–141.

Shapley, L. S., & Shubik, M. (1972). The assignment game I: The core. *International Journal of Game Theory 1,* 111–130.

Tijs, S. H., Parthasarathy, T., Potters, J. A. M., & Rajendra Prasad, V. (1984). Permutation games: Another class of totally balanced games. *Operations Research Spektrum, 6,* 119–123.

The game-theoretic literature on bargaining can be divided in two strands: the cooperative and the noncooperative approach. Here, the focus is on the cooperative approach, which was initiated by Nash (1950) and which is axiomatic in nature; see Sect. 10.1 for a first discussion. A seminal article on noncooperative bargaining is Rubinstein (1982). The basic idea of that paper is briefly repeated below, but see Sect. 6.7 for a more elaborate discussion. We conclude the chapter with a few remarks on games with nontransferable utility (NTU-games).

21.1 The Bargaining Problem

A *2-person bargaining problem* is a pair $(S, \mathbf{d})$ where S is a compact convex nonempty subset of $\mathbb{R}^2$ and $\mathbf{d}$ is an element of S such that $\mathbf{x} > \mathbf{d}$ for some $\mathbf{x} \in S$. The elements of S are called *outcomes* and $\mathbf{d}$ is the *disagreement outcome*. The interpretation of such a problem $(S, \mathbf{d})$ is as follows. Two bargainers, 1 and 2, have to agree on some outcome $\mathbf{x} \in S$, yielding utility x_i to bargainer i. If they fail to reach such an agreement, they end up with the disagreement utilities $\mathbf{d} = (d_1, d_2)$. B denotes the family of all 2-person bargaining problems.

A *(bargaining) solution* is a map $F : B \to \mathbb{R}^2$ such that $F(S, \mathbf{d}) \in S$ for all $(S, \mathbf{d}) \in B$. Nash (1950) proposed to characterize such a solution by requiring it to satisfy certain axioms. More precisely, he proposed the following axioms.[1]

Weak Pareto Optimality (WPO): $F(S, \mathbf{d}) \in W(S)$ for all $(S, \mathbf{d}) \in B$, where $W(S) := \{\mathbf{x} \in S \mid \forall \mathbf{y} \in \mathbb{R}^2 : \mathbf{y} > \mathbf{x} \Rightarrow \mathbf{y} \notin S\}$ is the *weakly Pareto optimal subset of S*.[2]

---

[1] See Fig. 10.2 for an illustration of these axioms. In Sect. 10.1 the stronger Pareto Optimality is imposed instead of Weak Pareto Optimality. In the diagram—panel (a)—that does not make a difference.

[2] The notation $\mathbf{y} > \mathbf{x}$ means $y_i > x_i$ for $i = 1, 2$.

© Springer-Verlag Berlin Heidelberg 2015
H. Peters, *Game Theory*, Springer Texts in Business and Economics,
DOI 10.1007/978-3-662-46950-7_21

Symmetry (SYM): $F_1(S, \mathbf{d}) = F_2(S, \mathbf{d})$ for all $(S, \mathbf{d}) \in B$ that are *symmetric*, i.e., $d_1 = d_2$ and $S = \{(x_2, x_1) \in \mathbb{R}^2 \mid (x_1, x_2) \in S\}$.

Scale Covariance (SC): $F(\mathbf{a}S + \mathbf{b}, \mathbf{a}\mathbf{d} + \mathbf{b}) = \mathbf{a}F(S, \mathbf{d}) + \mathbf{b}$ for all $(S, \mathbf{d}) \in B$, where $\mathbf{b} \in \mathbb{R}^2$, $\mathbf{a} \in \mathbb{R}^2_{++}$, $\mathbf{a}\mathbf{x} := (a_1 x_1, a_2 x_2)$ for all $\mathbf{x} \in \mathbb{R}^2$, $\mathbf{a}S := \{\mathbf{a}\mathbf{x} \mid \mathbf{x} \in S\}$, and $\mathbf{a}S + \mathbf{b} := \{\mathbf{a}\mathbf{x} + \mathbf{b} \mid \mathbf{x} \in S\}$.[3]

Independence of Irrelevant Alternatives (IIA): $F(S, \mathbf{d}) = F(T, \mathbf{e})$ for all $(S, \mathbf{d})$, $(T, \mathbf{e}) \in B$ with $\mathbf{d} = \mathbf{e}$, $S \subseteq T$, and $F(T, \mathbf{e}) \in S$.

Weak Pareto Optimality says that it should not be possible for both bargainers to gain with respect to the solution outcome. If a game is symmetric, then there is no way to distinguish between the bargainers, and a solution should not do that either: that is what Symmetry requires. Scale Covariance requires the solution to be covariant under positive affine transformations: the underlying motivation is that the *utility functions* of the bargainers are usually assumed to be of the von Neumann–Morgenstern type, which implies that they are representations of preferences unique only up to positive affine transformations (details are omitted here). Independence of Irrelevant Alternatives requires the solution outcome not to change when the set of possible outcomes shrinks, the original solution outcome still remaining feasible.

The *Nash (bargaining) solution* $N : B \to \mathbb{R}^2$ is defined as follows. For every $(S, \mathbf{d}) \in B$,

$$N(S, \mathbf{d}) = \operatorname{argmax}\{(x_1 - d_1)(x_2 - d_2) \mid \mathbf{x} \in S, \ \mathbf{x} \geq \mathbf{d}\} \ .$$

That the Nash bargaining solution is well defined, follows from Problem 21.3.

The following theorem shows that the four conditions above characterize the Nash bargaining solution.

Theorem 21.1 *Let $F : B \to \mathbb{R}^2$ be a bargaining solution. Then the following two statements are equivalent:*

(a) $F = N$.
(b) *F satisfies WPO, SYM, SC, IIA.*

Proof The implication (a)$\Rightarrow$(b) is the subject of Problem 21.4. For the implication (b)$\Rightarrow$(a), assume F satisfies WPO, SYM, SC, and IIA. Let $(S, \mathbf{d}) \in B$, and $\mathbf{z} := N(S, \mathbf{d})$. Note that $\mathbf{z} > \mathbf{d}$. Let $T := \{((z_1 - d_1)^{-1}, (z_2 - d_2)^{-1})(\mathbf{x} - \mathbf{d}) \mid \mathbf{x} \in S\}$. By SC,

$$F(T, \mathbf{0}) = \left(\frac{F_1(S, \mathbf{d})}{z_1 - d_1}, \frac{F_2(S, \mathbf{d})}{z_2 - d_2}\right) - \left(\frac{d_1}{z_1 - d_1}, \frac{d_2}{z_2 - d_2}\right) \tag{21.1}$$

---

[3] $\mathbb{R}^2_{++} = \{\mathbf{x} = (x_1, x_2) \in \mathbb{R}^2 \mid x_1, x_2 > 0\}$.

Fig. 21.1 Proof of
Theorem 21.1

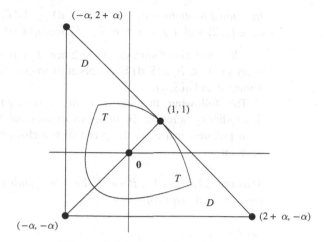

and

$$N(T, \mathbf{0}) = \left(\frac{z_1 - d_1}{z_1 - d_1}, \frac{z_2 - d_2}{z_2 - d_2} \right) = (1, 1) \,. \tag{21.2}$$

Hence, in order to prove $F(S, \mathbf{d}) = N(S, \mathbf{d})$, it is, in view of (21.1) and (21.2), sufficient to show that $F(T, \mathbf{0}) = (1, 1)$. By (21.2) and Problem 21.5, there is a supporting line of T at $(1, 1)$ with slope -1. So the equation of this supporting line is $x_1 + x_2 = 2$. Choose $\alpha > 0$ so large that $T \subseteq D := \text{conv}\{(-\alpha, -\alpha), (-\alpha, 2 + \alpha), (2 + \alpha, -\alpha)\}$. Cf. Fig. 21.1.

Then $(D, \mathbf{0}) \in B$, $(D, \mathbf{0})$ is symmetric, and $W(D) = \text{conv}\{(-\alpha, 2 + \alpha), (2 + \alpha, -\alpha)\}$. Hence by SYM and WPO of F:

$$F(D, \mathbf{0}) = (1, 1) \,. \tag{21.3}$$

Since $T \subseteq D$ and $(1, 1) \in T$, we have by IIA and (21.3): $F(T, \mathbf{0}) = (1, 1)$. This completes the proof. ∎

21.2   The Raiffa–Kalai–Smorodinsky Solution

Kalai and Smorodinsky (1975) replaced Nash's IIA (the most controversial axiom in Theorem 21.1) by the following condition. For a problem $(S, \mathbf{d}) \in B$,

$$u(S, \mathbf{d}) := (\max\{x_1 \mid \mathbf{x} \in S, \ \mathbf{x} \geq \mathbf{d}\}, \max\{x_2 \mid \mathbf{x} \in S, \ \mathbf{x} \geq \mathbf{d}\})$$

is called the *utopia point* of $(S, \mathbf{d})$.

Individual Monotonicity (IM): $F_j(S, \mathbf{d}) \leq F_j(T, \mathbf{e})$ for all $(S, \mathbf{d})$, $(T, \mathbf{e}) \in B$ and $i, j \in \{1, 2\}$ with $i \neq j$, $\mathbf{d} = \mathbf{e}$, $S \subseteq T$, and $u_i(S, \mathbf{d}) = u_i(T, \mathbf{e})$.

The *Raiffa–Kalai–Smorodinsky solution* $R : B \to \mathbb{R}^2$ is defined as follows. For every $(S, \mathbf{d}) \in B$, $R(S, \mathbf{d})$ is the point of intersection of $W(S)$ with the straight line joining $\mathbf{d}$ and $u(S, d)$.

The following theorem presents a characterization of the Raiffa–Kalai–Smorodinsky solution. In order to understand the proof it is recommended to draw pictures, just as in the proof of the characterization of the Nash bargaining solution.

Theorem 21.2 *Let $F : B \to \mathbb{R}^2$ be a bargaining solution. Then the following two statements are equivalent:*

(a) $F = R$.
(b) *F satisfies WPO, SYM, SC, and IM .*

Proof The implication (a)$\Rightarrow$(b) is the subject of Problem 21.7. For the converse implication, assume F has the four properties stated. Let $(S, \mathbf{d}) \in B$ and let $T :=$ $\{\mathbf{ax} + \mathbf{b} \mid \mathbf{x} \in S\}$ with $\mathbf{a} := ((u_1(S, \mathbf{d}) - d_1)^{-1}, (u_2(S, \mathbf{d}) - d_2)^{-1})$, $\mathbf{b} := -\mathbf{ad}$. By SC of R and F, $R(T, \mathbf{0}) = \mathbf{a}R(S, \mathbf{d}) + \mathbf{b}$ and $F(T, \mathbf{0}) = \mathbf{a}F(S, \mathbf{d}) + \mathbf{b}$. Hence, for $F(S, \mathbf{d}) = R(S, \mathbf{d})$, it is sufficient to prove that $R(T, \mathbf{0}) = F(T, \mathbf{0})$.

Since $u(T, \mathbf{0}) = (1, 1)$, $R(T, \mathbf{0})$ is the point of $W(T)$ with equal coordinates, so $R_1(T, \mathbf{0}) = R_2(T, \mathbf{0})$. If $R(T, \mathbf{0}) = (1, 1) = u(T, \mathbf{0})$, then let $L :=$ conv$\{(0, 0), (1, 1)\}$. Then by WPO, $F(L, \mathbf{0}) = (1, 1)$, so by IM, $F(T, \mathbf{0}) \geq F(L, \mathbf{0})$, hence $F(T, \mathbf{0}) = F(L, \mathbf{0}) = R(T, \mathbf{0})$.

Next assume $R(T, \mathbf{0}) < (1, 1)$. Let $\tilde{T} := \{\mathbf{x} \in \mathbb{R}^2 \mid \mathbf{y} \leq \mathbf{x} \leq \mathbf{z} \text{ for some } \mathbf{y}, \mathbf{z} \in T\}$. Clearly $T \subseteq \tilde{T}$ and $u(\tilde{T}, \mathbf{0}) = u(T, \mathbf{0}) = (1, 1)$ so by IM:

$$F(\tilde{T}, \mathbf{0}) \geq F(T, \mathbf{0}) , \tag{21.4}$$

and further, since $R(T, \mathbf{0}) \in W(T)$ and $R_1(\tilde{T}, \mathbf{0}) = R_2(\tilde{T}, \mathbf{0})$,

$$R(\tilde{T}, \mathbf{0}) = R(T, \mathbf{0}) . \tag{21.5}$$

Let $V := $ conv$\{\mathbf{0}, R(T, \mathbf{0}), (1, 0), (0, 1)\}$. By WPO and SYM, $F(V, \mathbf{0}) = R(T, \mathbf{0})$. By $V \subseteq \tilde{T}$, $u(V, \mathbf{0}) = u(\tilde{T}, \mathbf{0}) = (1, 1)$, and IM, we have $F(\tilde{T}, \mathbf{0}) \geq F(V, \mathbf{0}) = R(T, \mathbf{0})$, hence $F(\tilde{T}, \mathbf{0}) = R(T, \mathbf{0})$. Combined with (21.4), this implies $R(T, \mathbf{0}) \geq F(T, \mathbf{0})$, hence $R(T, \mathbf{0}) = F(T, \mathbf{0})$ by WPO and the fact $R(T, \mathbf{0}) < (1, 1)$. This completes the proof. ∎

21.3 The Egalitarian Solution

Consider the following two properties for a bargaining solution F.

Pareto Optimality (PO): $F(S, \mathbf{d}) \in P(S)$ for all $(S, \mathbf{d}) \in B$, where $P(S) := \{\mathbf{x} \in S \mid \forall \mathbf{y} \in S : \mathbf{y} \geq \mathbf{x} \Rightarrow \mathbf{y} = \mathbf{x}\}$ is the *Pareto optimal subset of* S.[4]

Monotonicity (MON): $F(S, \mathbf{d}) \leq F(T, \mathbf{e})$ for all $(S, \mathbf{d}), (T, \mathbf{e}) \in B$ with $S \subseteq T$ and $\mathbf{d} = \mathbf{e}$.

Clearly, $P(S) \subseteq W(S)$ for every $(S, \mathbf{d}) \in B$, and Pareto optimality is a stronger requirement than Weak Pareto Optimality. The Nash and Raiffa–Kalai–Smorodinsky solutions are Pareto optimal, and therefore WPO can be replaced by PO in Theorems 21.1 and 21.2. Monotonicity is much stronger than Individual Monotonicity or Restricted Monotonicity (see Problem 21.8 for the definition of the last axiom) and in fact it is inconsistent with Weak Pareto Optimality. (See Problem 21.10.)

Call a problem $(S, \mathbf{d}) \in B$ *comprehensive* if $\mathbf{z} \leq \mathbf{y} \leq \mathbf{x}$ implies $\mathbf{y} \in S$ for all $\mathbf{z}$, $\mathbf{x} \in S, \mathbf{y} \in \mathbb{R}^2$. By B^c we denote the subclass of comprehensive problems.

The *egalitarian solution* $E : B^c \rightarrow \mathbb{R}^2$ assigns to each problem $(S, \mathbf{d}) \in B^c$ the point $E(S, \mathbf{d}) \in W(S)$ with $E_1(S, \mathbf{d}) - d_1 = E_2(S, \mathbf{d}) - d_2$.

The following axiom is a weakening of Scale Covariance.

Translation Covariance (TC): $F(S + \mathbf{e}, \mathbf{d} + \mathbf{e}) = F(S, \mathbf{d}) + \mathbf{e}$ for all $(S, \mathbf{d}) \in B^c$ and all $\mathbf{e} \in \mathbb{R}^2$.

The following theorem gives a characterization of the egalitarian solution based on Monotonicity.

Theorem 21.3 *Let* $F : B^c \rightarrow \mathbb{R}^2$ *be a bargaining solution. Then the following two statements are equivalent:*

(a) $F = E$.
(b) F *satisfies WPO, MON, SYM, and TC.*

Proof The implication (a)$\Rightarrow$(b) is the subject of Problem 21.11. For the converse implication, let $(S, \mathbf{d}) \in B^c$. We want to show $F(S, \mathbf{d}) = E(S, \mathbf{d})$. In view of TC of F and E, we may assume $\mathbf{d} = \mathbf{0}$. Let $V := \{\mathbf{x} \in \mathbb{R}^2 \mid \mathbf{0} \leq \mathbf{x} \leq E(S, \mathbf{0})\}$. Clearly, $(V, \mathbf{0}) \in B^c$ is a symmetric problem, so $F(V, \mathbf{0}) = E(S, \mathbf{0})$ by SYM and WPO of F. By MON,

$$F(S, \mathbf{0}) \geq F(V, \mathbf{0}) = E(S, \mathbf{0}) . \tag{21.6}$$

---

[4]The notation $\mathbf{y} \geq \mathbf{x}$ means $y_i \geq x_i$ for $i = 1, 2$.

If $E(S, 0) \in P(S)$, then (21.6) implies $F(S, 0) = E(S, 0)$, so we are done. Now suppose $E(S, 0) \in W(S) \setminus P(S)$. Without loss of generality, assume $E_1(S, 0) = u_1(S, 0)$, i.e., $E_1(S, 0) = \max\{x_1 \mid \mathbf{x} \in S, \mathbf{x} \geq \mathbf{0}\}$. Hence, $E_1(S, 0) = F_1(S, 0)$ by (21.6).

Suppose $F_2(S, 0) > E_2(S, 0)$. The proof will be finished by contradiction. Let $\alpha > 0$ with $E_2(S, 0) < \alpha < F_2(S, 0)$. Let $T := \text{conv}(S \cup \{(\alpha, 0), (\alpha, \alpha)\})$. Then $(T, 0) \in B^c$ and $E(T, 0) = (\alpha, \alpha) \in P(T)$, so $F(T, 0) = (\alpha, \alpha)$ by our earlier argument [see the line below (21.6)]. On the other hand, by MON, $F_2(T, 0) \geq F_2(S, 0) > \alpha$, a contradiction.   ∎

An alternative characterization of the egalitarian solution can be obtained by considering the following axioms. For a bargaining problem $(S, \mathbf{d})$, denote $S_\mathbf{d} = \{\mathbf{x} \in S \mid \mathbf{x} > \mathbf{d}\}$.

Super-Additivity (SA): $F(S + T, \mathbf{d} + \mathbf{e}) \geq F(S, \mathbf{d}) + F(T, \mathbf{e})$ for all $(S, \mathbf{d}), (T, \mathbf{e}) \in B^c$. Here, $S + T := \{\mathbf{x} + \mathbf{y} \mid \mathbf{x} \in S, \mathbf{y} \in T\}$.

Independence of Non-Individually Rational Alternatives (INIR): $F(S, \mathbf{d}) = F(S_\mathbf{d}, \mathbf{d})$ for all $(S, \mathbf{d}) \in B^c$.

INIR is a stronger version of the following well-known axiom.

Individual Rationality (IR): $F(S, \mathbf{d}) \geq \mathbf{d}$ for all $(S, \mathbf{d}) \in B^c$.

Observe, indeed, that INIR implies IR.

Theorem 21.4 *Let* $F : B^c \to \mathbb{R}^2$ *be a bargaining solution. Then the following two statements are equivalent:*

(a) $F = E$.
(b) *F satisfies WPO, SA, SYM, INIR, and TC.*

Proof (a)$\Rightarrow$(b) follows from Theorem 21.3 and Problem 21.12. For the converse implication, let $(S, \mathbf{d}) \in B^c$. We wish to show $F(S, \mathbf{d}) = E(S, \mathbf{d})$. In view of TC of F and E we may assume $\mathbf{d} = \mathbf{0}$. For every $1 > \varepsilon > 0$ let $V^\varepsilon := \{\mathbf{x} \in \mathbb{R}^2 \mid \mathbf{0} \leq \mathbf{x} \leq (1 - \varepsilon)E(S, 0)\}$. Then $(V^\varepsilon, 0) \in B^c$ and $F(V^\varepsilon, 0) = E(V^\varepsilon, 0) = (1 - \varepsilon)E(S, 0)$ by WPO and SYM of F and E. Define $W^\varepsilon = \{\mathbf{x} - (1 - \varepsilon)E(S, 0) \mid \mathbf{x} \in S\}$ and $S^\varepsilon = W^\varepsilon + V^\varepsilon$ for every $1 > \varepsilon > 0$. Then $S_0 = S_0^\varepsilon$ and by SA we have

$$F(S^\varepsilon, 0) \geq (1 - \varepsilon)E(S, 0) + F(W^\varepsilon, 0) ,$$

hence by INIR

$$F(S, 0) \geq (1 - \varepsilon)E(S, 0) + F(W^\varepsilon, 0) \tag{21.7}$$

for all $1 > \varepsilon > 0$. Letting ε decrease to 0, we obtain by (21.7) and IR (which is implied by INIR):

$$F(S, \mathbf{0}) \geq E(S, \mathbf{0}) . \tag{21.8}$$

If $E(S, \mathbf{0}) \in P(S)$, then (21.8) implies $F(S, \mathbf{0}) = E(S, \mathbf{0})$ and we are done. Otherwise, suppose without loss of generality that $E_1(S, \mathbf{0}) = \max\{x_1 \mid \mathbf{x} \in S, \mathbf{x} \geq \mathbf{0}\}$. Let $\mathbf{z}$ be the point of $P(S)$ with $z_1 = E_1(S, \mathbf{0})$, hence $\alpha := E_2(S, \mathbf{0}) - z_2 < 0$ since, by assumption, $E(S, \mathbf{0}) \notin P(S)$. For $\varepsilon > 0$, let $R^\varepsilon := \text{conv}\{(0, \varepsilon), (0, \alpha), (\varepsilon, \alpha)\}$. Then $(R^\varepsilon, \mathbf{0}) \in B^c$. Further, let $T^\varepsilon := S + R^\varepsilon$. By construction, $E(T^\varepsilon, \mathbf{0}) \in P(T^\varepsilon)$, hence, as before, $F(T^\varepsilon, \mathbf{0}) = E(T^\varepsilon, \mathbf{0})$. If ε approaches 0, $F(T^\varepsilon, \mathbf{0})$ converges to $E(S, \mathbf{0})$ and by SA and IR, $F(T^\varepsilon, \mathbf{0}) \geq F(S, \mathbf{0})$. So $E(S, \mathbf{0}) \geq F(S, \mathbf{0})$. Combined with (21.8), this implies $F(S, \mathbf{0}) = E(S, \mathbf{0})$.   ∎

21.4 Noncooperative Bargaining

A different approach to bargaining is obtained by studying it a as strategic process. In this section we discuss the basics of the model of Rubinstein (1982) in an informal manner. See also Sect. 6.7 for a more elaborate treatment.

Point of departure is a bargaining problem $(S, \mathbf{d}) \in B$. Assume $\mathbf{d} = \mathbf{0}$ and write S instead of $(S, \mathbf{d})$. Suppose bargaining takes place over time, at moments $t = 0, 1, 2, \ldots$. At even moments, player 1 makes some proposal $\mathbf{x} = (x_1, x_2) \in P(S)$ and player 2 accepts or rejects it. At odd moments, player 2 makes some proposal $\mathbf{x} = (x_1, x_2) \in P(S)$ and player 1 accepts or rejects it. The game ends as soon as a proposal is accepted. If a proposal $\mathbf{x} = (x_1, x_2)$ is accepted at time t, then the players receive payoffs $(\delta^t x_1, \delta^t x_2)$. Here $0 < \delta < 1$ is a so called discount factor; it reflects impatience of the players, for instance because of foregone interest payments ('shrinking cake'). If no proposal is ever accepted, then the game ends with the disagreement payoffs of $(0, 0)$.

Suppose player 1 has in mind to make some proposal $\mathbf{y} = (y_1, y_2) \in P(S)$, and that player 2 has in mind to make some proposal $\mathbf{z} = (z_1, z_2) \in P(S)$. So player 1 offers to player 2 the amount y_2. Player 2 expects to get z_2 if he rejects $\mathbf{y}$, but he will get z_2 one round later. So player 1's proposal $\mathbf{y}$ will be rejected by player 2 if $y_2 < \delta z_2$; on the other hand, there is no need to offer strictly more than δz_2. This leads to the equation

$$y_2 = \delta z_2 . \tag{21.9}$$

By reversing in this argument the roles of players 1 and 2 one obtains

$$z_1 = \delta y_1 . \tag{21.10}$$

These two equations define unique points $\mathbf{y}$ and $\mathbf{z}$ in $P(S)$. The result of the Rubinstein bargaining approach is that player 1 starts by offering $\mathbf{y}$, player 2 accepts, and the game ends with the payoffs $\mathbf{y} = (y_1, y_2)$.

This description is informal. Formally, one defines a dynamic noncooperative game and looks for the (in this case) subgame perfect Nash equilibria of this game. It can be shown that all such equilibria result in the payoffs $\mathbf{y}$ (or in $\mathbf{z}$ if player 2 would start instead of player 1).

The surprising fact is that, although at first sight the Rubinstein approach is quite different from the axiomatic approach by Nash (Theorem 21.1) the resulting outcomes turn out to be closely related. From Eqs. (21.9) and (21.10) one derives easily that $y_1 y_2 = z_1 z_2$, i.e., the points $\mathbf{y}$ and $\mathbf{z}$ are on the same level curve of the function $\mathbf{x} = (x_1, x_2) \mapsto x_1 x_2$, which appears in the definition of the Nash bargaining solution. Moreover, if the discount factor δ approaches 1, the points $\mathbf{y}$ and $\mathbf{z}$ converge to one another on the curve $P(S)$, and hence to the Nash bargaining solution outcome. In words, as the players become more patient, the outcome of the Rubinstein model converges to the Nash bargaining solution outcome.

21.5   Games with Nontransferable Utility

Both TU-games and bargaining problems are special cases of NTU-games, games with nontransferable utility. In an NTU-game, the possibilities from cooperation for each coalition are described by a set, rather than a single number. For a TU-game (N, v) those sets can be defined as

$$V(S) = \{\mathbf{x} \in \mathbb{R}^S \mid x(S) \le v(S)\}$$

for every coalition S. For a two-person bargaining problem $(S, \mathbf{d})$ the set of feasible payoffs is S for the grand coalition $\{1, 2\}$ and $(-\infty, d_i]$ for each player i.

The core can be extended to NTU-games (for bargaining problems it is just the part of the Pareto optimal set weakly dominating the disagreement outcome). Also the balancedness concept can be extended; the main result here is that balanced games have a nonempty core, but the converse is not true. For further remarks and references see the Notes section.

21.6   Problems

21.1. *Anonymity and Symmetry*
Call a two-person bargaining solution *anonymous* if $F_1(S', \mathbf{d}') = F_2(S, \mathbf{d})$ and $F_2(S', \mathbf{d}') = F_1(S, \mathbf{d})$ whenever $(S, \mathbf{d})$, $(S', \mathbf{d}') \in B$ with $S' = \{(x_2, x_1) \in \mathbb{R}^2 \mid (x_1, x_2) \in S\}$ and $(d_1', d_2') = (d_2, d_1)$. Prove that Anonymity implies Symmetry but not vice versa.

21.2. *Revealed Preference*
Let $B_0 = \{(S, \mathbf{d}) \in B \mid \mathbf{d} = (0, 0)\}$. For $(S, \mathbf{d}) \in B_0$ write S instead of $(S, \mathbf{0})$. Let $\succeq$ be a binary relation on $\mathbb{R}^2$ and $F : B_0 \to \mathbb{R}^2$ a solution. Say that $\succeq$ *represents* F if for every $S \in B_0$:

$$\{F(S)\} = \{\mathbf{x} \in S \mid \mathbf{x} \succeq \mathbf{y} \text{ for every } \mathbf{y} \in S\},$$

i.e., if F uniquely maximizes $\succeq$ on S. Prove: F satisfies IIA if and only if F can be represented by a binary relation $\succeq$.

21.3. *The Nash Solution Is Well-Defined*
Show that N is well defined, i.e., that the function $(x_1 - d_1)(x_2 - d_2)$ takes its maximum on $\{\mathbf{x} \in S \mid \mathbf{x} \geq \mathbf{d}\}$ at a unique point.

21.4. $(a) \Rightarrow (b)$ *in Theorem 21.1*
Show that N satisfies the properties WPO, SYM, SC, and IIA.

21.5. *Geometric Characterization of the Nash Bargaining Solution*
Show that, for every $(S, \mathbf{d}) \in B$, $N(S, \mathbf{d}) = \mathbf{z} > \mathbf{d}$ if and only if there is a supporting line of S at $\mathbf{z}$ with slope the negative of the slope of the straight line through $\mathbf{d}$ and $\mathbf{z}$.

21.6. *Strong Individual Rationality*
Call a solution F *strongly individually rational* (SIR) if $F(S, \mathbf{d}) > \mathbf{d}$ for all $(S, \mathbf{d}) \in B$. The *disagreement* solution D is defined by $D(S, \mathbf{d}) := \mathbf{d}$ for every $(S, \mathbf{d}) \in B$. Show that the following two statements for a solution F are equivalent:

(a) $F = N$ or $F = D$.
(b) F satisfies IR, SYM, SC, and IIA.

Derive from this that N is the unique solution with the properties SIR, SYM, SC, and IIA. (Hint: For the implication $(b) \Rightarrow (a)$, show that, for every $(S, \mathbf{d}) \in B$, either $F(S, \mathbf{d}) = \mathbf{d}$ or $F(S, \mathbf{d}) \in W(S)$. Also show that, if $F(S, \mathbf{d}) = \mathbf{d}$ for *some* $(S, \mathbf{d}) \in B$, then $F(S, \mathbf{d}) = \mathbf{d}$ for *all* $(S, \mathbf{d}) \in B$.)

21.7. $(a) \Rightarrow (b)$ *in Theorem 21.2*
Show that the Raiffa–Kalai–Smorodinsky solution has the properties WPO, SYM, SC, and IM.

21.8. *Restricted Monotonicity*
Call a solution $F : B \to \mathbb{R}^2$ *restrictedly monotonic* (RM) if $F(S, \mathbf{d}) \leq F(T, \mathbf{e})$ whenever $(S, \mathbf{d}), (T, \mathbf{e}) \in B$, $\mathbf{d} = \mathbf{e}$, $S \subseteq T$, $u(S, \mathbf{d}) = u(T, \mathbf{e})$.

(a) Prove that IM implies RM.
(b) Show that RM does not imply IM.

21.9. *Global Individual Monotonicity*

For a problem $(S, \mathbf{d}) \in B$, $g(S) := (\max\{x_1 \mid \mathbf{x} \in S\}, \max\{x_2 \mid \mathbf{x} \in S\})$ is called the *global utopia point* of S. *Global Individual Monotonicity* (GIM) is defined in the same way as IM, with the condition "$u_i(S, \mathbf{d}) = u_i(T, \mathbf{e})$" replaced by: $g_i(S) = g_i(T)$. The solution $G : B \to \mathbb{R}^2$ assigns to each $(S, \mathbf{d}) \in B$ the point of intersection of $W(S)$ with the straight line joining $\mathbf{d}$ and $g(S)$. Show that G is the unique solution with the properties WPO, SYM, SC, and GIM.

21.10. *Monotonicity and (Weak) Pareto Optimality*

(a) Show that there is no solution satisfying MON and WPO.

(b) Show that, on the subclass B_0 introduced in Problem 21.2, there is no solution satisfying MON and PO. Can you find a solution on this class with the properties MON and WPO?

21.11. *The Egalitarian Solution (1)*

(a) Show that E satisfies MON, SYM, and WPO (on B^c).

(b) Show that E is translation covariant on B^c.

21.12. *The Egalitarian Solution (2)*

Show that the egalitarian solution is super-additive.

21.13. *Independence of Axioms*

In the characterization Theorems 21.1–21.4, show that none of the axioms used can be dispensed with.

21.14. *Nash and Rubinstein*

Suppose two players (bargainers) bargain over the division of one unit of a perfectly divisible good. Player 1 has utility function $u_1(\alpha) = \alpha$ and player 2 has utility function $u_2(\alpha) = 1 - (1 - \alpha)^2$ for amounts $\alpha \in [0, 1]$ of the good. If they do not reach an agreement on the division of the good they both receive nothing.

(a) Determine the set of feasible utility pairs. Make a picture.

(b) Determine the Nash bargaining solution outcome, in terms of utilities as well as of the physical distribution of the good.

(c) Suppose the players' utilities are discounted by a factor $\delta \in (0, 1)$. Calculate the Rubinstein bargaining outcome.

(d) Determine the limit of the Rubinstein bargaining outcome, for δ approaching 1, in two ways: by using the result of (b) and by using the result of (c).

21.7 Notes

The axiomatic study of bargaining problems was initiated by Nash (1950). For comprehensive surveys see Peters (1992) or Thomson (1994). Theorem 21.1 is due to Nash (1950).

The Raiffa–Kalai–Smorodinsky solution was introduced by Raiffa (1953) and axiomatized by Kalai and Smorodinsky (1975). Theorem 21.2 is a modified version of this characterization.

For the analysis of the noncooperative bargaining model in Sect. 21.4 see Rubinstein (1982) or Sutton (1986). For an elaborate discussion of noncooperative bargaining models see Muthoo (1999). The observation about the relation with the Nash bargaining solution is due to Binmore et al. (1986).

The fact that balanced NTU-games have a nonempty core is due to Scarf (1976). A complete characterization of NTU games with nonempty core based on a kind of local balancedness condition is provided by Predtetchinski and Herings (2004).

Most other solution concepts for NTU-games—in particular the Harsanyi (1963) and Shapley (1969) NTU-values, and the consistent value of Hart and Mas-Collel (1996)—extend the Nash bargaining solution as well as the Shapley value for TU-games. An exception are the monotonic solutions of Kalai and Samet (1985), which extend the egalitarian solution of the bargaining problem. See de Clippel et al. (2004) for an overview of various axiomatic characterizations of values for NTU-games, and see Peters (2003) for an overview of NTU-games in general. An extensive textbook treatment can be found in Peleg and Sudhölter (2003, Part II).

Most (though not all) results of this chapter on bargaining can be extended to the n-person case without too much difficulty. This is not true for the Rubinstein approach, the extension of which is not obvious. One possibility is presented by Hart and Mas-Collel (1996).

References

Binmore, K., Rubinstein, A., & Wolinsky, A. (1986). The Nash bargaining solution in economic modelling. *Rand Journal of Economics, 17*, 176–188.

de Clippel, G., Peters, H., & Zank, H. (2004). Axiomatizing the Harsanyi solution, the symmetric egalitarian solution, and the consistent solution for NTU-games. *International Journal of Game Theory, 33*, 145–158.

Harsanyi, J. C. (1963). A simplified bargaining model for the n-person cooperative game. *International Economic Review, 4*, 194–220.

Hart, S., & Mas-Collel, A. (1996). Bargaining and value. *Econometrica, 64*, 357–380.

Kalai, E., & Samet, D. (1985). Monotonic solutions to general cooperative games. *Econometrica, 53*, 307–327.

Kalai, E., & Smorodinsky, M. (1975). Other solutions to Nash's bargaining problem. *Econometrica, 43*, 513–518.

Muthoo, A. (1999). *Bargaining theory with applications*. Cambridge: Cambridge University Press.

Nash, J. F. (1950). The bargaining problem. *Econometrica, 18*, 155–162.

Peleg, B., & Sudhölter, P. (2003). *Introduction to the theory of cooperative games*. Boston: Kluwer Academic.

Peters, H. (1992). *Axiomatic bargaining game theory*. Dordrecht: Kluwer Academic.

Peters, H. (2003). NTU-games. In U. Derigs (Ed.), *Optimization and operations research. Encyclopedia of life support systems (EOLSS)*. Oxford: Eolss Publishers. http://www.eolss.net.

Predtetchinski, A., & Herings, P. J. J. (2004). A necessary and sufficient condition for non-emptiness of the core of a non-transferable utility game. *Journal of Economic Theory, 116*, 84–92.

Raiffa, H. (1953). Arbitration schemes for generalized two-person games. *Annals of Mathematics Studies, 28*, 361–387.

Rubinstein, A. (1982). Perfect equilibrium in a bargaining model. *Econometrica, 50*, 97–109.

Scarf, H. E. (1976). The core of an *n*-person game. *Econometrica, 35*, 50–69.

Shapley, L. S. (1969). Utility comparisons and the theory of games. In: G. Th. Guilbaud (Ed.), *La Decision* (pp. 251–263). Paris: CNRS.

Sutton, J. (1986). Non-cooperative bargaining theory: An introduction. *Review of Economic Studies, 53*, 709–724.

Thomson, W. (1994). Cooperative models of bargaining. In R. J. Aumann & S. Hart (Eds.), *Handbook of game theory with economic applications* (Vol. 2). Amsterdam: North-Holland.

Part IV

Tools, Hints and Solutions

Tools

22

This chapter collects some mathematical tools used in this book: (direct) convex separation results in Sects. 22.2 and 22.6; Lemmas of the Alternative, in particular Farkas' Lemma in Sect. 22.3; the Linear Duality Theorem in Sect. 22.4; the Brouwer and Kakutani Fixed Point Theorems in Sect. 22.5; the Krein–Milman Theorem and the Birkhoff–von Neumann Theorem in Sect. 22.6; and the Max Flow Min Cut Theorem of Ford and Fulkerson in Sect. 22.7.

22.1 Some Definitions

A subset $Z \subseteq \mathbb{R}^n$ is *convex* if with any two points $\mathbf{x}, \mathbf{y} \in Z$, also the line segment connecting $\mathbf{x}$ and $\mathbf{y}$ is contained in Z. Formally:

$$\forall \, \mathbf{x}, \mathbf{y} \in Z \; \forall \, 0 \le \lambda \le 1 : \lambda \mathbf{x} + (1 - \lambda) \mathbf{y} \in Z \, .$$

If Z is a closed set[1] then for convexity it is sufficient to check this condition for $\lambda = 1/2$ (see Problem 22.1). It is easy to see that a set $Z \subseteq \mathbb{R}^n$ is convex if and only if $\sum_{j=1}^{k} \lambda_j \mathbf{x}^j \in Z$ for all $\mathbf{x}^1, \ldots, \mathbf{x}^k \in Z$ and all nonnegative $\lambda_1, \ldots, \lambda_k \in \mathbb{R}$ with $\sum_{j=1}^{k} \lambda_j = 1$. Such a sum $\sum_{j=1}^{k} \lambda_j \mathbf{x}^j$ is called a *convex combination* of the $\mathbf{x}^j$. For an arbitrary subset $D \subseteq \mathbb{R}^n$, the *convex hull* of D is the set of all convex combinations of elements of D or, equivalently, the smallest (with respect to set inclusion) convex subset of $\mathbb{R}^n$ containing D.

For vectors $\mathbf{x} = (x_1, \ldots, x_n), \mathbf{y} = (y_1, \ldots, y_n) \in \mathbb{R}^n$,

$$\mathbf{x} \cdot \mathbf{y} := \sum_{i=1}^{n} x_i y_i$$

---

[1] A set $Z \subseteq \mathbb{R}^n$ is *closed* if it contains the limit of every converging sequence in Z.

© Springer-Verlag Berlin Heidelberg 2015
H. Peters, *Game Theory*, Springer Texts in Business and Economics,
DOI 10.1007/978-3-662-46950-7_22

denotes the *inner product* of **x** and **y**, and

$$||\mathbf{x} - \mathbf{y}|| := \sqrt{\sum_{i=1}^{n}(x_i - y_i)^2}$$

is the *Euclidean distance* between **x** and **y**. A set $C \subseteq \mathbb{R}^n$ is a (convex) *cone* if, with each $\mathbf{x}, \mathbf{y} \in C$ and $\lambda \in \mathbb{R}$, $\lambda \geq 0$, also $\lambda \mathbf{x} \in C$ and $\mathbf{x} + \mathbf{y} \in C$.

22.2   A Separation Theorem

In this section we derive the simplest version of a separation result, namely separating a point from a convex set.

Theorem 22.1 *Let $Z \subseteq \mathbb{R}^n$ be a closed convex set and let $\mathbf{x} \in \mathbb{R}^n \setminus Z$. Then there is a $\mathbf{y} \in \mathbb{R}^n$ with $\mathbf{y} \cdot \mathbf{z} > \mathbf{y} \cdot \mathbf{x}$ for every $\mathbf{z} \in Z$.*

Thus, this theorem states the geometrically obvious fact that a closed convex set and a point not in that set can be *separated* by a hyperplane (with normal **y**).

Proof of Theorem 22.1 Let $\mathbf{z}' \in Z$ such that $0 < ||\mathbf{x} - \mathbf{z}'|| \leq ||\mathbf{x} - \mathbf{z}||$ for all $\mathbf{z} \in Z$. Such a $\mathbf{z}'$ exists by the Theorem of Weierstrass, since the Euclidean distance from **x** is a continuous function on the set Z, and for the minimum of $\mathbf{z} \rightarrow ||\mathbf{x} - \mathbf{z}||$ on Z attention can be restricted to a compact (i.e., bounded and closed) subset of Z. Let $\mathbf{y} = \mathbf{z}' - \mathbf{x}$. Let $\mathbf{z} \in Z$. For any α, $0 \leq \alpha \leq 1$, convexity of Z implies $\mathbf{z}' + \alpha(\mathbf{z} - \mathbf{z}') \in Z$, and thus

$$||\mathbf{z}' + \alpha(\mathbf{z} - \mathbf{z}') - \mathbf{x}||^2 \geq ||\mathbf{z}' - \mathbf{x}||^2 \ .$$

Hence,

$$2\alpha(\mathbf{z}' - \mathbf{x}) \cdot (\mathbf{z} - \mathbf{z}') + \alpha^2 ||\mathbf{z} - \mathbf{z}'||^2 \geq 0 \ .$$

Thus, letting $\alpha \downarrow 0$, it follows that $(\mathbf{z}' - \mathbf{x}) \cdot (\mathbf{z} - \mathbf{z}') \geq 0$. From this, $(\mathbf{z}' - \mathbf{x}) \cdot \mathbf{z} \geq (\mathbf{z}' - \mathbf{x}) \cdot \mathbf{z}' = (\mathbf{z}' - \mathbf{x}) \cdot \mathbf{x} + (\mathbf{z}' - \mathbf{x}) \cdot (\mathbf{z}' - \mathbf{x}) > (\mathbf{z}' - \mathbf{x}) \cdot \mathbf{x}$.
    Because **z** was arbitrary, it follows that $\mathbf{y} \cdot \mathbf{z} > \mathbf{y} \cdot \mathbf{x}$ for every $\mathbf{z} \in Z$. ∎

Remark 22.2 A consequence of Theorem 22.1 is that there are real numbers α and β satisfying $\mathbf{y} \cdot \mathbf{z} > \alpha$ and $\mathbf{y} \cdot \mathbf{x} < \alpha$, and $\mathbf{y} \cdot \mathbf{z} > \beta$ and $\mathbf{y} \cdot \mathbf{x} = \beta$, for all $\mathbf{z} \in Z$ (notations as in the theorem). The last assertion is trivial. For the first assertion, note that in the proof of the theorem we have $\mathbf{y} \cdot \mathbf{z} \geq \mathbf{y} \cdot \mathbf{z}'$ for all $\mathbf{z} \in Z$, so $\mathbf{y} \cdot \mathbf{z}'$ is a lower bound for $\mathbf{y} \cdot \mathbf{z}$. Then take, for instance, $\alpha = \frac{1}{2}(\mathbf{y} \cdot \mathbf{z}' + \mathbf{y} \cdot \mathbf{x})$.

22.3 Lemmas of the Alternative

Theorem 22.1 can be used to derive several *lemmas of the alternative*. These lemmas have in common that they describe two systems of linear inequalities and equations, exactly one of which has a solution.

Lemma 22.3 (Theorem of the Alternative for Matrices) *Let A be an $m \times n$ matrix. Exactly one of the following two statements is true.*

(a) *There are $\mathbf{y} \in \mathbb{R}^n$ and $\mathbf{z} \in \mathbb{R}^m$ with $(\mathbf{y}, \mathbf{z}) \geq \mathbf{0}$, $(\mathbf{y}, \mathbf{z}) \neq \mathbf{0}$ and $A\mathbf{y} + \mathbf{z} = \mathbf{0}$.*
(b) *There is an $\mathbf{x} \in \mathbb{R}^m$ with $\mathbf{x} > \mathbf{0}$ and $\mathbf{x}A > \mathbf{0}$.*

Proof We leave it to the reader to prove that at most one of the systems in (a) and (b) has a solution (Problem 22.2).

Now suppose that (a) is not true. It is sufficient to prove that the system in (b) must have a solution. Observe that (a) implies that $\mathbf{0}$ is a convex combination of the columns of A and the set $\{\mathbf{e}^j \in \mathbb{R}^m \mid j = 1, \ldots, m\}$. This follows from dividing both sides of the equation $A\mathbf{y} + \mathbf{z} = 0$ by $\sum_{j=1}^n y_j + \sum_{i=1}^m z_i$. Hence, the assumption that (a) is not true means that $\mathbf{0} \notin Z$, where $Z \subseteq \mathbb{R}^m$ is the convex hull of the columns of A and the set $\{\mathbf{e}^j \in \mathbb{R}^m \mid j = 1, \ldots, m\}$. By Theorem 22.1 and Remark 22.2 there is an $\mathbf{x} \in \mathbb{R}^m$ and a number $\beta \in \mathbb{R}$ such that $\mathbf{x} \cdot \mathbf{p} > \beta$ for all $p \in Z$ and $\mathbf{x} \cdot \mathbf{0} = \beta$. Hence, $\beta = 0$ and, in particular, $\mathbf{x}A > \mathbf{0}$ and $\mathbf{x} > \mathbf{0}$ since the columns of A and all $\mathbf{e}^j$ for $j = 1, \ldots, m$ are elements of Z. Thus, (b) is true. ∎

Another lemma of the alternative is Farkas' Lemma below. In its proof we use the following result.

Lemma 22.4 *Let A be an $m \times n$ matrix and let*

$$Z = \{\mathbf{z} \in \mathbb{R}^n \mid \text{ there exists an } \mathbf{x} \in \mathbb{R}^m, \mathbf{x} \geq \mathbf{0} \text{ with } \mathbf{z} = \mathbf{x}A\}.$$

Then Z is closed.

Proof

(a) Suppose that the rank of the matrix A, $r(A)$, is equal to m. Then also $r(AA^T) = m$, where A^T is the transpose of A (Problem 22.3). Therefore, AA^T is invertible. Let $(\mathbf{z}^n)_{n \in \mathbb{N}}$ be a sequence in Z converging to $\mathbf{z} \in \mathbb{R}^n$. Let $\mathbf{z}^n = \mathbf{x}^n A$, $\mathbf{x}^n \geq \mathbf{0}$, for all $n \in \mathbb{N}$. Then $\mathbf{x}^n = \mathbf{x}^n AA^T (AA^T)^{-1}$ for every n, hence $\mathbf{x}^n A \to \mathbf{z}$ implies $\mathbf{x}^n \to \mathbf{z}A^T (AA^T)^{-1} =: \mathbf{x}$, and in particular $\mathbf{x} \geq \mathbf{0}$. Thus, since $\mathbf{x}^n A \to \mathbf{x}A$, we obtain $\mathbf{z} = \mathbf{x}A$, so that $\mathbf{z} \in Z$.

(b) Let $b \in Z \setminus \{\mathbf{0}\}$ and choose $\mathbf{x} \in \mathbb{R}^m$, $\mathbf{x} \geq \mathbf{0}$, with $\mathbf{b} = \mathbf{x}A$ such that $|S|$ is maximal, where $S := \{i \in \{1, \ldots, m\} \mid x_i > 0\}$. We show that the rows of A with numbers in S are linearly independent. If not, then there is a $\mu \in \mathbb{R}^m$ with $\mu A = \mathbf{0}$,

$\mu_j \neq 0$ for some $j \in S$, and $\mu_i = 0$ for all $i \notin S$. Then $(\mathbf{x} - t\mu)A = \mathbf{x}A = \mathbf{b}$ for every $t \in \mathbb{R}$. Choose $\hat{t}$ in such a way that $x_j - \hat{t}\mu_j \geq 0$ for all $j \in S$ and $x_j - \hat{t}\mu_j = 0$ for some $j \in S$. Then $\mathbf{b} = (\mathbf{x} - \hat{t}\mu)A$, $\mathbf{x} - \hat{t}\mu \geq \mathbf{0}$, and $|\{i \in \{1, \ldots, m\} \mid x_i - t\mu_i > 0\}| \leq |S| - 1$, a contradiction.

(c) In view of part (b) of the proof we can write

$$Z = \bigcup_B \{\mathbf{x}B \mid B \text{ a } k \times n \text{ submatrix of } A, r(B) = k \leq r(A), \mathbf{0} \leq \mathbf{x} \in \mathbb{R}^k\}.$$

By part (a), each of the sets at the right hand side of this equation is closed. Hence, Z is the union of finitely many closed sets, and therefore is closed itself.   ∎

Lemma 22.5 (Farkas' Lemma) *Let A be an $m \times n$ matrix and $\mathbf{b} \in \mathbb{R}^n$. Exactly one of the following two statements is true.*

(a) *There is an $\mathbf{x} \in \mathbb{R}^m$ with $\mathbf{x} \geq \mathbf{0}$ and $\mathbf{x}A = \mathbf{b}$.*
(b) *There is a $\mathbf{y} \in \mathbb{R}^n$ with $A\mathbf{y} \geq \mathbf{0}$ and $\mathbf{b} \cdot \mathbf{y} < 0$.*

Proof We leave it to the reader to show that at most one of the two systems in (a) and (b) can have a solution (Problem 22.4). Assume that the system in (a) does not have a solution. It is sufficient to prove that the system in (b) must have a solution.

The assumption that the system in (a) does not have a solution is equivalent to the statement $\mathbf{b} \notin Z$ where

$$Z = \{\mathbf{z} \in \mathbb{R}^n \mid \text{ there exists an } \mathbf{x} \in \mathbb{R}^m, \mathbf{x} \geq \mathbf{0} \text{ with } \mathbf{z} = \mathbf{x}A\}.$$

Clearly, the set Z is convex, and it is closed by Lemma 22.4. By Theorem 22.1 and Remark 22.2 it follows that there is a $\mathbf{y} \in \mathbb{R}^n$ and an $\alpha \in \mathbb{R}$ with $\mathbf{y} \cdot \mathbf{b} < \alpha$ and $\mathbf{y} \cdot \mathbf{z} > \alpha$ for all $\mathbf{z} \in Z$. Because $\mathbf{0} \in Z$ it follows that $\alpha < \mathbf{y} \cdot \mathbf{0} = 0$, hence $\mathbf{y} \cdot \mathbf{b} < \alpha < 0$. To prove that the system in (b) has a solution, it is sufficient to prove that $A\mathbf{y} \geq \mathbf{0}$. Suppose not, i.e., there is an i with $(A\mathbf{y})_i < 0$. Then $\mathbf{e}^i A\mathbf{y} < 0$, so $(M\mathbf{e}^i)A\mathbf{y} \to -\infty$ as $\mathbb{R} \ni M \to \infty$. Observe, however, that $(M\mathbf{e}^i)A \in Z$ for every $M > 0$, so that $(M\mathbf{e}^i)A\mathbf{y} > \alpha$ for every such M. This contradiction completes the proof of the lemma.   ∎

These lemmas can be interpreted geometrically. We show this for Farkas' Lemma in Fig. 22.1. Consider the row vectors $\mathbf{r}_i$ of A as points in $\mathbb{R}^n$. The set of all nonnegative linear combinations of the $\mathbf{r}_i$ forms a cone C. The statement that the system in (i) in Lemma 22.5 has no nonnegative solution means that the vector $\mathbf{b}$ does not lie in C. In this case, the lemma asserts the existence of a vector $\mathbf{y}$ which makes an obtuse angle with $\mathbf{b}$ and a nonobtuse angle with each of the vectors $\mathbf{r}_i$. This means that the hyperplane L orthogonal to $\mathbf{y}$ has the cone C on one side and the point $\mathbf{b}$ on the other.

Fig. 22.1 Geometric interpretation of Farkas' Lemma

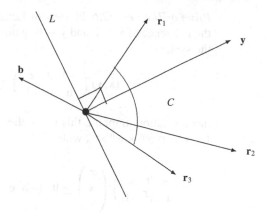

22.4 The Duality Theorem of Linear Programming

In this section we prove the following theorem.

Theorem 22.6 (Duality Theorem of Linear Programming) *Let A be an $n \times p$ matrix, $\mathbf{b} \in \mathbb{R}^p$, and $\mathbf{c} \in \mathbb{R}^n$. Suppose $V := \{\mathbf{x} \in \mathbb{R}^n \mid \mathbf{x}A \geq \mathbf{b}, \mathbf{x} \geq 0\} \neq \emptyset$ and $W := \{\mathbf{y} \in \mathbb{R}^p \mid A\mathbf{y} \leq \mathbf{c}, \mathbf{y} \geq 0\} \neq \emptyset$. Then $\min\{\mathbf{x} \cdot \mathbf{c} \mid \mathbf{x} \in V\} = \max\{\mathbf{b} \cdot \mathbf{y} \mid \mathbf{y} \in W\}$.*

To prove this theorem, we first prove the following variant of Farkas' Lemma.

Lemma 22.7 *Let A be an $m \times n$ matrix and $\mathbf{b} \in \mathbb{R}^n$. Exactly one of the following two statements is true.*

(a) *There is an $\mathbf{x} \in \mathbb{R}^m$ with $\mathbf{x}A \leq \mathbf{b}$ and $\mathbf{x} \geq 0$.*
(b) *There is a $\mathbf{y} \in \mathbb{R}^n$ with $A\mathbf{y} \geq 0$, $\mathbf{b} \cdot \mathbf{y} < 0$, and $\mathbf{y} \geq 0$.*

Proof Problem 22.5. ∎

The following two lemmas are further preparations for the proof of the Duality Theorem.

Lemma 22.8 *Let $\mathbf{x} \in V$ and $\mathbf{y} \in W$ (cf. Theorem 22.6). Then $\mathbf{x} \cdot \mathbf{c} \geq \mathbf{b} \cdot \mathbf{y}$.*

Proof $\mathbf{x} \cdot \mathbf{c} \geq \mathbf{x}A\mathbf{y} \geq \mathbf{b} \cdot \mathbf{y}$. ∎

Lemma 22.9 *Let $\hat{\mathbf{x}} \in V$, $\hat{\mathbf{y}} \in W$ with $\hat{\mathbf{x}} \cdot \mathbf{c} = \mathbf{b} \cdot \hat{\mathbf{y}}$. Then $\hat{\mathbf{x}} \cdot \mathbf{c} = \min\{\mathbf{x} \cdot \mathbf{c} \mid \mathbf{x} \in V\}$ and $\mathbf{b} \cdot \hat{\mathbf{y}} = \max\{\mathbf{b} \cdot \mathbf{y} \mid \mathbf{y} \in W\}$.*

Proof By Lemma 22.8, for every $\mathbf{x} \in V$: $\mathbf{x} \cdot \mathbf{c} \geq \mathbf{b} \cdot \hat{\mathbf{y}} = \hat{\mathbf{x}} \cdot \mathbf{c}$. Similarly, $\mathbf{b} \cdot \mathbf{y} \leq \hat{\mathbf{x}} \cdot \mathbf{c} = \mathbf{b} \cdot \hat{\mathbf{y}}$ for every $\mathbf{y} \in W$. ∎

Proof of Theorem 22.6 In view of Lemmas 22.8 and 22.9, it is sufficient to show the existence of $\hat{\mathbf{x}} \in V$ and $\hat{\mathbf{y}} \in W$ with $\hat{\mathbf{x}} \cdot \mathbf{c} \leq \mathbf{b} \cdot \hat{\mathbf{y}}$. So it is sufficient to show that the system

$$(\mathbf{x}, \mathbf{y}) \begin{pmatrix} -A & 0 & \mathbf{c} \\ 0 & A^T & -\mathbf{b} \end{pmatrix} \leq (-\mathbf{b}, \mathbf{c}, 0), \ \mathbf{x} \geq 0, \ \mathbf{y} \geq 0$$

has a solution. Suppose this is not the case. By Lemma 22.7, there exists a vector $(\mathbf{z}, \mathbf{w}, t) \in \mathbb{R}^p \times \mathbb{R}^n \times \mathbb{R}$ with

$$\begin{pmatrix} -A & 0 & \mathbf{c} \\ 0 & A^T & -\mathbf{b} \end{pmatrix} \begin{pmatrix} \mathbf{z} \\ \mathbf{w} \\ t \end{pmatrix} \geq 0, \ (-\mathbf{b}, \mathbf{c}, 0) \cdot (\mathbf{z}, \mathbf{w}, t) < 0, \ \mathbf{z} \geq 0, \ \mathbf{w} \geq 0, \ t \geq 0 .$$

Hence

$$Az \leq t\mathbf{c} \tag{22.1}$$

$$\mathbf{w}A \geq t\mathbf{b} \tag{22.2}$$

$$\mathbf{c} \cdot \mathbf{w} < \mathbf{b} \cdot \mathbf{z} . \tag{22.3}$$

If $t = 0$, then $A\mathbf{z} \leq 0$ and $\mathbf{w}A \geq 0$, hence, for $\mathbf{x} \in V$ and $\mathbf{y} \in W$:

$$\mathbf{b} \cdot \mathbf{z} \leq \mathbf{x}A\mathbf{z} \leq 0 \leq \mathbf{w}A\mathbf{y} \leq \mathbf{w} \cdot \mathbf{c}$$

contradicting (22.3). If $t > 0$, then by (22.1) and (22.2), $t^{-1}\mathbf{z} \in W$ and $t^{-1}\mathbf{w} \in V$. By (22.3), $\mathbf{b} \cdot (t^{-1}\mathbf{z}) > (t^{-1}\mathbf{w}) \cdot \mathbf{c}$, which contradicts Lemma 22.8. Hence, the first system above must have a solution. ∎

22.5 Some Fixed Point Theorems

Let $Z \subseteq \mathbb{R}^n$ be a nonempty convex and compact[2] set. Let $f : Z \to Z$ be a continuous function. A point $\mathbf{x}^* \in Z$ is a *fixed point* of f if $f(\mathbf{x}^*) = \mathbf{x}^*$.

If $n = 1$, then Z is a closed interval of the form $[a, b] \subseteq \mathbb{R}$, and then it is clear (by drawing a picture) that f must have a fixed point: formally, this is a straightforward implication of the intermediate-value theorem.

More generally, the following result holds.

---

[2]A set $Z \subseteq \mathbb{R}^n$ is *compact* if it is closed and bounded. A set $Z \subseteq \mathbb{R}^n$ is bounded if there is an $M > 0$ such that $||\mathbf{x}|| < M$ for all $\mathbf{x} \in Z$.

Theorem 22.10 (Brouwer Fixed Point Theorem) *Let $Z \subseteq \mathbb{R}^n$ be a nonempty compact and convex set and let $f : Z \to Z$ be a continuous function. Then f has a fixed point.*

A generalization of Brouwer's fixed point theorem is Kakutani's fixed point theorem. Let $F : Z \to Z$ be a *correspondence*, i.e., $F(\mathbf{x})$ is a nonempty subset of Z for every $\mathbf{x} \in Z$. Call F *convex-valued* if $F(\mathbf{x})$ is a convex set for every $\mathbf{x} \in Z$. Call F *upper semi-continuous* if the following holds: for every sequence $(\mathbf{x}^k)_{k \in \mathbb{N}}$ in Z converging to $\mathbf{x} \in Z$ and for every sequence $(\mathbf{y}^k)_{k \in \mathbb{N}}$ in Z converging to $\mathbf{y} \in Z$, if $\mathbf{y}^k \in F(\mathbf{x}^k)$ for every $k \in \mathbb{N}$, then $\mathbf{y} \in F(\mathbf{x})$. A point $\mathbf{x}^* \in Z$ is a *fixed point* of Z if $\mathbf{x}^* \in F(\mathbf{x}^*)$.

Theorem 22.11 (Kakutani Fixed Point Theorem) *Let $Z \subseteq \mathbb{R}^n$ be a nonempty compact and convex set and let $F : Z \to Z$ be an upper semi-continuous and convex-valued correspondence. Then F has a fixed point.*

22.6 The Birkhoff–von Neumann Theorem

Let C be a convex set in a linear space V. An element $e \in C$ is called an *extreme point* of C if for all $x, y \in C$ with $e = \frac{1}{2}(x + y)$ it holds that $x = y \, (= e)$. By $\text{ext}(C)$ the set of extreme points of C is denoted. See Problem 22.6 for alternative characterizations of extreme points.

An $n \times n$-matrix D is called *doubly stochastic* if $0 \leq d_{ij} \leq 1$ for all $i, j = 1, \ldots, n$, $\sum_{j=1}^{n} d_{ij} = 1$ for all i, and $\sum_{i=1}^{n} d_{ij} = 1$ for all j. If moreover $d_{ij} \in \{0, 1\}$ for all $i, j = 1, \ldots, n$, then D is called a *permutation matrix*. Let $D_{n \times n}$ denote the set of all $n \times n$ doubly stochastic matrices, and let $P_{n \times n}$ denote the set of all $n \times n$ permutation matrices. Note that $D_{n \times n}$ is a convex compact set, and that $P_{n \times n}$ is a finite subset of $D_{n \times n}$. The following theorem gives the exact relation.

Theorem 22.12 (Birkhoff–von Neumann)

(a) $\text{ext}(D_{n \times n}) = P_{n \times n}$
(b) $D_{n \times n} = \text{conv}(P_{n \times n})$.

Part (b) of Theorem 22.12 follows from the Theorem of Krein–Milman (Theorem 22.14 below). In the proof of the latter theorem the dimension of a subset of a linear space V plays a role. A subset of V of the form $a + L$ where $a \in V$ and L is a linear subspace of V, is called an *affine subspace*. Check that a subset A of V is affine if, and only if, with any two different elements x and y of A, also the straight line through x and y is contained in A (Problem 22.7). For an affine subspace $a + L$ of V the *dimension* is defined to be the dimension of the linear subspace L. For an arbitrary subset A of V, its *dimension* $\dim(A)$ is defined to be the dimension of the smallest affine subspace of V containing the set A.

The following separation lemma is used in the proof of the Theorem of Krein–Milman. By $\text{int}(C)$ and $\text{clo}(C)$ we denote the (topological) interior and closure of the set $C \subseteq \mathbb{R}^n$, respectively.

Lemma 22.13 *Let C be a nonempty convex subset of $\mathbb{R}^n$ and $\mathbf{a} \in \mathbb{R}^n \setminus \text{int}(C)$. Then there exists a $\mathbf{p} \in \mathbb{R}^n \setminus \{\mathbf{0}\}$ with $\mathbf{p} \cdot \mathbf{a} \le \mathbf{p} \cdot \mathbf{c}$ for every $\mathbf{c} \in C$.*

Proof We distinguish two cases: (a) $\mathbf{a} \notin \text{clo}(C)$ and (b) $\mathbf{a} \in \text{clo}(C)$.

(a) Suppose $\mathbf{a} \notin \text{clo}(C)$. Then the result follows from Theorem 22.1, with $\text{clo}(C)$ in the role of the set Z there.
(b) Suppose $\mathbf{a} \in \text{clo}(C)$. Because $\mathbf{a} \notin \text{int}(C)$, there is a sequence $\mathbf{a}^1, \mathbf{a}^2, \dots \in \mathbb{R}^n \setminus \text{clo}(C)$ converging to $\mathbf{a}$. By Theorem 22.1 again, for each k there is a $\mathbf{p}^k \in \mathbb{R}^n \setminus \{\mathbf{0}\}$ with $\mathbf{p}^k \cdot \mathbf{a}^k \le \mathbf{p}^k \cdot \mathbf{c}$ for all $\mathbf{c} \in \text{clo}(C)$, and we can take these vectors $\mathbf{p}^k$ such that $||\mathbf{p}^k|| = 1$ for every k ($|| \cdot ||$ denotes the Euclidean norm). Because $\{\mathbf{x} \in \mathbb{R}^n \mid ||\mathbf{x}|| = 1\}$ is a compact set, there exists a converging subsequence $\mathbf{p}^{k(1)}, \mathbf{p}^{k(2)}, \dots$ of $\mathbf{p}^1, \mathbf{p}^2, \dots$ with limit, say, $\hat{\mathbf{p}}$. Then $\hat{\mathbf{p}} \cdot \mathbf{a} = \lim_{\ell \to \infty} \mathbf{p}^{k(\ell)} \cdot \mathbf{a}^{k(\ell)} \le \lim_{\ell \to \infty} \mathbf{p}^{k(\ell)} \cdot \mathbf{c} = \hat{\mathbf{p}} \cdot \mathbf{c}$ for all $\mathbf{c} \in \text{clo}(C)$. ∎

Theorem 22.14 (Krein–Milman) *Let C be a nonempty compact and convex subset of $\mathbb{R}^n$. Then $\text{ext}(C) \neq \emptyset$ and $C = \text{conv}(\text{ext}(C))$.*

Proof

(1) Because C is compact and $\mathbf{x} \mapsto ||\mathbf{x}||$ (where $|| \cdot ||$ denotes the Euclidean norm) is continuous, there exists by the Theorem of Weierstrass an $\mathbf{e} \in C$ with $||\mathbf{e}|| = \max_{\mathbf{x} \in C} ||\mathbf{x}||$. Then $\mathbf{e} \in \text{ext}(C)$, which can be proved as follows. Suppose that $\mathbf{e} = \frac{1}{2}(\mathbf{x}^1 + \mathbf{x}^2)$ for some $\mathbf{x}^1, \mathbf{x}^2 \in C$. Then

$$||\mathbf{e}|| = ||\frac{1}{2}(\mathbf{x}^1 + \mathbf{x}^2)|| \le \frac{1}{2}||\mathbf{x}^1|| + \frac{1}{2}||\mathbf{x}^2|| \le \frac{1}{2}||\mathbf{e}|| + \frac{1}{2}||\mathbf{e}||$$

implies $||\mathbf{x}^1|| = ||\mathbf{x}^2|| = ||\frac{1}{2}(\mathbf{x}^1 + \mathbf{x}^2)||$. By definition of the Euclidean norm this is only possible if $\mathbf{x}^1 = \mathbf{x}^2 = \mathbf{e}$. This shows $\mathbf{e} \in \text{ext}(C)$. Hence, $\text{ext}(C) \neq \emptyset$.
(2) The second statement in the theorem will be proved by induction on $\dim(C)$.
  (a) If $\dim(C) = 0$, then $C = \{\mathbf{a}\}$ for some $\mathbf{a} \in \mathbb{R}^n$, so $\text{ext}(C) = \{\mathbf{a}\}$ and $\text{conv}(\text{ext}(C)) = \{\mathbf{a}\} = C$.
  (b) Let $k \in \mathbb{N}$, and suppose that $\text{conv}(\text{ext}(D)) = D$ for every nonempty compact and convex subset D of $\mathbb{R}^n$ with $\dim(D) < k$. Let C be a k-dimensional compact convex subset of $\mathbb{R}^n$. Obviously, $\text{conv}(\text{ext}(C)) \subseteq C$. So to prove is still: $C \subseteq \text{conv}(\text{ext}(C))$. Without loss of generality assume $\mathbf{0} \in C$ (otherwise, shift the whole set C). Let W be the smallest affine (hence, linear) subset of $\mathbb{R}^n$ containing C. Hence, $\dim(W) = k$. From part (1) of the proof there is an $\mathbf{e} \in \text{ext}(C)$. Let $\mathbf{x} \in C$. If $\mathbf{x} = \mathbf{e}$ then $\mathbf{x} \in \text{conv}(\text{ext}(C))$. If $\mathbf{x} \neq \mathbf{e}$ then the intersection of the straight line through $\mathbf{x}$ and $\mathbf{e}$ with C is a line segment of which one of the endpoints is $\mathbf{e}$. Let $\mathbf{b}$ be the other endpoint.

Then $\mathbf{b}$ is a boundary point of C. Then, by Lemma 22.13, there is a linear function $f : W \to \mathbb{R}$ with $f(\mathbf{b}) = \min\{f(\mathbf{c}) \mid \mathbf{c} \in C\}$ and $f \neq 0$ (check this). Let $D := \{\mathbf{y} \in C \mid f(\mathbf{y}) = f(\mathbf{b})\}$. Then D is a compact and convex subset of C. Because $f \neq 0$ it follows that $\dim(D) < k$. By the induction hypothesis, $D = \mathrm{conv}(\mathrm{ext}(D))$. Also, $\mathrm{ext}(D) \subseteq \mathrm{ext}(C)$, see Problem 22.8. Hence, $\mathbf{b} \in D = \mathrm{conv}(\mathrm{ext}(D)) \subseteq \mathrm{conv}(\mathrm{ext}(C))$. Further, $\mathbf{e} \in \mathrm{ext}(C)$. Because $\mathbf{x} \in \mathrm{conv}\{\mathbf{b}, \mathbf{e}\}$ it follows that $\mathbf{x} \in \mathrm{conv}(\mathrm{ext}(C))$. So $C \subseteq \mathrm{conv}(\mathrm{ext}(C))$. ∎

Proof of Theorem 22.12 Because $D_{n \times n}$ is compact and convex, part (b) follows from part (a) and Theorem 22.14. So only (a) still has to be proved.

(1) We first prove that $P_{n \times n} \subseteq \mathrm{ext}(D_{n \times n})$. Let $P = [p_{ij}]_{i,j=1}^{n}$ be a permutation matrix with $P = \frac{1}{2}(A+B)$ for some $A, B \in D_{n \times n}$. Then $p_{ij} = \frac{1}{2}(a_{ij}+b_{ij})$ and $p_{ij} \in \{0, 1\}$ for all $i, j \in \{1, 2, \ldots, n\}$. If $p_{ij} = 0$ then $a_{ij} = b_{ij} = 0$ because $a_{ij}, b_{ij} \geq 0$. If $p_{ij} = 1$ then $a_{ij} = b_{ij} = 1$ because $a_{ij}, b_{ij} \leq 1$. Hence, $A = B$, so that $P \in \mathrm{ext}(D_{n \times n})$.

(2) Let now $D = [d_{ij}] \in D_{n \times n}$ such that D is not a permutation matrix. The proof is complete if we show that D is not an extreme point. For this, it is sufficient to show that there exists an $n \times n$-matrix $C \neq [0]$ with
   (i) $c_{ij} = 0$ whenever $d_{ij} = 0$ or $d_{ij} = 1$,
   (ii) $\sum_{i=1}^{n} c_{ij} = 0$ for all $j \in \{1, 2, \ldots, n\}$ with $d_{ij} \neq 1$ for every i,
   (iii) $\sum_{j=1}^{n} c_{ij} = 0$ for all $i \in \{1, 2, \ldots, n\}$ with $d_{ij} \neq 1$ for every j.
   For in that case, for $\varepsilon > 0$ sufficiently small, the matrices $D + \varepsilon C$ and $D - \varepsilon C$ are two different doubly stochastic matrices with $D = \frac{1}{2}((D+\varepsilon C)+(D-\varepsilon C))$, implying that $D \notin \mathrm{ext}(D_{n \times n})$.

   We are left to construct C. In order to satisfy (i), for those rows or columns of D that contain a 1 the corresponding rows or columns of C contain only zeros. Suppose there are k rows (and hence columns) of D that do not contain a 1. Because D is not a permutation matrix, $2 \leq k \leq n$. In these k rows there are at least $2k$ elements unequal to 0 and 1. The corresponding $2k$ or more elements of C are to be chosen such that they satisfy the system of $2k$ homogeneous linear equations described in (ii) and (iii). Without loss of generality assume that these equations correspond to the first k rows and the first k columns. Then, if $\sum_{j=1}^{k} c_{ij} = 0$ for all $i \in \{1, \ldots, k-1\}$ and $\sum_{i=1}^{k} c_{ij} = 0$ for all $j \in \{1, \ldots, k\}$, we have $\sum_{j=1}^{k} c_{kj} = 0$ as well, so that this last equation is redundant. Thus, we have a system with less than $2k$ independent equations and at least $2k$ variables. Hence, it has a nonzero solution, which gives the required $C \neq [0]$. ∎

22.7   The Max-Flow Min-Cut Theorem

A *capacitated network* is a triple (V, E, k), where V is a finite set containing at least two distinguished elements $q, s \in V$ called *source* (q) and *sink* (s); E is a subset of $V \times V$ such that $v \neq w$, $v \neq s$, and $w \neq q$ for all $(v, w) \in E$; and $k : E \to \mathbb{R}_+$. Elements of V are called *vertices* and elements of E are called *edges*. The number

Fig. 22.2 A capacitated
network

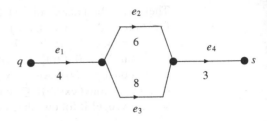

$k(e)$ is the *capacity* of the edge e; if $e = (v, w)$ then $k(e)$ is interpreted as the maximal amount that can flow from v to w through edge e. The source has only outgoing and the sink only incoming edges. See Fig. 22.2 for an example.

A *flow* in this network is a map $f : E \to \mathbb{R}_+$ with $f(e) \leq k(e)$ and such that for all $v \in V \setminus \{q, s\}$

$$\sum_{w \in V:\ (w,v) \in E} f(w, v) = \sum_{w \in V:\ (v,w) \in E} f(v, w) .$$

In other words, a flow satisfies the capacity constraints and for all vertices (except source and sink) the 'inflow' equals the 'outflow'.

The *value* of a flow f is defined as the inflow in the sink, i.e., as the number

$$\sum_{v \in V:\ (v,s) \in E} f(v, s) .$$

A flow is called *maximal* if it has maximal value among all possible flows. Intuitively, the value of a maximal flow is determined by the 'bottlenecks' in the network. In order to formalize this, define a *cut* in the network to be a partition of V into two subsets V_1 and V_2 such that $q \in V_1$ and $s \in V_2$. Such a cut is denoted by (V_1, V_2). The *capacity* of a cut is the number

$$k(V_1, V_2) := \sum_{v \in V_1, w \in V_2:(v,w) \in E} k(v, w) ,$$

i.e., the total capacity along edges going from V_1 to V_2. A cut is called *minimal* if it has minimal capacity among all possible cuts.

In the example in Fig. 22.2, a minimal cut has only the sink in V_2, and its capacity is equal to 3. Obviously, this is also the value of a maximal flow, but such a flow is not unique.

Flows and cuts are, first of all, related as described in the following lemma.

Lemma 22.15 *Let f be a flow in the capacitated network (V, E, k), and let $\varphi : E \to \mathbb{R}$ be an arbitrary function. Then:*

(a) $\displaystyle\sum_{v \in V} \sum_{(w,v):\ (w,v) \in E} \varphi(w, v) = \sum_{v \in V} \sum_{(v,w):\ (v,w) \in E} \varphi(v, w) .$

(b) $\displaystyle\sum_{(q,v):\,(q,v)\in E} f(q,v) = \sum_{(v,s):\,(v,s)\in E} f(v,s)$.

(c) *For every cut* (V_1, V_2) *the value of the flow* f *is equal to*

$$\sum_{(v,w):\,(v,w)\in E,v\in V_1,w\in V_2} f(v,w) - \sum_{(v,w):\,(v,w)\in E,v\in V_2,w\in V_1} f(v,w) \ .$$

Proof (a) follows because summation at both sides is taken over the same sets. Part (a) moreover implies

$$\sum_{(q,v):\,(q,v)\in E} f(q,v) + \sum_{(v,w)\in E:\,v\neq q} f(v,w) =$$

$$\sum_{(v,s):\,(v,s)\in E} f(v,s) + \sum_{(v,w)\in E:\,w\neq s} f(v,w)$$

which implies (b) because $\sum_{(v,w)\in E:\,v\neq q} f(v,w) = \sum_{(v,w)\in E:\,w\neq s} f(v,w)$ by definition of a flow ('inflow' equals 'outflow' at every vertex that is not the source and not the sink). For part (c), let (V_1, V_2) be a cut of the network. Then

$$\sum_{(v,w)\in E:\,v\in V_1,w\in V_2} f(v,w) = \sum_{(v,w)\in E:\,v\in V_1} f(v,w) - \sum_{(v,w)\in E:\,v,w\in V_1} f(v,w)$$

$$= \sum_{(v,w)\in E:\,v=q} f(v,w) + \sum_{(v,w)\in E:\,w\in V_1} f(v,w)$$

$$- \sum_{(v,w)\in E:\,v,w\in V_1} f(v,w)$$

$$= \sum_{(v,w)\in E:\,v=q} f(v,w) + \sum_{(v,w)\in E:\,v\in V_2,w\in V_1} f(v,w)$$

$$= \sum_{(v,w)\in E:\,w=s} f(v,w) + \sum_{(v,w)\in E:\,v\in V_2,w\in V_1} f(v,w) \ ,$$

where the second equality follows since everything that leaves from a node of V_1 also has entered that node, except for q; and the fourth equality follows from (b). This implies part (c) of the lemma. ∎

The following theorem is the famous Max Flow Min Cut Theorem.

Theorem 22.16 *Let* (V, E, k) *be a capacitated network. Then the value of a maximal flow is equal to the capacity of a minimal cut.*

Proof Let f be a maximal flow. (Note that f is an optimal solution of a feasible bounded linear program, so that existence of f is guaranteed.) Part (c) of

Lemma 22.15 implies that the value of *any* flow f is smaller than or equal to

$$\sum_{(v,w):\ (v,w)\in E, v\in V_1, w\in V_2} k(v,w) - \sum_{(v,w):\ (v,w)\in E, v\in V_2, w\in V_1} f(v,w)$$

$$\leq \sum_{(v,w):\ (v,w)\in E, v\in V_1, w\in V_2} k(v,w)$$

for *any* cut (V_1, V_2), so that it is sufficient to find a cut of which the capacity is equal to the value of f.

For points v, w in the network define a *path* as a sequence of different non-directed edges starting in v and ending in w; 'non-directed' means that for any edge $(x, y) \in E$, (x, y) as well as (y, x) may be used in this path. Such a path may be described by a sequence $v = x_1, x_2, \ldots, x_m = w$ with $(x_i, x_{i+1}) \in E$ or $(x_{i+1}, x_i) \in E$ for every $i = 1, \ldots, m - 1$. Call such a path *non-satiated* if for every $i = 1, \ldots,$ $m - 1$ it holds that $f(x_i, x_{i+1}) < k(x_i, x_{i+1})$ if $(x_i, x_{i+1}) \in E$, and $f(x_{i+1}, x_i) > 0$ if $(x_{i+1}, x_i) \in E$. In other words, the flow is below capacity in edges that are traversed in the 'right' way, and positive in edges that are traversed in the 'wrong' way.

Define V_1 to be the set of vertices x for which there is a non-satiated path from q to x, together with the vertex q, and let V_2 be the complement of V_1 in V. Then $s \in V_2$ because otherwise there would be a non-satiated path from q to s, implying that f would not be maximal; the flow f could be increased by increasing it in edges on this path that are traversed in the right way and decreasing it in edges along the path that are traversed in the wrong way, without violating the capacity constraints or the inflow-outflow equalities. Hence (V_1, V_2) is a cut in the network.

Let $(x, y) \in E$ with $x \in V_1$ and $y \in V_2$. Then $f(x, y) = k(x, y)$ because otherwise there would be a non-satiated path from q to a vertex in V_2. Similarly, $f(x', y') = 0$ whenever $(x', y') \in E$ with $x' \in V_2$ and $y' \in V_1$. By Lemma 22.15, part (c), the value of the flow f is equal to

$$\sum_{(v,w):\ (v,w)\in E, v\in V_1, w\in V_2} f(v,w) - \sum_{(v,w):\ (v,w)\in E, v\in V_2, w\in V_1} f(v,w)$$

hence to $\sum_{(v,w):\ (v,w)\in E, v\in V_1, w\in V_2} k(v,w)$, which is by definition the capacity of the cut (V_1, V_2). This completes the proof.   ∎

Observe that the proof of Theorem 22.16 suggests an algorithm to determine a maximal flow, by starting with an arbitrary flow, looking for a non-satiated path, and improving this path. By finding an appropriate cut, maximality of a flow can be checked. Theorem 22.16 is actually a (linear) duality result, but the above proof is elementary.

22.8 Problems

22.1. *Convex Sets*
Prove that a closed set $Z \subseteq \mathbb{R}^n$ is convex if and only if $\frac{1}{2}\mathbf{x} + \frac{1}{2}\mathbf{y} \in Z$ for all $\mathbf{x}, \mathbf{y} \in Z$.

22.2. *Proof of Lemma 22.3*
Prove that at most one of the systems in Lemma 22.3 has a solution.

22.3. *Rank of AA^T*
Let A be an $m \times n$ matrix with rank k. Prove that the rank of AA^T (where A^T is the transpose of A) is also equal to k.

22.4. *Proof of Lemma 22.5*
Prove that at most one of the systems in Lemma 22.5 has a solution.

22.5. *Proof of Lemma 22.7*
Prove Lemma 22.7.

22.6. *Extreme Points*
Let C be a convex set in a linear space V and let $e \in C$. Prove that the following three statements are equivalent.

(a) $e \in \text{ext}(C)$.
(b) For all $0 < \alpha < 1$ and all $x, y \in C$, if $x \neq y$ then $e \neq \alpha x + (1 - \alpha)y$.
(c) $C \setminus \{e\}$ is a convex set.

22.7. *Affine Subspaces*
Prove that a subset A of a linear space V is affine if, and only if, with any two different elements x and y of A, also the straight line through x and y is contained in A.

22.8. *The Set of Sup-points of a Linear Function on a Convex Set*
Let $f : \mathbb{R}^n \to \mathbb{R}^n$ be a linear function. Let C be a convex subset of $\mathbb{R}^n$ and $\alpha := \sup\{f(\mathbf{x}) \mid \mathbf{x} \in C\}$, $D := \{\mathbf{x} \in C \mid f(\mathbf{x}) = \alpha\}$. Show that D is convex and that $\text{ext}(D) \subseteq \text{ext}(C)$.

22.9 Notes

Theorem 22.10 is due to Brouwer (1912). For proofs, see also e.g. Scarf (1973) or Vohra (2005).

Theorem 22.11 is from Kakutani (1941). One way to prove this theorem is to derive it from the Brouwer Fixed Point Theorem: see, e.g., Hildenbrand and Kirman (1976).

A good source for results on convexity is Rockafellar (1970).

The Max Flow Min Cut Theorem (Theorem 22.16) is due to Ford and Fulkerson (1956).

References

Brouwer, L. E. J. (1912). Über Abbildung von Mannigfaltigkeiten. *Mathematische Annalen, 71*, 97–115.

Ford, L. R., & Fulkerson, D. R. (1956). Maximal flow through a network. *Canadian Journal of Mathematics, 8*, 399–404.

Hildenbrand, H., & Kirman, A. P. (1976). *Introduction to equilibrium analysis*. Amsterdam: North-Holland.

Kakutani, S. (1941). A generalization of Brouwer's fixed point theorem. *Duke Mathematical Journal, 8*, 457–459.

Rockafellar, R. T. (1970). *Convex analysis*. Princeton: Princeton University Press

Scarf, H. E. (1973). *The computation of economic equilibria*. New Haven: Cowles Foundation, Yale University Press.

Vohra, R. V. (2005). *Advanced mathematical economics*. New York: Routledge.

This chapter contains Review Problems to Chaps. 2–10, organized per chapter.

Chapter 2

RP 1 *Matrix Games (1)*

(a) Give the maximin rows and minimax columns in the following matrix game:

$$A = \begin{pmatrix} 0 & 1 & 1 \\ 2 & 2 & 0 \\ 0 & 0 & 2 \end{pmatrix}.$$

What can you infer from this about the value of the game?

(b) Calculate all (mixed) maximin and minimax strategies and the value of the following matrix game:

$$A = \begin{pmatrix} 2 & 2 & 3 \\ 3 & 1 & 4 \\ 2 & 0 & 2 \end{pmatrix}.$$

(c) Answer the same questions as in (a) and (b) for the following game:

$$A = \begin{pmatrix} 1 & 0 & 2 & 3 \\ 4 & 2 & 1 & 3 \\ 3 & 1 & 3 & 3 \end{pmatrix}.$$

© Springer-Verlag Berlin Heidelberg 2015
H. Peters, *Game Theory*, Springer Texts in Business and Economics,
DOI 10.1007/978-3-662-46950-7_23

RP 2 *Matrix Games (2)*

(a) Consider the following matrix game:

$$A = \begin{pmatrix} 4 & 2 \\ 1 & 3 \\ 0 & 5 \\ 5 & 1 \end{pmatrix}.$$

Determine all maximin rows and minimax columns, and also all saddlepoint(s), if any. What can you conclude from this about the value of this game?

(b) Consider the six different 2×2-matrix games that can be obtained by choosing two rows from A, as follows:

$$A_1 = \begin{pmatrix} 4 & 2 \\ 1 & 3 \end{pmatrix} \quad A_2 = \begin{pmatrix} 4 & 2 \\ 0 & 5 \end{pmatrix} \quad A_3 = \begin{pmatrix} 4 & 2 \\ 5 & 1 \end{pmatrix}$$

$$A_4 = \begin{pmatrix} 1 & 3 \\ 0 & 5 \end{pmatrix} \quad A_5 = \begin{pmatrix} 1 & 3 \\ 5 & 1 \end{pmatrix} \quad A_6 = \begin{pmatrix} 0 & 5 \\ 5 & 1 \end{pmatrix}.$$

Determine the values of all these games. Which one must be equal to the value of A? (Give an argument for your answer!)

(c) Determine all (possibly mixed) maximin and minimax strategies of A. [Hint: Use your answer to (b).]

RP 3 *Matrix Games (3)*

(a) Consider the following matrix game:

$$A = \begin{pmatrix} 8 & 16 & 12 \\ 9 & 8 & 4 \\ 12 & 4 & 16 \end{pmatrix}.$$

Determine all maximin rows and minimax columns, and also all saddlepoint(s), if any. What can you conclude from this about the value of this game?

(b) Give an argument why player 1 (the row player) will never put probability on the second row in a maximin strategy. Give also an argument why, consequently, player 2 will not put probability on the third column in a minimax strategy.

(c) Determine all maximin and minimax strategies and the value of A. [Hint: Use part (b).]

Chapter 3

RP 4 *Bimatrix Games (1)*
Consider the following bimatrix game:

$$
\begin{array}{c c c c}
 & L & C & R \\
U & 0,0 & 1,1 & 2,2 \\
M & 3,1 & 0,0 & 0,1 \\
D & 1,1 & 0,0 & 1,0
\end{array}.
$$

(a) Simplify this game by iterated elimination of strictly dominated strategies. Each time you eliminate a pure strategy, say by which pure or mixed strategy the eliminated strategy is strictly dominated and why.
(b) Find all Nash equilibria in the remaining game by computing the best reply functions and making a diagram.
(c) Determine all Nash equilibria in the original game.

RP 5 *Bimatrix Games (2)*

(a) Consider the following bimatrix game:

$$(A, B) = \begin{pmatrix} x,1 & x,0 \\ 0,0 & 2,1 \end{pmatrix}.$$

For every $x \in \mathbb{R}$, determine all Nash equilibria (in mixed and pure strategies) of this game.
(b) Consider the following bimatrix game:

$$
\begin{array}{c c c c c}
 & e & f & g & h \\
a & 8,0 & 0,1 & 0,2 & 2,10 \\
b & 1,7 & 3,2 & 2,0 & 3,1 \\
c & 3,0 & 0,3 & 4,5 & 4,0 \\
d & 0,2 & 1,1 & 1,3 & 7,7
\end{array}.
$$

Show that strategy f for player 2 is strictly dominated by a mixed strategy, but not by a pure strategy. Which strategies survive iterated elimination of strictly dominated strategies in this game? Find all Nash equilibria in pure and mixed strategies of this game.

RP 6 *Voting*
Two political parties, *I* and *II*, each have four votes that they can distribute over two party-candidates each. A committee is to be elected, consisting of three members. Each political party would like to see as many as possible of their own candidates elected in the committee. Of the total of four candidates those three that have most

of the votes will be elected; in case of ties, tied candidates are drawn with equal probabilities.

(a) Model this situation as a bimatrix game between the two parties.
(b) Determine all (possibly mixed) Nash equilibria in this game.

Chapter 4

RP 7 *A Bimatrix Game*
Consider the following bimatrix game, where a can be any real number:

$$
\begin{array}{c}
 & \begin{array}{cc} L & \quad R \end{array} \\
\begin{array}{c} T \\ B \end{array} & \left(\begin{array}{cc} a,0 & 0,1 \\ 0,1 & a,0 \end{array} \right)
\end{array}
$$

(a) Determine the set of (mixed and pure) Nash equilibria of this game for every value of a.
    Suppose now that player 1 (the row player) chooses first, and that player 2 (the column player) observes this move and chooses next.
(b) Write down the extensive form of this game, determine the associated strategic form, and for every $a \in \mathbb{R}$ determine all subgame perfect equilibria (in pure strategies) in this game.

RP 8 *An Ice-cream Vendor Game*
Three ice-cream vendors choose a location on the beach. This beach has the following form:

A	B
C	D

Each of the four regions has 300 customers. Each customer goes to the nearest ice-cream vendor, but can only move vertically or horizontally (hence not diagonally). In case of ties customers are distributed equally over vendors.

(a) Suppose the three vendors simultaneously and independently choose one of the four regions. Determine the Nash equilibrium or equilibria of this game, if any.
(b) Suppose vendor 1 chooses first, vendor 2 observes this and chooses next, and vendor 3 observes both choices and chooses last. Determine the subgame perfect Nash equilibrium or equilibria of this game, if any.

RP 9 *A Repeated Game*
Consider the three player game where player 1 chooses between U and D, player 2 between L and R, and player 3 between A and B. Choices are made

simultaneously and independently, and only pure strategies are considered. The payoffs are given by:

$$A: \begin{array}{c} \\ U \\ D \end{array} \begin{pmatrix} L & R \\ 2,2,0 & 5,5,5 \\ 8,6,8 & 0,7,4 \end{pmatrix} \qquad B: \begin{array}{c} \\ U \\ D \end{array} \begin{pmatrix} L & R \\ 4,4,1 & 4,2,8 \\ 0,2,9 & 4,2,5 \end{pmatrix}.$$

Hence, if player 3 plays A the left matrix applies and if player 3 plays B the right matrix applies. The triples are the payoffs for players 1, 2, and 3, respectively.

(a) Find the (pure strategy) Nash equilibria of this game, if any.
(b) Consider the twice repeated game with as payoffs the sums of the payoffs in the two periods. Suppose the players play a subgame perfect Nash equilibrium. What can you say about the actions chosen in the second period?
(c) Consider again the twice repeated game of (b). Is there a subgame perfect Nash equilibrium of the game in which (U, R, A) is played in the first period? If not, argue why not; otherwise, give a complete description of such an equilibrium.
(d) Answer the same question as in (b) for the combination (U, R, B).

RP 10 *Locating a Pub*

Two pub owners, player 1 and player 2, must choose a location for their pubs in a long street. In fact, player 1 wants to open two new pubs in the street, but player 2 only one. The possible locations are A, B and C, where A is right at the beginning of the street, B is somewhere in between A and C, and C is at the very end. Between A and B there are 200 potential customers (uniformly distributed), and between B and C there are 300 potential customers (uniformly distributed as well). Every customer will always go to the pub that is closest to his house.
The two pub owners have agreed on the following procedure. First, player 1 chooses a location for his first pub. Next, player 2 observes player 1's choice and chooses a location for his pub. Player 2 must choose a different location than player 1. Finally, player 1 obtains the remaining location for his second pub.
The objective for both pub owners is to maximize the number of customers.

(a) Formulate this situation as an extensive form game. How many strategies does player 1 have? And player 2?
(b) Find the unique subgame perfect equilibrium for this game.

Suppose now that they change the procedure as follows. After player 1 has chosen the location for his first pub, players 1 and 2 simultaneously choose one of the two remaining locations. If they happen to choose the same location, then they will enter into a fierce fight and the overall utility for both pub owners is 0.

(c) Formulate this situation as an extensive form game. How many strategies does player 1 have? And player 2?

(d) Find a subgame perfect equilibrium in which player 1 starts by choosing A. Find another subgame perfect equilibrium in which player 1 starts by choosing B. Find yet another subgame perfect equilibrium in which player 1 starts by choosing C.

RP 11 *A Two-stage Game*
Consider the two bimatrix games

$$G_1 = \begin{array}{c} U \\ D \end{array} \begin{pmatrix} \overset{L}{3,3} & \overset{R}{0,4} \\ 4,0 & 1,1 \end{pmatrix}, \quad G_2 = \begin{array}{c} T \\ M \\ B \end{array} \begin{pmatrix} \overset{X}{3,1} & \overset{Y}{0,0} & \overset{Z}{0,0} \\ 0,0 & 3,3 & 0,0 \\ 0,0 & 0,0 & 1,3 \end{pmatrix}.$$

(a) Compute all the pure strategy Nash equilibria in G_1 and all the pure strategy Nash equilibria in G_2.

Now suppose that G_1 is played first, then the players learn the outcome of G_1 and next play the game G_2. For each player the payoff is the sum of the payoffs of each game separately.

(b) In this two-stage game, how many (pure) strategies does each player have?
(c) Is there a subgame perfect Nash equilibrium in which (U, L) is played in G_1? If so, then describe such an equilibrium. If not, explain why.
(d) Now suppose that the order of play is reversed: first G_2 is played, then G_1. Determine all subgame perfect Nash equilibria of this two-stage game.

Chapter 5

RP 12 *Job Market Signaling*[1]
A worker (W) has private information about his level of ability. With probability $\frac{1}{3}$ he is a high type (H) and with probability $\frac{2}{3}$ he is a low type (L). After observing his own type, the worker decides whether to obtain a costly education (E) or not (N); think of E as getting a degree. The firm (F) observes the worker's education decision but not the worker's ability. The firm then decides whether to employ the worker in an important managerial job (M) or in a much less important clerical job (C). The payoffs are as follows. If the worker gets job M then he receives a payoff of 10 but has to subtract the cost of education if he has chosen E: this is 4 for a worker of type H but 7 for a worker of type L. If the worker gets job C then he receives a payoff of 4 minus the cost of education if he has chosen E. The firm has a payoff of 10 if it offers the job M to a H type worker but 0 if it offers M to an L type worker; and it has payoff 4 if it offers job C to any type worker.

---

[1]From Watson (2002).

(a) Represent this game in extensive form.

(b) Find all the pure strategy Nash equilibria of this game, if any. (You may but do not have to use the strategic form.)

(c) Which one(s) is (are) perfect Bayesian? Pooling or separating? Provide the associated beliefs and discuss whether or not the intuitive criterion is satisfied.

RP 13 *Second-hand Cars (1)*

We consider the following variation on the 'lemons' (second hand car) market. A seller wants to sell his car. He knows the quality, which can be good or bad with equal probabilities. There is also a buyer, who does not know the quality. The seller has the choice between offering a guarantee certificate or not, next the buyer has the choice between buying the car or not, having observed the action of the seller. The car has a fixed price of 15. The value of a bad or a good car for the buyer equals 10 or 20, respectively (some time after the sale, the buyer will discover the quality of the car). The car does not have any value for the seller. Offering a guarantee certificate represents an expected transfer from the seller to the buyer of 0 if the car is good, and of 10 if the car is bad (if the buyer buys the car). If the buyer does not buy the car then payoffs are 0 to both.

(a) Model this situation as a game in extensive form. Is this a signaling game? Why or why not?

(b) Determine the strategic form of this game, and all Nash equilibria in pure strategies, if any.

(c) Determine all perfect Bayesian Nash equilibria. Which ones are pooling or separating, if any?

RP 14 *Second-hand Cars (2)*

You have seen a nice-looking second-hand car, and you are considering buying it. The problem is that you do not know the precise value of the car. Suppose that there is a 25 % chance that the value of the car is 2,000, and there is a 75 % chance that the value is 4,000. You are aware of these probabilities. The car seller, on the other hand, knows the precise value of the car. You and the car seller simultaneously name a price, and a transaction only takes place if your price is higher or equal than the price named by the seller. In that case, the eventual price to be paid will be the average of your price and the price named by the seller.

For convenience, assume that both you and the seller can only name prices 1,000, 3,000 and 5,000. If you buy the car, your utility would be the value of the car minus the price paid. For the seller the utility would be the price received minus the value of the car. If no transaction takes place, the utility for both would be zero.

(a) Model this situation as a game with incomplete information.

(b) How many types and how many strategies do you have? What about the seller?

(c) One of your strategies is strictly dominated. Which one? Find a pure strategy, or mixed strategy, for you that strictly dominates it.

(d) Show that there is no Nash equilibrium in which you choose a price of 3,000.
(e) Find the Nash equilibrium, or equilibria, in pure strategies of this game. Will
    you eventually buy the car?

RP 15 *Signaling Games*

Compute the strategic form and all the pure strategy equilibria in the games in (a)
and (b) below. Also determine all perfect Bayesian equilibria in pure strategies.
Which ones are pooling or separating, if any? Give the corresponding beliefs. Apply
the intuitive criterion.

(a)

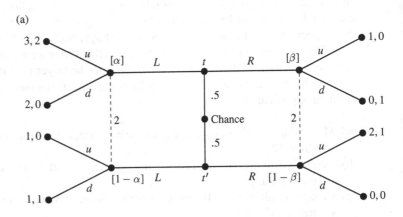

(b)

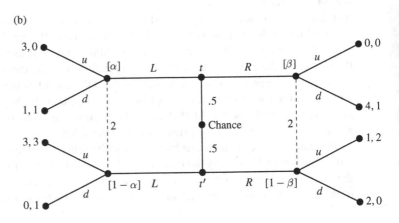

RP 16 *A Game of Incomplete Information*
Consider the following two games G_1 and G_2.

$$G_1 = \begin{matrix} & C & D & E \\ A & \begin{pmatrix} 5,2 & 3,2 & 2,3 \\ B & 5,2 & 3,2 & 2,3 \end{pmatrix} \end{matrix} \quad G_2 = \begin{matrix} & C & D & E \\ A & \begin{pmatrix} 1,2 & 0,1 & 3,0 \\ B & 3,3 & 2,4 & 0,0 \end{pmatrix} \end{matrix}.$$

At the beginning, a chance move determines whether the payoffs are as in G_1 or as in G_2. Both events happen with probability 0.5. Player 1 knows whether G_1 is played or G_2, but player 2 does not know which of the two games is played. Finally, both players simultaneously choose an action.

(a) Draw the extensive form representation of this game. (That is, draw the game tree.)
(b) Compute the strategic form of this game with incomplete information, and find all Nash equilibria in pure strategies.

Suppose now that player 2 knows whether G_1 or G_2 is played, but that player 1 does not know.

(c) Find all Nash equilibria in pure strategies of this new game with incomplete information. Use reasoning to find these equilibria, without making the strategic form.

RP 17 *A Bayesian Game*[2]
Two persons are involved in a dispute, and each person can either fight (F) or yield (Y). Each person has a payoff of 0 from yielding, regardless of the other person's action, and a payoff of 1 from fighting if the other person yields. If both persons fight, then the payoffs are -1 to person 1 and 1 to person 2. (This reflects the fact that person 2 is *strong*.)

(a) Represent this game in bimatrix from and compute all the Nash equilibria (so, in pure and mixed strategies).

Now assume that person 2 could also be *weak*, in which case the payoffs from both persons fighting are equal to 1 for person 1 and -1 for person 2; all other payoffs stay the same. Suppose that person 2 knows whether he is weak or strong, but person 1 only knows that person 2 is strong with probability α.

(b) Represent this game in extensive form.
(c) Determine the strategic form of the game in (b) and compute all Nash equilibria in pure strategies, if any. (Your answer may depend on the value of α, so you may have to distinguish cases.)

---

[2]From Osborne (2004).

RP 18 *Entry as a Signaling Game*[3]
A challenger contests an incumbent's market. The challenger is *strong* with
probability $1/4$ and *weak* with probability $3/4$; it knows its type, but the incumbent
does not. The challenger may either *prepare* itself for battle, or remain *unprepared*
(it does not have the option of staying out). The incumbent observes whether the
challenger is prepared or not, but not its type, and chooses whether to *fight* or
acquiesce. An unprepared challenger's payoff is 5 if the incumbent acquiesces to
its entry. Preparations cost a strong challenger 1 unit of payoff and a weak one 3
units, and fighting entails a loss of 2 units for each type. The incumbent prefers to
fight (payoff 2) rather than to acquiesce to (payoff 0) a weak challenger (who is
quickly dispensed with), and prefers to acquiesce to (payoff 2) rather than to fight
(payoff 0) a strong one.

(a) Represent this game in extensive form.
(b) Find all the pure strategy Nash equilibria of this game, if any. (You may but do
    not have to use the strategic form.)
(c) Which one(s) is (are) perfect Bayesian? Pooling or separating? Provide the
    associated beliefs and discuss whether or not the intuitive criterion is satisfied.

Chapter 6

RP 19 *Bargaining (1)*
Consider the following two-player bargaining game. Player 1 owns an object which
has worth 0 to him. The object has worth $v \in [0, 1]$ to player 2: player 2 knows v
but player 1 only knows that v has been drawn according to the uniform distribution
over $[0, 1]$. There are two periods. In period 1, player 1 makes a price offer p_1 and
player 2 responds with "yes" or "no". If player 2 rejects player 1's offer, then player
2 makes the price offer p_2 in the second period, to which player 1 can say "yes" or
"no". Agreement in the first period yields the payoff p_1 to player 1 and $v - p_1$ to
player 2. Agreement in the second period yields δp_2 to player 1 and $\delta(v - p_2)$ to
player 2 ($0 < \delta < 1$ is a discount factor). No agreement yields 0 to both.

(a) Explain that this is a game of incomplete information. What are the type sets of
    the players? Describe the strategy set of player 1.
(b) Describe the strategy set of player 2.
(c) Suppose player 1 offers a price p_1 in period 1. Give the best response of player
    2 (assume that players 1 and 2 accept in case of indifference).
(d) Compute the perfect Bayesian Nash equilibrium of this game.

RP 20 *Bargaining (2)*
Suppose two players (bargainers) bargain over the division of one unit of a
perfectly divisible good. Player 1 has utility function $u_1(\alpha) = \alpha$ and player 2

---

[3]From Osborne (2004).

has utility function $u_2(\beta) = 1 - (1 - \beta)^2$ for amounts $\alpha, \beta \in [0, 1]$ of the good.

(a) Determine the set of feasible utility pairs. Make a picture.
(b) Determine the Nash bargaining solution outcome, in terms of utilities as well as of the physical distribution of the good.
(c) Suppose the players' utilities are discounted by a factor $\delta \in [0, 1)$. Calculate the Rubinstein bargaining outcome.
(d) Determine the limit of the Rubinstein bargaining outcome, for δ approaching 1, in two ways: by using the result of (b) and by using the result of (c).

RP 21 *Bargaining (3)*

Suppose two players (bargainers) bargain over the division of one unit of a perfectly divisible good. Assume that utilities are just equal to the amounts of the good obtained, discounted by a common discount factor $0 < \delta < 1$. The players play a finite Rubinstein bargaining game over three periods $t = 0, 1, 2$, player 1 starting to make an offer at time $t = 0$, player 2 making an offer at time $t = 1$ (if reached), and the game ending at time $t = 2$ (if reached) with equal split $(\frac{1}{2}, \frac{1}{2})$.

(a) Calculate the backwards induction outcome of this game. Argue that player 1 has a beginner's advantage.

Suppose again that two players (bargainers) bargain over the division of one unit of a perfectly divisible good. Assume that player 1 has utility function $u(x)$ $(0 \le x \le 1)$ and player 2 has utility function $v(x) = 2u(x)$ $(0 \le x \le 1)$.

(b) Determine the physical distribution of the good according to the Nash bargaining solution. Can you say something about the resulting utilities?

RP 22 *Ultimatum Bargaining*

Consider the ultimatum bargaining game where player 1 offers a division $(1 - m, m)$ of one Euro, and player 2 can accept (receiving m whereas player receives $1 - m$) or reject (in which both players receive zero). Suppose that player 1 only cares about how much money he receives, but player 2 also cares about the division: specifically, if the game results in monetary payoffs (x_1, x_2), then the (utility) payoff to player 2 is $x_2 + a(x_2 - x_1)$, where a is some positive constant.

(a) Represent this game in extensive form, writing the payoffs in terms of m and a.
(b) Determine the subgame perfect Nash equilibrium.
(c) What happens to the equilibrium monetary split as a becomes large? What is the explanation for this?

RP 23 *An Auction (1)*

Two players (bidders) have valuations $v_1 > 0$ and $v_2 > 0$ for an object. They simultaneously and independently announce bids $b_1, b_2 \in [0, \infty)$. The player with the highest bid obtains the object and has a payoff $v_i - b_i$, the other player has

payoff 0. If the bids are equal, then each obtains the object and associated payoff
with probability $\frac{1}{2}$.

(a) Does this game have perfect or imperfect information? Complete or incomplete
information?
(b) Suppose that $v_1 = v_2$. Determine all Nash equilibria (in pure strategies), if any.
(c) Suppose that $v_1 > v_2$. Determine all Nash equilibria (in pure strategies), if any.
(d) Suppose that $v_1 = 1$ and $v_2 = 3$. Also suppose that only integer numbers up
to 3 are allowed as bids: $b_1, b_2 \in \{0, 1, 2, 3\}$. Represent this game as a bimatrix
game and solve for all pure Nash equilibria.

RP 24 *An Auction (2)*
$n \geq 4$ players participate in an auction for an object for which their evaluations are
$v_1 > v_2 > \ldots > v_n$. It is a sealed-bid auction, and the highest bidder obtains the
object and pays the fourth-highest bid. In case of a tie among highest bidders, the
player with the lowest number among the highest bidders obtains the object.

(a) Show that for any player i the bid of v_i weakly dominates any lower bid but does
not weakly dominate any higher bid.
(b) Show that a strategy profile in which each player bids his true valuation is not a
Nash equilibrium.
(c) Find all Nash equilibria in which all players submit the same bid.

RP 25 *An Auction (3)*
Two individuals (players) participate in the auction of a painting. The painting
has worth 6 for player 1 and 4 for player 2. The individuals simultaneously and
independently submit their bids b_1 and b_2, where these bids can only be whole
numbers: $b_1, b_2 \in \{0, \ldots, 6\}$ (higher bids do not make sense). The highest bidder
wins the auction, receives the painting and pays his bid, whereas the other player
receives and pays nothing. In case of a draw player 1 wins. Hence, the payoff to
player 1 is $6 - b_1$ if $b_1 \geq b_2$ and 0 if $b_1 < b_2$, and the payoff to player 2 is $4 - b_2$ if
$b_2 > b_1$ and 0 if $b_2 \leq b_1$.

(a) Compute the best reply functions of both players, and draw a diagram.
(b) Compute the Nash equilibrium or equilibria of this game, if any.

RP 26 *Quantity Versus Price Competition*
Suppose, in the Cournot model, that the two firms produce heterogenous goods,
which have different market prices. Specifically, suppose that these market prices
are given by

$$p_1 = \max\{4 - 2q_1 - q_2, 0\}, \quad p_2 = \max\{4 - q_1 - 2q_2, 0\}. \tag{*}$$

These are the prices of the goods of firms 1 and 2, respectively. The firms compete
in quantities. Both fixed and marginal costs are assumed to be zero.

(a) Write down the payoff (= profit) functions of the two firms and compute the Nash equilibrium quantities.
(b) Use (*) to show that

$$q_1 = \max\{\tfrac{1}{3}(p_2 - 2p_1 + 4), 0\}, \quad q_2 = \max\{\tfrac{1}{3}(p_1 - 2p_2 + 4), 0\}. \qquad (**)$$

Assume now that the firms compete in prices, with demands given by (**).

(c) Write down the payoff (= profit) functions of the two firms in terms of prices and compute the Nash equilibrium prices.
(d) Compare the equilibria found under (a) and (c). Specifically, compare the quantities and (associated) prices under (a) to the (associated) quantities and prices under (c). What about the associated profits? Which is tougher: quantity or price competition?

RP 27 *An Oligopoly Game (1)*
Three oligopolists operate in a market with inverse demand function given by $P(Q) = a - Q$, where $Q = q_1 + q_2 + q_3$ and q_i is the quantity produced by firm i. Each firm has a constant marginal cost of production, $0 < c < a$, and no fixed cost. The firms choose their quantities as follows: (1) firm 1 chooses $q_1 \geq 0$; (2) firms 2 and 3 observe q_1 and then simultaneously choose q_2 and q_3, respectively.

(a) Draw a picture of the extensive form of this game. Also give the strategic form: describe the strategy spaces of the players and the associated payoff functions.
(b) Determine the subgame perfect Nash equilibrium of this game.

RP 28 *An Oligopoly Game (2)*
Consider the Cournot model with three firms. Each firm $i = 1, 2, 3$ offers $q_i \geq 0$, and the market price of the good is $P(q_1, q_2, q_3) = 10 - q_1 - q_2 - q_3$ (or zero if this amount should be negative). Firms 1 and 2 have marginal costs equal to 0 while firm 3 has marginal cost equal to 1. We assume that the firms are involved in Cournot quantity competition.

(a) Derive the reaction functions of the three firms.
(b) Compute the Nash equilibrium of this game.
(c) Determine the maximal joint profit the three firms can achieve by making an agreement on the quantities q_1, q_2, and q_3. Is such an agreement unique?

RP 29 *A Duopoly Game with Price Competition*
Two firms (1 and 2) sell one and the same good. They engage in price competition à la Bertrand. The demand for the good at price p is $100 - p$. The firm that sets the lower price gains the whole market; in case of equal prices the market is split evenly.

The unit marginal cost for firm 1 is $c_1 = 30$, whereas for firm 2 it is $c_2 = 50$. Prices are in whole units, i.e., $p \in \{0, 1, \ldots\}$. (So, e.g., a price of 20.5 is not allowed.)

(a) Write down the profit functions of both firms, and compute the monopoly prices.
(b) Derive the reaction functions of both firms.
(c) Compute the Nash equilibrium in which the price of firm 1 is minimal among all Nash equilibria.
(d) Compute the Nash equilibrium in which the price of firm 1 is maximal among all Nash equilibria.

RP 30 *Contributing to a Public Good*

Three people simultaneously decide whether or not to contribute to a public good. At least two contributions are needed in order to provide the public good. Suppose that the public good has a value of 8 units to every person if it is provided, and that the contribution is fixed at 3 units. If a person decides to contribute, then the contribution must be paid also if the good is not provided.

(a) Find all Nash equilibria in pure strategies of this game.
(b) Suppose now that players 2 and 3 use the same mixed strategy, in which they contribute with probability p. Show that player 1's expected payoff of contributing is equal to

$$16p - 8p^2 - 3 .$$

(c) Compute the two symmetric mixed strategy Nash equilibria of this game.

RP 31 *A Demand Game*

Three players divide one perfectly divisible Euro. The players simultaneously submit their demands, x_1, x_2, and x_3, respectively, where $0 \leq x_i \leq 1$ for each player $i = 1, 2, 3$. If the sum of these demands is smaller than or equal to 1, i.e., $x_1 + x_2 + x_3 \leq 1$, then each player i obtains his demand x_i; otherwise, i.e., if $x_1 + x_2 + x_3 > 1$, then each player obtains 0.
First, for each of the following amounts, either exhibit a Nash equilibrium of this game with sum of demands equal to that amount, or show that such a Nash equilibrium does not exist.

(a) 0.9; (b) 1.2; (c) 1.5; (d) 1.8

(e) Determine all Nash equilibria of this game, i.e., all triples (x_1, x_2, x_3) with $0 \leq x_1, x_2, x_3 \leq 1$ so that no player can do better given the demands of the other players.

Chapter 7

RP 32 *A Repeated Game (1)*
Consider the following bimatrix game

$$G = \begin{array}{c} \\ T \\ B \end{array} \begin{array}{cc} L & R \\ \left(\begin{array}{cc} 16,24 & 0,25 \\ 0,18 & 16,16 \end{array} \right) \end{array}.$$

(a) Which set of payoffs can be reached as the long run average payoffs in subgame perfect Nash equilibria of the infinitely repeated discounted game $G^\infty(\delta)$ for suitable choices of δ?

(b) Same question as under (a), but now for Nash equilibrium (not necessarily subgame perfect).

(c) Describe a Nash equilibrium of $G^\infty(\delta)$ resulting in the long run average payoffs $(16, 20)$. Is there any value of δ for which this equilibrium is subgame perfect? Why or why not?

RP 33 *A Repeated Game (2)*
Consider the following stage game G:

$$\begin{array}{c} \\ U \\ M \\ D \end{array} \begin{array}{ccc} L & C & R \\ \left(\begin{array}{ccc} 8,8 & 0,9 & 4,1 \\ 9,1 & 2,1 & 4,2 \\ 10,3 & 2,4 & 4,4 \end{array} \right) \end{array}.$$

Player 1 is the row player and player 2 the column player.

(a) Compute all the pure strategy Nash equilibria of G.

(b) Compute all mixed-strategy Nash equilibria of the game G.

(c) Suppose that the game G is played twice, with as payoffs the sums of the payoffs of the two stages. Determine the number of (pure) strategies of each player. Is there a subgame perfect (pure strategy) Nash equilibrium of the twice repeated game in which (U, L) is played in the first stage? (Describe such an equilibrium or argue that it cannot exist.)

(d) Consider the infinite repetition of the game G in which the players use the discounted sums of payoffs to evaluate the outcome. The players have a common discount factor $0 < \delta < 1$. Describe a subgame perfect Nash equilibrium of the repeated game in which always (U, L) is played, using trigger strategies. Also compute the minimal value of δ for which this equilibrium exists.

RP 34 *A Repeated Game (3)*
Consider the following stage game G:

$$\begin{array}{c} \\ U \\ M \\ D \end{array} \begin{array}{c} \begin{array}{ccc} L & C & R \end{array} \\ \begin{pmatrix} 0,0 & 11,0 & 6,1 \\ 0,12 & 10,10 & 5,5 \\ 1,6 & 12,4 & 6,6 \end{pmatrix} \end{array}.$$

Player 1 is the row player and player 2 the column player.

(a) Compute all the pure strategy Nash equilibria of G.
(b) Compute all mixed-strategy Nash equilibria of the game G.
(c) Suppose that the game G is played twice, with as payoffs the sums of the payoffs of the two stages. Determine the number of (pure) strategies of each player. Is there a subgame perfect (pure strategy) Nash equilibrium of the twice repeated game in which (M, C) is played in the first stage? (Describe such an equilibrium or argue that it cannot exist.)
(d) Consider the infinite repetition of the game G in which the players use the discounted sums of payoffs to evaluate the outcome. The players have a common discount factor $0 < \delta < 1$. Describe a subgame perfect Nash equilibrium of the repeated game in which always (M, C) is played, using trigger strategies. Also compute the minimal value of δ for which this equilibrium exists.

RP 35 *A Repeated Game (4)*
Consider the bimatrix game

$$G = \begin{array}{c} \\ T \\ B \end{array} \begin{array}{c} \begin{array}{cc} L & R \end{array} \\ \begin{pmatrix} 4,10 & 3,9 \\ 5,5 & 3,4 \end{pmatrix} \end{array}.$$

Suppose this game is played twice. After each play of the game the players learn the outcome. The total payoff is the sum of the payoffs of each of the two plays of the game.

(a) Describe the subgame perfect equilibrium or equilibria in pure strategies of this twice repeated game, if any. Also give the associated outcome(s) and payoffs.
(b) Is there a Nash equilibrium of the twice repeated game in which the combination (T, L) is played the first time? If so, describe such an equilibrium and the associated outcome and payoffs. If not, explain why.

Now assume that the game is repeated infinitely many times. Payoffs are the discounted sums of stage payoffs, with common discount factor $0 < \delta < 1$.

(c) Which long run average payoffs can be obtained in a subgame perfect equilibrium in (pure) trigger strategies of this infinitely repeated game for a suitably chosen value of δ? Give an example of such a subgame perfect equilibrium, and also give the associated outcome, payoffs, and long-run average payoffs, as well as the values of δ for which this is an equilibrium.

RP 36 *A Repeated Game (5)*
Consider the following bimatrix game:

$$
\begin{array}{c c}
 & \begin{array}{c c} L & \quad R \end{array} \\
\begin{array}{c} T \\ B \end{array} &
\begin{pmatrix} 2,1 & 5,0 \\ 0,6 & 1,1 \end{pmatrix}.
\end{array}
$$

(a) Determine all Nash equilibria (in pure or mixed strategies) in this game.

Now suppose that the game is infinitely repeated, at times $t = 0, 1, 2, \ldots$, and that the players learn the outcome after each play of the game. There is a common discount factor $0 < \delta < 1$, and the payoffs are the discounted streams of payoffs.

(b) Which pairs of payoffs can be reached as long-run average payoffs in a subgame perfect equilibrium (in trigger strategies) in this game, assuming that we can take δ as close to 1 as desired?

(c) Give a subgame perfect equilibrium of the infinitely repeated game resulting in the average payoffs $(2\frac{1}{2}, 3)$. Give the values of δ for which your strategies indeed form a subgame perfect equilibrium.

Chapter 8

RP 37 *An Evolutionary Game*
Consider the following matrix

$$
A = \begin{array}{c c}
 & \begin{array}{c c} C & \quad D \end{array} \\
\begin{array}{c} C \\ D \end{array} &
\begin{pmatrix} 0 & 2 \\ 3 & 1 \end{pmatrix}.
\end{array}
$$

(a) Suppose this matrix represents an evolutionary game between animals of the same species. Give an interpretation of a mixed strategy $(p, 1 - p)$.
(b) Determine the replicator dynamics, rest points and stable rest points, and evolutionary stable strategies. Include a phase diagram for the replicator dynamics.
(c) For the evolutionary stable strategy or strategies, show directly (that is, without using the replicator dynamics) that the strategy (or strategies) is (or are) evolutionary stable.

Chapter 9

RP 38 *An Apex Game*

A voting committee consists of five members. Player 1 is a major player, called the *apex player*. The other players are called *minor players*. In order to pass a decision one needs the consent of either the apex player and at least one minor player, or of the four minor players. Therefore, a coalition that contains either the apex player and at least one minor player, or all minor players, is called *winning*. We model this situation as a so-called *simple game*: winning coalitions obtain worth 1, all other coalitions worth 0.

(a) Is the core of this game empty or not? If not, then compute it.
(b) Compute the Shapley value of this game. (Use the symmetry.)
(c) Compute the nucleolus of this game. (Use the symmetry.)

RP 39 *A Three-person Cooperative Game (1)*
A three-person cooperative game with player set $\{1, 2, 3\}$ is described in the following table:

S	$\emptyset$	$\{1\}$	$\{2\}$	$\{3\}$	$\{1, 2\}$	$\{1, 3\}$	$\{2, 3\}$	$\{1, 2, 3\}$
$v(S)$	0	0	1	0	3	3	a	10

(a) Determine all values of a for which the core of this game is not empty.
(b) Determine the Shapley value of this game. For which values of a is it in the core?
(c) Determine all values of a, if any, for which the vector $(\frac{16-2a}{3}, \frac{7+a}{3}, \frac{7+a}{3})$ is the nucleolus of this game.

RP 40 *A Three-person Cooperative game (2)*
For each $a \in \mathbb{R}$ the three-person cooperative game $(\{1, 2, 3\}, v_a)$ is given by $v_a(1) = a$, $v_a(2) = v_a(3) = 0$, $v_a(12) = 2$, $v_a(13) = 3$, $v_a(23) = 4$, and $v_a(N) = v_a(123) = 5$.

(a) For which values of a is the core of v_a non-empty? For these values, compute the core.
(b) Compute the Shapley value of v_a. For which values of a is it in the core of v_a?
(c) For which values of a are the maximal excesses at the nucleolus of v_a reached by the two-person coalitions? For those values, compute the nucleolus.

RP 41 *Voting*
Suppose in Parliament there are four parties A, B, C, D with numbers of votes equal to $40, 30, 20, 10$, respectively. To pass any law an absolute majority ($>50\%$) is needed.

(a) Formulate this situation as a four-person cooperative game where winning coalitions (coalitions that have an absolute majority) have worth 1 and losing coalitions worth 0. Determine the Shapley value of this game.

For every party X in $\{A, B, C, D\}$ let p_X denote the number of coalitions, containing X, that are winning but would be losing without X. Define $\beta(X) = \frac{p_X}{p_A + p_B + p_C + p_D}$ for every party X.

(b) Compute $\beta(X)$ for every $X \in \{A, B, C, D\}$.

Consider now a Parliament with three parties A, B, C and numbers of votes equal to 20, 10, 10, respectively. To pass any law a two-third majority is needed.

(c) Answer the same questions as in (a) and (b) for this Parliament. (Of course, now $p_X = \frac{p_X}{p_A + p_B + p_C}$ for $X \in \{A, B, C\}$.)

RP 42 *An Airport Game*
Three airline companies share the cost of a runway. To serve the planes of company $i \in \{1, 2, 3\}$ the length of the runway must be c_i, which is also the cost of a runway of that length. The airline companies can form coalitions, and the cost of a coalition is the cost of the smallest runway long enough to serve the planes of all companies in the coalition. The costs c_i are given by $c_i = i$ for each company $i \in \{1, 2, 3\}$, and we assume $c_1 \leq c_2 \leq c_3$.

(a) Model this situation as a three-player cost savings (TU) game.
(b) Compute the core of this game.
(c) Compute the Shapley value of this game. Is it in the core?
(d) Suppose that at the nucleolus of this game the excesses of the two-person coalitions are equal and maximal. What does this imply for c_1, c_2, and c_3?

RP 43 *A Glove Game*
There are five players. Players 1 and 2 each possess a right-hand glove, while players 3, 4, and 5 each possess a left-hand glove. The players can form coalitions, and the worth of each coalition is equal to the number of pairs of gloves that the coalition can make.

(a) Compute the Shapley value of this game.
(b) Compute the core of this game.
(c) Compute the nucleolus of this game.

RP 44 *A Four-person Cooperative Game*
Consider the four-person game (N, v) with $N = \{1, 2, 3, 4\}$, $v(\{1, 2\}) = v(\{3, 4\}) = 2$, $v(\{1, 3\}) = 3$, $v(N) = 4$, and $v(S) = 0$ for all other coalitions S.

(a) Compute the core of this game. Plot the possible core payoffs of players 1 and 3 in a two-dimensional diagram with the payoffs of player 1 on the horizontal axis and the payoffs of player 3 on the vertical axis.
(b) Compute the Shapley value of this game.

Chapter 10

RP 45 *A Matching Problem*

Let $X = \{x_1, x_2, x_3, x_4\}$ and $Y = \{y_1, y_2, y_3, y_4\}$ be two groups of people, and let their preferences concerning possible partners be given by the following table.

x_1 x_2 x_3 x_4	y_1 y_2 y_3 y_4
y_3 y_3 y_3 y_4	x_2 x_3 x_2 x_1
y_4 y_2 y_1 y_1	x_1 x_4 x_1 x_4
y_1 y_4 y_2 y_2	x_3 x_2 x_3 x_2
y_2 y_1 y_4 y_3	x_4 x_1 x_4 x_3

(a) Apply the deferred acceptance procedure with proposals by members of X. Which matching do you obtain?

(b) Apply the deferred acceptance procedure with proposals by members of Y. Which matching do you obtain?

(c) Explain why the matching $(x_1, y_1), (x_2, y_3), (x_3, y_2), (x_4, y_4)$ is not in the core.

(d) How many matchings are in the core? Explain your answer.

RP 46 *House Exchange*

Player i owns house h_i, $i = 1, 2, 3$. The preferences of the players over the houses are given by the following table:

Player 1	Player 2	Player 3
h_2	h_3	h_1
h_1	h_2	h_2
h_3	h_1	h_3

(a) Compute all core allocations of this game.

(b) Which of these allocations are in the strong core?

(c) Give a new preference of player 1 such that the new game has a unique core allocation.

RP 47 *A Marriage Market*

Consider a marriage market with four men (m_1, m_2, m_3, and m_4) and four women (w_1, w_2, w_3, and w_4).

(a) Suppose that every woman has the same preference $m_1 > m_2 > m_3 > m_4$ over the men. Argue that there is a unique matching in the core, and describe this matching.

For (b)–(d) assume that the preferences are given by the following table.

m_1	m_2	m_3	m_4	w_1	w_2	w_3	w_4
w_1	w_2	w_3	w_4	m_4	m_3	m_2	m_1
w_2	w_1	w_4	w_3	m_3	m_4	m_1	m_2
w_3	w_4	w_1	w_2	m_2	m_1	m_4	m_3
w_4	w_3	w_2	w_1	m_1	m_2	m_3	m_4

(b) Compute the core matching that is optimal from the point of view of the men.
(c) Compute the core matching that is optimal from the point of view of the women.
(d) Find a core matching in which m_1 and w_2 are coupled.

References

Osborne, M. J. (2004). *An introduction to game theory*. New York: Oxford University Press.
Watson, J. (2002). *Strategy, an introduction to game theory*. New York: Norton.

Hints, Answers and Solutions

24

Problems of Chapter 1

1.2 *Variant of Matching Pennies*
There are saddlepoint(s) if and only if $x \leq -1$.

1.3 *Mixed Strategies*

(b) $(3/4, 1/4)$.
(c) $(1/2, 1/2)$.
(d) By playing $(3/4, 1/4)$ player 1 obtains $10/4 = 2.5$ for sure (independent of what player 2 does). Similarly, by playing $(1/2, 1/2)$, player 2 is sure to pay 2.5. So 2.5 is the value of this game. Given a rational opponent, no player can hope to do better by playing differently.

1.4 *Sequential Cournot*

(b) $q_1 = 1/3$ and $q_2 = 1/6$.

1.6 *Glove Game*

(a) $(0, 0, 1)$ is the unique vector in the core of the glove game.

1.7 *Dentist Appointments*
The Shapley value $(9\frac{1}{2}, 6\frac{1}{2}, 8)$ is *not* in the core of this game. The nucleolus is in the core of the game.

© Springer-Verlag Berlin Heidelberg 2015
H. Peters, *Game Theory*, Springer Texts in Business and Economics,
DOI 10.1007/978-3-662-46950-7_24

1.8 *Nash Bargaining*

(a) The problem to solve is $\max_{0\leq\alpha\leq1} \alpha\sqrt{1-\alpha}$. Obviously, the solution must be interior: $0 < \alpha < 1$. Set the first derivative equal to 0, solve, and check that the second derivative is negative.

(b) In terms of utilities, the Nash bargaining solution is $(2/3, (1/3)\sqrt{3})$.

1.9 *Variant of Glove Game*
The worth of a coalition S depends on the minimum of the numbers of right-hand and left-hand players in the coalition.

Problems of Chapter 2

2.1 *Solving Matrix Games*

(a) The optimal strategies are $(5/11, 6/11)$ for player 1 and $(5/11, 6/11)$ for player 2. The value of the game is $30/11$. In the original game the optimal strategies are $(5/11, 6/11, 0)$ for player 1 and $(5/11, 6/11, 0)$ for player 2.

(b) The value of the game is 0. The unique maximin strategy is $(0, 1, 0)$. The minimax strategies are $(0, q, 1 - q, 0)$ for any $0 \leq q \leq 1$.

(c) The value of the game is 1, the unique minimax strategy is $(1/2, 0, 1/2)$, and the maximin strategies are: $(p, (1 - p)/2, (1 - p)/2)$ for $0 \leq p \leq 1$.

(d) The value of the game is 9 and player 1's maximin strategy is $(1/2, 1/2, 0, 0)$. The set of all minimax strategies is $\{(\alpha, (7 - 14\alpha)/10, (3 + 4\alpha)/10) \in \mathbb{R}^3 \mid 0 \leq \alpha \leq 1/2\}$.

(e) The value is $8/5$. The unique maximin strategy is $(2/5, 3/5)$ and the unique minimax strategy is $(0, 4/5, 1/5, 0)$.

(f) The value is equal to 1, player 2 has a unique minimax strategy namely $(0, 1, 0)$, and the set of maximin strategies is $\{(0, p, 1 - p) \mid 0 \leq p \leq 1\}$.

2.2 *Saddlepoints*

(b) There are saddlepoints at $(1, 4)$ and at $(4, 1)$.

2.3 *Maximin Rows and Minimax Columns*

(c) The unique maximin strategy is $(\frac{4}{7}, 0, \frac{3}{7}, 0)$ and the unique minimax strategy is $(\frac{4}{7}, 0, \frac{3}{7})$.

2.4 *Subgames of Matrix Games*

(c) The unique minimax strategy is $(0, 4/5, 1/5, 0)$ and the unique maximin strategy is $(2/5, 3/5)$.

2.5 *Rock-Paper-Scissors*
The associated matrix game is:

$$
\begin{array}{c}
\quad\quad R \quad P \quad S \\
\begin{array}{c} R \\ P \\ S \end{array}
\left(
\begin{array}{ccc}
0 & -1 & 1 \\
1 & 0 & -1 \\
-1 & 1 & 0
\end{array}
\right).
\end{array}
$$

Problems of Chapter 3

3.1 *Some Applications*

(a) Let Smith be the row player and Brown the column player, then the bimatrix game is:

$$
\begin{array}{c}
\quad\quad L \quad\quad\quad S \\
\begin{array}{c} L \\ S \end{array}
\left(
\begin{array}{cc}
2,2 & -1,-1 \\
-1,-1 & 1,1
\end{array}
\right).
\end{array}
$$

(b) Let the government be the row player and the pauper the column player. The bimatrix game is:

$$
\begin{array}{c}
\quad\quad\quad work \quad\quad not \\
\begin{array}{c} aid \\ not \end{array}
\left(
\begin{array}{cc}
3,2 & -1,3 \\
-1,1 & 0,0
\end{array}
\right).
\end{array}
$$

(c) This game has two pure strategy Nash equilibria and one other (mixed strategy) Nash equilibrium.

(e, f) This situation can be modelled as a 3×3 bimatrix game. For (e), the expected numbers of candidates in the committee can be taken as payoffs; for (f), the payoff is the expected utility (using $\sqrt{c}$) of the lottery determining the number of candidates in the committee.

3.2 *Matrix Games*

(a) You should find the same solution, namely $(5/11, 6/11)$ for player 1 and $(5/11, 6/11)$ for player 2, as the unique Nash equilibrium.

(b) If player 2 plays a minimax strategy then 2's payoff is at least $-v$, where v is the value of the game. Hence, any strategy that gives player 1 at least v is a best reply. So a maximin strategy is a best reply. Similarly, a minimax strategy is a best reply against a maximin strategy, so any pair consisting of a maximin and a minimax strategy is a Nash equilibrium.

Conversely, in a Nash equilibrium the payoffs must be $(v, -v)$ otherwise one of the players could improve by playing an optimal (maximin or minimax) strategy. But then player 1's strategy must be a maximin strategy since otherwise player 2 would have a better reply, and player 2's strategy must be a minimax strategy since otherwise player 1 would have a better reply.

(c) The appropriate definition for player 2 would be: a maximin strategy *for player 2* in B, since now B represents the payoffs to player 2, and not what player 2 has to pay to player 1.

The Nash equilibrium of Problem 3.1(b), for instance, does not consist of maximin strategies of the players. The maximin strategy of player 1 in A is $(1/5, 4/5)$, which is not part of a (the) Nash equilibrium. The maximin strategy of player 2 (!) in B is $(1, 0)$, which is not part of a (the) Nash equilibrium.

3.3 Strict Domination

(c) There are three Nash equilibria: $((1, 0), (1, 0, 0, 0))$, $((0, 1), (0, 0, 1, 0))$, and $((3/7, 4/7), (1/3, 0, 2/3, 0))$.

3.4 Iterated Elimination (1)

(b) The unique equilibrium is (B, Y).

3.5 Iterated Elimination (2)
The Nash equilibria are $((1/3, 2/3, 0), (2/3, 0, 1/3))$, $((0, 1, 0), (1, 0, 0))$, and $((1, 0, 0), (0, 0, 1))$.

3.6 Weakly Dominated Strategies

(b) Consecutive deletion of Z, C, A results in the Nash equilibria (B, X) and (B, Y). Consecutive deletion of C, Y, B, Z results in the Nash equilibrium (A, X).

3.7 A Parameter Game
Distinguish three cases: $a > 2$, $a = 2$, and $a < 2$.

3.8 Equalizing Property of Mixed Equilibrium Strategies

(a) Check by substitution.
(b) Suppose the expected payoff (computed by using $\mathbf{q}^*$) of row i played with positive probability (p_i^*) in a Nash equilibrium $(\mathbf{p}^*, \mathbf{q}^*)$, hence the number $e^i A \mathbf{q}^*$, would not be maximal. Then player 1 would improve by adding the probability p_i^* to some row j with higher expected payoff $e^j A \mathbf{q}^* > e^i A \mathbf{q}^*$, and in this way increase his payoff, a contradiction. A similar argument can be made for player 2 and the columns.

3.9 *Voting*

(a, b, c) Set the total number of voters equal to 10 (in order to avoid fractions). We
   obtain a 6×6 bimatrix game in which the sum of the payoffs per entry is
   always 10. This game has four Nash equilibria in pure strategies.
   (d) Now we obtain a unique Nash equilibrium in pure strategies.
   (e) In both games, subtract 5 from all payoffs.

3.10 *Guessing Numbers*

(d) The unique Nash equilibrium is the one where each player chooses each number
   with equal probability.
(e) The value of this game is $\frac{1}{K}$.

3.11 *Bimatrix Games*

(b) $e < a, b < d, c < g, h < f$. Then there is a unique Nash equilibrium.

Problems of Chapter 4

4.1 *Counting Strategies*
White has 20 possible opening moves, and therefore also 20 possible strategies.
Black has many more strategies.

4.2 *Extensive Versus Strategic Form*
For the game with perfect information, start with a decision node of player 1. For
the game with imperfect information, start with a decision node of player 2.

4.4 *Choosing Objects*

(c) In any subgame perfect equilibrium the game is played as follows: player 1 picks
   O_3, then player 2 picks O_2 or O_1, and finally player 1 picks O_4. These are the
   (two) subgame perfect equilibrium outcomes of the game. Due to ties (of player
   2) there is more than one subgame perfect equilibrium, namely eight in total.
   All subgame perfect equilibria result in the same distribution of the objects.
(d) There is a Nash equilibrium in which player 1 obtains the objects O_4 and O_2.

4.5 *A Bidding Game*

(c) Due to ties, there are four different subgame perfect equilibria. They all result
   in the same outcome.

4.6 *An Extensive Form Game*
There is a unique pure strategy Nash equilibrium, which is also subgame perfect.
This equilibrium is perfect Bayesian for an appropriate choice of player 2's belief.

4.7 *Another Extensive Form Game*
There is a unique Nash equilibrium (in pure strategies). This equilibrium is not perfect Bayesian.

4.8 *Still Another Extensive Form Game*

(b) There are three Nash equilibria in pure strategies.
(c) There are two subgame perfect Nash equilibria in pure strategies.
(d) Both equilibria in (c) are perfect Bayesian.

4.9 *A Centipede Game*

(b) Consider any strategy combination. The last player who has continued when playing his strategy could have improved by stopping if possible. Hence, in equilibrium the play of the game must have stopped immediately.
  To exhibit a non-subgame perfect Nash equilibrium, assume that player 1 always stops, and that player 2 also always stops except at his second decision node. Check that this is a Nash equilibrium. [One can also write down the strategic form, which is an 8×8 bimatrix game.]

4.10 *Finitely Repeated Prisoners' Dilemma*

(a) There are five subgames, including the entire game.
(b) There is a unique subgame perfect equilibrium.

4.11 *A Twice Repeated 2×2 Bimatrix Game*

(b) Player 1: play B at the first stage; if (B, L) was played at the first stage play B at the second stage, otherwise play T at the second stage.

4.12 *Twice Repeated 3×3 Bimatrix Games*

(a) There are ten subgames, including the entire game.
(b) Player 1: play T at the first stage. Player 2: play L at the first stage. Second stage play is given by the following diagram:

$$
\begin{array}{c}
\\ T \\ C \\ B
\end{array}
\begin{array}{ccc}
L & M & R \\
\left(\begin{array}{ccc}
B,R & C,R & C,R \\
B,M & B,R & B,R \\
B,M & B,R & B,R
\end{array} \right).
\end{array}
$$

For instance, if first stage play results in (C, L), then player 1 plays B and player 2 plays M at stage 2. Verify that this defines a subgame perfect equilibrium in which (T, L) is played at the second stage. (Other solutions are possible, as long as players 1 and 2 are punished for unilateral deviations at stage 1.)

Problems of Chapter 5

5.1 *Battle-of-the-Sexes*
The strategic form is a 4×4 bimatrix game. List the strategies of the players as in the text. We can then compute the expected payoffs. For example, if the first row corresponds to strategy FF of player 1 and strategies FF, FB, BF, and BB of player 2, then the payoffs are, respectively, $1/6$ times $(8,3)$, $(6,9)$, $(6,0)$, and $(4,6)$. The (pure strategy) Nash equilibria are (FF, FB) and (BF, BB).

5.2 *A Static Game of Incomplete Information*
There are three pure Nash equilibria: (TT, L), (TB, R), and (BB, R). (The first letter in a strategy of player 1 applies to Game 1, the second letter to Game 2.)

5.3 *Another Static Game of Incomplete Information*

(b) The unique pure strategy Nash equilibrium is: t_1 and t_1' play B, t_2 and t_2' play R.

5.4 *Job-Market Signaling*

(b) There is a separating perfect Bayesian equilibrium. There is another Nash equilibrium in which no worker takes education, but this is not perfect Bayesian.

5.5 *A Joint Venture*

(c) There is a unique Nash equilibrium (even in mixed strategies). This is also subgame perfect and perfect Bayesian.

5.6 *Entry Deterrence*
For $x \leq 100$ the strategy combination where the entrant always enters and the incumbent colludes is a perfect Bayesian equilibrium. For $x \geq 50$, the combination where the entrant always stays out and the incumbent fights is a perfect Bayesian equilibrium if the incumbent believes that, if the entrant enters, then fighting yields 0 with probability at most $1 - \frac{50}{x}$. IC applies only to the second equilibrium, which survives it.

5.7 *The Beer-Quiche Game*

(b) There are two perfect Bayesian equilibria, both of which are pooling. In the first one, player 1 always eats quiche. This equilibrium does not survive IC. In the second one, player 1 always drinks beer. This equilibrium does survive IC.

5.8 *Issuing Stock*

(b) There is a pooling equilibrium in which the manager never proposes to issue new stock, and such a proposal would not be approved of by the existing shareholders since they believe that this proposal signals a good state with high enough probability. [The background of this is that a new stock issue would dilute the value of the stock of the existing shareholders in a good state of the world, see the original article Myers and Majluf (1984) for details.] This equilibrium survives the intuitive criterion.

There is also a separating equilibrium in which a stock issue is proposed in the bad state but not in the good state. If a stock issue is proposed, then it is approved of.

Finally, there is a separating equilibrium in which a stock issue is proposed in the good state but not in the bad state. If a stock issue is proposed, then it is not approved of.

(c) In this case, a stock issue proposal would always be approved of, so the 'bad news effect' of a stock issue vanishes. The reason is that the investment opportunity is now much more attractive.

5.9 *More Signaling Games*

(a) IC does not apply, since it would rule out both types of player 1.
(b) There is a unique, pooling perfect Bayesian equilibrium. This equilibrium does not survive the intuitive criterion.
(c) There are two strategy combinations that are perfect Bayesian.

Problems of Chapter 6

6.1 *Cournot with Asymmetric Costs*
The Nash equilibrium is $q_1 = (a - 2c_1 + c_2)/3$ and $q_2 = (a - 2c_2 + c_1)/3$, given that these amounts are nonnegative.

6.2 *Cournot Oligopoly*

(b) The reaction function of player i is: $\beta_i(q_1, \ldots, q_{i-1}, q_{i+1}, \ldots, q_n) = (a - c - \sum_{j \neq i} q_j)/2$ if $\sum_{j \neq i} q_j \leq a - c$, and $\beta_i(q_1, \ldots, q_{i-1}, q_{i+1}, \ldots, q_n) = 0$ otherwise.
(c) One should compute the point of intersection of the n reaction functions. This amounts to solving a system of n linear equations in n unknowns $q_1, \ldots, q_n$. Alternatively, one may guess that there is a solution $q_1 = q_2 = \ldots = q_n$. Then $q_1 = (a - c - (n - 1)q_1)/2$, resulting in $q_1 = (a - c)/(n + 1)$. Hence, each firm producing $(a - c)/(n + 1)$ is a Nash equilibrium. If the number of firms becomes large then this amount converges to 0, which is no surprise since demand is bounded by a.
(d) To show that this equilibrium is unique, it is sufficient to show that the determinant of the coefficient matrix associated with the system of n linear equations in n unknowns (the reaction functions) is unequal to zero.

6.3 *Quantity Competition with Heterogenous Goods*

(a) $\Pi_i(q_1, q_2) = q_i p_i(q_1, q_2) - cq_i$ for $i = 1, 2$.

(b) The equilibrium is: $q_1 = (21 - 4c)/33, q_2 = (13 - 3c)/22, p_1 = (21 + 7c)/11,$ $p_2 = (39 + 13c)/22$. From this the profits are easily computed.

(c) $q_1 = (57 - 10c)/95, q_2 = (38 - 10c)/95, p_1 = (228 + 50c)/95, p_2 = (228 + 45c)/95$. From this the profits are easily computed.

(d) $q_1 = \max\{1 - \frac{1}{2}p_1 + \frac{1}{3}p_2, 0\}, q_2 = \max\{1 - \frac{1}{2}p_2 + \frac{1}{4}p_1\}$. The profit functions are now $\Pi_1(p_1, p_2) = p_1 q_1 - cq_1$ and $\Pi_2(p_1, p_2) = p_2 q_2 - cq_2$, with q_1 and q_2 as given.

(e) The equilibrium is $p_1 = (16 + 8c)/11, p_2 = (30 + 15c)/22$. Note that these prices are different from the ones in (c). Profits under price competition will be lower than those under quantity competition.

(f) These are the same prices and quantities as under (c).

(g) See the answers to (e) and (f).

6.4 *A Numerical Example of Cournot Competition with Incomplete Information*
$q_1 = 18/48, q_H = 9/48, q_L = 15/48$. In the complete information case with low cost we have $q_1 = q_2 = 16/48$, with high cost it is $q_1 = 20/48$ and $q_2 = 8/48$. Note that the low cost firm 'suffers' from incomplete information since firm 1 attaches some positive probability to firm 2 having high cost and therefore has higher supply. For the high cost firm the situation is reversed: it 'benefits' from incomplete information.

6.5 *Cournot Competition with Two-Sided Incomplete Information*
Similar to (6.3) we derive:

$$q_\ell = q_\ell(q_H, q_L) = \frac{a - c_\ell - \vartheta q_H - (1 - \vartheta)q_L}{2},$$

$$q_h = q_h(q_H, q_L) = \frac{a - c_h - \vartheta q_H - (1 - \vartheta)q_L}{2},$$

$$q_L = q_L(q_h, q_\ell) = \frac{a - c_L - \pi q_h - (1 - \pi)q_\ell}{2},$$

$$q_H = q_H(q_h, q_\ell) = \frac{a - c_H - \pi q_h - (1 - \pi)q_\ell}{2}.$$

Here, q_ℓ and q_h correspond to the low and high cost types of firm 1 and q_L, and q_H correspond to the low and high cost types of firm 2. The (Bayesian) Nash equilibrium follows by solving these four equations in the four unknown quantities.

6.6 *Incomplete Information About Demand*
The equilibrium is: $q_1 = (\vartheta a_H + (1 - \vartheta)a_L - c)/3$, $q_H = (a_H - c)/3 + ((1 - \vartheta)/6)(a_H - a_L)$, $q_L = (a_L - c)/3 - (\vartheta/6)(a_H - a_L)$. (Assume that all these quantities are positive.)

6.7 *Variations on Two-Person Bertrand*

(a) If $c_1 < c_2$ then there is no Nash equilibrium. (Write down the reaction functions or—easier—consider different cases.)
(b) In both cases (i) and (ii), there are two equilibria.

6.8 *Bertrand with More Than Two Firms*
A strategy combination is a Nash equilibrium if and only if at least two firms charge a price of c and the other firms charge prices higher than c.

6.9 *Variations on Stackelberg*

(a) With firm 1 as a leader we have $q_1 = (1/2)(a - 2c_1 + c_2)$ and $q_2 = (1/4)(a + 2c_1 - 3c_2)$. With firm 2 as a leader we have $q_2 = (1/2)(a - 2c_2 + c_1)$ and $q_1 = (1/4)(a + 2c_2 - 3c_1)$.
(b) The leader in the Stackelberg game can always play the Cournot quantity: since the follower plays the best reply, this results in the Cournot outcome. Hence, the Stackelberg equilibrium—where the leader maximizes—can only give a higher payoff. (This argument holds for an arbitrary game where one player moves first and the other player moves next, having observed the move of the first player.)
(c) $q_i = (1/2^i)(a - c)$ for $i = 1, 2, \ldots, n$.

6.10 *First-Price Sealed-Bid Auction*

(b) Suppose that in some Nash equilibrium player i wins with valuation $v_i < v_1$. Then the winning bid b_i must be at most v_i otherwise player i makes a negative profit and therefore can improve by bidding (e.g.) v_i. But then player 1 can improve by bidding higher than b_i (and win) but lower than v_1 (and make positive profit). Other Nash equilibria: $(v_1, v_1, 0, 0, \ldots, 0)$, $(b, b, b, \ldots, b)$ with $v_1 \geq b \geq v_2$, etc.
(d) If not, then there would be a Nash equilibrium in which—in view of (c)—all players bid below their valuations. By (b) a player with the highest valuation wins the auction, so this must be player 1 if each player bids below his true valuation. But then player 1 can improve if $b_1 \geq v_2$ and player 2 can improve if $b_1 < v_2$.

6.11 *Second-Price Sealed-Bid Auction*

(d) Also $(v_1, 0 \ldots, 0)$ is a Nash equilibrium.
(e) The equilibria are: $\{(b_1, b_2) \mid b_2 \geq v_1,\ 0 \leq b_1 \leq v_2\} \cup \{(b_1, b_2) \mid b_1 \geq v_2,\ b_2 \leq b_1\}$.

6.12 Third-Price Sealed-Bid Auction

(b) Suppose $v_1 > v_2 > v_3 > \ldots$, then bidder 2 could improve by bidding higher than v_1.

(c) Everybody bidding the highest valuation v_1 is a Nash equilibrium. Also everybody bidding the second highest valuation v_2 is a Nash equilibrium. (There are many more!)

6.13 n-Player First-Price Sealed-Bid Auction with Incomplete Information

Suppose every player $j \neq i$ plays s_j^*. If player i's type is v_i and he bids b_i (which can be assumed to be at most $1 - 1/n$ since no other bidder bids higher than this) then the probability of winning the auction is equal to the probability that very bid $b_j, j \neq i$, is at most b_i (including equality since this happens with zero probability). In turn, this is equal to the probability that $v_j \leq n/(n-1)b_i$ for every $j \neq i$. Since the players's valuations are independently drawn from the uniform distribution, the probability that player i wins the auction is equal to $((n/(n-1))b_i)^{n-1}$, hence player i should maximize the expression $(v_i - b_i)((n/(n-1))b_i)^{n-1}$, resulting in $b_i = (1 - 1/n)v_i$.

6.14 Double Auction

(b) The probability of trade given that $v_s \leq v_b$ is equal to $2x(1-x)$. Note that this is maximal for $x = 1/2$, and then it is equal to $1/2$.

(c) $p_b(v_b) = (2/3)v_b + 1/12$ and $p_s(v_s) = (2/3)v_s + 1/4$.

(d) Observe that no trade occurs if $v_s > v_b$. Suppose $v_s \leq v_b$. Then the (conditional) probability that trade occurs is $9/16$. Observe that this is larger than the maximal probability in (b).

6.15 Mixed Strategies and Objective Uncertainty

(a) $((1/2, 1/2), (2/5, 3/5))$.

6.16 Variations on Finite Horizon Bargaining

(a) Adapt Table 6.1 for the various cases.

(b) The subgame perfect equilibrium *outcome* is: player 1 proposes $(1 - \delta_2 + \delta_1\delta_2, \delta_2 - \delta_1\delta_2)$ at $t = 0$ and player 2 accepts.

(c) The subgame perfect equilibrium *outcome* in shares of the good is: player 1 proposes $(1 - \delta_2^2 + \delta_1\delta_2^2, \delta_2^2 - \delta_1\delta_2^2)$ at $t = 0$ and player 2 accepts.

(d) The subgame perfect equilibrium *outcome* is: player 1 proposes $(1 - \delta + \delta^2 - \ldots + \delta^{T-1} - \delta^T s_1, \delta - \delta^2 + \ldots - \delta^{T-1} + \delta^T s_1)$ at $t = 0$ and player 2 accepts.

(e) The limits are $(1/(1 + \delta), \delta/(1 + \delta))$, independent of s.

6.17 *Variations on Infinite Horizon Bargaining*

(a) Conditions (6.10) are replaced by $x_2^* = \delta_2 y_2^*$ and $y_1^* = \delta_1 x_1^*$. This implies $x_1^* = (1 - \delta_2)/(1 - \delta_1 \delta_2)$ and $y_1^* = (\delta_1 - \delta_1 \delta_2)/(1 - \delta_1 \delta_2)$. In the strategies (σ_1^*) and (σ_2^*), replace δ by δ_1 and δ_2, respectively. The equilibrium outcome is that player 1's proposal x^* at $t = 0$ is accepted.

(b) Nothing essential changes. Player 2's proposal y^* is accepted at $t = 0$.

(c) Nothing changes compared to the situation in the text, since s is only obtained at $t = \infty$.

(e) Let p denote the probability that the game ends. Then p is also the probability that the game ends given that it does not end at $t = 0$. Hence, $p = (1 - \delta) + \delta p$, so that $p = 1$.

6.18 *A Principal-Agent Game*

(a) This is a game of complete information.

(b) The subgame perfect equilibrium can be found by backward induction. Distinguish two cases: (i) strategy h is optimal for the worker and (ii) strategy l is optimal for the worker. Show that it is optimal for the employer to induce high effort by a wage combination (w_H, w_L) with $8w_H + 2w_L = 50$ and $w_H - w_L \geq 5$.

6.19 *The Market for Lemons*

(b) There are many subgame perfect equilibria: the buyer offers $p \leq 5{,}000$ and the seller accepts any price of at least 5,000 if the car is bad and of at least 15,000 if the car is good. All these equilibria result in expected payoff of zero for both. There are no other subgame perfect equilibria.

6.20 *Corporate Investment and Capital Structure*

(b) Suppose the investor's belief that $\pi = L$ after observing s is equal to q. Then the investor accepts s if and only if

$$s[qL + (1 - q)H + R] \geq I(1 + r) . \tag{*}$$

The entrepreneur prefers to receive the investment if and only if

$$s \leq R/(\pi + R) , \tag{**}$$

for $\pi \in \{L, H\}$.

In a pooling equilibrium, $q = p$. Note that (**) is more difficult to satisfy for $\pi = H$ than for $\pi = L$. Thus, (*) and (**) imply that a pooling equilibrium exists only if

$$\frac{I(1 + r)}{pL + (1 - p)H + R} \leq \frac{R}{H + R}.$$

A separating equilibrium always exists. The low-profit type offers $s = I(1 + r)/(L+R)$, which the investor accepts, and the high-profit type offers $s < I(1 + r)/(H + R)$, which the investor rejects.

6.21 A Poker Game

(a) The strategic form of this game is as follows:

$$\begin{array}{cccc} & aa & aq & ka & kq \\ \text{believe} & \begin{pmatrix} -1,1 & -1/3,1/3 & -2/3,2/3 & 0,0 \\ 2/3,-2/3 & 1/3,-1/3 & 0,0 & -1/3,1/3 \end{pmatrix} \end{array}.$$

Here, 'believe' and 'show' are the strategies of player I. The first letter in any strategy of player II is what player II says if the dealt card is a King, the second letter is what II says if the dealt card is a Queen—if the dealt card is an Ace player II has no choice.

(b) Player I has a unique optimal (maximin) strategy and player 2 has a unique optimal (minimax) strategy. The value of the game is $-2/9$.

6.22 A Hotelling Location Problem

(a) $x_1 = x_2 = \frac{1}{2}$.
(c) $x_1 = x_2 = \frac{1}{2}$.

6.23 Median Voting

(a) The strategy set of each player is the interval $[0, 30]$. If each player i plays x_i, then the payoff to each player i is $-|((x_1 + \ldots + x_n)/n) - t_i|$. A Nash equilibrium always exists.

(b) The payoff to player i is now $-|\operatorname{med}(x_1, \ldots, x_n) - t_i|$, where $\operatorname{med}(\cdot)$ denotes the median. For each player, proposing a temperature different from his true ideal temperature either leaves the median unchanged or moves the median farther away from the ideal temperature, whatever the proposals of the other players. Hence, proposing one's ideal temperature is a weakly dominant strategy.

6.24 The Uniform Rule

(b) $M = 4 : (1, 3/2, 3/2)$, $M = 5 : (1, 2, 2)$, $M = 5.5 : (1, 2, 5/2)$, $M = 6 : (1, 2, 3)$, $M = 7 : (2, 2, 3)$, $M = 8 : (5/2, 5/2, 3)$, $M = 9 : (3, 3, 3)$.

(c) If player i reports t_i and receives $s_i > t_i$ then, apparently the total reported quantity is above M and thus, player i can only further increase (hence, worsen) his share by reporting a different quantity. If player i reports t_i and receives $s_i < t_i$ then, apparently the total reported quantity is below M and thus, player i can only further decrease (hence, worsen) his share by reporting a different quantity.

There exist other Nash equilibria, but they do not give different outcomes (shares). For example, if $M > \sum_{j=1}^{n} t_j$, then player 1 could just as well report 0 instead of t_1.

6.25 *Reporting a Crime*

(b) $p = 1 - (c/v)^{1/(n-1)}$.
(c) The probability of the crime being reported in this equilibrium is $1 - (1-p)^n = 1 - (c/v)^{n/(n-1)}$. This converges to $1 - (c/v)$ for n going to infinity. Observe that both p and the the probability of the crime being reported decrease if n becomes larger.

6.26 *Firm Concentration*
Let, in equilibrium, n firms locate downtown and m firms in the suburbs, with $n = 6$ and $m = 4$.

6.27 *Tragedy of the Commons*

(d) Suppose, to the contrary, $G^* \leq G^{**}$. Then $v(G^*) \geq v(G^{**})$ since $v' < 0$, and $0 > v'(G^*) \geq v'(G^{**})$ since $v'' < 0$. Also, $G^*/n < G^{**}$. Hence

$$v(G^*) + (1/n)G^* v'(G^*) - c > v(G^{**}) + G^{**} v'(G^{**}) - c ,$$

a contradiction since both sides should be zero.

Problems of Chapter 7

7.1 *Nash and Subgame Perfect Equilibrium in a Repeated Game (1)*

(a) $v(A) = 1$ and the minimax strategy in A is R; $v(-B) = -1$ and the maximin strategy in $-B$ is D.
(d) Player 1 plays always U but after a deviation switches to D forever. Player 2 always plays L but after a deviation switches to R forever. We need $\delta \geq 1/2$ to make this a Nash equilibrium of $G^\infty(\delta)$.

7.2 Nash and Subgame Perfect Equilibrium in a Repeated Game (2)

(a) The limiting average payoffs $(2, 1)$, $(1, 2)$, and $(2/3, 2/3)$, resulting from playing, respectively, the Nash equilibria (U, L), (D, R), and $((2/3, 1/3), (1/3, 2/3))$ at every stage; and all payoffs (x_1, x_2) with $x_1, x_2 > 2/3$.

(b) $v(A) = 2/3$ and $-v(-B) = 2/3$. Hence, all payoffs (x_1, x_2) with $x_1, x_2 > 2/3$.

(c) The players play (U, L) at even times and (D, R) at odd times. Since at each time they play a Nash equilibrium of the stage game, no trigger strategies (describing punishment after a deviation) are needed.

(d) In this case a trigger strategy is needed. The players alternate between (U, L), (D, L), and (D, R).

7.3 Nash and Subgame Perfect Equilibrium in a Repeated Game (3)

(a) The stage game has a unique Nash equilibrium.

(b) $v(A) = 4$ since (D, L) is a saddlepoint in A. The minimax strategy of player 2 is L. The value of $-B$ is -1 and the maximin strategy of player 1 is $(1/2, 1/2)$.

(c) These limit average payoffs are obtained, for instance, by letting the players play (U, L) at even times and (D, R) at odd times. After any deviation the players switch to playing (D, L) (or $((1/2, 1/2), L)$) forever.

7.4 Subgame Perfect Equilibrium in a Repeated Game

(c) Alternate between (T, L) and (M, C).

7.5 The Strategies Tr_1^* and Tr_2^*
An optimal moment for player 1 to deviate would be $t = 1$. For player 2 it would be $t = 3$.

7.6 Repeated Cournot and Bertrand

(a) Each player offers half of the monopoly quantity (half of $(a-c)/2$) at each time, but if a deviation from this occurs, then each player offers the Cournot quantity $(a - c)/3$ forever. This is a subgame perfect equilibrium for $\delta \geq 9/17$.

(b) In this case, each player asks the monopoly price $(a + c)/2$ at each time; if a deviation from this occurs, each player switches to the Bertrand equilibrium price $p = c$ forever. This is a subgame perfect equilibrium for $\delta \geq 1/2$.

7.7 Repeated Duopoly

(b) The Nash equilibrium prices are $p_1 = p_2 = 6$.

(c) Joint profit is maximized at $p_1 = p_2 = 5$.

(d) Ask prices $p_1 = p_2 = 5$, but after a deviation switch to the equilibrium prices $p_1 = p_2 = 6$. This is a subgame perfect equilibrium for $\delta \geq 25/49$.

7.8 *On Discounting*
See the solution to Problem 6.17(e).

7.9 *On Limit Average*
A sequence like $1, 3, 5, 7, \ldots$ has a limit average of infinity. More interestingly, one may construct a sequence containing only the numbers $+1$ and -1 of which the finite averages 'oscillate', e.g, below $-1/2$ and above $+1/2$, so that the limit does not exist.

Problems of Chapter 8

8.1 *Symmetric Games*

(a) $(0, 1)$ is the only *ESS*.
(b) Both $(1, 0)$ and $(0, 1)$ are *ESS*. The (Nash equilibrium) strategy $(1/3, 2/3)$ is not an *ESS*.

8.2 *More Symmetric Games*

(a) The replicator dynamics is $\dot{p} = p(p-1)(p-1/2)$, with rest points $p = 0, 1, 1/2$, of which only $p = 1/2$ is stable. The game (A, A^T) has a unique symmetric Nash equilibrium, namely $((1/2, 1/2), (1/2, 1/2))$. The unique *ESS* is $(1/2, 1/2)$.

8.3 *Asymmetric Games*

(b) The replicator dynamics is given by the equations $\dot{p} = pq(1 - p)$ and $\dot{q} = pq(1 - q)$. There is one stable rest point, namely $p = q = 1$, corresponding to the unique strict Nash equilibrium $((1, 0), (1, 0))$ of the game. The other rest points are all points in the set

$$\{(p, q) \mid p = 0 \text{ and } 0 \leq q \leq 1 \text{ or } q = 0 \text{ and } 0 \leq p \leq 1\}.$$

8.4 *More Asymmetric Games*

(a) The replicator dynamics are $dx/dt = x(1 - x)(2 - 3y)$ and $dy/dt = 2y(1 - 2x)(y - 1)$. There are no stable rest points.
(b) The replicator dynamics are $dx/dt = x(x-1)(2y-1)$ and $dy/dt = y(y-1)(2x-1)$.

8.5 *Frogs Call for Mates*
Note that for (a) and (b) Proposition 8.5 can be used. Similarly, for (c) we can use Proposition 8.8, by stating the conditions under which each of the four pure strategy combinations is a strict Nash equilibrium: if $z_1 < P + m - 1$ and $z_2 < P + m - 1$ then (Call, Call) is a strict Nash equilibrium, etc.

8.6 *Video Market Game*

There are four rest points, of which only one is stable.

Problems of Chapter 9

9.2 *Computing the Core*

(a) $\{(0,0,1)\}$; (b) polygon with vertices $(15,5,4)$, $(9,5,10)$, $(14,6,4)$, and $(8,6,10)$.

9.4 *The Core of the General Glove Game*

The Shapley value is in the core.

9.6 *Non-monotonicity of the Core*

(b) The core of (N, v') is the set $\{(0,0,1,1)\}$ [use the fact that $C(N, v') \subseteq C(N, v)$].
Hence, player 1 can only obtain less in the core although the worth of coalition $\{1,3,4\}$ has increased.

9.7 *Efficiency of the Shapley Value*

Consider an order $i_1, i_2, \ldots, i_n$ of the players. The sum of the coordinates of the associated marginal vector is

$$
\begin{aligned}
&[v(\{i_1\}) - v(\emptyset)] \\
&+[v(\{i_1, i_2\}) - v(\{i_1\})] \\
&+[v(\{i_1, i_2, i_3\}) - v(\{i_1, i_2\})] \\
&+ \ldots \\
&+[v(N) - v(N \setminus \{i_n\})] \\
&= v(N) - v(\emptyset) = v(N) \, .
\end{aligned}
$$

Hence, every marginal vector is efficient, so the Shapley value is efficient since it is the average of the marginal vectors.

9.8 *Computing the Shapley Value*

(a) $\Phi(N, v) = (1/6, 1/6, 2/3) \notin C(N, v)$; (b) $(9\frac{1}{2}, 6\frac{1}{2}, 8)$, not in the core.

9.9 *The Shapley Value and the Core*

(a) $a = 3$ (use Problem 9.5).
(b) $(2.5, 2, 1.5)$.
(c) The Shapley value is $(a/3 + 1/2, a/3, a/3 - 1/2)$. The minimal value of a for which this is in the core is $15/4$.

9.10 *Shapley Value in a Two-Player Game*
$\Phi(N,v) = (v(\{1\})+(v(\{1,2\})-v(\{1\})-v(\{2\}))/2, v(\{2\})+(v(\{1,2\})-v(\{1\})-v(\{2\}))/2)$.

9.11 *Computing the Nucleolus*

(a) $(0,0,1)$.
(b) $(11.5, 5.5, 7)$.
(c) $(1/5, 1/5, 1/5, 1/5, 1/5, 0, \ldots, 0) \in \mathbb{R}^{15}$.
(d) In (N,v): $(1/2, 1/2, 1/2, 1/2)$; in (N, v'): $(0, 0, 1, 1)$.

9.12 *Nucleolus of Two-Player Games*
The nucleolus is $(v(\{1\}) + (v(\{1,2\}) - v(\{1\}) - v(\{2\}))/2, v(\{2\}) + (v(\{1,2\}) - v(\{1\}) - v(\{2\}))/2)$.

9.13 *Computing the Core, the Shapley Value, and the Nucleolus*

(a) The nucleolus and Shapley value coincide and are equal to $(1.5, 2, 2.5)$.
(c) The maximal value of $v(\{1\})$ is 2. For that value the core is the line segment with endpoints $(2, 1, 3)$ and $(2, 3, 1)$.

9.14 *Voting (1)*

(a) The Shapley value is $\Phi(N, v) = (7/12, 3/12, 1/12, 1/12)$.
(b) The nucleolus is $(1, 0, 0, 0)$.

9.15 *Voting (2)*

(c) $\Phi(N, v) = (1/60)(9, 9, 14, 14, 14)$.
(d) The nucleolus is $(0, 0, 1/3, 1/3, 1/3)$.
(e) The nucleolus is not in the core (e.g., $v(\{1, 3, 4\} = 1 > 2/3)$, so the core must be empty. This can also be seen directly.

9.16 *Two Buyers and a Seller*

(c) $\Phi(N, v) = (1/6, 4/6, 7/6)$.
(d) The nucleolus is $(0, 1/2, 3/2)$.

9.17 *Properties of the Shapley Value*

(a) In $\Phi_i(N, v)$ the term $v(S \cup \{i\}) - v(S)$ occurs the same number of times as the term $v(S \cup \{j\}) - v(S)$ in $\Phi_j(N, v)$, for every coalition $S \subseteq N \setminus \{i, j\}$. Let S be a coalition with $i \in S$ and $j \notin S$. Then $v(S \setminus \{i\} \cup \{j\}) = v(S \setminus \{i\} \cup \{i\})$, so that

$$v(S \cup \{j\}) - v(S) = v((S \setminus \{i\} \cup \{j\}) \cup \{i\}) - v((S \setminus \{i\}) \cup \{i\})$$
$$= v((S \setminus \{i\} \cup \{j\}) \cup \{i\}) - v((S \setminus \{i\}) \cup \{j\}),$$

and also these expressions occur the same umber of times. Similarly for coalitions S that contain j but not i.

(b) This is obvious from Definition 9.4.

(c) Observe that it is sufficient to show $\sum_{S:i\notin S} \frac{|S|!(n-|S|-1)!}{n!} = 1$. To show this, note that

$$\frac{|S|!(n-|S|-1)!}{n!} = \frac{1}{n}\binom{n-1}{|S|}^{-1}, \text{ so that}$$

$$\sum_{S:i\notin S} \frac{|S|!(n-|S|-1)!}{n!} = \frac{1}{n}\sum_{s=0,1,\dots,n-1}\binom{n-1}{s}\binom{n-1}{s}^{-1}$$

$$= \frac{1}{n}\cdot n = 1\,.$$

Problems of Chapter 10

10.1 *A Division Problem (1)*

(b) In terms of utilities: $(\frac{1}{3}\sqrt{3}, \frac{2}{3})$, in terms of distribution: $(\frac{1}{3}\sqrt{3}, 1 - \frac{1}{3}\sqrt{3})$.

(c) The Rubinstein outcome is x^* where $x_1^* = \sqrt{\frac{1}{1+\delta+\delta^2}}$ and $x_2^* = 1 - \frac{1}{1+\delta+\delta^2}$.

(d) $\lim_{\delta\to 1} x_1^* = \frac{1}{3}\sqrt{3}$, consistent with what was found under (a).

10.2 *A Division Problem (2)*

Use symmetry, Pareto optimality and covariance of the Nash bargaining solution.

10.3 *A Division Problem (3)*

(a) The distribution of the good is $\left(2\frac{1-\delta^3}{1-\delta^4}, 2 - 2\frac{1-\delta^3}{1-\delta^4}\right)$. In utility terms this is $\left(\frac{1-\delta^3}{1-\delta^4}, \sqrt[3]{2 - 2\frac{1-\delta^3}{1-\delta^4}}\right)$.

(b) By taking the limit for $\delta \to 1$ in (b), we obtain $(1.5, 0.5)$ as the distribution assigned by the Nash bargaining solution. In utilities: $(0.75, \sqrt[3]{0.5})$.

10.4 *An Exchange Economy*

(a) $x_1^A(p_1, p_2) = (3p_2 + 2p_1)/2p_1$, $x_2^A = (4p_1 - p_2)/2p_2$, $x_1^B = (p_1 + 6p_2)/2p_1$, $x_2^B = p_1/2p_2$.

(b) $(p_1, p_2) = (9, 5)$ (or any positive multiple thereof); the equilibrium allocation is $((33/18, 31/10), (39/18, 9/10))$.

(c) The (non-boundary part of the) contract curve is given by the equation $x_2^A = (17x_1^A + 5)/(2x_1^A + 8)$. The core is the part of this contract curve such that $\ln(x_1^A + 1) + \ln(x_2^A + 2) \geq \ln 4 + \ln 3 = \ln 12$ (individual rationality constraint for A) and $3\ln(5-x_1^A)+\ln(5-x_2^A)\geq 3\ln 2+\ln 4=\ln 12$ (individual rationality constraint for B).

(d) The point $\mathbf{x}^A=(33/18,31/10)$ satisfies the equation $x_2^A=(17x_1^A+5)/(2x_1^A+8)$.

(e) For the disagreement point $\mathbf{d}$ one can take the point $(\ln 12,\ln 12)$. The set S contains all points $\mathbf{u}\in\mathbb{R}^2$ that can be obtained as utilities from any distribution of the goods that does not exceed total endowments $\mathbf{e}=(4,4)$. Unlike the Walrasian equilibrium allocation, the allocation obtained by applying the Nash bargaining solution is not independent of arbitrary monotonic transformations of the utility functions. It is a 'cardinal' concept, in contrast to the Walrasian allocation, which is 'ordinal'.

10.5 *The Matching Problem of Table 10.1 Continued*

(a) The resulting matching is (w_1,m_1), (w_2,m_2), w_3 and m_3 remain single.

10.6 *Another Matching Problem*

(a) With the men proposing: (m_1,w_1), (m_2,w_2), (m_3,w_3). With the women proposing: (m_1,w_1), (m_2,w_3), (m_3,w_2).

(b) Since in any stable matching we must have (m_1,w_1), the matchings found in (a) are the only stable ones.

(c) Obvious: every man weakly or strongly prefers the men proposing matching in (a); and vice versa for the women.

10.7 *Yet Another Matching Problem: Strategic Behavior*

(b) There are no other stable matchings.

(c) The resulting matching is (m_1,w_1), (m_2,w_3), (m_3,w_2). This is clearly better for w_1.

10.8 *Core Property of Top Trading Cycle Procedure*

All players in a top trading cycle get their top houses, and thus none of these players can be a member of a blocking coalition, say S. Omitting these players and their houses from the problem, by the same argument none of the players in a top trading cycle in the second round can be a member of S: the only house that such a player may prefer is no longer available in S; etc.

10.9 *House Exchange with Identical Preferences*

Without loss of generality, assume that each player has the same preference $h_1 h_2\ldots h_n$. Show that in a core allocation each player keeps his own house.

10.10 *A House Exchange Problem*

There are three core allocations namely: (i) $1:h_3$, $2:h_4$, $3:h_1$, $4:h_2$; (ii) $1:h_2$, $2:h_4$, $3:h_1$, $4:h_3$; (iii) $1:h_3$, $2:h_1$, $3:h_4$, $4:h_2$. Allocation (i) is in the strong core.

10.11 *Cooperative Oligopoly*
(a)–(c) Analogous to Problems 6.1, 6.2. Parts (d) and (f) follow directly from (c).
For parts (e) and (g) use the methods of Chap. 9.

Problems of Chapter 11

11.1 *Preferences*

(a) If $a \neq b$ and aRb and bRa then neither aPb nor bPa, so P is not necessarily complete.
(b) I is not complete unless aRb for all $a, b \in A$. I is only antisymmetric if R is a linear order.

11.2 *Pairwise Comparison*

(a) $C(r)$ is reflexive and complete but not antisymmetric.
(c) There is no Condorcet winner in this example.

11.3 *Independence of the Conditions in Theorem 11.1*
The social welfare function based on the Borda scores is Pareto efficient but does not satisfy IIA and is not dictatorial (cf. Sect. 11.1). The social welfare function that assigns to each profile of preferences the reverse preference of agent 1 satisfies IIA and is not dicatorial but also not Pareto efficient.

11.4 *Independence of the Conditions in Theorem 11.2*
A constant social welfare function (i.e., always assigning the same fixed alternative) is strategy-proof and nondictatorial but not surjective. The social welfare function that always assigns the bottom element of agent 1 is surjective, nondictatorial, and not strategy-proof.

11.5 *Independence of the Conditions in Theorem 11.3*
A constant social welfare function (i.e., always assigning the same fixed alternative) is monotonic and nondictatorial but not unanimous. A social welfare function that assigns the common top alternative to any profile where all agents have the same top alternative, and a fixed constant alternative to any other profile, is unanimous and nondictatorial but not monotonic.

11.6 *Copeland Score and Kramer Score*

(a) The Copeland ranking is a preference. The Copeland ranking is not antisymmetric. It is easy to see that the Copeland ranking is Pareto efficient. By Arrow's Theorem therefore, it does not satisfy IIA.
(b) The Kramer ranking is a preference. The Kramer ranking is not antisymmetric and not Pareto efficient. It violates IIA.

11.7 *Two Alternatives*

Consider the social welfare function based on majority rule, i.e., it assigns aPb if $|N(a, b, r)| > |N(b, a, r)|$; bPa if $|N(a, b, r)| < |N(b, a, r)|$; and aIb if $|N(a, b, r)| = |N(b, a, r)|$.

Problems of Chapter 12

12.1 *Solving a Matrix Game*

(c) $v(A) = 12/5$ and the unique optimal strategies of players 1 and 2 are, respectively, $(0, 4/5, 1/5, 0)$ and $(0, 2/5, 0, 3/5)$.

(d) The answer is independent of y and the same as in (c).

12.3 2×2 *Games*

(a) To have no saddlepoints we need $a_{11} > a_{12}$ or $a_{11} < a_{12}$. By assume the first, the other inequalities follow.

(b) For optimal strategies $\mathbf{p} = (p, 1-p)$ and $\mathbf{q} = (q, 1-q)$ we must have $0 < p < 1$ and $0 < q < 1$. Then use that p should be such that player 2 is indifferent between the two columns and q such that player 1 is indifferent between the two rows.

12.4 *Symmetric Games*

Let $\mathbf{x}$ be optimal for player 1. Then $\mathbf{x}A\mathbf{y} \geq v(A)$ for all $\mathbf{y}$; hence $\mathbf{y}A\mathbf{x} = -\mathbf{x}A\mathbf{y} \leq -v(A)$ for all $\mathbf{y}$; hence (take $\mathbf{y} = \mathbf{x}$) $v(A) \leq -v(A)$, so $v(A) \leq 0$. Similarly, derive the converse inequality by considering an optimal strategy for player 2.

12.5 *The Duality Theorem Implies the Minimax Theorem*

Let A be an $m \times n$ matrix game. Without loss of generality assume that all entries of A are positive. Consider the associated LP as in Sect. 12.2.

Consider the vector $\bar{\mathbf{x}} = (1/m, \ldots, 1/m, \eta) \in \mathbb{R}^{m+1}$ with $\eta > 0$. Since all entries of A are positive it is straightforward to check that $\bar{\mathbf{x}} \in V$ if $\eta \leq \sum_{i=1}^{m} a_{ij}/m$ for all $j = 1, \ldots, n$. Since $\bar{\mathbf{x}} \cdot c = -\eta < 0$, it follows that the value of the LP must be negative.

Let $\mathbf{x} \in O_{\min}$ and $\mathbf{y} \in O_{\max}$ be optimal solutions of the LP. Then $-x_{m+1} = -y_{n+1} < 0$ is the value of the LP. We have $x_i \geq 0$ for every $i = 1, \ldots, m$, $\sum_{i=1}^{m} x_i \leq 1$, and $(x_1, \ldots, x_m)Ae^j \geq x_{m+1}$ (> 0) for every $j = 1, \ldots, n$. Optimality in particular implies $\sum_{i=1}^{m} x_i = 1$, so that $v_1(A) \geq (x_1, \ldots, x_m)Ae^j \geq x_{m+1}$ for all j, hence $v_1(A) \geq x_{m+1}$. Similarly, it follows that $v_2(A) \leq y_{n+1} = x_{m+1}$, so that $v_2(A) \leq v_1(A)$. The Minimax Theorem now follows.

12.6 *Infinite Matrix Games*

(a) A is an infinite matrix game with for all $i, j \in \mathbb{N}$: $a_{ij} = 1$ if $i > j$, $a_{ij} = 0$ if $i = j$, and $a_{ij} = -1$ if $i < j$.

(b) Fix a mixed strategy $\mathbf{p} = (p_1, p_2, \ldots)$ for player 1 with $p_i \geq 0$ for all $i \in \mathbb{N}$ and $\sum_{i=1}^{\infty} p_i = 1$. If player 2 plays pure strategy j, then the expected payoff for player 1 is equal to $-\sum_{i=1}^{j-1} p_i + \sum_{i=j+1}^{\infty} p_i$. Since $\sum_{i=1}^{\infty} p_i = 1$, this expected payoff converges to -1 as j approaches ∞. Hence, $\inf_{\mathbf{q}} \mathbf{p} A \mathbf{q} = -1$, so $\sup_{\mathbf{p}} \inf_{\mathbf{q}} \mathbf{p} A \mathbf{q} = -1$. Similarly, one shows $\inf_{\mathbf{q}} \sup_{\mathbf{p}} \mathbf{p} A \mathbf{q} = 1$, hence the game has no 'value'.

12.7 Equalizer Theorem

Assume, without loss of generality, $v = 0$. It is sufficient to show that there exists $\mathbf{q} \in \mathbb{R}^n$ with $\mathbf{q} \geq 0$, $A\mathbf{q} \leq 0$, and $q_n = 1$. The required optimal strategy is then obtained by normalization.

This is equivalent to existence of a vector $(\mathbf{q}, \mathbf{w}) \in \mathbb{R}^{n+m}$ with $\mathbf{q} \geq 0$, $\mathbf{w} \geq 0$, such that

$$\begin{pmatrix} A & I \\ \mathbf{e}^n & 0 \end{pmatrix} \begin{pmatrix} \mathbf{q} \\ \mathbf{w} \end{pmatrix} = \begin{pmatrix} 0 \\ 1 \end{pmatrix} ,$$

where row vector $\mathbf{e}^n \in \mathbb{R}^n$, I is the $m \times m$ identity matrix, 0 is an $1 \times m$ vector on the left hand side and an $m \times 1$ vector on the right hand side. Thus, we have to show that the vector $\mathbf{x} := (0, 1) \in \mathbb{R}^{m+1}$ is in the cone spanned by the columns of the $(m+1) \times (n+m)$ matrix on the left hand side. Call this matrix B and call this cone Z. Assume $\mathbf{x} \notin Z$ and derive a contradiction using Theorem 22.1.

Problems of Chapter 13

13.1 Existence of Nash Equilibrium Using Brouwer

(c) Let $\sigma^* \in \prod_{i \in N} \Delta(S_i)$. If σ^* is a Nash equilibrium of G then

$$\sigma_i^*(s_i) = \frac{\sigma_i^*(s_i) + \max\{0, u_i(s_i, \sigma_{-i}^*) - u_i(\sigma^*)\}}{1 + \sum_{s_i' \in S_i} \max\{0, u_i(s_i', \sigma_{-i}^*) - u_i(\sigma^*)\}} \tag{*}$$

for all $i \in N$ and $s_i \in S_i$, so that σ^* is a fixed point of f. Conversely, let σ^* be a fixed point of f. Then (*) holds for all $i \in N$ and $s_i \in S_i$. Hence

$$\sigma_i^*(s_i) \sum_{s_i' \in S_i} \max\{0, u_i(s_i', \sigma_{-i}^*) - u_i(\sigma^*)\} = \max\{0, u_i(s_i, \sigma_{-i}^*) - u_i(\sigma^*)\} .$$

Multiply both sides of this equation by $u_i(s_i, \sigma_{-i}^*) - u_i(\sigma^*)$ and next sum over all $s_i \in S_i$.

13.2 *Existence of Nash Equilibrium Using Kakutani*
For upper semi-continuity of β, take a sequence σ^k converging to σ, a sequence $\tau^k \in \beta(\sigma^k)$ converging to τ, and show $\tau \in \beta(\sigma)$.

13.3 *Lemma 13.2*
The only-if direction is straightforward from the definition of best reply.

13.4 *Lemma 13.3*
Take i such that $e^i A\mathbf{q} \geq e^k A\mathbf{q}$ for all $k = 1, \ldots, m$. Then, clearly, $e^i A\mathbf{q} \geq \mathbf{p}'A\mathbf{q}$ for all $\mathbf{p}' \in \Delta^m$, so $e^i \in \beta_1(\mathbf{q})$. The second part is analogous.

13.5 *Dominated Strategies*

(b) Denote by $NE(A, B)$ the set of Nash equilibria of (A, B). Then

$$(\mathbf{p}^*, \mathbf{q}^*) \in NE(A, B) \Leftrightarrow (\mathbf{p}^*, (\mathbf{q}', 0)) \in NE(A, B) \text{ where } (\mathbf{q}', 0) = \mathbf{q}^*$$
$$\Leftrightarrow \forall \mathbf{p} \in \Delta^m, \mathbf{q} \in \Delta^{n-1}[\mathbf{p}^* A(\mathbf{q}', 0) \geq \mathbf{p}A(\mathbf{q}', 0),$$
$$\mathbf{p}^* B(\mathbf{q}', 0) \geq \mathbf{p}^* B(\mathbf{q}, 0)]$$
$$\Leftrightarrow \forall \mathbf{p} \in \Delta^m, \mathbf{q} \in \Delta^{n-1}[\mathbf{p}^* A' \mathbf{q}' \geq \mathbf{p}A'\mathbf{q}',$$
$$\mathbf{p}^* B' \mathbf{q}' \geq \mathbf{p}^* B' \mathbf{q}]$$
$$\Leftrightarrow (\mathbf{p}^*, \mathbf{q}') \in NE(A', B') .$$

Note that the first equivalence follows by part (a).

13.6 *A 3×3 Bimatrix Game*

(c) The unique Nash equilibrium is $((0, 0, 1), (0, 0, 1))$.

13.7 *A 3×2 Bimatrix Game*
The set of Nash equilibria is $\{(\mathbf{p}, \mathbf{q}) \in \Delta^3 \times \Delta^2 \mid p_1 = 0, q_1 \geq \frac{1}{2}\} \cup \{((1, 0, 0), (0, 1))\}$.

13.8 *The Nash Equilibria in Example 13.18*

(a) Let $\mathbf{p} = (p_1, p_2, p_3)$ be the strategy of player 1. We distinguish two cases: (i) $p_2 = 0$ (ii) $p_2 > 0$.
   In case (i), reduce the game to

$$\begin{array}{c} \\ p_1 \\ p_3 \end{array} \begin{array}{ccc} q_1 & q_2 & q_3 \\ \left(\begin{array}{ccc} 1,1 & 0,0 & 2,0 \\ 0,0 & 1,1 & 1,1 \end{array} \right) \end{array}$$

where $\mathbf{q} = (q_1, q_2, q_3)$ is player 2's strategy. Solve this game graphically. As long as player 1 gets at least 1 (the payoff from playing M) theobtained equilibria are also equilibria of the original game G.

In case (ii), R gives a lower expected payoff to player 2 than C, so the game can be reduced to

$$
\begin{array}{c}
 & \begin{array}{cc} q_1 & q_2 \end{array} \\
\begin{array}{c} p_1 \\ p_2 \\ p_3 \end{array} & \left(\begin{array}{cc} 1,1 & 0,0 \\ 1,2 & 1,2 \\ 0,0 & 1,1 \end{array} \right).
\end{array}
$$

Solve this game graphically and extend to G.

(b) Consider again the perturbed games $G(\varepsilon)$ as in Example 13.18. For $q = 0$ consider the strategy combination $((\varepsilon, 1 - 2\varepsilon, \varepsilon), (\varepsilon, 1 - 2\varepsilon, \varepsilon))$ in $G(\varepsilon)$. For $q = 1$ consider, similarly, $((\varepsilon, 1 - 2\varepsilon, \varepsilon), (1 - 2\varepsilon, \varepsilon, \varepsilon))$ in $G(\varepsilon)$; for $0 < q < 1$ consider $((\varepsilon, 1 - 2\varepsilon, \varepsilon), (q - \varepsilon/2, 1 - q - \varepsilon/2, \varepsilon))$.

13.9 *Proof of Theorem 13.8*

'If': conditions (13.1) are satisfied and $f = 0$, which is optimal since $f \leq 0$ always.

'Only-if': clearly we must have $a = \mathbf{p}A\mathbf{q}$ and $b = \mathbf{p}B\mathbf{q}$ (otherwise $f < 0$ which cannot be optimal). From the conditions (13.1) we have $\mathbf{p}'A\mathbf{q} \leq a = \mathbf{p}A\mathbf{q}$ and $\mathbf{p}B\mathbf{q}' \leq b = \mathbf{p}B\mathbf{q}$ for all $\mathbf{p}' \in \Delta^m$ and $\mathbf{q}' \in \Delta^n$, which implies that $(\mathbf{p}, \mathbf{q})$ is a Nash equilibrium.

13.10 *Matrix Games*
This is a repetition of the proof of Theorem 12.5. Note that the solutions of program (13.3) give exactly the value of the game a and the optimal (minimax) strategies of player 2. The solutions of program (13.4) give exactly the value of the game $-b$ and the optimal (maximin) strategies of player 1.

13.11 *Tic-Tac-Toe*

(a) Start by putting a cross in the center square. Then player 2 has essentially two possibilities for the second move, and it is easy to see that in each of the two cases player 1 has a forcing third move. After this, it is equally easy to see that player 1 can always enforce a draw.
(b) If player 1 does not start at the center, then player 2 can put his first circle at the center and then can place his second circle in such a way that it becomes forcing. If player 1 starts at the center then either a pattern as in (a) is followed, leading to a draw, or player 2's second circle becomes forcing, also resulting in a draw.

13.12 *Iterated Elimination in a Three-Player Game*
The resulting strategy combination is (D, l, L).

13.13 *Never a Best Reply and Domination*
First argue that strategy Y is not strictly dominated. Next assume that Y is a best reply to strategies $(p, 1 - p)$ of player 1 and $(q, 1 - q)$ of player 2, and derive a contradiction.

13.15 *A 3-Player Game with an Undominated But Not Perfect Equilibrium*

(a) First observe that the set of Nash equilibria is $\{((p, 1 - p), l, L) \mid 0 \le p \le 1\}$, where p is the probability with which player 1 plays U.

13.16 *Existence of Proper Equilibrium*
Tedious but straightforward.

13.17 *Strictly Dominated Strategies and Proper Equilibrium*

(a) The only Nash equilibria are $(U, l(, L))$ and $(D, r(, L))$. Obviously, only the first one is perfect and proper.
(b) (D, r, L), is a proper Nash equilibrium.

13.18 *Strictly Perfect Equilibrium*

(a) Identical to the proof of Lemma 13.16, see Problem 13.14: note that any sequence of perturbed games converging to the given game must eventually contain any given completely mixed Nash equilibrium σ.
(c) The set of Nash equilibria is $\{((p, 1 - p), L) \mid 0 \le p \le 1\}$, where p is the probability on U. Every Nash equilibrium of the game (A, B) is perfect and proper. No Nash equilibrium is strictly perfect.

13.19 *Correlated Equilibria in the Two-Driver Example (1)*
Use inequalities (13.5) and (13.6) to derive the conditions: $p_{11} + p_{12} + p_{21} + p_{22} = 1$, $p_{ij} \ge 0$ for all $i, j \in \{1, 2\}$, $p_{11} \le \frac{3}{5} \min\{p_{12}, p_{21}\}$, $p_{22} \le \frac{5}{3} \min\{p_{12}, p_{21}\}$.

13.20 *Nash Equilibria are Correlated*
Check that (13.5) and (13.6) are satisfied for P.

13.21 *The Set of Correlated Equilibria is Convex*
Let P and Q be correlated equilibria and $0 \le t \le 1$. Check that (13.5) and (13.6) are satisfied for $tP + (1 - t)Q$.

13.22 *Correlated vs. Nash Equilibrium*

(a) The Nash equilibria are: $((1, 0), (0, 1))$, $((0, 1), (1, 0))$, and $((2/3, 1/3), (2/3, 1/3))$.

13.23 *Correlated Equilibria in the Two-Driver Example (2)*
The matrix C is:

$$
\begin{array}{c}
 \\
(1,1') \\
(1,2') \\
(2,1') \\
(2,2')
\end{array}
\begin{array}{cccc}
(1,2) & (2,1) & (1',2') & (2',1') \\
\left(\begin{array}{cccc}
-10 & 0 & -10 & 0 \\
6 & 0 & 0 & 10 \\
0 & 10 & 6 & 0 \\
0 & -6 & 0 & -6
\end{array} \right)
\end{array}.
$$

13.24 *Finding Correlated Equilibria*
There is a unique correlated equilibrium

$$
P = \begin{pmatrix} \frac{1}{3} & \frac{1}{3} \\ \frac{1}{6} & \frac{1}{6} \end{pmatrix}.
$$

13.25 *Nash, Perfect, Proper, Strictly Perfect, and Correlated Equilibria*

(d) $\{((0,1,0,(q_1,q_2,q_3)) \mid q_3 \leq 2q_1 \leq 4q_3\} \cup \{((p_1,p_2,p_3),(0,1,0)) \mid 0 < p_1 \leq 2p_3 \leq 4p_1\}$.
(e) The only perfect equilibrium is $((0,1,0),(0,1,0))$.
(f) $((0,1,0),(0,1,0))$ is also the only proper and strictly perfect equilibrium.
(h) For instance $\beta = \gamma = 3/8, \alpha = \delta = 1/8$.

13.26 *Independence of the Axioms in Corollary 13.40*
Not OPR: take the set of all strategy combinations in every game. Not CONS: in games with maximal player set take all strategy combinations, in other games take the set of Nash equilibria. Not COCONS: drop a Nash equilibrium in some game with maximal player set, but otherwise always take the set of all Nash equilibria.

13.27 *Inconsistency of Perfect Equilibria*
First show that the perfect equilibria in G_0 are all strategy combinations where player 2 plays L, player 3 plays D, and player 1 plays any mixture between T and B. Next consider the reduced game by fixing player 3's strategy at D.

Problems of Chapter 14

14.2 *An Extensive Form Structure without Perfect Recall*

(a) The paths $\{(x_0,x_1)\}$ and $\{(x_0,x_2)\}$ contain different player 1 actions.

14.3 *Consistency Implies Bayesian Consistency*
With notations as in Definition 14.13, for $h \in H$ with $\mathbb{P}_b(h) > 0$ and $x \in h$ we have: $\beta_h(x) = \lim_{m \to \infty} \beta_h^m(x) = \lim_{m \to \infty} \mathbb{P}_{b^m}(x)/\mathbb{P}_{b^m}(h) = \mathbb{P}_b(x)/\mathbb{P}_b(h)$. Here, the second equality follows from Bayesian consistency of the (b^m, β^m).

14.4 *(Bayesian) Consistency in Signaling Games*
The idea of the proof is as follows. Let (b, β) be a Bayesian consistent assessment. This means that β is determined on every information set of player 2 that is reached with positive probability, given b_1. Take $m \in \mathbb{N}$. Assign the number $1/m^2$ to action a of a type i of player 1 if that type does not play a but some other type of player 1 plays a with positive probability. Assign the number $1/m^2$ also to action a of type i if no type of player 1 plays a and player 2 attaches zero belief probability to type i conditional on player 1 having played a. To every other action of a of player 1, assign the number $\beta(i, a)/m$, where $\beta(i, a)$ is the (positive) belief that player 2 attaches to player 1 being of type i conditional on having played a. Next, normalize all these numbers to behavioral strategies b_1^m of player 1. For player 2, just take completely mixed behavioral strategies b_2^m converging to b_2. Then $(b^m, \beta^m) \to (b, \beta)$, where the β^m are determined by Bayesian consistency.

14.5 *Sequential Equilibria in a Signaling Game*
There is one pure and one completely mixed sequential equilibrium.

14.6 *Computation of Sequential Equilibrium (1)*
The unique sequential equilibrium consists of the behavioral strategies where player 1 plays B with probability 1 and C with probability $1/2$, and player 2 plays L with probability $1/2$; and player 1 believes that x_3 and x_4 are equally likely.

14.7 *Computation of Sequential Equilibrium (2)*

(b) The Nash equilibria are (L, l), and $(R, (\alpha, 1 - \alpha))$ for all $\alpha \leq 1/2$, where α is the probability with which player 2 plays l.

(c) Let π be the belief player 2 attaches to node y_1. Then the sequential equilibria are: (L, l) with belief $\pi = 1$; (R, r) with belief $\pi \leq 1/2$; and $(R, (\alpha, 1 - \alpha))$ for any $\alpha \leq 1/2$ with belief $\pi = 1/2$.

14.8 *Computation of Sequential Equilibrium (3)*

(b) The Nash equilibria are $(R, (q, 1 - q))$ with $1/3 \leq q \leq 2/3$. (The conditions on q keep player 1 from deviating to L or M.)

14.9 *Computation of Sequential Equilibrium (4)*
The Nash equilibria in this game are: $(R, (q_1, q_2, q_3))$ with $q_3 \leq 1/3$ and $q_1 \leq 1/2 - (3/4)q_3$, where q_1, q_2, q_3 are the probabilities put on l, m, r, respectively; and $((1/4, 3/4, 0), (1/4, 0, 3/4))$ (probabilities on L, M, R and l, m, r, respectively).

Let π be the belief attached by player 2 to y_1. Then with $\pi = 1/4$ the equilibrium $((1/4, 3/4, 0), (1/4, 0, 3/4))$ becomes sequential. The first set of equilibria contains no equilibrium that can be extended to a sequential equilibrium.

14.10 *Computation of Sequential Equilibrium (5)*
The Nash equilibria are: (DB, r); $((R, (s, 1 - s)), (q, 1 - q))$ with $0 \leq s \leq 1$ and $q \geq 1/3$, where s is the probability on A and q is the probability on l. The subgame perfect equilibria are: (DB, r); (RA, l); $((R, (3/4, 1/4)), (3/5, 2/5))$. The first one

becomes sequential with $\beta = 0$; the second one with $\beta = 1$; and the third one with $\beta = 3/5$.

Problems of Chapter 15

15.1 *Computing ESS in 2×2 Games (1)*
$ESS(A)$ can be computed using Proposition 15.4.
(a) $ESS(A) = \{e^2\}$. (b) $ESS(A) = \{e^1, e^2\}$. (c) $ESS(A) = \{(2/3, 1/3)\}$.

15.2 *Computing ESS in 2×2 Games (2)*
Case (1): $ESS(A') = \{e^2\}$; case (2): $ESS(A') = \{e^1, e^2\}$; case (3): $ESS(A') = \{\hat{\mathbf{x}}\} = \{(a_2/(a_1 + a_2), a_1/(a_1 + a_2))\}$.

15.3 *Rock-Paper-Scissors (1)*
The unique Nash equilibrium is $((1/3, 1/3, 1/3), (1/3, 1/3, 1/3))$, which is symmetric. But $(1/3, 1/3, 1/3)$ is not an ESS.

15.4 *Uniform Invasion Barriers*
Case (1), e^2: maximal uniform invasion barrier is 1.
Case (2), e^1: maximal uniform invasion barrier is $a_1/(a_1 + a_2)$.
Case (2), e^2: maximal uniform invasion barrier is $a_2/(a_1 + a_2)$.
Case (3), $\hat{\mathbf{x}}$: maximal uniform invasion barrier is 1.

15.5 *Replicator Dynamics in Normalized Game (1)*
Straightforward computation.

15.6 *Replicator Dynamics in Normalized Game (2)*
The replicator dynamics can be written as $\dot{x} = [x(a_1 + a_2) - a_2]x(1 - x)$, where $\dot{x} = \dot{x}_1$. So $x = 0$ and $x = 1$ are always stationary points. In case (1) the graph of $\dot{x}$ on $(0, 1)$ is below the horizontal axis. In case (2) there is another stationary point, namely at $x = a_2/(a_1 + a_2)$; on $(0, a_2/(a_1 + a_2))$ the function $\dot{x}$ is negative, on $(a_2/(a_1 + a_2), 1)$ it is positive. In case (3) the situation of case (2) is reversed: the function $\dot{x}$ is positive on $(0, a_2/(a_1 + a_2))$ and negative on $((a_2/(a_1 + a_2)), 1)$.

15.7 *Weakly Dominated Strategies and Replicator Dynamics*
(b) The stationary points are e^1, e^2, e^3, and all points with $x_3 = 0$. Except e^3, all stationary points are Lyapunov stable. None of these points is asymptotically stable. Also, e^3 is strictly dominated (by e^1)). [One can also derive $d(x_1/x_2)/dt = x_1 x_3/x_2 > 0$ at completely mixed strategies, i.e., at the interior of Δ^3. Hence, the share of subpopulation 1 grows faster than that of 2 but this difference goes to zero if x_3 goes to zero (e^2 is weakly dominated by e^1).]

15.8 *Stationary Points and Nash Equilibria (1)*

(a) $NE(A) = \{(\alpha, \alpha, 1 - 2\alpha) \mid 0 \le \alpha \le 1/2\}$.
(b) By Proposition 15.18 and (a) it follows that $\{(\alpha, \alpha, 1 - 2\alpha) \mid 0 \le \alpha \le 1/2\} \cup \{e^1, e^2, e^3\} \subseteq ST(A)$, and that possibly other stationary points must be boundary

points of Δ^3. By considering the replicator dynamics it follows that there are no additional stationary points. All stationary points except $\mathbf{e}^1$ and $\mathbf{e}^2$ are Lyapunov stable, but no point is asymptotically stable.

15.9 *Stationary Points and Nash Equilibria (2)*

(a) The Nash equilibrium strategies are: $(0, 1, 0)$, $(1/2, 0, 1/2)$, $(0, 2/3, 1/3)$, and $(0, 0, 1)$.
(b) Use Proposition 15.18. This implies that $(1, 0, 0)$, $(0, 1, 0)$, $(0, 0, 1)$, $(1/2, 0, 1/2)$, and $(0, 2/3, 1/3)$ all are stationary states. Any other stationary state must be on the boundary of Δ^3 and have exactly one zero coordinate. Using this it can be shown that there are no other stationary states.
(c) The state $(0, 0, 1)$ is asymptotically stable. All other stationary states are not Lyapunov stable.

15.10 *Lyapunov Stable States in 2×2 Games*
Case (1): $\mathbf{e}^2$; case (2): $\mathbf{e}^1$ and $\mathbf{e}^2$; case (3): $\hat{\mathbf{x}}$. (Cf. Problem 15.6.)

15.11 *Nash Equilibrium and Lyapunov Stability*
$NE(A) = \{\mathbf{e}^1\}$. If we start at a completely mixed strategy close to $\mathbf{e}^1$, then first x_3 increases, and we can make the solution trajectory pass $\mathbf{e}^3$ as closely as desired. This shows that $\mathbf{e}^1$ is not Lyapunov stable.

15.12 *Rock-Paper-Scissors (2)*

(e) Follows from (d). If $a > 0$ then any trajectory converges to the maximum point of $x_1 x_2 x_3$, i.e. to $(1/3, 1/3, 1/3)$. If $a = 0$ then the trajectories are orbits ($x_1 x_2 x_3$ constant) around $(1/3, 1/3, 1/3)$. If $a < 0$ then the trajectories move outward, away from $(1/3, 1/3, 1/3)$.

Problems of Chapter 16

16.1 *Imputation Set of an Essential Game*
Note that $I(v)$ is a convex set and $\mathbf{f}^i \in I(v)$ for every $i = 1, \ldots, n$. Thus, $I(v)$ contains the convex hull of $\{\mathbf{f}^i \mid i \in N\}$. Now let $\mathbf{x} \in I(v)$, and write $\mathbf{x} = (v(1), \ldots, v(n)) + (\alpha_1, \ldots, \alpha_n)$, where $\sum_{i \in N} \alpha_i = v(N) - \sum_{i \in N} v(i) =: \alpha$.

16.2 *Convexity of the Domination Core*
First prove the following claim: For each $\mathbf{x} \in I(v)$ and $\emptyset \neq S \subseteq N$ we have

$$\exists \mathbf{z} \in I(v) : \mathbf{z} \, \mathrm{dom}_S \mathbf{x} \Leftrightarrow x(S) < v(S) \text{ and } x(S) < v(N) - \sum_{i \notin S} v(i) .$$

Use this claim to show that $I(v) \setminus D(S)$ is a convex set. Finally, conclude that $DC(v)$ must be convex.

16.3 Dominated Sets of Imputations

(b) In both games, $D(ij) = \{\mathbf{x} \in I(v) \mid x_i + x_j < v(ij)\}$, $i, j \in \{1, 2, 3\}$, $i \neq j$.

16.7 A Glove Game

(b) The core and the domination core are both equal to $\{(0, 1, 0)\}$, cf. Theorem 16.12.

16.11 Core and D-Core
Condition (16.1) is not a necessary condition for equality of the core and the D-core. To find a counterexample, first note that if $C(v) \neq \emptyset$ then (16.1) must hold. Therefore, a counterexample has to be some game with empty core and D-core.

16.12 Strategic Equivalence
Straightforward using the definitions.

16.13 Proof of Theorem 16.20
Write $B = \begin{pmatrix} A \\ -A \end{pmatrix}$. Then

$$
\begin{aligned}
\max\{\mathbf{b} \cdot \mathbf{y} \mid A\mathbf{y} = \mathbf{c}, \mathbf{y} \geq \mathbf{0}\} &= \max\{\mathbf{b} \cdot \mathbf{y} \mid B\mathbf{y} \leq (\mathbf{c}, -\mathbf{c}), \mathbf{y} \geq \mathbf{0}\} \\
&= \min\{(\mathbf{c}, -\mathbf{c}) \cdot (\mathbf{x}, \mathbf{z}) \mid (\mathbf{x}, \mathbf{z})B \geq \mathbf{b}, (\mathbf{x}, \mathbf{z}) \geq \mathbf{0}\} \\
&= \min\{\mathbf{c} \cdot (\mathbf{x} - \mathbf{z}) \mid (\mathbf{x} - \mathbf{z})A \geq \mathbf{b}, (\mathbf{x}, \mathbf{z}) \geq \mathbf{0}\} \\
&= \min\{\mathbf{c} \cdot \mathbf{x}' \mid \mathbf{x}'A \geq \mathbf{b}\} \ .
\end{aligned}
$$

The second equality follows from Theorem 22.6.

16.14 Infeasible Programs in Theorem 16.20
Follow the hint.

16.15 Proof of Theorem 16.22 Using Lemma 22.5
Follow the hint and investigate (b) of Lemma 22.5.

16.17 Minimum of Balanced Games
Follows by using the definition of balancedness or by Theorem 16.22.

16.18 Balanced Simple Games
Let (N, v) be a simple game.

Suppose i is a veto player. Let B be a balanced collection with balanced map λ. Then

$$
\sum_{S \in B} \lambda(S)v(S) = \sum_{S \in B: i \in S} \lambda(S)v(S) \leq 1 = v(N) \ ,
$$

since i is a veto player. Hence, v is balanced.

For the converse, suppose v is balanced, and distinguish two cases:

Case 1: There is an i with $v(\{i\}) = 1$. Show that i is a veto player.
Case 2: $v(\{i\}) = 0$ for every $i \in N$. Show that also in this case v has veto players.

Problems of Chapter 17

17.1 *The Games* 1_T

(c) For $i \in T$: $\Phi_i(1_T) = \frac{(|T|-1)!(n-|T|)!}{n!}$.

17.2 *Unanimity Games*

(a) Suppose $\sum_{T \neq \emptyset} \alpha_T u_T = 0$, where 0 means the zero-game, for some $\alpha_T \in \mathbb{R}$. Show that all α_T are zero by induction, starting with one-person coalitions.
(b) Let $W \in 2^N$, then show

$$\sum_{T \neq \emptyset} c_T u_T(W) = v(W) + \sum_{S:\, S \subsetneq W} v(S) \sum_{T:\, S \subseteq T \subseteq W} (-1)^{|T|-|S|}.$$

It is sufficient to show that the second term of the last expression is equal to 0, hence that $\sum_{T:\, S \subseteq T \subseteq W} (-1)^{|T|-|S|} = 0$.

17.3 *If-Part of Theorem 17.4*
EFF, NP and ADD are straightforward. SYM needs more attention. Let i, j be symmetric in v. Note that for $S \subseteq N$ with $i \notin S$ and $j \in S$ we have $v((S \cup i) \setminus j) = v(S)$ by symmetry of i and j, since $v((S \cup i) \setminus j) = v((S \setminus j) \cup i)$ and $v(S) = v((S \setminus j) \cup j)$. Use this to show $\Phi_i(v) = \Phi_j(v)$ by collecting terms in a clever way.

17.4 *Dummy Player Property and Anonymity*
That DUM implies NP and the Shapley value satisfies DUM is straightforward. AN implies SYM: Let i and j be symmetric players, and let the value ψ satisfy AN. Then consider the permutation σ with $\sigma(i) = j$, $\sigma(j) = i$, and $\sigma(k) = k$ otherwise.

17.5 *Shapley Value, Core, and Imputation Set*
In the case of two players the core and the imputation set coincide. If the core is not empty then the Shapley value is in the core, cf. Example 17.2. In general, consider any game with $v(1) = 2$, $v(N) = 3$, and $v(S) = 0$ otherwise. Then the Shapley value is not even in the imputation set as soon as $n \geq 3$.

17.6 *Shapley Value as a Projection*
If a is an additive game then $\Phi(a) = (a(1), a(2), \ldots, a(n))$. For a general game v let a^v be the additive game generated by $\Phi(v)$. Then $\Phi(a^v) = (a^v(1), \ldots, a^v(n)) = \Phi(v)$.

17.7 *Shapley Value of Dual Game*
Follow the hint, or give a direct proof by using (17.4).

17.8 *Multilinear Extension*

(b) Let g be another multilinear extension of $\tilde{v}$ to $[0,1]^n$, say $g(\mathbf{x}) = \sum_{T \subseteq N} b_T \left(\prod_{i \in T} x_i \right)$. Show $b_T = c_T$ for all T by induction, starting with one-player coalitions.

17.9 *The Beta-Integral Formula*
Apply partial integration.

17.10 *Path Independence of* Φ
Use Theorem 17.12(c).

17.11 *An Alternative Characterization of the Shapley Value*
The Shapley value satisfies all these conditions. Conversely, (b)–(d) imply standardness for two-person games, so the result follows from Theorem 17.18.

Problems of Chapter 18

18.1 *Marginal Vectors and Dividends*

(b) For each $i \in N$, $m_i^\pi = \sum_{T \subseteq P_\pi(i) \cup i,\ i \in T} \Delta_v(T)$.

18.2 *Convexity and Marginal Vectors*
Use Theorems 18.3 and 18.6.

18.3 *Strictly Convex Games*
Let π and σ be two different permutations and suppose that $k \geq 1$ is the minimal number such that $\pi(k) \neq \sigma(k)$. Then show that $m_{\pi(k)}^\pi(v) < m_{\pi(k)}^\sigma(v)$. Hence, $m^\pi \neq m^\sigma$.

18.4 *Sharing Profits*

(a) For the landlord: $\Phi_0(v) = \frac{1}{n+1} \left[\sum_{s=0}^n f(s) \right]$.
(c) Extend f to a piecewise linear function on $[0, n]$. Then v is convex if and only if f is convex.

18.5 *Sharing Costs*

(a) For every nonempty coalition S, $v(S) = \sum_{i \in S} c_i - \max\{c_i \mid i \in S\}$. If we regard $c = (c_1, \ldots, c_2)$ as an additive game we can write $v = c - c_{\max}$, where $c_{\max}(S) = \max\{c_i \mid i \in S\}$.

18.6 *Independence of the Axioms in Theorem 18.8*

(a) Consider the value which, for every game v, gives each dummy player his individual worth and distributes the rest, $v(N) - \sum_{i \in D} v(i)$ where D is the subset

of dummy players, evenly among the players. This value satisfies all axioms except LIN.
(b) Consider the value which, for every game v, distributes $v(N)$ evenly among all players. This value satisfies all axioms except DUM.
(c) The value which gives each player his individual worth satisfies all axioms except EFF.
(d) Consider any set of weights $\{\alpha_\pi \mid \pi \in \Pi(N)\}$ with $\alpha_\pi \in \mathbb{R}$ for all π and $\sum_{\pi \in \Pi(N)} \alpha_\pi = 1$. The value $\sum_{\pi \in \Pi(N)} \alpha_\pi m^\pi$ satisfies LIN, DUM and EFF, but not MON unless all weights are nonnegative.

18.7 Null-Player in Theorem 18.8
Check that the dummy axiom in the proof of this theorem is only used for unanimity games. In those games, dummy players are also null-players, so it is sufficient to require NP. Alternatively, one can show that DUM is implied by ADD (and, thus, LIN), EFF and NP.

18.8 Characterization of Weighted Shapley Values
Check that every weighted Shapley value satisfies the Partnership axiom. Conversely, let ψ be a value satisfying the Partnership axiom and the four other axioms. Let $S^1 := \{i \in N \mid \psi_i(u_N) > 0\}$ and for every $i \in S^1$ let $\omega_i := \psi_i(u_N)$. Define, recursively, $S^k := \{i \in N \setminus (S^1 \cup \ldots \cup S^{k-1}) \mid \psi_i(u_{N\setminus(S^1\cup\ldots\cup S^{k-1})}) > 0\}$ and for every $i \in S^k$ let $\omega_i := \psi_i(u_{N\setminus(S^1\cup\ldots\cup S^{k-1})})$. This results in a partition $(S^1, \ldots, S^m)$ of N. Now define the weight system w by the partition $(S_1, \ldots, S_m)$ with $S_1 := S^m$, $S_2 := S^{m-1}, \ldots, S_m := S^1$, and the weights ω. Then it is sufficient to prove that for each coalition S and each player $i \in S$ we have $\psi_i(u_S) = \Phi_i^w(u_S)$. Let $h := \max\{j \mid S \cap S_j \neq \emptyset\}$, then with $T = N \setminus (S_{h+1} \cup \ldots \cup S_m)$ we have by the Partnership axiom: $\psi_i(u_S) = \frac{1}{\psi(u_T)(S)}\psi_i(u_T)$. If $i \notin S_h$ then $\psi_i(u_T) = 0$, hence $\psi_i(u_S) = 0 = \Phi_i^w(u_S)$. If $i \in S_h$ then $\psi_i(u_S) = \frac{\omega_i}{\sum_{j\in S\cap S_h}\omega_j} = \Phi_i^w(u_S)$.

18.9 Core and Weighted Shapley Values in Example 18.2
First write the game as a sum of unanimity games:

$$v = u_{\{1,2\}} + u_{\{1,3\}} - u_{\{2,3\}} + 2u_N .$$

Then consider all possible ordered partitions of N, there are 13 different ones, and associated weight vectors. This results in a description of all payoff vectors associated with weighted Shapley values, including those in the core of the game.

Problems of Chapter 19

19.1 Binary Relations
Not (4): $\succeq$ on $\mathbb{R}$ defined by $x \succeq y \Leftrightarrow x^2 \geq y^2$.
Not (3): $\geq$ on $\mathbb{R}^2$.

Not (2): $\succeq$ on $\mathbb{R}$ defined by: for all $x, y, x \geq y$, let $x \succeq y$ if $x - y \geq 1$, and let $y \succeq x$
    if $x - y < 1$.
Not (1): $>$ on $\mathbb{R}$.

The ordering on $\mathbb{R}$, defined by $[x \succeq y] \Leftrightarrow [x = y$ or $0 \leq x, y \leq 1]$ is reflexive and
transitive but not complete and not anti-symmetric.

19.2 *Linear Orders*
If $x \succ y$ then by definition $x \succeq y$ and not $y \succeq x$: hence $x \neq y$ since otherwise $y \succeq x$
by reflexivity.
If $x \succeq y$ and $x \neq y$ then not $y \succeq x$ since otherwise $x = y$ by anti-symmetry. Hence
$x \succ y$.

19.3 *The Lexicographic Order (1)*
Check (1)–(4) in Sect. 19.2 for $\succeq_{\text{lex}}$. Straightforward.

19.4 *The Lexicographic Order (2)*
This is the set $\{(x_1, x_2) \in \mathbb{R}^2 \mid [x_1 = 3, x_2 \geq 1]$ or $[x_1 > 3]\}$. This set is not closed.

19.5 *Representability of Lexicographic Order (1)*
Consider Problem 19.4. Since $(\alpha, 0) \succeq_{\text{lex}} (3, 1)$ for all $\alpha > 3$, we have $u(\alpha, 0) \geq$
$u(3, 1)$ for all $\alpha > 3$ and hence, by continuity of u, $\lim_{\alpha \downarrow 3} u(\alpha, 0) \geq u(3, 1)$.
Therefore $(3, 0) \succeq_{\text{lex}} (3, 1)$, a contradiction.

19.6 *Representability of Lexicographic Order (2)*
Suppose that u represents $\succeq_{\text{lex}}$ on $\mathbb{R}^2$, that is, $\mathbf{x} \succeq_{\text{lex}} \mathbf{y}$ if and only if $u(\mathbf{x}) \geq u(\mathbf{y})$
for all $\mathbf{x}, \mathbf{y} \in \mathbb{R}^2$. Then for any $t \in \mathbb{R}$ let $q(t)$ be a rational number in the interval
$[u(t, 0), u(t, 1)]$. Since $(t, \alpha) \succ_{\text{lex}} (s, \beta)$ and hence $u(t, \alpha) > u(s, \beta)$ for all $t > s$ and
all $\alpha, \beta \in [0, 1]$, we have $[u(t, 0), u(t, 1)] \cap [u(s, 0), u(s, 1)] = \emptyset$ for all $t \neq s$. Hence,
$q(t) \neq q(s)$ for all $t \neq s$. Therefore, we have found uncountably many different
rational numbers, a contradiction.

19.7 *Single-Valuedness of the Pre-nucleolus*
Consider the pre-nucleolus on a suitable compact convex subset and apply Theorem 19.3.

19.8 *(Pre-)Nucleolus and Core*
Use the fact that core elements have all excesses non-positive.

19.9 *Kohlberg Criterion for the Nucleolus*
First observe that the following modification of Theorem 19.4 holds:
*Theorem 19.4' Let (N, v) be a game and $\mathbf{x} \in I(N, v)$. Then the following two
statements are equivalent.*

(1) $\mathbf{x} = v(N, v)$.
(2) *For every α such that $\mathcal{D}(\alpha, \mathbf{x}, v) \neq \emptyset$ and for every side-payment $\mathbf{y}$ with $y(S) \geq$
    0 for every $S \in \mathcal{D}(\alpha, \mathbf{x}, v)$ and with $y_i \geq 0$ for all $i \in N$ with $x_i = v(i)$ we have
    $y(S) = 0$ for every $S \in \mathcal{D}(\alpha, \mathbf{x}, v)$.*

The proof of this theorem is almost identical to the proof of Theorem 19.4. In the second sentence of the proof, note that $\mathbf{z}_\varepsilon \in I(N, v)$ for ε small enough. In the second part of the proof, $(2)\Rightarrow(1)$, note that $y_i = z_i - x_i \geq 0$ whenever $x_i = v(i)$.

For the 'if'-part of the statement in this problem, let $\mathbf{x} \in I(N, v)$, $\mathcal{D}(\alpha, \mathbf{x}, v) \neq \emptyset$, and $\mathcal{E}(\alpha, \mathbf{x}, v)$ such that $\mathcal{D}(\alpha, \mathbf{x}, v)\cup\mathcal{E}(\alpha, \mathbf{x}, v)$ is balanced. Consider a side-payment $\mathbf{y}$ with $y(S) \geq 0$ for every $S \in \mathcal{D}(\alpha, \mathbf{x}, v)$ and $y_i \geq 0$ for every i with $x_i = v(i)$ [hence in particular for every i with $\{i\} \in \mathcal{E}(\alpha, \mathbf{x}, v)$]. The argument in the first part of the proof of Theorem 19.5 now applies to $\mathcal{D}(\alpha, \mathbf{x}, v) \cup \mathcal{E}(\alpha, \mathbf{x}, v)$, and Theorem 19.4′ implies $\mathbf{x} = v(N, v)$.

For the 'only-if' part, consider the program (19.4) in the second part of the proof of Theorem 19.5 but add the constraints $-y_i \leq 0$ for every $i \in N$ with $x_i = v(i)$. Theorem 19.4′ implies that the dual of this program is feasible, that is, there are $\lambda(S) \geq 0$, $S \in \mathcal{D}(\alpha, \mathbf{x}, v)$, $\lambda(\{i\}) \geq 0$, i such that $x_i = v(i)$, and $\lambda(N) \in \mathbb{R}$ such that

$$- \sum_{i\in N:\ x_i=v(i)} \lambda(\{i\})\mathbf{e}^{\{i\}} - \sum_{S\in\mathcal{D}(\alpha,\mathbf{x},v)} \lambda(S)\mathbf{e}^S + \lambda(N)\mathbf{e}^N = \sum_{S\in\mathcal{D}(\alpha,\mathbf{x},v)} \mathbf{e}^S .$$

Hence $\lambda(N)\mathbf{e}^N = \sum_{S\in\mathcal{D}(\alpha,\mathbf{x},v)}(1+\lambda(S))\mathbf{e}^S + \sum_{i\in N:\ x_i=v(i)} \lambda(\{i\})\mathbf{e}^{\{i\}}$. Let $\mathcal{E}(\alpha, \mathbf{x}, v)$ consist of those one-person coalitions $\{i\}$ with $x_i = v(i)$ and $\lambda(\{i\}) > 0$, then $\mathcal{D}(\alpha, \mathbf{x}, v) \cup \mathcal{E}(\alpha, \mathbf{x}, v)$ is balanced.

19.10 *Proof of Theorem 19.5*
To formulate the dual program, use for instance the formulation in Theorem 16.20. For instance, the primal (19.4) can be converted to the minimization problem in Theorem 16.20; then the dual corresponds to the maximization problem in Theorem 16.20. Feasibility of the dual follows from Problem 16.14.

19.11 *Nucleolus of a Three-Person Game (1)*
The nucleolus is $(5, 4, 3)$.

19.12 *Nucleolus of a Three-Person Game (2)*
The (pre-)nucleolus is $(5, 3, 2)$.

19.14 *Individual Rationality Restrictions for the Nucleolus*
The nucleolus is $(1, 0, 0)$. The pre-nucleolus is $(5/3, -1/3, -1/3)$.

19.15 *Example 19.7*
The set $\mathcal{B}_1 = \{123, 124, 34\}$ is balanced with weights all equal to $1/2$. The set $\mathcal{B}_1 \cup \mathcal{B}_2 = \{123, 124, 34, 134, 234\}$ is balanced with weights, respectively, equal to $5/12, 5/12, 3/12, 2/12, 2/12$.

19.16 *(Pre-)Nucleolus of a Symmetric Game*

(a) $v(v) = v^*(v) = (v(N)/n)\mathbf{e}^N$.

19.17 *COV and AN of the Pre-nucleolus*

Covariance of the pre-nucleolus follows since applying a transformation as in the definition of this property changes all excesses (only) by the same positive (multiplicative) factor.

Anonymity of the pre-nucleolus follows since a permutation of the players does not change the ordered vectors $\theta(\mathbf{x})$, but only permutes the coalitions to which the excesses correspond.

19.18 Apex Game

The (pre-)nucleolus is $(3/7, 1/7, 1/7, 1/7, 1/7)$. This can easily be verified using the Kohlberg criterion.

19.19 Landlord Game

(a) By anonymity, each worker is assigned $\frac{1}{2}[f(n) - f(n - 1)]$. By computing the excesses, it follows that among all coalitions containing the landlord, with this payoff vector the maximal excesses are reached by the coalitions containing $n - 1$ workers, and further also by the coalitions consisting of a single worker and not the landlord. By the Kohlberg criterion this immediately implies that the given vector is the (pre-)nucleolus. For the Shapley value, see Problem 18.4.

(b) Compute the excesses for the payoff vector $\frac{f(n)}{n+1}\mathbf{e}^{\{0,1,...,n\}}$, and apply the Kohlberg criterion.

19.20 Game in Sect. 19.1

The first linear program is: minimize α subject to the constraints $x_i + \alpha \geq 4$ for $i = 1, 2, 3$, $x_1 + x_2 + \alpha \geq 8$, $x_1 + x_3 + \alpha \geq 12$, $x_2 + x_3 + \alpha \geq 16$, $x_1 + x_2 + x_3 = 24$. The program has optimal value $\alpha = -2$, reached for $x_1 = 6$ and $x_2, x_3 \geq 6$.

In the second program x_1 has been eliminated. This program reduces to: minimize α subject to $x_2 + \alpha \geq 4$, $x_2 \leq 12 + \alpha$, $x_2 + x_3 = 18$. This has optimal value $\alpha = -4$, reached for $x_2 = 8$, $x_3 = 10$.

19.21 The Prekernel

For $i, j \in N$ denote by $\mathcal{T}_{ij}$ those coalitions that contain player i and not player j. For a payoff vector $\mathbf{x}$ denote by $s_{ij}(\mathbf{x}, v)$ the maximum of $e(S, \mathbf{x}, v)$ over all $S \in \mathcal{T}_{ij}$.

Let now $\mathbf{x}$ be the pre-nucleolus and suppose, contrary to what has to be proved, that there are two distinct players k, ℓ such that $s_{k\ell}(\mathbf{x}, v) > s_{\ell k}(\mathbf{x}, v)$. Let $\delta = (s_{k\ell}(\mathbf{x}, v) - s_{\ell k}(\mathbf{x}, v))/2$ and define $\mathbf{y}$ by $y_k = x_k + \delta$, $y_\ell = x_\ell - \delta$, and $y_i = x_i$ for all $i \neq k, \ell$. Denote $\mathcal{S} = \{S \in 2^N \setminus \mathcal{T}_{k\ell} \mid e(S, \mathbf{x}, v) \geq s_{k\ell}(\mathbf{x}, v)\}$ and $s = |\mathcal{S}|$. Then $\theta_{s+1}(\mathbf{x}) = s_{k\ell}(\mathbf{x}, v)$. For $S \in 2^N \setminus (\mathcal{T}_{k\ell} \cup \mathcal{T}_{\ell k})$, we have $e(S, \mathbf{x}, v) = e(S, \mathbf{y}, v)$. For $S \in \mathcal{T}_{k\ell}$ we have $e(S, \mathbf{y}, v) = e(S, \mathbf{x}, v) - \delta$. Finally, for $S \in \mathcal{T}_{\ell k}$ we have

$$e(S, \mathbf{y}, v) = e(S, \mathbf{x}, v) + \delta \leq s_{\ell k}(\mathbf{x}, v) + \delta = s_{k\ell}(\mathbf{x}, v) - \delta .$$

Thus, $\theta_t(\mathbf{y}) = \theta_t(\mathbf{x})$ for all $t \leq s$ and $\theta_{s+1}(\mathbf{y}) < s_{k\ell}(\mathbf{x}, v) = \theta_{s+1}(\mathbf{x})$. Hence $\theta(\mathbf{x}) \succ_{\text{lex}} \theta(\mathbf{y})$, a contradiction.

Problems of Chapter 20

20.2 *Example 20.3*
Argue that $a_{12} = a_{13} = 3$ if v were an assignment game. Use this to derive a contradiction.

20.3 *Subgames of Permutation Games*
That a subgame of a permutation game is again a permutation game follows immediately from the definition: in (20.3) the worth $v(S)$ depends only on the numbers k_{ij} for $i, j \in S$. By a similar argument [consider (20.1)] this also holds for assignment games.

20.4 *A Flow Game*

(c) $(1, 1, 0, 0)$, corresponding to the minimum cut through e_1 and e_2; $\{(0, 0, 1 + \alpha, 1 - \alpha) \mid 0 \le \alpha \le 1\}$, corresponding to the minimum cut through e_3 and e_4.

20.5 *Every Nonnegative Balanced Game is a Flow Game*
Let v be a nonnegative balanced game, and write (following the hint to the problem) $v = \sum_{r=1}^{k} \alpha_r w_r$, where $\alpha_r > 0$ and w_r a balanced simple game for each $r = 1, \ldots, k$. Consider the controlled capacitated network with two vertices, the source and the sink, and k edges connecting them, where each edge e_r has capacity α_r and is controlled by w_r. Then show that the associated flow game is v.

20.6 *On Theorem 20.6 (1)*

(a) This follows straightforwardly from the proof of Theorem 20.6.
(b) For example, each player receiving $5\frac{1}{4}$ is a core element.

20.7 *On Theorem 20.6 (2)*
In any core element, player should 1 receive at least 1 and player 2 also, but $v(N) = 1$. Hence the game has an empty core.

20.8 *Totally Balanced Flow Games*
This follows immediately from Theorem 20.6, since every dictatorial game is balanced, i.e., has veto players.

20.9 *If-Part of Theorem 20.9*
We show that the Banzhaf value satisfies 2-EFF (the other properties are obvious). With notations as in the formulation of 2-EFF, we have

$$\psi_p(v_p) = \sum_{S \subseteq (N \setminus p) \cup \{p\}: \, p \notin S} \frac{1}{2^{|N|-2}} \left[v_p(S \cup \{p\}) - v_p(S) \right]$$

$$= \sum_{S \subseteq N \setminus \{i,j\}} \frac{1}{2^{|N|-2}} \left[v(S \cup \{ij\}) - v(S) \right]$$

$$= \sum_{S \subseteq N \setminus \{i,j\}} \frac{1}{2^{|N|-1}} \left[2v(S \cup \{ij\}) - 2v(S)\right] .$$

The term in brackets can be written as

$$[v(S \cup \{i,j\}) - v(S \cup \{i\}) + v(S \cup \{j\}) - v(S)]$$
$$+ [v(S \cup \{i,j\}) - v(S \cup \{j\}) + v(S \cup \{i\}) - v(S)] ,$$

hence $\psi_p(v_p) = \psi_j(v) + \psi_i(v)$.

Show that DUM cannot be weakened to NP by finding a different value satisfying 2-EFF, SYM, NP, and SMON.

Problems of Chapter 21

21.1 *Anonymity and Symmetry*
An example of a symmetric but not anonymous solution is as follows. To symmetric problems, assign the point in $W(S)$ with equal coordinates; otherwise, assign the point of S that is lexicographically (first player 1, then player 2) maximal.

21.3 *The Nash Solution is Well-Defined*
The function $\mathbf{x} \mapsto (x_1 - d_1)(x_2 - d_2)$ is continuous on the compact set $\{\mathbf{x} \in S \mid \mathbf{x} \geq \mathbf{d}\}$ and hence attains a maximum on this set. We have to show that this maximum is attained at a unique point. In general, consider two points $\mathbf{z}, \mathbf{z}' \in \{\mathbf{x} \in S \mid \mathbf{x} \geq \mathbf{d}\}$ with $(z_1 - d_1)(z_2 - d_2) = (z_1' - d_1)(z_2' - d_2) = \alpha$. Then one can show that at the point $\mathbf{w} = \frac{1}{2}(\mathbf{z} + \mathbf{z}') \in S$ one has $(w_1 - d_1)(w_2 - d_2) > \alpha$. This implies that the maximum is attained at a unique point.

21.4 *(a)* $\Rightarrow$ *(b) in Theorem 21.1*
WPO and IIA are straightforward by definition, and SC follows from an easy computation. For SYM, note that if $N(S, \mathbf{d}) = \mathbf{z}$ for a symmetric problem $(S, \mathbf{d})$, then also $(z_2, z_1) = N(S, \mathbf{d})$ by definition of the Nash bargaining solution. Hence, $z_1 = z_2$ by uniqueness.

21.5 *Geometric Characterization of the Nash Bargaining Solution*
Let $(S, \mathbf{d}) \in B$ and $N(S, \mathbf{d}) = \mathbf{z}$. The slope of the tangent line ℓ to the graph of the function $x_1 \mapsto (z_1 - d_1)(z_2 - d_2)/(x_1 - d_1) + d_2$ (which describes the level set of $\mathbf{x} \mapsto (x_1 - d_1)(x_2 - d_2)$ through $\mathbf{z}$) at $\mathbf{z}$ is equal to $-(z_2 - d_2)/(z_1 - d_1)$, i.e., the negative of the slope of the straight line through d and z. Clearly, ℓ supports S at $\mathbf{z}$: this can be seen by invoking a separating hyperplane theorem, but also as follows. Suppose there were some point $\mathbf{z}'$ of S on the other side of ℓ than $\mathbf{d}$. Then there is a point $\mathbf{w}$ on the line segment connecting $\mathbf{z}'$ and $\mathbf{z}$ (hence, $\mathbf{w} \in S$) with $(w_1 - d_1)(w_2 - d_2) > (z_1 - d_1)(z_2 - d_2)$, contradicting $\mathbf{z} = N(S, \mathbf{d})$. The existence of such a point $\mathbf{w}$ follows since otherwise the straight line through $\mathbf{z}'$ and $\mathbf{z}$ would also be a tangent line to the level curve of the Nash product at $\mathbf{z}$.

For the converse, suppose that at a point $\mathbf{z}$ there is a supporting line of S with slope $-(z_2 - d_2)/(z_1 - d_1)$. Clearly, this line is tangent to the graph of the function $x_1 \mapsto (z_1 - d_1)(z_2 - d_2)/(x_1 - d_1) + d_2$ at $\mathbf{z}$. It follows that $\mathbf{z} = N(S, \mathbf{d})$.

21.6 *Strong Individual Rationality*
The implication $(a) \Rightarrow (b)$ is straightforward. For $(b) \Rightarrow (a)$, if F is also weakly Pareto optimal, then $F = N$ by Theorem 21.1. So it is sufficient to show that, if F is not weakly Pareto optimal then $F = D$. Suppose that F is not weakly Pareto optimal. Then there is an $(S, \mathbf{d}) \in B$ with $F(S, \mathbf{d}) \notin W(S)$. By IR, $F(S, \mathbf{d}) \geq \mathbf{d}$. Suppose $F(S, \mathbf{d}) \neq \mathbf{d}$. By SC, we may assume w.l.o.g. $\mathbf{d} = (0, 0)$. Let $\alpha > 0$ be such that $F(S, (0, 0)) \in W((\alpha, \alpha)S)$. Since $F(S, (0, 0)) \notin W(S)$, $\alpha < 1$. So $(\alpha, \alpha)S \subseteq S$. By IIA, $F((\alpha, \alpha)S, (0, 0)) = F(S, (0, 0))$, so by SC, $F((\alpha, \alpha)S, (0, 0)) = (\alpha, \alpha)F(S, (0, 0)) = F(S, (0, 0))$, contradicting $\alpha < 1$. So $F(S, (0, 0)) = (0, 0)$. Suppose $F(T, (0, 0)) \neq (0, 0)$ for some $(T, (0, 0)) \in B$. By SC we may assume $(0, 0) \neq F(T, (0, 0)) \in S$. By IIA applied twice, $(0, 0) = F(S \cap T, (0, 0)) = F(T, (0, 0)) \neq (0, 0)$, a contradiction. Hence, $F = D$.

21.7 $(a) \Rightarrow (b)$ *in Theorem 21.2*
Straightforward. Note in particular that in a symmetric game the utopia point is also symmetric, and that the utopia point is 'scale covariant'.

21.8 *Restricted Monotonicity*

(a) Follows from applying IM twice.
(b) For $(S, \mathbf{d})$ with $\mathbf{d} = (0, 0)$ and $u(S, \mathbf{d}) = (1, 1)$, let $F(S, \mathbf{d})$ be the lexicographically (first player 1, then player 2) maximal point of $S \cap \mathbb{R}^2_+$. Otherwise, let F be equal to R. This F satisfies RM but not IM.

21.9 *Global Individual Monotonicity*
It is straightforward to verify that G satisfies WPO, SYM, SC, and GIM. For the converse, suppose that F satisfies these four axioms, let $(S, \mathbf{d}) \in B$ and $\mathbf{z} := G(S, \mathbf{d})$. By SC, w.l.o.g. $\mathbf{d} = (0, 0)$ and $g(S) = (1, 1)$. Let $\alpha \leq 0$ such that $S \subseteq \tilde{S}$ where $\tilde{S} := \{\mathbf{x} \in \mathbb{R}^2 \mid (\alpha, \alpha) \leq \mathbf{x} \leq \mathbf{y}$ for some $\mathbf{y} \in S\}$. In order to prove $F(S, (0, 0)) = G(S, (0, 0))$ it is sufficient to prove that $F(\tilde{S}, (0, 0)) = G(\tilde{S}, (0, 0))$ (in view of GIM and WPO). Let $T = \text{conv}\{\mathbf{z}, (\alpha, g_2(\tilde{S})), (g_1(\tilde{S}), \alpha)\} = \text{conv}\{\mathbf{z}, (\alpha, 1), (1, \alpha)\}$. By SYM and WPO, $F(T, (0, 0)) = \mathbf{z}$. By GIM, $F(\tilde{S}, (0, 0)) \geq F(T, (0, 0)) = \mathbf{z} = G(S, (0, 0)) = G(\tilde{S}, (0, 0))$, so by WPO: $F(\tilde{S}, (0, 0)) = G(\tilde{S}, (0, 0))$. (Make pictures. Note that this proof is analogous to the proof of Theorem 21.2.)

21.10 *Monotonicity and (Weak) Pareto Optimality*

(a) Consider problems of the kind $(\text{conv}\{\mathbf{d}, \mathbf{a}\}, \mathbf{d})$ for some $\mathbf{a} > \mathbf{d}$.
(b) The egalitarian solution E satisfies MON and WPO on B_0.

21.11 *The Egalitarian Solution (1)*
Straightforward.

21.12 *The Egalitarian Solution (2)*
Let $z := E(S, \mathbf{d}) + E(T, \mathbf{e})$. Then it is straightforward to derive that $z_1 - (d_1 + e_1) = z_2 - (d_2 + e_2)$. Since $E(S + T, d + e)$ is the maximal point $\mathbf{x}$ such that $x_1 - (d_1 + e_1) = x_2 - (d_2 + e_2)$, it follows that $E(S + T, d + E) \geq \mathbf{z}$.

21.13 *Independence of Axioms*
Theorem 21.1:
WPO, SYM, SC: $F = R$; WPO, SYM, IIA: $F = L$, where $L(S, \mathbf{d})$ is the point of $P(S)$ nearest to the point $\mathbf{z} \geq \mathbf{d}$ with $z_1 - d_1 = z_2 - d_2$ measured along the boundary of S; WPO, SC, IIA: $F = D^1$, where $D^1(S, \mathbf{d})$ is the point of $\{\mathbf{x} \in P(S) \mid \mathbf{x} \geq \mathbf{d}\}$ with maximal first coordinate; SYM, SC, IIA: $F = D$ (disagreement solution).
Theorem 21.2:
WPO, SYM, SC: $F = N$; WPO, SYM, IM: $F = L$; WPO, SC, IM: if $\mathbf{d} = (0, 0)$ and $u(S, \mathbf{d}) = (1, 1)$, let F assign the point of intersection of $W(S)$ and the line segment connecting $(1/4, 3/4)$ and $(1, 1)$ and, otherwise, let F be determined by SC; SYM, SC, IM: $F = D$.
Theorem 21.3:
WPO, MON, SYM: $F(S, \mathbf{d})$ is the maximal point of S on the straight line through $\mathbf{d}$ with slope $1/3$ if $\mathbf{d} = (1, 0)$, $F(S, \mathbf{d}) = E(S, \mathbf{d})$ otherwise; WPO, MON, TC: $F(S, \mathbf{d})$ is the maximal point of S on the straight line through $\mathbf{d}$ with slope $1/3$; WPO, SYM, TC: $F = N$; MON, SYM, TC: $F = D$.

21.14 *Nash and Rubinstein*

(b) The Nash bargaining solution outcome is $(\frac{1}{3}\sqrt{3}, \frac{2}{3})$, hence $(\frac{1}{3}\sqrt{3}, 1 - \frac{1}{3}\sqrt{3})$ is the resulting distribution of the good.

(c) The Rubinstein bargaining outcome is $\left(\sqrt{\frac{1-\delta}{1-\delta^3}}, \frac{\delta - \delta^3}{1 - \delta^3} \right)$.

(d) The outcome in (c) converges to the outcome in (b) if δ converges to 1.

Problems of Chapter 22

22.1 *Convex Sets*
The only-if part is obvious. For the if-part, for any two vectors $\mathbf{x}$ and $\mathbf{y}$ in Z the condition implies that $\frac{k}{2^m}\mathbf{x} + \frac{2^m - k}{2^m}\mathbf{y} \in Z$ for every $m \in \mathbb{N}$ and $k \in \{0, 1, \dots, 2^m\}$. By closedness of Z, this implies that $\text{conv}\{\mathbf{x}, \mathbf{y}\} \subseteq Z$, hence Z is convex.

22.2 *Proof of Lemma 22.3*
Suppose that both systems have a solution, say $(\mathbf{y}, \mathbf{z}) \geq \mathbf{0}$, $(\mathbf{y}, \mathbf{z}) \neq \mathbf{0}$, $A\mathbf{y} + \mathbf{z} = \mathbf{0}$, $\mathbf{x} > \mathbf{0}$, $\mathbf{x}A > \mathbf{0}$. Then $\mathbf{x}A\mathbf{y} + \mathbf{x} \cdot \mathbf{z} = \mathbf{x}(A\mathbf{y} + \mathbf{z}) = 0$, hence $\mathbf{y} = \mathbf{0}$ and $\mathbf{z} = \mathbf{0}$ since $\mathbf{x} > \mathbf{0}$, $\mathbf{x}A > \mathbf{0}$. This contradicts $(\mathbf{y}, \mathbf{z}) \neq \mathbf{0}$.

22.3 *Rank of AA^T*
This follows from basic linear algebra. Prove that the null spaces of A and $A^T A$ are equal and use the Rank Theorem to conclude that the rank of $A^T A$, and thus that of AA^T, is equal to the rank of A.

22.4 *Proof of Lemma 22.5*
Suppose that both systems have a solution, say $\mathbf{x} > 0$, $\mathbf{x}A = \mathbf{b}$, $A\mathbf{y} \geq 0$, $\mathbf{b} \cdot \mathbf{y} < 0$.
Then $\mathbf{x}A\mathbf{y} < 0$, contradicting $\mathbf{x} > 0$ and $A\mathbf{y} \geq 0$.

22.5 *Proof of Lemma 22.7*

(a) If $\mathbf{x} \geq 0$, $\mathbf{x}A \leq \mathbf{b}$, $\mathbf{y} \geq 0$ and $\mathbf{b} \cdot \mathbf{y} < 0$ then $\mathbf{x}A\mathbf{y} \leq \mathbf{b} \cdot \mathbf{y} < 0$, so $A\mathbf{y} \not\geq 0$. This
    shows that at most one of the two systems has a solution.
(b) Suppose the system in (a) has no solution. Then also the system $\mathbf{x}A + zI = \mathbf{b}$,
    $\mathbf{x} \geq 0$, $z \geq 0$ has no solution. Hence, by Farkas' Lemma the system $\begin{pmatrix} A \\ I \end{pmatrix} \mathbf{y} \geq 0$,
    $\mathbf{b} \cdot \mathbf{y} < 0$ has a solution. Therefore, the system in (b) has a solution.

22.6 *Extreme Points*
The implication (b) $\Rightarrow$ (a) follows by definition of an extreme point.
For the implication (a) $\Rightarrow$ (c), let $x, y \in C \setminus \{e\}$ and $0 < \lambda < 1$. Let $z = \lambda x + (1-\lambda)y$.
If $z \neq e$ then $z \in C \setminus \{e\}$ by convexity of C. Suppose now that $z = e$. W.l.o.g. let
$\lambda \geq 1/2$. Then $e = \lambda x + (1-\lambda)y = (1/2)x + (1/2)[\mu x + (1-\mu)y]$ for $\mu = 2\lambda - 1$.
Since $\mu x + (1-\mu)y \in C$, this implies that e is not an extreme point of C. This proves
the implication (a) $\Rightarrow$ (c).
For the implication (c) $\Rightarrow$ (b), let $x, y \in C$, $x \neq y$, and $0 < \alpha < 1$. If $x = e$ or $y = e$
then clearly $\alpha x + (1-\alpha)y \neq e$. If $x \neq e$ and $y \neq e$ then $\alpha x + (1-\alpha)y \in C \setminus \{e\}$
by convexity of $C \setminus \{e\}$, hence $\alpha x + (1-\alpha)y \neq e$ as well.

22.7 *Affine Subspaces*
Let $A = a + L$ be an affine subspace, $x, y \in A$, and $\lambda \in \mathbb{R}$. Write $x = a + \bar{x}$ and
$y = a + \bar{y}$ for $\bar{x}, \bar{y} \in L$, then $\lambda x + (1-\lambda)y = a + \lambda\bar{x} + (1-\lambda)\bar{y} \in A$ since
$\lambda\bar{x} + (1-\lambda)\bar{y} \in L$ (L is a linear subspace).
Conversely, suppose that A contains the straight line through any two of its elements.
Let a be an arbitrary element of A and let $L := \{x - a \mid x \in A\}$. Then it follows
straightforwardly that L is a linear subspace of V, and thus $A = a + L$ is an affine
subspace.

22.8 *The Set of Sup-points of a Linear Function on a Convex Set*
In general, linearity of f implies that, if $f(\mathbf{x}) = f(\mathbf{y})$, then $f(\lambda\mathbf{x} + (1-\lambda)\mathbf{y}) = f(\mathbf{x}) = f(\mathbf{y})$ for any two points of C and $0 < \lambda < 1$. It follows, in particular, that the set D
is convex.
Let $\mathbf{e} \in \text{ext}(D)$ and suppose $\mathbf{e} = (1/2)\mathbf{x} + (1/2)\mathbf{y}$ for some $\mathbf{x}, \mathbf{y} \in C$. Then by
linearity of f, $f(\mathbf{e}) = (1/2)f(\mathbf{x}) + (1/2)f(\mathbf{y})$, hence $\mathbf{x}, \mathbf{y} \in D$ since $\mathbf{e} \in D$. So
$\mathbf{e} = \mathbf{x} = \mathbf{y}$ since $\mathbf{e}$ is an extreme point of D. Thus, $\mathbf{e}$ is also an extreme point of C.

Problems of Chapter 23

RP 1 *Matrix Games (1)*

(a) All rows are (pure) maximin strategies (with minimum 0) and all columns are pure minimax strategies (with maximum 2). The value of the game is between 0 and 2 (which is obvious anyway in this case).
(b) The third column is strictly dominated by the second column and the third row is strictly dominated by the second row. Entry $(1,2)$ is a saddlepoint, hence the value of the game is 2. The unique maximin strategy is $(1,0,0)$, and the minimax strategies are the strategies in the set $\{(q, 1-q, 0) \mid 0 \le q \le 1/2\}$.
(c) The second and third rows are the maximin rows. The second column is the unique minimax column. From this we can conclude that the value of the game is between 1 and 2. The first and fourth columns are strictly dominated by the second. Next, the first row is strictly dominated by the last row. The unique maximin strategy is $(0, 2/3, 1/3)$ and the unique minimax strategy is $(0, 2/3, 1/3, 0)$. The value of the game is $5/3$.

RP 2 *Matrix Games (2)*

(a) The first row is the unique maximin row (with minimum 2) and both columns are minimax columns (with maximum 5). So the value is between 2 and 5. The game has no saddlepoint.
(b) $v(A_1) = 5/2$, $v(A_2) = 20/7$, $v(A_3) = 2$ (saddlepoint), $v(A_4) = 1$ (saddlepoint), $v(A_5) = 7/3$, $v(A_6) = 25/9$. Since player 1 can pick rows, the value must be the maximum of these amounts, hence $20/7$, the value of A_2.
(c) The unique maximin strategy is $(5/7, 0, 2/7, 0)$ and the unique minimax strategy is $(3/7, 4/7)$.

RP 3 *Matrix Games (3)*

(a) The unique maximin row is the first row, with minimum 8. The unique minimax column is the first column, with maximum 12. So the value of the game is between 8 and 12. The game has no saddlepoint.
(b) The second row is strictly dominated by for instance putting probability $1/2$ on the first row and $1/2$ on the third row. After eliminating the second row, the third column is strictly dominated by the first column.
(c) The unique maximin strategy is $(1/2, 0, 1/2)$ and the unique minimax strategy is $(3/4, 1/4, 0)$. The value of the game is 10.

RP 4 *Bimatrix Games (1)*

(a) D is strictly dominated by $3/5 \cdot U + 2/5 \cdot M$. Next, C is strictly dominated by R.
(b) In the reduced (two by two) game, the best reply function of player 1 is: play U if player 2 puts less than probability $2/5$ on L, play M if player 2 puts more

than probability $2/5$ on L, and play any combination of U and M if player 2 puts probability $2/5$ on L. The best reply function of player 2 is: play R if player 1 puts positive probability on U, and play any combination of L and R if player 1 plays M. The set of Nash equilibria is: $\{((1,0),(0,1))\} \cup \{((0,1),(q,1-q)) \mid 1 \geq q \geq 2/5\}$.

(c) The set of Nash equilibria in the original game is: $\{((1,0,0),(0,0,1))\} \cup \{((0,1,0),(q,0,1-q)) \mid 1 \geq q \geq 2/5\}$.

RP 5 *Bimatrix Games (2)*

(a) For $x > 2$: $\{((1,0),(1,0))\}$. For $x = 2$: $\{((1,0),(1,0))\} \cup \{((p,1-p),(0,1)) \mid 0 \leq p \leq 1/2\}$. For $0 < x < 2$: $\{((1/2,1/2),((2-x)/2,x/2))\}$. For $x = 0$: $\{((0,1),(0,1))\} \cup \{((p,1-p),(1,0)) \mid 1 \geq p \geq 1/2\}$. For $x < 0$: $\{((0,1),(0,1))\}$.

(b) f is strictly dominated by $1/3 \cdot e + 2/3 \cdot g$. Next: b is strictly dominated by c, e by g, a by d. The remaining two by two game has a unique Nash equilibrium. In the original game the unique Nash equilibrium is $((0,0,4/9,5/9),(0,0,1/2,1/2))$.

RP 6 *Voting*

(a)

$$
\begin{array}{c}
\quad\quad\quad (4,0)\quad\quad (3,1)\quad\quad (2,2)\\
\begin{array}{c}(4,0)\\(3,1)\\(2,2)\end{array}
\left(\begin{array}{ccc}
3/2,3/2 & 1,2 & 1,2\\
2,1 & 3/2,3/2 & 1,2\\
2,1 & 2,1 & 3/2,3/2
\end{array}\right).
\end{array}
$$

(b) By iterated elimination of strictly dominated strategies it follows that the unique Nash equilibrium in this game is $((2,2),(2,2))$. (This is a constant sum game: $(2,2)$ is the optimal strategy for each party.)

RP 7 *A Bimatrix Game*

(a) For $a \neq 0$ the unique Nash equilibrium is $((1/2,1/2),(1/2,1/2))$. For $a = 0$ the set of Nash equilibria is $\{((p,1-p),(0,1)) \mid 1 \geq p > 1/2\} \cup \{((p,1-p),(1,0)) \mid 0 \leq p < 1/2\} \cup \{((1/2,1/2),(q,1-q)) \mid 0 \leq q \leq 1\}$.

(b) The strategic form of this game is

$$
\begin{array}{c}
\quad\quad LL\quad LR\quad RL\quad RR\\
\begin{array}{c}T\\B\end{array}
\left(\begin{array}{cccc}
a,0 & a,0 & 0,1 & 0,1\\
0,1 & a,0 & 0,1 & a,0
\end{array}\right).
\end{array}
$$

There are two subgame perfect equilibria in pure strategies: player 1 plays T and player 2 plays RL (i.e., R after T and L after B); and player 1 plays B and player 2 plays RL.

RP 8 *An Ice-Cream Vendor Game*

(a) There are four different situations: (i) all vendors in the same location: each gets 400; (ii) two in the same location and the third vendor in a neighboring location: the first two get 300 and the third gets 600; (iii) two in the same location and the third vendor in the opposite location: the first two get 300 and the third gets 600; and (iv) all vendors in different locations: the middle one gets 300 and the others get 450 each. From this it is easily seen that (iii) and (iv) are Nash equilibria but (i) and (ii) are not Nash equilibria.

(b) There are many subgame perfect Nash equilibria, but they can be reduced to three types: (i) player 1 chooses arbitrarily, player 2 chooses the opposite location of player 1, and player 3 chooses a remaining optimal open location; (ii) player 1 chooses arbitrarily, player 2 chooses one of the neighboring locations of player 1, and player 3 chooses the opposite location of player 2 if that is unoccupied, and otherwise the same location as player 2; (iii) player 1 chooses arbitrarily, player 2 chooses the same location as player 1, and player 3 chooses the opposite location of player 1.

RP 9 *A Repeated Game*

(a) (U, L, B) and (D, R, B).

(b) In the second period, after each action combination of the first period, one of the two equilibria in (a) has to be played.

(c) In the first period player 1 plays U, player 2 plays R, and player 3 plays A. In the second period, if the first period resulted in (U, R, A) then player 1 plays D, player 2 plays R, and player 3 plays B; in all other cases, player 1 plays U, player 2 plays L, and player 3 plays B.

(d) In the first period player 1 plays U, player 2 plays R, and player 3 plays B. In the second period, if the first period resulted in (U, R, B) then player 1 plays U, player 2 plays L, and player 3 plays B; in all other cases, player 1 plays D, player 2 plays R, and player 3 plays B.

RP 10 *Locating a Pub*

(a) Player 1 has three pure strategies and player 2 has eight pure strategies.

(b) Player 1 chooses B. Player 2 chooses B, C, B, if player 1 chooses A, B, C respectively.

(c) Player 1 has 24 pure strategies and player 2 has 8 pure strategies.

(d) (i) Player 1 plays A; after A the subgame equilibrium (B, C) is played, after B the subgame equilibrium (A, C), and after C the subgame equilibrium (A, B).
(ii) Player 1 plays B; after A the subgame equilibrium (B, C) is played, after B the subgame equilibrium (C, A), and after C the subgame equilibrium (A, B).
(iii) Player 1 plays C; after A the subgame equilibrium (B, C) is played, after B the subgame equilibrium (C, A), and after C the subgame equilibrium (B, A).

RP 11 *A Two-Stage Game*

(a) In G_1: (D,R); in G_2: (T,X), (M,Y), and (B,Z).

(b) Each player has $2 \cdot 3^4 = 162$ pure strategies.

(c) In G_1 player 1 plays U and player 2 plays L. In G_2 the players play as follows. If (U,L) was played, then player 1 plays M and player 2 plays Y. If (D,L) was played, then player 1 plays B and player 2 plays Z. If (U,R) was played, then player 1 plays T and player 2 plays X. If (D,R) was played, then player 1 plays M and player 2 plays Y.

(d) In the second stage (in G_1) always (U,L) has to be played. Hence, there are three subgame perfect equilibria, corresponding to the three Nash equilibria of G_2.

RP 12 *Job Market Signaling*

(a)

(a)

(b) The Nash equilibria are: (i) type H plays E, type L plays N, F plays M after E and C after N; (ii) both types play N, F always plays C.

(c) The equilibrium in (i) is separating with (forced) beliefs $\alpha = 1$ and $\beta = 0$. The equilibrium in (ii) is pooling with $\beta = 1/3$ (forced) and $\alpha \leq 2/5$. According to the intuitive criterion we must have $\alpha = 1$, so that the intuitive criterion is not satisfied by the latter equilibrium. (It does not apply to the first equilibrium.)

RP 13 *Second-Hand Cars (1)*

(a, b) The extensive form of this signaling game is as follows:

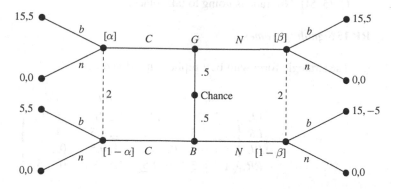

The strategic form is:

$$
\begin{array}{c c c c c}
 & bb & bn & nb & nn \\
CC & \begin{pmatrix} 10,\underline{5} & \underline{10,5} & 0,0 & \underline{0,0} \\ 15,0 & 7.5,\underline{2.5} & 7.5,-2.5 & \underline{0,0} \\ 10,\underline{5} & 2.5,2.5 & 7.5,2.5 & \underline{0,0} \\ \underline{15,0} & 0,\underline{0} & \underline{15,0} & \underline{0,0} \end{pmatrix} \\
CN \\
NC \\
NN
\end{array}.
$$

The Nash equilibria are: (CC, bn), (NN, bb), (NN, nb), (NN, nn).

(c) (CC, bn) is pooling with $\beta \leq 1/2$, (NN, bb) is pooling for all α. The other two equilibria are not perfect Bayesian, since player 2 will play b after C.

RP 14 *Second-Hand Cars (2)*

(a) This is a static game of incomplete information, represented by the pair G_1, G_2:

$$
G_1 = \begin{array}{c} 1 \\ 3 \\ 5 \end{array}
\begin{array}{ccc}
1 & 3 & 5 \\
\begin{pmatrix} 1,-1 & 0,0 & 0,0 \\ 0,0 & -1,1 & 0,0 \\ -1,1 & -2,2 & -3,3 \end{pmatrix}
\end{array}
\qquad
G_2 = \begin{array}{c} 1 \\ 3 \\ 5 \end{array}
\begin{array}{ccc}
1 & 3 & 5 \\
\begin{pmatrix} 3,-3 & 0,0 & 0,0 \\ 2,-2 & 1,-1 & 0,0 \\ 1,-1 & 0,0 & -1,1 \end{pmatrix}
\end{array}
$$

where G_1 is played with probability 25 % and G_2 with probability 75 %. (The numbers should be multiplied by 1,000, the buyer is the row and the seller the column player.)

(b) The buyer has one type and three pure strategies, the seller has two types and nine pure strategies.

(c) Strategy "5" is strictly dominated by strategy "3".

(d) Against strategy "3" of the buyer the best reply of the seller is the combination $(3, 5)$, but against this combination the best reply of the buyer is "1".

(e) Against strategy "1" of the buyer the seller has four best replies: $(3,3)$, $(3,5)$, $(5,3)$, and $(5,5)$. In turn, (only) against $(3,5)$ and $(5,5)$ is "1" a best reply. Hence there are two Nash equilibra in pure stategies: (i) $(1,(3,5))$ and (ii) $(1,(5,5))$. No trade is going to take place.

RP 15 *Signaling Games*

(a) The strategic form with best replies underlined is:

	uu	*ud*	*du*	*dd*
LL	2,$\underline{1}$	$\underline{2},\underline{1}$	1.5,0.5	$\underline{1.5},0.5$
LR	$\underline{2.5},\underline{1.5}$	1.5,1	$\underline{2},0.5$	1,0
RL	1,0	0.5,0.5	1,0.5	0.5,$\underline{1}$
RR	1.5,$\underline{0.5}$	0,$\underline{0.5}$	1.5,$\underline{0.5}$	0,$\underline{0.5}$

(LR, uu) is a separating perfect Bayesian equilibrium with beliefs $\alpha = 1$ and $\beta = 0$. (LL, ud) is a pooling Bayesian equilibrium with beliefs $\alpha = 1/2$ and $\beta \geq 1/2$. For the latter, the intuitive criterion requires $\beta = 0$, so that this equilibrium does not satisfy it.

(b) The strategic form with best replies underlined is:

	uu	*ud*	*du*	*dd*
LL	$\underline{3},\underline{1.5}$	$\underline{3},\underline{1.5}$	0.5,1	0.5,1
LR	2,1	2.5,0	$\underline{1},\underline{1.5}$	1.5,0.5
RL	1.5,1.5	$\underline{3.5},\underline{2}$	0,0.5	2,1
RR	0.5,$\underline{1}$	3,0.5	0.5,$\underline{1}$	$\underline{3},0.5$

(LL, uu) is a pooling perfect Bayesian equilibrium with beliefs $\alpha = 1/2$ and $\beta \leq 2/3$. The intuitive criterion requires $\beta = 1$, so this pooling equilibrium does not satisfy it. (LR, du) is a separating perfect Bayesian equilibrium with beliefs $\alpha = 1$ and $\beta = 0$, and (RL, ud) is a separating perfect Bayesian equilibrium with beliefs $\alpha = 0$ and $\beta = 1$.

RP 16 *A Game of Incomplete Information*

(a) Start with the decision node of player 1. Player 1 has four actions/strategies: AA, AB, BA, BB. All these actions lead to one and the same information set of player 2, who has three actions/strategies: C, D, E.

(b) The strategic form is:

$$
\begin{array}{c}
 & C & D & E \\
AA & 3,\underline{2} & 1.5,1.5 & \underline{2.5},1.5 \\
AB & \underline{4},2.5 & \underline{2.5},3 & 1,1.5 \\
BA & 3,\underline{2} & 1.5,1.5 & \underline{2.5},1.5 \\
BB & \underline{4},2.5 & \underline{2.5},3 & 1,1.5
\end{array}
$$

The Nash equilibria in pure strategies are (AB,D) and (BB,D).

(c) Player 1 has now two pure strategies, namely A and B. If player 1 plays A then the best reply of player 2 is EC. Against EC, the payoff of A is 1.5 and the payoff of B is 2.5, so that A is not a best reply against EC. Against B, the best reply of player 2 is ED. In turn, B is player 1's best reply against ED (yields 2 whereas A only yields 1). So the unique Nash equilibrium in pure strategies is (B,ED).

RP 17 *A Bayesian Game*

(a) This is the game

$$
\begin{array}{c}
 & F & Y \\
F & \begin{pmatrix} -1,1 & 1,0 \\ 0,1 & 0,0 \end{pmatrix}
\end{array}
$$

with (Y,F) as unique Nash equilibrium (also in mixed strategies).

(b) Start with the decision node for player 1, who has two actions/strategies: F and Y. Player 2 has a singe information set and four actions/strategies: FF, FY, YF, YY.

(c) The strategic form is:

$$
\begin{array}{c}
 & FF & FY & YF & YY \\
F & \begin{pmatrix} 1-2\alpha,2\alpha-1 & 1-2\alpha,\alpha & 1,\alpha-1 & 1,0 \\ 0,1 & 0,\alpha & 0,1-\alpha & 0,0 \end{pmatrix} \\
Y &
\end{array}
$$

For $\alpha = 0$ the Nash equilibria in pure strategies are (F,FY) and (F,YY). For $0 < \alpha < 1/2$: (F,FY). For $\alpha = 1/2$: (F,FY) and (Y,FF). For $1/2 < \alpha < 1$: (Y,FF). For $\alpha = 1$: (Y,FF) and (Y,FY).

RP 18 *Entry as a Signaling Game*

(a) The extensive form of this signaling game is:

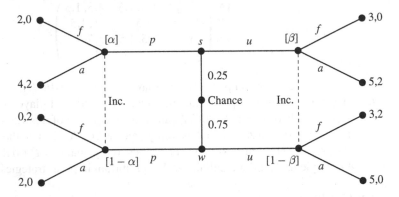

(b, c) The strategy combination (pu, af) (strong type p, incumbent a after p) is a Nash equilibrium. It is a separating perfect Bayesian equilibrium for $\alpha = 1$ and $\beta = 0$. Also (uu, ff) is a Nash equilibrium. It is pooling perfect Bayesian for $\beta = 1/2$ and $\alpha \leq 1/2$. It does not satisfy the intuitive criterion since that requires $\alpha = 1$.

RP 19 *Bargaining (1)*

(a) Player 1 has only one type. Player 2 has infinitely many types, namely each $v \in [0, 1]$ is a possible type of player 2. A typical strategy of player 1 consists of a price $p_1 \in [0, 1]$ and a yes/no decision depending on the price p_2 of player 2 if that player rejects p_1—in principle, the yes/no decision may also depend on p_1.

(b) A typical strategy of player 2 consists, for every type $v \in [0, 1]$, of a yes/no decision depending on the price p_1 asked by player 1 and a price p_2 in case the decision was 'no'. In principle, p_2 may also depend on p_1 (not only via the yes/no decision).

(c) Player 2 accepts if $v - p_1 \geq \delta v$ (noting that he can offer $p_2 = 0$ if he does not accept the price p_1 of player 1); rejects and offers $p_2 = 0$ if $v - p_1 < \delta v$.

(d) Using (c) player 1 asks the price p_1 that maximizes $p_1 \cdot Pr[p_1 \leq (1 - \delta)v]$, i.e., his expected payoff—note that his payoff is 0 if player 2 rejects. Hence, player 1 solves $\max_{p_1 \in [0,1]} p_1 \cdot [1 - p_1/(1 - \delta)]$, which has solution $p_1 = (1 - \delta)/2$. So the equilibrium is, that player 1 asks this price and accepts any price of player 2; and player 2 accepts any price at most $(1 - \delta)/2$, and rejects any higher price and then offers $p_2 = 0$.

RP 20 *Bargaining (2)*

(a) The (Pareto) boundary of the feasible set consists of all pairs $(x, 1 - x^2)$ for $x \in [0, 1]$.

(b) The Nash bargaining solution outcome is found by maximizing the expression $x(1 - x^2)$ over all $x \in [0, 1]$. The solution is $((1/3)\sqrt{3}, 2/3)$. In distribution of the good: $((1/3)\sqrt{3}, 1 - (1/3)\sqrt{3})$.

(c, d) Let $(x, 1 - x^2)$ be the proposal of player 1 and $(y, 1 - y^2)$ that of player 2. Then the equations $1 - x^2 = \delta(1 - y^2)$ and $y = \delta x$ hold for the Rubinstein outcome. This results in $x = 1/\sqrt{1 + \delta + \delta^2}$; taking the limit for $\delta \to 1$ gives $(1/3)\sqrt{3}$, which is indeed the Nash bargaining solution outcome for player 1.

RP 21 *Bargaining (3)*

(a) Player 1 proposes $(1 - \delta + (1/2)\delta^2, \delta - (1/2)\delta^2)$ at $t = 0$ and player 2 accepts. Note that $1 - \delta + (1/2)\delta^2 > \delta - (1/2)\delta^2$, so the beginning player has an advantage.

(b) If the utility function of player 2 were the same as that of player 1, then the Nash bargaining solution would result in equal split. This is still the case if player 2's utility function is multiplied by 2, as is the case here: the maximum of $u(x) \cdot 2u(1-x)$ is attained at the same point as the maximum of $u(x) \cdot u(1 - x)$. So the division of the good is $(1/2, 1/2)$. In terms of utilities, this gives $(u(1/2), 2u(1/2))$. (The Nash bargaining solution is symmetric, Pareto optimal, and scale covariant: see Chap. 10.)

RP 22 *Ultimatum Bargaining*

(a) Player 1 chooses an action/strategy $(1-m, m)$. Player 2 decides for each strategy of player 1 whether to accept or reject the offer. If he accepts, the payoffs are $(1 - m, m + a(2m - 1))$, otherwise the payoffs are $(0, 0)$.

(b) Player 1 proposes $(1-a/(1+2a), a/(1+2a))$, and player 2 accepts $(1-m, m)$ if and only if $m \geq a/(1+2a)$. Hence, the outcome is $(1-a/(1+2a), a/(1+2a))$.

(c) If a becomes large, then this outcome converges to equal split: this is because then player 2 cares mainly about the division and not so much about what he gets.

RP 23 *An Auction (1)*

(a) The game has imperfect but complete information.

(b) The unique Nash equilibrium is each bidder bidding $v_1 = v_2$.

(c) There is no Nash equilibrium.
(d) The associated bimatrix game is:

$$
\begin{array}{c@{\quad}cccc}
 & 0 & 1 & 2 & 3 \\
0 & 1/2,3/2 & 0,2 & 0,1 & 0,0 \\
1 & 0,0 & 0,1 & 0,1 & 0,0 \\
2 & -1,0 & -1,0 & -1/2,1/2 & 0,0 \\
3 & -2,0 & -2,0 & -2,0 & -1,0
\end{array}
$$

The Nash equilibria are $(0,1)$, $(1,1)$, and $(1,2)$.

RP 24 *An Auction (2)*

(a) Let $b_i < v_i$. If b_i wins then v_i is equally good. If b_i loses and the winning bid is below v_i then v_i is a strict improvement. If b_i loses and the winning bid is at least v_i then v_i is at least as good. If, on the other hand, $b_i > v_i$, then, if b_i wins, the fourth-highest bid is below v_i and the second highest bid is above v_i, then bidding v_i results in zero instead of positive payoff.
(b) For instance, player 2 can improve by any bid above v_1.
(c) All bidders bid $\tilde{v}$ where $\tilde{v} \in [v_2, v_1]$.

RP 25 *An Auction (3)*

(a) The best reply function b_2 of player 2 is given by: $b_2(0) = \{1\}$, $b_2(1) = \{2\}$, $b_2(2) = \{3\}$, $b_2(3) = b_2(4) = \{0,\ldots,4\}$, $b_2(5) = \{0,\ldots,5\}$, $b_2(6) = \{0,\ldots,6\}$. The best reply function b_1 of player 1 is given by: $b_1(0) = \{0\}$, $b_1(1) = \{1\}$, $b_1(2) = \{2\}$, $b_1(3) = \{3\}$, $b_1(4) = \{4\}$, $b_1(5) = \{5\}$, $b_1(6) = \{0,\ldots,6\}$.
(b) The Nash equilibria are: $(3,3)$, $(4,4)$, $(5,5)$, and $(6,6)$.

RP 26 *Quantity Versus Price Competition*

(a) The profit functions are $q_1(4 - 2q_1 - q_2)$ and $q_2(4 - q_1 - 2q_2)$ respectively (or zero in case an expression is negative). The first-order conditions (best reply functions) are $q_1 = (4 - q_2)/4$ and $q_2 = (4 - q_1)/4$ (or zero) and the equilibrium is $q_1 = q_2 = 4/5$ with associated prices equal to $8/5$ and profits equal to $32/25$.
(b) Follows by substitution.
(c, d) The profit functions are $(1/3)p_1(p_2 - 2p_1 + 4)$ and $(1/3)p_2(p_1 - 2p_2 + 4)$ (or zero) respectively. The first-order conditions (best reply functions) are $p_1 = (p_2 + 4)/4$ and $p_2 = (p_1 + 4)/4$. The equilibrium is $p_1 = p_2 = 4/3$ with associated quantities $q_1 = q_2 = 8/9$ and profits equal to $32/27$. Price competition is tougher.

RP 27 *An Oligopoly Game (1)*

(a, b) Player 1 chooses $q_1 \geq 0$. Players 2 and 3 then choose q_2 and q_3 simultaneously, depending on q_1. The best reply functions of players 2 and 3 in the subgame following q_1 are $q_2 = (a - q_1 - q_3 - c)/2$ and $(a - q_1 - q_2 - c)/2$ (or zero), and the equilibrium in the subgame is $q_2 = q_3 = (a - q_1 - c)/3$. Player 1 takes this into account and maximizes $q_1(a - c - q_1 - 2(a - q_1 - c)/3)$, which gives $q_1 = (a - c)/2$. Hence, the subgame perfect equilibrium is: player 1 plays $q_1 = (a - c)/2$; players 2 and 3 play $q_2 = q_3 = (a - q_1 - c)/3$. The outcome is player 1 playing $(a - c)/6$ and players 2 and 3 playing $(a - c)/6$.

RP 28 *An Oligopoly Game (2)*

(a) The best-reply functions are $q_1 = (10 - q_2 - q_3)/2$, $q_2 = (10 - q_1 - q_3)/2$, $q_3 = (9 - q_1 - q_2)/2$.
(b) The equilibrium is $q_1 = q_2 = 11/4$, $q_3 = 7/4$.
(c) To maximize joint profit, $q_3 = 0$ and $q_1 + q_2 = 5$. (This follows by using intuition: firm 3 has higher cost, or by solving the problem as a maximization problem under nonnegativity constraints.)

RP 29 *A Duopoly Game with Price Competition*

(a) The monopoly price of firm 1 is $p_1 = 65$ and the monopoly price of player 2 is $p_2 = 75$.
(b)

$$p_1(p_2) = \begin{cases} \{x \mid x > p_2\} & \text{if } p_2 < 30 \\ \{x \mid x \geq 30\} & \text{if } p_2 = 30 \\ \{31\} & \text{if } p_2 = 31 \\ \{p_2 - 1\} & \text{if } p_2 \in [32, 65] \\ \{65\} & \text{if } p_2 \geq 66 \end{cases} \qquad p_2(p_1) = \begin{cases} \{x \mid x > p_1\} & \text{if } p_1 < 50 \\ \{x \mid x \geq 50\} & \text{if } p_1 = 50 \\ \{51\} & \text{if } p_1 = 51 \\ \{p_1 - 1\} & \text{if } p_1 \in [52, 75] \\ \{75\} & \text{if } p_1 \geq 76 \end{cases}$$

(c) $(p_1, p_2) = (31, 32)$.
(d) $(p_1, p_2) = (50, 51)$.

RP 30 *Contributing to a Public Good*

(a) The Nash equilibria in pure strategies are all strategy combinations where exactly two persons contribute.
(b) The expected payoff of contributing is equal to $-3 + 8(1 - (1 - p)^2)$, which in turn is equal to $16p - 8p^2 - 3$.
(c) A player should be indifferent between contributing or not if the other two players contribute, hence $16p - 8p^2 - 3 = 8p^2$. This holds for $p = 1/4$ and for $p = 3/4$.

RP 31 *A Demand Game*

(a) Not possible: each player can gain by raising his demand by 0.1. (b) Not possible: at least one player has $x_i > 0.2$ and can gain by decreasing his demand by 0.2. (c) The unique Nash equilibrium is $(0.5, 0.5, 0.5)$. (d) A Nash equilibrium is for instance $(0.6, 0.6, 0.6)$.

(e) All triples with sum equal to one, and all triples such that the sum of each pair is at least one.

RP 32 *A Repeated Game (1)*

(a) The unique Nash equilibrium in the stage game is $((2/3, 1/3), (1/2, 1/2))$, with payoffs $(8, 22)$. Therefore, all payoffs pairs in the quadrangle with vertices $(16, 24)$, $(0, 25)$, $(0, 18)$, and $(16, 16)$ which are strictly larger than $(8, 22)$, as well as $(8, 22)$, can be reached as long run average payoffs in a subgame perfect equilibrium in the repeated game, for suitable choices of δ.

(b) Write $G = (A, B)$, then $v(A) = 8$ and $-v(-B) = 18$. Therefore, all payoffs pairs in the quadrangle with vertices $(16, 24)$, $(0, 25)$, $(0, 18)$, and $(16, 16)$ which are strictly larger than $(8, 20)$, can be reached as long run average payoffs in a Nash equilibrium in the repeated game, for suitable choices of δ.

(c) The players alternate between (T, L) and (B, R). Player 1 has no incentive to deviate, but uses the eternal punishment strategy B to keep player 2 from deviating. Player 2 will not deviate provided

$$25 + 18\delta/(1 - \delta) \leq 24/(1 - \delta^2) + 16\delta/(1 - \delta^2)$$

and

$$18 + 18\delta/(1 - \delta) \leq 16/(1 - \delta^2) + 24\delta/(1 - \delta^2).$$

The first inequality is satisfied if δ is at least (approximately) 0.55, and the second inequality if $\delta \geq 1/3$. Hence, this is a Nash equilibrium for $\delta \geq 0.55$. It is not subgame perfect since player 2 can obtain 22 by playing the stage game equilibrium strategy.

RP 33 *A Repeated Game (2)*

(a) (D, C), (D, R), and (M, R).

(b) Let $((p_1, p_2, p_3), (q_1, q_2, q_3))$ be a Nash equilibrium. First consider the case $q_3 < 1$. Then $p_1 = 0$ and therefore $q_1 = 0$. If $p_2 > 0$ then $q_2 = 0$ and $q_3 = 1$, a contradiction. Hence, $p_2 = 0$, and then $p_3 = 1$. We obtain the set of Nash equilibria $\{((0, 0, 1), (0, q_2, q_3)) \mid q_2, q_3 \geq 0, q_2 + q_3 = 1, q_3 < 1\}$.

Next, consider the case $q_3 = 1$. Then $9p_1 + p_2 + 4p_3 \leq p_1 + 2p_2 + 4p_3$, hence $8p_1 \leq p_2$. We obtain another set of Nash equilibria $\{((p_1, p_2, p_3), (0, 0, 1)) \mid p_1 \geq 0, 8p_1 \leq p_2, p_1 + p_2 = 1\}$.

(c) Each player has $3 \times 3^9 = 3^{10}$ pure strategies. In the first stage the players play (U, L) and in the second stage they play (for instance) according to the table

$$
\begin{array}{c}
\quad\quad\quad L \quad\quad C \quad\quad R \\
\begin{array}{c} U \\ M \\ D \end{array}
\left(
\begin{array}{ccc}
D, R & M, R & D, R \\
D, C & D, R & D, R \\
D, C & D, R & D, R
\end{array}
\right).
\end{array}
$$

(d) Always play (U, L) but after a deviation by player 1, player 2 reverts to C forever, to which player 1 replies by D, and after a deviation by player 2, player 1 reverts to M forever, to which player 2 replies by R. This is a subgame perfect equilibrium provided that

$$
10 + 2\delta/(1 - \delta) \le 8/(1 - \delta) \Leftrightarrow \delta \ge 1/4
$$

and

$$
9 + 2\delta/(1 - \delta) \le 8/(1 - \delta) \Leftrightarrow \delta \ge 1/7
$$

hence if $\delta \ge 1/4$.

RP 34 *A Repeated Game (3)*

(a) (D, L), (U, R), and (D, R).
(b) The second row and next the second column can be deleted by iterated elimination of strictly dominated strategies. This results in the sets of Nash equilibria $\{((0, 0, 1), (q_1, 0, q_3)) \mid q_1, q_3 \ge 0, q_1 + q_3 = 1\}$ and $\{((p_1, 0, p_3), (0, 0, 1)) \mid p_1, p_3 \ge 0, p_1 + p_3 = 1\}$.
(c) In the first stage the players play (M, C) and in the second stage they play (for instance) according to the table

$$
\begin{array}{c}
\quad\quad\quad L \quad\quad C \quad\quad R \\
\begin{array}{c} U \\ M \\ D \end{array}
\left(
\begin{array}{ccc}
D, R & D, L & D, R \\
U, R & D, R & D, R \\
D, R & D, L & D, R
\end{array}
\right).
\end{array}
$$

(d) Always play (M, C) but after a deviation by player 1 player 2 reverts to L forever, to which player 1 replies by D, and after a deviation by player 2 player 1 reverts to U, to which player 2 replies by R. This is a subgame perfect equilibrium provided that

$$
12 + \delta/(1 - \delta) \le 10/(1 - \delta)
$$

which holds for $\delta \ge 2/11$.

RP 35 *A Repeated Game (4)*

(a) Player 1 plays B and player 2 plays L in both stages.
(b) They play (T, L) in the first stage. If player 1 would deviate to B, then player 2 plays R in the second stage, otherwise L. Player 1 plays B in the second stage.
(c) Since (B, L) is the unique Nash equilibrium in the stage game and there are no payoff pairs better for both players, the only possibility is that player 1 plays B and player 2 plays L forever. This is a subgame perfect equilibrium for any value of δ, with long run average payoffs $(5, 5)$.

RP 36 *A Repeated Game (5)*

(a) Only (T, L).
(b) The payoff pair $(2, 1)$, and all payoff pairs larger for both players in the triangle with vertices $(5, 0)$, $(0, 6)$, and $(1, 1)$.
(c) At even times play (B, L) and at odd times play (T, R). After a deviation revert to T (player 1) and L (player 2) forever. This is a subgame perfect Nash equilibrium provided that

$$2 + 2\delta/(1 - \delta) \leq 5\delta/(1 - \delta^2)$$

and

$$1 + \delta/(1 - \delta) \leq 6\delta/(1 - \delta^2)$$

which is equivalent to $\delta \geq \max\{2/3, 1/5\} = 2/3$.

RP 37 *An Evolutionary Game*

(a) The species consists of $100p\%$ animals of type C and $100(1 - p)\%$ animals of type D.
(b) $\dot{p} = p(0p + 2(1-p) - 2p(1-p) - 3(1-p)p - (1-p)^2)$ which after simplification yields $\dot{p} = 4p(p - 1)(p - 1/4)$. Hence the rest points are $p = 0, 1/4, 1$ and $p = 1/4$ is stable.
(c) The unique symmetric Nash equilibrium strategy is $(1/4, 3/4)$. One has to check that $(1/4, 3/4)A(q, 1-q) > (q, 1-q)A(q, 1-q)$ for all $q \neq 1/4$, which follows readily by computation.

RP 38 *An Apex Game*

(a) Suppose $(x_1, \ldots, x_5)$ is in the core. Since $x_1 + x_2 \geq 1$, and all x_i are nonnegative and sum to one, we must have $x_3 = x_4 = x_5 = 0$. Similarly, $x_2 = 0$, but this contradicts $x_2 + \ldots + x_5 \geq 1$. So the core is empty.
(b) $\Phi_2(v) = 1!3!/5! + 3!1!/5! = 1/10$, hence $\Phi(v) = (6/10, 1/10, 1/10, 1/10, 1/10)$.

(c) Let $(1 - 4a, a, a, a, a)$ be the nucleolus of this game. The relevant (maximal) excesses to consider are $1-(1-4a)-a = 3a$ (e.g., $\{1, 2\}$) and $1-4a$ ($\{2, \ldots, 5\}$). Equating these yields $a = 1/7$.

RP 39 *A Three-person Cooperative Game (1)*

(a) For $a > 10$ the core is empty. For $a = 10$, a core element is for instance $(0, 5, 5)$. Hence, $a \leq 10$.
(b) The Shapley value is $((25 - 2a)/6, (19 + a)/6, (16 + a)/6)$. By writing down the core constraints, it follows that this is in the core for $-13 \leq a \leq 8.75$.
(c) At this vector, the excesses of the three two-player coalitions are equal, namely to $(a - 14)/3$. For this to be the nucleolus we need that the excesses of the one-person coalitions are not larger than this, i.e.,

$$(2a - 16)/3 \leq (a - 14)/3, (-a - 4)/3 \leq (a - 14)/3, (-a - 7)/3 \leq (a - 14)/3$$

and it is straightforward to check that this is true for no value of a.

RP 40 *A Three-person Cooperative Game (2)*

(a) The core is nonempty for $a \leq 1$. In that case, the core is the quadrangle (or line segment if $a = 1$) with vertices $(1, 2, 2)$, $(a, 2, 3 - a)$, $(1, 1, 3)$, and $(a, 2 - a, 3)$.
(b) The Shapley value is $((2a + 7)/6, (10 - a)/6, (13 - a)/6)$, which is in the core for $-2 \leq a \leq -1/2$.
(c) By equating the excesses of the two-person coalitions we obtain the vector $(2/3, 5/3, 8/3)$ with excess $-1/3$. This is the nucleolus if $a - 2/3 \leq -1/3$, hence if $a \leq 1/3$.

RP 41 *Voting*

(a) The winning coalitions (omitting set braces) are AB, AC, ABC, ABD, ACD, $ABCD$, and BCD. Then $\Phi_A(v) = 1!2!/4! + 1!2!/4! + 2!1!/4! + 2!1!/4! + 2!1!/4! = 5/12$. Similarly, one computes the other values to obtain $\Phi(v) = (1/12)(5, 3, 3, 1)$. (In fact, it is sufficient to compute $\Phi_B(v)$ and $\Phi_C(v)$.)
(b) $p_A = 5, p_B = 3, p_C = 3, p_D = 1$; $\beta(A) = 5/12, \beta(B) = 3/12, \beta(C) = 3/12$, $\beta(D) = 1/12$.
(c) The winning coalitions are AB, AC, ABC. The Shapley value is $(2/3, 1/6, 1/6)$. Further, $p_A = 3, p_B = p_C = 1$; $\beta(A) = 3/5, \beta(B) = \beta(C) = 1/5$.

RP 42 *An Airport Game*

(a) $v(1) = v(2) = v(3) = 0$, $v(12) = v(13) = c_1$, $v(23) = c_2$, and $v(N) = c_1 + c_2$.
(b) The core is the quadrangle with vertices $(c_1, c_2, 0)$, $(0, c_2, c_1)$, $(0, c_1, c_2)$, and $(c_1, 0, c_2)$.

(c) $\Phi(v) = (1/6)(4c_1, 3c_2 + c_1, 3c_2 + c_1)$. This is a core element (check the constraints).

(d) The nucleolus is of the form $(a, (c_1 + c_2 - a)/2, (c_1 + c_2 - a)/2)$. By equating the excesses of the two-person coalitions it follows that $a = (3c_1 - c_2)/3$, hence the nucleolus would be $((3c_1 - c_2)/3, 2c_2/3, 2c_2/3)$ and the excess of the two-person coalitions is then $-c_2/3$. We need that the excesses of the one-person coalitions are not larger, that is, $-(3c_1 - c_2)/3 \leq -c_2/3$ and $-(2/3)c_2 \leq -c_2/3$. This results in the condition $c_1 \geq 2c_2/3$.

RP 43 *A Glove Game*

(a) By straightforward computation, $\Phi(v) = (1/60)(39, 39, 14, 14, 14)$: note that it is sufficient to compute one of these values.
(b) $C(v) = \{(1, 1, 0, 0, 0)\}$.
(c) By (b) and the fact that the nucleolus is in the core whenever the core is nonempty, the nucleolus is $(1, 1, 0, 0, 0,)$.

RP 44 *A Four-Person Cooperative Game*

(a) $C(v) = \{\mathbf{x} \in \mathbb{R}^4 \mid x_i \geq 0 \, \forall i, x_1 + x_2 = x_3 + x_4 = 2, x_1 + x_3 \geq 3\}$. In the intended diagram, the core is a triangle with vertices $(2, 1)$, $(2, 2)$, and $(1, 2)$.
(b) $\Phi(v) = (1/4)(5, 3, 5, 3)$ (it is sufficient to compute one of these values).

RP 45 *A Matching Problem*

(a) The resulting matching is $(x_1, y_4), (x_2, y_3), (x_3, y_2), (x_4, y_1)$.
(b) The resulting matching is $(x_1, y_4), (x_2, y_3), (x_3, y_1), (x_4, y_2)$.
(c) x_1 prefers y_4 over y_1 and y_4 prefers x_1 over y_4.
(d) In any core matching, x_2 and y_3 have to be paired, since they are each other's top choices. Given this, x_1 and y_4 have to be paired. This leaves only the two matchings in (a) and (b).

RP 46 *House Exchange*

(a) There are two core allocations: $1 : h_1, 2 : h_3, 3 : h_2$ and $1 : h_2, 2 : h_3, 3 : h_1$.
(b) The unique top trading cycle is $1, 2, 3$, with allocation $1 : h_2, 2 : h_3, 3 : h_1$.
(c) Take preference h_1, h_2, h_3 with unique core allocation $1 : h_1, 2 : h_3, 3 : h_2$.

RP 47 *A Marriage Market*

(a) m_1 must be paired to his favorite woman in the core. Next, m_2 must be paired to his favorite of the remaining women, etc.
(b) $(m_1, w_1), (m_2, w_2), (m_3, w_3), (m_4, w_4)$.
(c) $(m_1, w_4), (m_2, w_3), (m_3, w_2), (m_4, w_1)$.

(d) (m_1, w_2), (m_2, w_1), (m_3, w_3), (m_4, w_4) (one can reason about this but also just try the six possibilities).

Reference

Myers, S. C., & Majluf, N. S. (1984). Corporate financing and investment decisions when firms have information that investors do not have. *Journal of Financial Economics, 13*, 187–221.

References

Akerlof, G. (1970). The market for lemons: Quality uncertainty and the market mechanism. *Quarterly Journal of Economics, 84*, 488–500.

Arrow, K. J. (1963). *Social choice and individual values* (2nd ed.). New York: Wiley.

Arrow, K. J., Sen, A. K., & Suzumura, K. (Eds.), (2002). *Handbook of social choice and welfare* (Vol. 1). Amsterdam: North-Holland.

Arrow, K. J., Sen, A. K., & Suzumura, K. (Eds.), (2011). *Handbook of social choice and welfare* (Vol. 2). Amsterdam: North-Holland.

Aumann, R. J. (1974). Subjectivity and correlation in randomized strategies. *Journal of Mathematical Economics, 1*, 67–96.

Aumann, R. J., & Hart, S. (Eds.), (1992/1994/2002). *Handbook of game theory with economic applications* (Vols. 1, 2 and 3). Amsterdam: North-Holland.

Aumann, R. J., & Maschler, M. (1985). Game theoretic analysis of a bankruptcy problem from the Talmud. *Journal of Economic Theory, 36*, 195–213.

Axelrod, R. (1984). *The evolution of cooperation*. New York: Basic Books.

Benoit, J.-P., & Krishna, V. (1985). Finitely repeated games. *Econometrica, 53*, 905–922.

Bernheim, B. (1984). Rationalizable strategic behavior. *Econometrica, 52*, 1007–1028.

Bertrand, J. (1883). Review of Walras's 'Théorie mathématique de la richesse sociale' and Cournot's 'Recherches sur les principes mathématiques de la théorie des richesses'. *Journal des Savants*, 499–508 [translated by M. Chevaillier and reprinted in Magnan de Bornier (1992), 646–653].

Binmore, K., Rubinstein, A., & Wolinsky, A. (1986). The Nash bargaining solution in economic modelling. *Rand Journal of Economics, 17*, 176–188.

Bird, C. G. (1976). On cost allocation for a spanning tree: A game theoretic approach. *Networks, 6*, 335–350.

Black, D. (1969). Lewis Carroll and the theory of games. In *American Economic Review, Proceedings*

Bomze, I. (1986). Non-cooperative two-person games in biology: A classification. *International Journal of Game Theory, 15*, 31–57.

Bondareva, O. N. (1962). Theory of the core in the *n*-person game. *Vestnik Leningradskii Universitet, 13*, 141–142 (in Russian).

Brams, S. J. (1980). *Biblical games: A strategic analysis of stories in the Old Testament*. Cambridge: MIT Press.

Brouwer, L. E. J. (1912). Über Abbildung von Mannigfaltigkeiten. *Mathematische Annalen, 71*, 97–115.

Cho, I. K., & Kreps, D. M. (1987). Signalling games and stable equilibria. *Quarterly Journal of Economics, 102*, 179–221.

Churchill, W. (1983). *Second world war*. Boston: Houghton Mifflin.

© Springer-Verlag Berlin Heidelberg 2015
H. Peters, *Game Theory*, Springer Texts in Business and Economics,
DOI 10.1007/978-3-662-46950-7

Cournot, A. (1838). *Recherches sur les principes mathématiques de la théorie des richesses* [English translation (1897) Researches into the mathematical principles of the theory of wealth]. New York: Macmillan.

Curiel, I. (1997). *Cooperative game theory and applications: Cooperative games arising from combinatorial optimization problems*. Boston: Kluwer Academic.

Curiel, I., Derks, J., & Tijs, S. H. (1986). On balanced games and games with committee control. *Operations Research Spektrum, 11*, 83–88.

Dantzig, G. B. (1963). *Linear programming and extensions*. Princeton: Princeton University Press.

Davis, M., & Maschler, M. (1965). The kernel of a cooperative game. *Naval Research Logistics Quarterly, 12*, 223–259.

de Borda, J. C. (1781). *Mémoires sur les élections au scrutin*. Paris: Histoire de l'Académie Royale des Sciences.

de Clippel, G., Peters, H., & Zank, H. (2004). Axiomatizing the Harsanyi solution, the symmetric egalitarian solution, and the consistent solution for NTU-games. *International Journal of Game Theory, 33*, 145–158.

de Condorcet, M. (1785). *Essai sur l'application de l'analyse à la probabilité des décisions rendues à la pluralité des voix*. Paris: Imprimerie Royale.

Derks, J. (1987). Decomposition of games with non-empty core into veto-controlled simple games. *Operations Research Spektrum, 9*, 81–85.

Derks, J. (1992). A short proof of the inclusion of the core in the Weber set. *International Journal of Game Theory, 21*, 149–150.

Derks, J., Haller, H., & Peters, H. (2000). The Selectope for cooperative games. *International Journal of Game Theory, 29*, 23–38.

Dimand, M. A., & Dimand, R. W. (1996). *The history of game theory, volume 1: From the beginnings to 1945*. London: Routledge.

Dodgson, C. L. (1884). *The principles of parliamentary representation*. London: Harrison & Sons.

Dresher, M., Shapley, L. S., & Tucker, A. W. (Eds.), (1964). *Advances in game theory. Annals of mathematics studies* (Vol. 52). Princeton: Princeton University Press.

Dresher, M., Tucker. A. W., & Wolfe, P. (Eds.), (1957). *Contributions to the theory of games III. Annals of mathematics studies* (Vol. 39). Princeton: Princeton University Press.

Feyerabend, P. K. (1974). *Against method*. London: New Left Books.

Ford, L. R., & Fulkerson, D. R. (1956). Maximal flow through a network. *Canadian Journal of Mathematics, 8*, 399–404.

Friedman, J. W. (1985). Cooperative equilibria in finite horizon noncooperative supergames. *Journal of Economic Theory, 35*, 390–398.

Fudenberg, D., & Maskin, E. (1986). The folk theorem in repeated games with discounting or with incomplete information. *Econometrica, 54*, 533–556.

Fudenberg, D., & Tirole, J. (1991a). *Game theory*. Cambridge: MIT Press.

Fudenberg, D., & Tirole, J. (1991b). Perfect Bayesian equilibrium and sequential equilibrium. *Journal of Economic Theory, 53*, 236–260.

Gale, D., & Shapley, L. S. (1962). College admissions and the stability of marriage. *American Mathematical Monthly, 69*, 9–15.

Gardner, R. (1995). *Games for business and economics*. New York: Wiley.

Gehrlein, W. V. (2006). *Condorcet's paradox*. Berlin: Springer.

Gibbard, A. (1973). Manipulation of voting schemes: A general result. *Econometrica, 41*, 587–601.

Gibbons, R. (1992). *A primer in game theory*. Hertfordshire: Harvester Wheatsheaf.

Gillies, D. B. (1953). *Some theorems on n-person games*, Ph.D. Thesis. Princeton: Princeton University Press.

Hardin, G. (1968). The tragedy of the commons. *Science, 162*, 1243–1248.

Harsanyi, J. C. (1959). A bargaining model for the cooperative *n*-person game. In *Contributions to the theory of games* (Vol. 4, pp. 324–356). Princeton: Princeton University Press.

Harsanyi, J. C. (1963). A simplified bargaining model for the *n*-person cooperative game. *International Economic Review, 4*, 194–220.

Harsanyi, J. C. (1967/1968). Games with incomplete information played by "Bayesian" players I, II, and III. *Management Science, 14*, 159–182, 320–334, 486–502.

Harsanyi, J. C. (1973). Games with randomly disturbed payoffs: A new rationale of mixed strategy equilibrium points. *International Journal of Game Theory, 2*, 1–23.

Hart, S., & Mas-Colell, A. (1989). Potential, value, and consistency. *Econometrica, 57*, 589–614.

Hart, S., & Mas-Collel. A. (1996). Bargaining and value. *Econometrica, 64*, 357–380.

Hildenbrand, H., & Kirman, A. P. (1976). *Introduction to equilibrium analysis.* Amsterdam: North-Holland.

Hirsch, M., & Smale, S. (1974). *Differential equations, dynamical systems, and linear algebra.* San Diego: Academic.

Hofbauer, J., Schuster, P., & Sigmund, K. (1979). A note on evolutionary stable strategies and game dynamics. *Journal of Theoretical Biology, 81*, 609–612.

Hofbauer, J., & Sigmund, K. (1988). *The theory of evolution and dynamical systems.* Cambridge: Cambridge University Press.

Ichiishi, T. (1981). Super-modularity: Applications to convex games and to the greedy algorithm for LP. *Journal of Economic Theory, 25*, 283–286.

Jehle, G. A., & Reny, P. J. (2001). *Advanced microeconomic theory.* Boston: Addison Wesley.

Kakutani, S. (1941). A generalization of Brouwer's fixed point theorem. *Duke Mathematical Journal, 8*, 457–459.

Kalai, E., & Samet, D. (1985). Monotonic solutions to general cooperative games. *Econometrica, 53*, 307–327.

Kalai, E., & Samet, D. (1987). On weighted Shapley values. *International Journal of Game Theory, 16*, 205–222.

Kalai, E., & Smorodinsky, M. (1975). Other solutions to Nash's bargaining problem. *Econometrica, 43*, 513–518.

Kalai, E., & Zemel, E. (1982). Totally balanced games and games of flow. *Mathematics of Operations Research, 7*, 476–478.

Kohlberg, E. (1971). On the nucleolus of a characteristic function game. *SIAM Journal of Applied Mathematics, 20*, 62–66.

Kreps, D. M., & Wilson, R. B. (1982). Sequential equilibria. *Econometrica, 50*, 863–894.

Krishna, V. (2002). *Auction theory.* San Diego: Academic.

Kuhn, H. W. (1953). Extensive games and the problem of information. In H. W. Kuhn & A. W. Tucker (Eds.), *Contributions to the theory of games II. Annals of mathematics studies* (Vol. 28, pp. 193–216). Princeton: Princeton University Press.

Kuhn, H. W., & Tucker, A. W. (Eds.), (1950). *Contributions to the theory of games I. Annals of mathematics studies* (Vol. 24). Princeton: Princeton University Press.

Kuhn, H. W., & Tucker, A. W. (Eds.), (1953). *Contributions to the theory of games II. Annals of mathematics studies* (Vol. 28). Princeton: Princeton University Press.

Lemke, C. E., & Howson, J. T. (1964). Equilibrium points of bimatrix games. *Journal of the Society for Industrial and Applied Mathematics, 12*, 413–423.

Littlechild, S. C., & Owen, G. (1974). A simple expression for the Shapley value in a special case. *Management Science, 20*, 370–372.

Lucas, W. F. (1969). The proof that a game may not have a solution. *Transactions of the American Mathematical Society, 136*, 219–229.

Luce, R. D., & Raiffa, H. (1957). *Games and decisions: Introduction and critical survey.* New York: Wiley.

Luce, R. D., & Tucker, A. W. (Eds.), (1958). *Contributions to the theory of games IV. Annals of mathematics studies* (Vol. 40). Princeton: Princeton University Press.

Magnan de Bornier, J. (1992). The 'Cournot-Bertrand debate': A historical perspective. *History of Political Economy, 24*, 623–656.

Mailath, G. J., & Samuelson, L. (2006). *Repeated games and reputations.* New York: Oxford University Press.

Mangasarian, O. L., & Stone, H. (1964). Two-person nonzero-sum games and quadratic programming. *Journal of Mathematical Analysis and Applications, 9*, 348–355.

Maschler, M., Solan, E., & Zamir, S. (2013). *Game theory*. Cambridge: Cambridge University Press.

Maskin, E. (1999). Nash equilibrium and welfare optimality. *Review of Economic Studies, 66*, 23–38.

Maynard Smith, J., & Price, G. R. (1973). The logic of animal conflict. *Nature, 246*, 15–18.

Milgrom, P. (2004). *Putting auction theory to work*. Cambridge: Cambridge University Press.

Monderer, D., Samet, D., & Shapley, L. S. (1992). Weighted values and the core. *International Journal of Game Theory, 21*, 27–39.

Morris, P. (1994). *Introduction to game theory*. New York: Springer.

Moulin, H. (1980). On strategy-proofness and single-peakedness. *Public Choice, 35*, 437–455.

Moulin, H. (1988). *Axioms of cooperative decision making*. Cambridge: Cambridge University Press.

Moulin, H. (1995). *Cooperative microeconomics; a game-theoretic introduction*. Hemel Hempstead: Prentice Hall/Harvester Wheatsheaf.

Muller, E., & Satterthwaite, M. A. (1977). The equivalence of strong positive association and strategy-proofness. *Journal of Economic Theory, 14*, 412–418.

Muthoo, A. (1999). *Bargaining theory with applications*. Cambridge: Cambridge University Press.

Myers, S. C., & Majluf, N. S. (1984). Corporate financing and investment decisions when firms have information that investors do not have. *Journal of Financial Economics, 13*, 187–221.

Myerson, R. B. (1978). Refinements of the Nash equilibrium concept. *International Journal of Game Theory, 7*, 73–80.

Myerson, R. B. (1991). *Game theory, analysis of conflict*. Cambridge: Harvard University Press.

Nasar, S. (1998). *A beautiful mind*. London: Faber and Faber Ltd.

Nash, J. F. (1950). The bargaining problem. *Econometrica, 18*, 155–162.

Nash, J. F. (1951). Non-cooperative games. *Annals of Mathematics, 54*, 286–295.

Nash, J. F. (1953). Two-person cooperative games. *Econometrica, 21*, 128–140.

Norde, H., Potters, J., Reijnierse, H., & Vermeulen, D. (1996). Equilibrium selection and consistency. *Games and Economic Behavior, 12*, 219–225.

Nowak, A. S. (1997). On an axiomatization of the Banzhaf value without the additivity axiom. *International Journal of Game Theory, 26*, 137–141.

Okada, A. (1981). On stability of perfect equilibrium points. *International Journal of Game Theory, 10*, 67–73.

Osborne, M. J. (2004). *An introduction to game theory*. New York: Oxford University Press.

Owen, G. (1972). Multilinear extensions of games. *Management Science, 18*, 64–79.

Owen, G. (1995). *Game theory* (3rd ed.). San Diego: Academic.

Pearce, D. (1984). Rationalizable strategic behavior and the problem of perfection. *Econometrica, 52*, 1029–1050.

Peleg, B., & Sudhölter, P. (2003). *Introduction to the theory of cooperative games*. Boston: Kluwer Academic.

Peleg, B., & Tijs, S. (1996). The consistency principle for games in strategic form. *International Journal of Game Theory, 25*, 13–34.

Perea, A. (2001). *Rationality in extensive form games*. Boston: Kluwer Academic.

Perea, A. (2012). *Epistemic game theory: Reasoning and choice*. Cambridge: Cambridge University Press.

Peters, H. (1992). *Axiomatic bargaining game theory*. Dordrecht: Kluwer Academic.

Peters, H. (2003). NTU-games. In U. Derigs (Ed.), *Optimization and operations research*. Encyclopedia of life support systems (EOLSS). Oxford: Eolss Publishers. http://www.eolss.net.

Predtetchinski, A., & Herings, P. J. J. (2004). A necessary and sufficient condition for nonemptiness of the core of a non-transferable utility game. *Journal of Economic Theory, 116*, 84–92.

Raghavan, T. E. S. (1994). *Zero-sum two-person games*, Chap. 20. In R. J. Aumann, & S. Hart (Eds.), *Handbook of game theory with economic applications* (Vol. 2). Amsterdam: North-Holland.

Raiffa, H. (1953). Arbitration schemes for generalized two-person games. *Annals of Mathematics Studies, 28*, 361–387.

Rasmusen, E. (1989). *Games and information: An introduction to game theory* (2nd ed., 1994/1995). Oxford: Basil Blackwell.

Rawls, J. (1971). *A theory of justice*. Cambridge: Harvard University Press.

Reny, P. J. (2001). Arrow's theorem and the Gibbard-Satterthwaite theorem: A unified approach. *Economics Letters, 70*, 99–105.

Rockafellar, R. T. (1970). *Convex analysis*. Princeton: Princeton University Press.

Roth, A. E. (Ed.), (1988). *The Shapley value, essays in honor of Lloyd S. Shapley*. Cambridge: Cambridge University Press.

Rubinstein, A. (1982). Perfect equilibrium in a bargaining model. *Econometrica, 50*, 97–109.

Samuelson, L., & Zhang, J. (1992). Evolutionary stability in asymmetric games. *Journal of Economic Theory, 57*, 363–391.

Satterthwaite, M. A. (1975). Strategy-proofness and Arrow's conditions: Existence and correspondence theorems for voting procedures and social welfare functions. *Journal of Economic Theory, 10*, 187–217.

Scarf, H. E. (1973). *The computation of economic equilibria*. New Haven: Cowles Foundation, Yale University Press.

Scarf, H. E. (1976). The core of an n-person game. *Econometrica, 35*, 50–69.

Schelling, T. C. (1960). *The strategy of conflict*. Cambridge: Harvard University Press.

Schmeidler, D. (1969). The nucleolus of a characteristic function game. *SIAM Journal on Applied Mathematics, 17*, 1163–1170.

Selten, R. (1965). Spieltheoretische Behandlung eines Oligopolmodels mit Nachfragezeit. *Zeitschrift für Gesammte Staatswissenschaft, 121*, 301–324.

Selten, R. (1975). Reexamination of the perfectness concept for equilibrium points in extensive games. *International Journal of Game Theory, 4*, 25–55.

Selten, R. (1980). A note on evolutionary stable strategies in asymmetric animal conflicts. *Journal of Theoretical Biology, 84*, 93–101.

Selten, R. (1983). Evolutionary stability in extensive-form two-person games. *Mathematical Social Sciences, 5*, 269–363.

Shapley, L. S. (1953). A value for n-person games. In A. W. Tucker & H. W. Kuhn (Eds.), *Contributions to the theory of games* (Vol. II, pp. 307–317). Princeton: Princeton University Press.

Shapley, L. S. (1967). On balanced sets and cores. *Naval Research Logistics Quarterly, 14*, 453–460.

Shapley, L. S. (1969). Utility comparisons and the theory of games. In G.Th. Guilbaud (Ed.), *La Decision* (pp. 251–263). Paris: CNRS.

Shapley, L. S. (1971). Cores of convex games. *International Journal of Game Theory, 1*, 11–26.

Shapley, L. S., & Shubik, M. (1972). The assignment game I: The core. *International Journal of Game Theory 1*, 111–130.

Snijders, C. (1995). Axiomatization of the nucleolus. *Mathematics of Operations Research, 20*, 189–196.

Sobolev, A. I. (1975). Characterization of the principle of optimality through functional equations. *Mathematical Methods in the Social Sciences, 6*, 92–151.

Spence, A. M. (1973). Job market signalling. *Quarterly Journal of Economics, 87*, 355–374.

Sprumont, Y. (1991). The division problem with single-peaked preferences: A characterization of the uniform allocation rule. *Econometrica, 59*, 509–520.

Sun Tzu (1988). *The art of war* [Translated by Thomas Cleary]. Boston: Shambala.

Sutton, J. (1986). Non-cooperative bargaining theory: An introduction. *Review of Economic Studies, 53*, 709–724.

Tan, T., & Werlang, S. R. C. (1988). The Bayesian foundations of solution concepts of games. *Journal of Economic Theory, 45*, 370–391.

Taylor, P., & Jonker, L. (1978). Evolutionary stable strategies and game dynamics. *Mathematical Biosciences, 40*, 145–156.

Thomas, L. C. (1986). *Games, theory and applications*. Chichester: Ellis Horwood Limited.

Thomson, W. (1994). Cooperative models of bargaining. In R. J. Aumann & S. Hart (Eds.), *Handbook of game theory with economic applications* (Vol. 2). Amsterdam: North-Holland.

Tijs, S. H., Parthasarathy, T., Potters, J. A. M., & Rajendra Prasad, V. (1984). Permutation games: Another class of totally balanced games. *Operations Research Spektrum, 6*, 119–123.

Tirole, J. (1988). *The theory of industrial organization*. Cambridge: MIT Press.

van Damme, E. C. (1991). *Stability and perfection of Nash equilibria*. Berlin: Springer.

Vickers, G., & Cannings, C. (1987). On the definition of an evolutionarily stable strategy. *Journal of Theoretical Biology, 132*, 387–408.

Vickrey, W. (1961). Counterspeculation, auctions, and competitive sealed tenders. *Journal of Finance, 16*, 8–37.

Vohra, R. V. (2005). *Advanced mathematical economics*. New York: Routledge.

von Neumann. J. (1928). Zur Theorie der Gesellschaftsspiele. *Mathematische Annalen, 100*, 295–320.

von Neumann, J., & Morgenstern, O. (1944/1947). *Theory of games and economic behavior*. Princeton: Princeton University Press.

von Stackelberg, H. F. (1934). *Marktform und Gleichgewicht*. Wien: Julius Springer.

von Stengel, B. (2002). Computing equilibria for two-person games. In R. Aumann & S. Hart (Ed.), *Handbook of game theory with economic applications* (Vol. 3). Amsterdam: North-Holland.

Walker, M., & Wooders, J. (2001). Minimax play at Wimbledon. *American Economic Review, 91*, 1521–1538.

Watson, J. (2002). *Strategy, an introduction to game theory*. New York: Norton.

Weber, R. J. (1988). Probabilistic values for games. In A. E. Roth (Ed.), *The Shapley value. Essays in honor of Lloyd S. Shapley* (pp. 101–119). Cambridge: Cambridge University Press.

Weibull, J. W. (1995). *Evolutionary game theory*. Cambridge: MIT Press.

Young, H. P. (1985). Monotonic solutions of cooperative games. *International Journal of Game Theory, 14*, 65–72.

Young, H. P. (2004). *Strategic learning and its limits*. Oxford: Oxford University Press.

Zermelo, E. (1913). Über eine Anwendung der Mengenlehre auf die Theorie des Schachspiels. In *Proceedings Fifth International Congress of Mathematicians* (Vol. 2, pp. 501–504).

Index

© Springer-Verlag Berlin Heidelberg 2015
H. Peters, *Game Theory*, Springer Texts in Business and Economics,
DOI 10.1007/978-3-662-46950-7

Printed in the United States
By Bookmasters